From Pigments
to Perception
Advances in Understanding
Visual Processes

NATO ASI Series

Advanced Science Institutes Series

A series presenting the results of activities sponsored by the NATO Science Committee, which aims at the dissemination of advanced scientific and technological knowledge, with a view to strengthening links between scientific communities.

The series is published by an international board of publishers in conjunction with the NATO Scientific Affairs Division

A	**Life Sciences**	Plenum Publishing Corporation
B	**Physics**	New York and London
C	**Mathematical and Physical Sciences**	Kluwer Academic Publishers
D	**Behavioral and Social Sciences**	Dordrecht, Boston, and London
E	**Applied Sciences**	
F	**Computer and Systems Sciences**	Springer-Verlag
G	**Ecological Sciences**	Berlin, Heidelberg, New York, London,
H	**Cell Biology**	Paris, Tokyo, Hong Kong, and Barcelona
I	**Global Environmental Change**	

Recent Volumes in this Series

Series A: Life Sciences

From Pigments
to Perception

Advances in Understanding
Visual Processes

Edited by

Arne Valberg

University of Oslo
Oslo, Norway

and

Barry B. Lee

Max-Planck-Institute for Biophysical Chemistry
Gottingen, Germany

Plenum Press
New York and London
Published in cooperation with NATO Scientific Affairs Division

Proceedings of a NATO Advanced Research Workshop on
Advances in Understanding Visual Processes:
Convergence of Neurophysiological and Psychophysical Evidence,
held August 6–10, 1990,
in Røros, Norway

QP
475
.N32
1990

Library of Congress Cataloging in Publication Data

NATO Advanced Research Workshop on Advances in Understanding Visual Pro-
cesses: Convergence of Neurophysiological and Psychophysical Evidence (1990:
Røros, Norway)
   From pigments to perception: advances in understanding visual processes /
edited by Arne Valberg and Barry B. Lee.
      p.   cm.—(NATO ASI series. Series A, Life sciences: v. 203)
   "Proceedings of a NATO Advanced Research Workshop on Advances in
Understanding Visual Processes: Convergence of Neurophysiological and
Psychophysical Evidence, held August 6–10, 1990, in Røros, Norway"—T.p.
verso.
   Includes bibliographical references and index.
   ISBN 0-306-43905-0
   1. Visual pathways—Congresses. 2. Psychophysics—Congresses. 3. Photore-
ceptors—Congresses. 4. Visual evoked response—Congresses. 5. Visual per-
ception—Congresses. I. Valberg, Arne. II. Lee, Barry B. III. Title. IV. Series.
   [DNLM: 1 Color Perception—physiology—congresses. 2. Evoked Potentials,
Visual—congresses. 3. Models, Theoretical—congresses. 4. Photoreceptors—
physiology—congresses. 5. Space Perception—physiology—congresses. 6.
Visual Cortex—physiology—congresses. 7. Visual Pathways—physiology—
congresses. 8. Visual Perception—physiology—congresses. WW 105 N2785f
1990]
QP475.N32   1990
152.14—dc20
DNLM/DLC   91-3529
for Library of Congress   CIP

© 1991 Plenum Press, New York
A Division of Plenum Publishing Corporation
233 Spring Street, New York, N.Y. 10013

Printed in the United States of America

"Les Newtoniens mettent les couleurs uniquement dans les rayons, qu'ils distinguent en rouges, jaunes, verts, bleus et.violets; et ils disent, qu'un corps nous paroît de telle ou telle couleur, lorsqu'il réfléchit des rayons de cette espèce. D'autres, auxquels ce sentiment paroît trop grossier, prétendent que les couleurs n'existent que dans nous-mêmes. C'est un excellent moyens de cacher son ignorance; autrement le peuple pourroit croire, que le savant ne connoit pas mieux la nature des couleurs que lui."

Euler, Leonhard, 1760
From 'Lettres à une Princesse d'Allemagne', XXVIII Lettre

PREFACE

The nature of the visual process has been disputed since modern theories of light and colour began to be developed in the eighteenth and nineteenth centuries. The citation above exemplifies the conflicts which still arise when relating the physical nature of light to visual perception. This book constitutes a record of lectures and discussions held in a workshop from the 6th to the 10th of August 1990 in Røros, Norway. Its objective was an attempt to resolve such conflicts, by providing better links between visual psychophysics and neurophysiology.

With "psychophysical processes", Gustav Fechner, the father of psychophysics, seems to have meant the neural processes intervening between physical stimuli and sensation. He distinguished between an inner and an outer psychophysics, and in his "Elements of Psychophysics" (1860) he said: "... there can be no development of outer psychophysics without constant regard to inner psychophysics, in view of the fact that the body's external world is functionally related to the mind only by the mediation of the body's internal world." And later he continues: "To the extent that lawful relationships between sensation and stimulus can be found, they must include lawful relationships between stimulus and this inner physical activity ..." Much in the same sense, the physicist James Clerk Maxwell wrote some years later (1872) that: "... if the sensation which we call colour has any laws, it must be something in our own nature which determines the form of these laws".

We hoped to encourage participants to consider these processes and the problems involved in multidisciplinary approaches to vision. We wished to focus not only on detection tasks involving correlations between visual thresholds and cellular sensitivities, but also on the current status of 'neuron doctrines', and the relationships to be found between sensations and supra-threshold neuronal processes. Thus, we were seeking a common ground where Fechner's inner and outer psychophysics, or, as one would say today, neurophysiology and psychophysics, might cross-fertilize each other.

There are, of course, severe methodological problems attached to this kind of inter-disciplinary approach. Looking across the borders between disciplines is a fascinating and a popular subject for informal discussion, but is often regarded by the scientific community as somewhat improper when carried out in public. Early attempts in this direction frequently seem naive today, and this has engendered an attitude in favour of keeping experimental work entirely within one's own discipline, either within physiology (the 'inside') or psychophysics (the 'outside'). However, attempts to understand vision without such psycho-physiological linking concepts resemble research without a hypothesis. In recent years, we have begun to see how different neural systems share the task of mediating information about the visual world. In particular, recent research has indicated a division of labour between parvocellular (P) and magnocellular (M) systems. Where this division lies is not yet settled, and several viewpoints will be found in these pages. Clearly, progress is dependent upon studying the system as a whole, involving interaction between psychophysics and physiology, and cooperation between physicists, physiologists, and psychologists.

Such interdisciplinary activity relies heavily on various conceptual assumptions and linking hypotheses. An attempt to better define the nature of these hypotheses was made by Giles Brindley in his distinction between class A and class B observations. Some of the ways linking hypotheses are used are apparent from the contributions. It will be seen that logical or causal connections between physiology and psychophysics are sometimes difficult to pin down; analogies are often as important as logical deduc-tions. In particular, problems arise if one does not wish to restrict oneself to detection paradigms and hence to specification of sensitivity for some particular task. In this context, one promising possibility is to develop models of neural coding, which formally and quantitatively can combine data from related psychophysical and neurophysiological experiments.

It is instructive to compare this volume with the Proceedings of the Freiburg Symposium organized by Richard Jung and Hans Kornhuber in 1960. Similar problems were discussed then as now, but formerly parallels were often drawn between the visual system of humans with those of cats, fishes, turtles, and a wide variety of other species. Today we are able to make a more relevant comparison, in that we can concentrate more on the primate as a model for human vision.

The contributions here are related to physiology and psychophysics at several lev-els. One primary stimulus to the workshop was to collate various viewpoints as to the functions of the P- and M-cell systems. Analysis at the receptor level provides informa-tion as to the input to these two systems, whereas a natural extension is the interactions between the diverging, parallel, and converging streams of information processing in the cortex. More general, conceptual questions frequently arose in discussion of the inter-pretation of cell responses at the higher levels of processing, and their relationships to executive systems. For instance, how do cell systems process different dimensions of objects in time and space? How are these dimensions coded, and how are features extracted and combined to generate new attributes? Cues and qualitative dimensions, for example colour, texture, depth, orientation and movement, are presumably created by combining inputs of diverging and converging neural pathways, and then integrated, for instance into the perception of shape, and made available for the preparation for appropriate action. How far neural processes and sensations can be regarded as identical remains another, philosophical, question related to the mind-body problem.

We strove to make poster and discussion sessions an integral part of the meeting. We present an edited version of these discussions, hoping it reflects some of the vigour with which they were conducted, and the diversity of subjects covered. We apologise if, in the interests of brevity, we have misinterpreted the views of some commentators.

We would like to thank the other members of the Organising Committee, Svein Magnussen, Joel Pokorny, and Steven Shevell, for help in shaping the program. We also thank Discussants and Chairmen for their assistance in making the meeting a success, the Secretariat of Anne-Sophie Andresen and Anne-Grethe Gulbrandsen for valuable help, and Sissel Knudsen for transcribing the discussions. We also acknowledge the technical assistance of Lars Rune Bjørnevik, Stein Dyrnes, Thorstein Seim, and Jan-Henrik Wold. Finally, we thank Jan Kremers for editorial assistance in preparing this volume for press.

This meeting was made possible by financial support from NATO, and also by grants from the National Science Foundation (U.S.A.), the Norwegian Council for Science and the Humanities, the Fridjof Nansen Foundation, the University of Oslo, the Norwegian Physical Society, Norwegian Telecom, and the Scandinavian Airline System. Their generous support permitted attendance by some students and younger scientists.

Arne Valberg and Barry B. Lee

October, 1990

CONTENTS

VISUAL EVOKED POTENTIALS

CORTICAL PROCESSING AND PSYCHOPHYSICAL MEASUREMENT

PSYCHOPHYSICAL STUDIES AND POST-RECEPTORAL PROCESSES

VISUAL PIGMENTS AND COLOUR VISION IN PRIMATES

J. K. Bowmaker

Department of Visual Science
Institute of Ophthalmology
University of London
Judd Street
London WC1H 9QS

INTRODUCTION

There is now considerable information from behavioural, electro-physiological and microspectrophotometric studies, concerning the forms of colour vision in primates, and the spectral location of the underlying visual pigments. These investigations give some insight into the evolution of colour vision in the primates, from dichromacy to trichromacy, and this has been amplified by the recent analysis of the molecular genetics of human colour vision (Nathans et al., 1986). However, there are no pub-lished results relating to the molecular genetics of either Old World or New World monkeys, and a number of questions still remain concerning the number and spectral location of visual pigments in man, both in normal observers and in anomalous trichromats.

In terms of colour vision, the primates can be considered in three distinct groups: a) those considered the most primitive, the *Prosimii*, b) New World monkeys, the *Platyrrhinae*, and c) Old World monkeys and man, the *Catarrhinae*.

PROSIMIANS

Unfortunately little data are available on either the forms of colour vision or the visual pigments within this group. However there is some information concerning the *Lemuridae* (lemurs), the *Lorisidae* (*Galago*, bush-baby) and the *Tupaiidae* (tree shrews), though the status of the tree shrew as a primitive primate has been much disputed (see Luckett, 1980).

In the tree shrew, behavioural and erg flicker photometric studies (Jacobs and Neitz, 1986) have established that their colour vision is di-chromatic, subserved by two classes of cone with peak sensitivities (λ_{max}) at about 556 and 444 nm. We have confirmed these values by microspectro-photometry (Bowmaker et al., 1990) and our data are in broad agreement with

From Pigments to Perception, Edited by A. Valberg and
B.B. Lee, Plenum Press, New York, 1991

those of Petry and Hárosi (1987, 1990). The retina of *Tupaia* is dominated by cones with the rods constituting from 1 to 14% of the photoreceptor population, depending on retinal location. The majority of the cones are long-wave sensitive with short-wave cones constituting only from 4 and 10% of the total cone population (Muller and Peichl, 1989). From a micro-spectrophotometric study of a single male *Tupaia glis*, we obtained 18 analyzable records from long-wave cones that formed a single distribution with a mean λ_{max} of 555 nm (identical with that of Petry and Hárosi) and a further single record from a short-wave sensitive cone with λ_{max} between 440 and 445 nm. Petry and Hárosi (1990) report λ_{max} ranging from 410 to 442 nm for 8 short-wave sensitive cones with a mean λ_{max} at 428 nm. We also recorded from two rods with maximum absorbance at about 498 nm, in good agreement with the value of 496 nm suggested from a single rod by Petry and Hárosi.

In the lemurs the situation is less clear. Although their retinas are dominated by rods, cones are present and a number of species are diurnal. Behavioural data from ringed-tailed lemurs (*Lemur catta*) show clearly that they have colour vision. Blakeslee and Jacobs (1985) have demonstrated that they can discriminate wavelengths over a large portion of the spectrum and are able to set a distinct Rayleigh match. This would suggest that they are trichromats, but they require considerably more red light when matching a red and green mixture to a yellow than do humans and could perhaps be classified in human terms as 'severely protanomalous'. The basis for their colour vision is not clear, but Blakeslee and Jacobs (1985) appear to favour trichromacy based on two types of cone interacting with rods. The severe protanomaly implies that their long-wave cone has a λ_{max} considerably shorter than that of humans. From recent microspectro-photometric data from the black lemur (*Lemur macaco*), we have identified rods with λ_{max} at 501 nm, but we have been able to record from only three cones and these appear to form a single population at longer wavelengths with λ_{max} at about 543 nm (Figure 1).

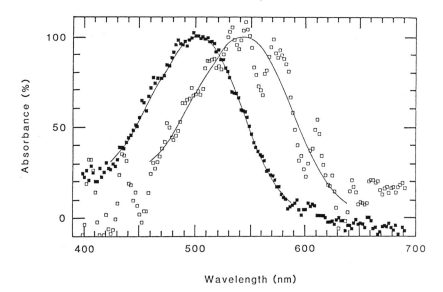

Figure 1. Mean absorbance spectra of 9 rods (solid symbols) and 3 cones (open symbols) from *Lemur macaco*. The solid lines show the best-fitting rhodopsin templates with λ_{max} at 501 and 543 nm.

The bushbaby, *Galago crassicaudatus*, is nocturnal and has often been thought to have a pure rod retina (Dartnall et al., 1965). However, Dodt (1967) has demonstrated a Purkinje shift that suggests the presence of a cone population maximally sensitive at about 552 nm. A recent microspectrophotometric examination of the retina of the bushbaby revealed only rods with λ_{max} near 501 nm (Petry and Hárosi, 1990), though this cannot exclude the possibility of a small cone population.

NEW WORLD MONKEYS

New World monkeys are basically dichromatic, but a polymorphism of the long-wave pigments results in a much more complex picture, with some two thirds of females having the benefit of trichromacy (Mollon et al., 1984; Jacobs and Neitz, 1985, 1987a; Travis et al., 1988). Both the *Callitrichidae* and the *Cebidae* have three cone pigments available in the red/green region of the spectrum, in addition to a short-wave pigment. In individual males only one of the three long-wave pigments is expressed and the animals are dichromatic, but in females either one or a combination of any two of the long-wave pigments may be present.

The most likely explanation for this diversity of pigment distribution is that these species have only a single gene locus on the X-chromosome that determines the long-wave pigment, but that three alleles of the locus are available, coding for slightly different opsins that are expressed as cone pigments with slightly different peak sensitivities. Females homozygous for the locus will be dichromats whereas those heterozygous for the locus will be trichromats, since they can express two long-wave pigments. X-chromosome inactivation is necessary to ensure the segregation of the two pigments into separate classes of cone.

In the *Callitrichidae*, the common marmoset (*Callithrix jacchus*) and the saddle-backed tamarin (*Saguinus fusicollis*) have long-wave pigments with λ_{max} at about 543, 556 and 563 nm (Jacobs et al., 1987; Travis et al., 1988), whereas in the *Cebidae*, the squirrel monkey (*Saimiri sciureus*), has pigments with maxima at about 536, 549 and 563 nm (Mollon et al., 1984, Jacobs and Neitz, 1987a). These pigment locations also appear to occur in the tufted capuchin (*Cebus apella*) and the dusky titi (*Callicebus moloch*): Jacobs and Neitz (1987b) report pigments with λ_{max} at about 549 and 562 nm and we have recorded a P535 in a male *Cebus* (Bowmaker et al., 1983).

We have recently studied, both behaviourally and microspectrophotometrically, the inheritance of cone visual pigments within family groups of marmosets (Tovée, 1990). In one family, the parents and four off-spring, two sets of twins, were studied. The mother was presumed to be trichromatic, since microspectrophotometry showed that she possessed both the P556 and P563 pigments, whereas the father possessed only the P556 and was shown behaviourally to be dichromatic. In one pair of twins, a male and female, the male was behaviourally dichromatic and possessed the P556, whereas the female was shown behaviourally to be trichromatic and microspectrophotometry demonstrated the presence of the P556 and P563 (Figures 2). The inheritance of the pigments in these twins appears to follow the single gene hypothesis: the son inheriting the P556 from his mother, and the daughter inheriting the paternal P556 and the maternal P563. It is interesting to note that the female possessing long- and middle-wave cones separated by only about 7 nm clearly showed opponent processes in the red/green spectral region, with a distinct notch in the spectral sensitivity function at about 580-590 nm.

The short-wave cones of the prosimians and the platyrrhines also vary. Those of the squirrel monkey have maximum absorbance at about 430 nm (Mollon et al., 1984), those of the marmoset absorb maximally at about 423 nm (Travis et al., 1988), whereas that of the tree shrew probably peaks close to 444 nm (Petry and Hárosi, 1990; Bowmaker et al, 1990).

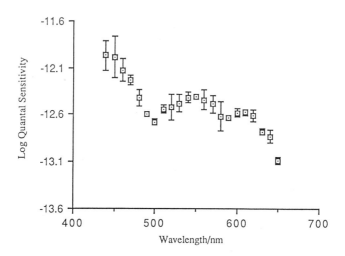

Figure 2. Increment threshold spectral sensitivity function, determined on an achromatic background, for a female marmoset. The behavioural task was a three-alternative forced-choice problem. Note the three peaks at about 450, 550 and 620 nm and the two dips at about 500 and 580-590 nm. Microspectrophotometry demonstrated the presence of two cone pigments at longer wavelengths with λ_{max} at about 556 and 563 nm (from Tovée, 1990).

OLD WORLD MONKEYS

Until recently our knowledge of the colour vision and cone pigments of catarrhine monkeys was based almost solely on macaques (e.g. Bowmaker et al., 1978; Hárosi, 1987; Baylor et al., 1987). We now have microspectro-photometric data from seven other species of Old world monkey, all from Africa (Bowmaker et al., 1990). These include the baboon, *Papio papio*, patus monkey (*Erythrocebus patus*), and five species of *Cercopithecus*. These monkeys come from different geographical regions and from a wide range of ecological niches. They differ considerably in their body weight and colouring, some are exclusively arboreal and others predominantly terrestrial. Nevertheless, all have long-wave and middle-wave cone pigments with absorbance maxima close to 565 and 535 nm with an inter-specific variation of about 5 nm in each case (Table 1). The short-wave cones have λ_{max} around 430 nm.

Table 1. λ_{max} of middle- and long-wave cones from
Old World monkeys, lemur and tree-shrew.

Species	Sex	Middle-wave cones	Long-wave cones
Cercopithecus aethiops	M	535.0 ± 2.7 (13)*	566.3 ± 1.9 (11)
Cercopithecus diana	F	530.7 ± 3.9 (11)	565.9 ± 2.4 (18)
Cercopithecus ascanius	M	533.9 ± 2.8 (11)	562.7 ± 3.2 (8)
Cercopithecus cephus	F	533.0 ± 1.7 (21)	565.3 ± 2.9 (17)
Erythrocebus patus	M	533.0 ± 2.3 (11)	566.3 ± 1.9 (20)
Cercopithecus talapoin	M	533.3 ± 2.8 (20)	564.0 ± 3.9 (10)
Macaca mulatta	F	536.0 ± 1.8 (12)	561.5 ± 5.8 (6)
Papio papio	3M & 1F	536.2 ± 3.0 (32)	565.6 ± 3.0 (22)
Lemur macao	2M	543 (3)	
Tupaia glis	M		554.5 ± 3.5 (18)

* Numbers in parentheses are the number of cones analyzed.
(From Bowmaker et al., 1990)

Table 2. Numbers of middle- and long-wave cones
obtained from catarrhine monkeys.

Species	Sex	Middle-wave cones	Long-wave cones	Ratio
Cercopithecus aethiops	M	22	19	1.16
Cercopithecus diana	F	12	30	0.40
Cercopithecus ascanius	M	19	21	0.90
Cercopithecus cephus	F	26	22	1.18
Erythocephus patus	M	25	35	0.71
Cercopithecus talapoin	M	27	28	0.96
Macaca mulatta	F	23	13	1.77
Papio papio	3M & 1F	72	48	1.71
Total		226	216	1.05

The Table includes all records obtained, whereas the numbers given
in Table 1 refer only to those cells yielding absorbance spectra
for which a peak wavelength could be precisely given.

The numerosity of middle- and long-wave cones has often been debated
(e.g. Cicerone and Nerger, 1989) and Table 2 gives the total number of
cones identified in our retinal samples (taken from within 1-2 mm of the
fovea) from catarrhine monkeys (Bowmaker et al., 1990). Although the ratio
of middle- to long-wave cones varies considerable between individual
species and the sample of cells for each species is small, the overall
ratio (based on 442 cones) is close to unity and clearly different from the
ratio of 1:2 postulated from psychophysical studies of human foveal vision
(Vos and Walraven, 1971; Cicerone and Nerger, 1989).

MAN

The spectral location of the middle-wave and long-wave cone pigments
in man appear to be well established with maxima at about 530 and 565 nm

(Dartnall et al., 1983; Schnapf et al., 1987). Our best estimate from microspectrophotometry puts the λ_{max} at 531 and 563 nm (Bowmaker, 1990). These values are very similar to those of catarrhine monkeys, though the middle-wave cone pigment appears to be somewhat shorter. The short-wave cone pigment has not been so clearly specified and microspectrophotometry suggests an absorbance maximum close to 420 nm, distinctly shorter than that in Old World monkeys.

Two major questions remain. Is there more than one long-wave and one middle-wave cone pigment in humans with normal colour vision (Neitz and Jacobs, 1986, 1990; Jordan and Mollon, 1989)? How many, and what are the spectral locations of the spectrally displaced visual pigments in anomalous observers (Neitz et al., 1989)?

Neitz and Jacobs (1986, 1990) show that within a relatively large population of young males with normal colour vision, Rayleigh matches vary and fall into four distinct groups. If the average match point (R/R+G) is made equal to 0.5 then the four groups have match points at about 0.43, 0.47, 0.53 and 0.57. Most of the subjects fall within the two centre groups whereas a small number of outliers make up the two more extreme groups. In contrast, a similar sized group of females show a large number with a Rayleigh match of 0.5, in addition to the two main distributions at about 0.47 and 0.53.

Neitz and Jacobs (1990) suggest that these discrete variations in Rayleigh matches are produced by discrete variations in the spectal positions of both the middle- amd long-wave cone mechanisms and that the cone mechanisms are separated by about 3 nm. They further suggest that the variation in cone mechanisms can be explained by a polymorphism of both the middle- and long-wave pigments and propose that within the normal population there can be four different genes on the X-chromosome coding for pigments in the red/green spectral range. Thus there would be two middle-wave cone pigments separated by about 6 nm and two long-wave cone pigments also separated by approximately 6 nm. Individuals carrying both genes for one pigment class would have a cone mechanism maximally sensitive mid-way between the λ_{max} of the two pigments, thus explaining the 3 nm separation in cone mechanisms.

The separation of 6 nm between the pairs of pigments is similar to that reported for human middle- and long-wave cones from a microspectrophotometric study of human cones from the eyes of seven persons (Dartnall et al., 1983) where we estimated λ_{max} at about 528 and 534 nm for the middle-wave sensitive cones and 554 and 563 nm for the long-wave cones. Such bimodality was not apparent in our more recent study (Bowmaker, 1990).

VARIATION AND EVOLUTION OF VISUAL PIGMENTS

Two striking features emerge from the distribution of primate cone visual pigments sensitive in the red/green spectral region. First, the polymorphism of pigments in the New World monkeys and secondly, the contrasting conservatism of the spectral locations of the pigments in Old World monkeys. The polymorphism in New World monkeys, primarily dichromatic species, is maintained presumably because the trichromacy it confers on two thirds of the females offers behavioural advantages over dichromacy. It is not clear why there should be a difference in λ_{max} between the pigments in the cebids and the callitrichids, but given the range of spectral locations available to the platyrrhines (536, 543, 549, 556, 563 nm), it is perhaps surprising that more variation has not found in the spectral locations of the middle- and long-wave pigments in catarrhines.

It has been proposed (see Knowles and Dartnall, 1977) that the opsins of visual pigments can evolve relatively rapidly and that, at least in man, the homology and tandem arrangement of the middle- and long-wave opsin genes on the X-chromosome should allow frequent opportunity for the formation of hybrid genes that code for pigments with intermediate λ_{max} (Nathans et al., 1986). However, the apparently fixed positions of the Old World cone pigments implies that either there is a functional or ecological constraint that maintains the peak sensitivities at about 535 and 563 nm, or else it must be more difficult than thought previously for the structure of opsins to be modified so as to displace the absorbance spectra of visual pigments.

In terms of the evolution of cone pigments within the primates, the ancestral simian was presumably dichromatic. In New World monkeys the gene locus for the longer-wave pigment, located on the X-chromosome, has mutated to produce at least three alleles (with the polymorphism being maintained because of the heterozygous advantage of trichromacy), whereas in Old World monkeys the gene must have duplicated as a consequence of unequal crossing over during cell division. Mutation of one or both loci would cause the expression of two spectrally distinct pigments conferring trichromacy on both males and females. The only common spectral location of a cone pigment within the simians is the longest i.e. close to 563 nm, and this might be thought to reflect the spectral location of an ancestral pigment. However, such a long-wave pigment has not been found amongst the prosimians and indeed has not been identified in any other mammal, and it would seem more likely that the longer-wave pigment of the ancestral simian had λ_{max} at an intermediate wavelength between 535 and 563 nm.

Within the New World and Old World monkeys there appear to be only five or six spectral locations for the middle-wave and long-wave pigments: at about 528, 536, 543, 549, 556 and 563 nm, and it is interesting that the longer-wave pigments of the lemur (P543) and tree shrew (P555) are located at spectrally similar positions to two of those of the *Callitrichidae*. This spectral 'clustering' at intervals of about 6 or 7 nm, is reminiscent of the clustering of the λ_{max} of extracted pigments from rods noted by Dartnall and Lythgoe (1965) and also appears to occur within the rhodopsins of deep-sea fish (Partridge et al., 1989). Unfortunately, insufficient data are available from short-wave pigments to establish whether clustering occurs at these wavelengths, though it should be noted that the short-wave cones of man and marmosets have λ_{max} around 420-424 nm whereas those of Old World monkeys and the squirrel monkey have λ_{max} at about 430nm.

It is perhaps not surprising that it is somewhat difficult to be precise about the exact spectral locations of the 'clusters' in the red/green spectral range, since the published λ_{max} for cone pigments from different species have often been determined by different methods; micro-spectrophotometry, electroretinography and suction eletrode techniques. Nevertheless, if the clustering is real, it must reflect constraints on the structure of opsin: substitutions of non-polar and polar amino-acids close to the relevant regions of the chromophore cannot yield an infinite range of visual pigments with different spectral maxima. This would appear to be the case for the anomalous pigments in man (Pokorny et al., 1973; Nagy et al., 1985; Neitz et al., 1989) and the comparison of opsin structures from prosimians, platyrrhines and catarrhines with those from man will be of interest.

REFERENCES

Baylor, D.A., Nunn, B.J. and Schnapf, J.L. (1987) Spectral sensitivity of cones of the monkey *Macaca fascicularis*. J. Physiol. (Lond.) 390, 145-160.

Blakeslee, B. and Jacobs, G.H. (1985) Color vision in the ringed-tailed lemur (*Lemur catta*). Brain Behav. Evol. 26: 154-166.

Bowmaker, J.K. (1990) Cone visual pigments in man and monkeys, in "Frontiers of Visual Science: proceedings of the 1988 symposium". National Acad. Press, Washington.

Bowmaker, J.K., Astell, S., Hunt, D.M. and Mollon, J.D. (1990) Photo-sensitive and photostable pigments in the retinae of Old World monkeys. J. Exptl. Biol. In press.

Bowmaker, J.K., Dartnall, H.J.A., Lythgoe, J.N. and Mollon, J.D. (1978) The visual pigments of rods and cones in the Rhesus monkey, *Macaca mulatta*. J. Physiol. (Lond.) 274, 329-348.

Bowmaker, J.K., Mollon, J.D. and Jacobs, G.H. (1983) microspectrophoto-metric results for Old and New World primates, in: "Colour Vision", J.D. Mollon and L. T. Sharpe, eds., Academic Press, London.

Cicerone, C.M. and Nerger, J.L. (1989) The relative numbers of long-wavelength-sensitive to middle-wavelength-sensitive cones in the human fovea centalis. Vision Res., 29, 115-128.

Dartnall, H.J.A., Arden, G.B., Ikeda, G.B., Luck, C.P., Rosenberg, M.E., Pedler, C.M.H. and Tansley, K. (1965) Anatomical, electrophysiological and pigmentary aspects of vision in the bush baby: an interpretative study. Vision Res. 5, 399-424.

Dartnall, H.J.A., Bowmaker, J.K. and Mollon, J.D. (1983) Human visual pigments: microspectrophotometric results from the eyes of seven persons. Proc. Roy. Soc. (Lond.) B, 220, 115-130.

Dartnall, H.J.A. and Lythgoe, J.N. (1965) The spectral clustering of visual pigments. Vision Res. 5, 81-100.

Dodt, E. (1967). Purkinje shift in the rod eye of the bush baby, *Galago crassicaudatus*. Vision Res. 7, 509-517.

Hárosi, F.I. (1987) Cynomolgus and Rhesus monkey visual pigments. J. gen. Physiol. 89, 717-743.

Jacobs, G.H. and Neitz, J. (1986) Spectral mechanisms of color vision in the tree-shrew (*Tupaia belangeri*). Vision Res. 26, 291-298.

Jacobs, G.H. and Neitz, J. (1987a) Inheritance of colour vision in a New World monkey (*Saimiri sciureus*). Proc. natn. Acad. Sci. 84, 2545-2549.

Jacobs, G.H. and Neitz, J. (1987b) Polymorphism of the middle wavelength cone in two species of South American monkey: *Cebus apella* and *Callicebus moloch*. Vision Res. 27, 1263-1268.

Jacobs, G.H., Neitz, J. and Crognale, M. (1987) Color vision polymorphism and its photopigment basis in a callitrichid monkey (*Saguinus fusicollis*). Vision Res. 27, 2089-2100.

Jordan, G. and Mollon, J.D. (1989) On the relationship between Rayleigh matches and 'unique green'. Perception 18, 530-531.

Knowles A. and Dartnall, H.J.A. (1977) "The photobiology of vision", Vol 2B of "The Eye", H. Davson, ed., Academic Press, New York.

Luckett, W.P. (1990) "Comparative biology and evolutionary relationships of tree shrews", Plenum, New York.

Mollon, J.D., Bowmaker, J.K. and Jacobs, G.H. (1984) Variations of colour vision in a New World primate can be explained by polymorphism of retinal photopigments. _Proc. Roy. Soc. (Lond.)_ B, 222, 373-399.

Muller, B and Peichl, L. (1989) Topography of cones and rods in the tree shrew retina. _J. Comp. Neurol_. 282, 581-594.

Nagy, A.L., Purl, K.F. and Houston, J.S (1985) Cone mechanisms underlying the color discrimination of deutan color deficients. _Vision Res_. 25, 661-669.

Nathans, J., Thomas, D. and Hogness, D.S. (1986) Molecular genetics of human color vision: the genes encoding blue, green and red pigments. _Science_ 232, 193-203.

Nathans, J., Piantanida, T.P., Eddy, R.L., Shows, T.B. and Hogness, D.S. (1986) Molecular genetics of inherited variations in human color vision. _Science_ 232, 203-210.

Neitz, J. and Jacobs, G.H. (1986) Polymorphism of the long-wavelength cone in normal human colour vision. _Nature (Lond.)_ 323, 623-625.

Neitz, J., Neitz, M. and Jacobs, G.H. (1989) Analysis of fusion gene and encoded photopigment of colour-blind humans. _Nature (Lond.)_ 342, 679-682.

Neitz, J. and Jacobs, G.H. (1990) Polymorphism in normal human colour vision and its mechanism. _Vision Res_. 30, 621-636.

Partridge, J.C., Shand, J., Archer, S.N., Lythgoe, J.N. and van Groningen-Luyben, W.A.H.M. (1989) Interspecific variation in the visual pigments of deep-sea fishes. _J. Comp. Physiol_. 164, 513-529

Petry, H.M. and Hárosi, F.I. (1987) Tree shrew visual pigments by microspectrophotometry. _Ann. N. Y. Acad. Sci_. 494, 250-252.

Petry, H.M. and Hárosi, F.I. (1990) Visual pigments of the tree shrew (_Tupaia belangeri_) and greater galago (_Galago crassicaudatus_): a microspectrophotometric investigation. _Vision Res_. 30, 839-851.

Pokorny, J., Smith, V.C. and Katz, I. (1973) Derivation of the photopigment absorption spectra in anomalous trichromats. _J. Opt. Soc. Am_. 63, 232-237.

Schnapf, J.L., Kraft, T.W. and Baylor, D.A. (1987) Spectral sensitivity of human cone photoreceptors. _Nature (Lond.)_ 325, 439-441.

Tovée, M.J. (1990) A polymorphism of the middle- to long-wave cone photopigments in the common marmoset (_Callithrix jacchus jacchus_): a behavioural and microspectrophotometric study. Ph.D. Thesis, University of Cambridge.

Travis, D.S., Bowmaker, J.K. and Mollon, J.D. (1988) Polymorphism of visual pigments in a callitrichid monkey. _Vision Res_. 28, 481-490.

Vos, J.J. and Walraven, P.L. (1971) On the derivation of the foveal receptor primaries. _Vision Res._, 11, 799-818.

THE COST OF TRICHROMACY FOR SPATIAL VISION

David R. Williams, Nobutoshi Sekiguchi, William Haake,
David Brainard, and Orin Packer

Center for Visual Science
University of Rochester
Rochester, NY 14627

INTRODUCTION

An ideal trichromatic visual system would provide three different spectral samples at each and every point in the retinal image. One way to achieve this would be to construct a three-tiered cone mosaic with each class of photoreceptor in a different tier, much like color film. Each tier could then have the same spatial resolution as a single-tiered monochromatic mosaic and trichromacy could be incorporated without a loss of spatial resolution. The human cone mosaic, on the other hand, has its three cone submosaics interleaved in a single tier. This means that on a small spatial scale the retina is not trichromatic since there can be only one spectral sample at each point in the retinal image. This chapter explores the cost for vision of incorporating trichromacy in a single-tiered cone mosaic.

Discrete spatial sampling by the cone mosaic has a potential cost for spatial vision in the form of aliasing distortion in the representation of high spatial frequencies (Yellott, Jr., 1982; Williams, 1985a; Williams, 1986; Coletta & Williams, 1987; Williams, 1988; Williams, 1990, Williams, in press). Aliasing arises because sampling discards information about the behavior of the retinal image between sample points. All images that are the same at the sample points, but differ between them, are aliases of one another and cannot be discriminated by post-receptoral processes. An alternative description of aliasing is that it is the formation of moiré patterns between the image and the sampling array. These moiré patterns have the property that they could be substituted for the original stimulus, without changing the quantum catch in any of the receptors.

Aliasing in foveal vision first appears near the cone Nyquist frequency of about 60 c/deg, and higher spatial frequencies than this are seen as flickering, lower frequency patterns that resemble zebra stripes. In the fovea (though not in the extrafovea), aliasing can only be seen with interference fringes. Under ordinary viewing conditions, the optics of the eye are well-matched to cone spacing, blurring just those high spatial frequencies that would otherwise produce cone aliasing. On the other hand, the sampling density of each submosaic is lower than that of the entire mosaic, so one might expect each submosaic to reveal aliasing effects over and above those produced by the mosaic as a whole. Furthermore, the lower sampling density of each submosaic means that the protective effect of the eye's optics is reduced, suggesting that aliasing by a single cone submosaic might be visible in ordinary vision without the use of interference fringes.

Fig. 1 shows an idealized example in which the submosaics of a dichromatic retina produce this kind of aliasing, which we will refer to as "chromatic aliasing". The two types of cones are indicated in gray and white. The period of the white square wave grating imaged on the mosaic is slightly different than the horizontal spacing of cones indicated in gray, so that the bright bars of the grating slide in and out of register with these columns of cones across the retina. This modifies the local ratio of excitation of the two cone types,   and it is this ratio that determines the hue that would

From Pigments to Perception, Edited by A. Valberg and
B.B. Lee, Plenum Press, New York, 1991

Fig. 1. Idealized illustration of chromatic aliasing showing a crystalline cone mosaic composed of two cone classes, indicated as gray and white disks, sampling a square wave grating to which both cone classes are equally sensitive. As the grating moves in and out of register with the submosaic indicated in gray, the local ratio of excitation of the two cone classes smoothly changes, so that an alias of low spatial frequency is produced that is defined by variation in color.

be seen by an observer with such a mosaic. This produces a low spatial frequency, chromatic moiré pattern, one full cycle of which can be seen (in shades of gray instead of color in the example.) This chromatic pattern is an alias in the sense that it is one of many possible interpretations of the observed distribution of quantum catches. The ambiguity that characterizes chromatic aliasing is a potential cost of incorporating trichromacy into a single-tiered cone mosaic. Next we examine evidence for chromatic aliasing in human vision.

SPATIAL SAMPLING BY THE S CONE SUBMOSAIC

In the human retina, the low sampling density of the S cone mosaic makes it particularly susceptible to chromatic aliasing, at least in the laboratory. Anatomical evidence shows that S cones form a sparse mosaic, accounting for only about 10% of the cone population, and that the mosaic is somewhat disordered though by no means random (Marc & Sperling, 1977; DeMonasterio et al.., 1981; Szél et al.., 1988; Curcio et al., 1989). This low density makes it possible to map individual S cones in the living human eye (Williams et al, 1981). The visual resolution of the S cone pathway is correspondingly low, with maximum resolution no greater than 10-14 c/deg (Green, 1968; Williams & Collier, 1983). Disordered mosaics are expected to produce aliasing effects that resemble two-dimensional spatial noise for gratings that exceed the Nyquist frequency (Yellott, Jr., 1983; Williams & Coletta, 1987; Coletta & Williams, 1987; Coletta et al., in press; Tiana et al., in press). If the chromatic aberration of the eye is corrected, gratings seen by the S cone pathway with spatial frequencies above about 10 c/deg have the noisy appearance expected from aliasing by the S cone mosaic (Williams & Collier, 1983; Williams et al., 1983).

A striking example of S cone aliasing arises from their absence at the foveal center. Williams, MacLeod, and Hayhoe (1981a, b) showed that most observers (though not all) have a foveal region roughly 20-25' of arc that lacks S cones. More recently, Curcio et al. (1989) provided anatomical support for this finding, staining human S cones with an antibody to the opsin of the S photopigment. The largest disk that could fit within the tritanopic area they identified was 21' of arc, in good agreement with the psychophysical estimate. Thus normal observers are reduced to dichromats at a retinal location that is the most critical for most visual tasks. This leads to the filling-in phenomenon

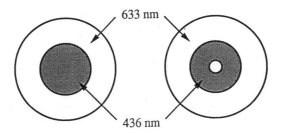

633 nm

436 nm

Fig. 2. A violet disk and a violet annulus, each super-imposed on a long wavelength background designed to isolate the S cones. When each field is centrally fixated, the two stimuli are chromatic aliases, provided the inner diameter of the annulus is smaller than the tritanopic area.

reported by Williams, MacLeod, & Hayhoe (1981b) and illustrated in Fig. 2. In their experiment, observers fixated either of two violet fields, a disc and an annulus, each superimposed on a long wavelength background bright enough to isolate the S cone mechanism. If the inner diameter of the annular field was smaller than the tritanopic area, these two fields were indistinguishable. The hole in the violet annulus was filled in perceptually and the field looked like a uniform disc. If an eye movement was made that was large enough to move the hole in the annulus outside the tritanopic area, the hole immediately became apparent.

Recall that two patterns are aliases of one another if they produce identical distributions of quantum catches in the photoreceptors. The disk and the annulus are therefore aliases for the S cone mosaic, and they provide a clear example of aliasing with stimuli other than gratings. Of all the different images that are consistent with the data available to the S cones surrounding the tritanopic area, the visual system adopts the simple interpretation that it is uniform. Just how the visual system would cope with more complicated light distributions is largely unexplored, and the sophistication of the interpolation mechanisms that perform this task has yet to be revealed.

Under ordinary viewing conditions, the eye suffers from considerable axial chromatic aberration particularly at the short wavelengths to which the S cones are sensitive (Wyszecki and Stiles, 1982). Chromatic aberration is effective at reducing chromatic aliasing for the S cone submosaic because the spatial frequencies that would alias (above about 10 c/deg) are severely blurred in the retinal image (Yellott, Jr., et al., 1984). Thus the cost of sprinkling the retina with a small number of S cones is apparently quite small, and the benefit of a third degree of freedom for color vision can be had without a loss in spatial resolution for the system as a whole.

SPATIAL SAMPLING BY THE M AND L CONE SUBMOSAICS

The M and L cone submosaics, on the other hand, do not receive the protection afforded the S cones by chromatic aberration, and these two cone classes form the basis for the most acute spatial vision we enjoy. Several theoretical investigations have suggested that aliasing by the M and L sub-mosaics should be visible (Ahumada, 1986; Brainard et al., 1989; Packer et al., 1989; Sekiguchi et al., in press; Williams, in press), but there has been surprisingly little evidence for it. Some encouragement that these effects should be visible with grating stimuli comes from observations of the appearance of point sources. Holmgren (1884) reported that the color appearance of stars varies from instant to instant, which he attributed to selective excitation of different cone classes as the point image moved across the mosaic (see also Krauskopf, 1964; 1978; Krauskopf and Srebro, 1965; Cicerone and Nerger, 1989; Vimal et al., 1989).

But previous attempts to uncover M and L cone chromatic aliasing with interference fringes have come up empty handed. Packer et al. (1989) made an extensive but unsuccessful search for chromatic aliasing in the M and L cones. They isolated each of the cone types with an interference fringe of one wavelength superimposed on a background of another, and then varied the fringe spatial frequency and orientation to search for chromatic aliases. Sekiguchi, Williams, and Packer (in press) described "secondary zebra stripes"; a very faint pattern observed with interference fringes in

Fig. 3. A stimulus for producing Brewster's colors. With steady fixation and viewing at normal reading distance, most observers report desaturated splotches of color that are several times larger than the period of the grating.

the fovea. The pattern is seen at the foveal cone Nyquist frequency rather than at twice the Nyquist frequency where achromatic aliasing effects (or "primary zebra stripes") are most prominent (Williams, 1985a; Williams, 1988). These secondary zebra stripes sometimes have a reddish and greenish appearance and both these properties made them candidates for chromatic aliasing by M and L cones. However, several lines of evidence argue against this view, favoring an alternative explanation involving nonlinear distortion in the retina.

BREWSTER'S COLORS

We introduce the argument here that chromatic aliasing not only by the S cones but also by the M and L cones can be seen with simple patterns other than laser interference fringes and that this phenomenon, though not its proper explanation, has been known for more than 150 years. Brewster (1832) observed that a fine pattern of black and white lines, such as those used on maps to indicate the sea, sometimes appear to be covered with a splotchy pattern of desaturated colors. Colored effects with high contrast achromatic patterns have been described repeatedly since then (Skinner, 1932; Luckiesh & Moss, 1933; Erb & Dallenbach, 1939), though they are usually associated with Benham's colors, a point we will return to later. This effect, which we will call Brewster's colors, can be seen with black and white gratings of high contrast and high spatial frequency. Periodic arrays of small white dots are also suitable. The range of spatial frequencies over which the effect can be seen is perhaps 10-40 c/deg, but this range is not very clearly defined. Fig. 3 shows a stimulus that produces Brewster's colors when viewed steadily at normal reading distance. The chromatic splotches are several times larger than the grating period. Many observers report that a fine black and white grating is seen at the center of fixation, but that Brewster's colors are visible outside this central zone of highest resolution. When the spatial frequency of the test pattern is increased, Brewster's colors encroach on the center of fixation. We asked 43 observers to view patterns such as these and 81% described Brewster's colors. The remaining 19% did not, even with coaching about what to look for.

Even for those observers who report Brewster's colors, it is a subtle and fleeting phenomenon. However, we were able to obtain reliable measurements of their hues with a matching technique. Two observers with normal color vision were used. With the right eye, each observer viewed a test field that produced Brewster's colors. The test stimulus was a photographic transparency mounted on a diffuser, back illuminated with an array of fluorescent lights. For observer NS, the test stimulus was a white square wave grating of unity contrast oriented 45 deg clockwise from vertical. For observer DRW, the pattern was a triangular array of white dots on a black background, where each dot subtended 1.3' of arc. In both cases, the fundamental spatial frequency of the pattern was 20 c/deg. The square test field was viewed at a distance of 2 meters, subtended 5 deg, and had a mean retinal illuminance of 377 Td for observer NS and 670 Td for DRW. It was viewed through an artificial pupil (3 mm for NS, 4 mm for DRW) that was aligned to minimize the chromatic fringes caused by axial chromatic aberration. As we will see shortly, the refractive state of the eye has a strong impact on the particular colors that are seen. Trial lenses were used to manipulate refractive state in 0.25 diopter steps. Accommodation was paralyzed with cyclopentolate hydrochloride (1%).

The left eye viewed a calibrated color monitor that displayed a pattern of chromatic splotches designed to emulate Brewster's colors. The hue and contrast of this pattern could be adjusted to match the true Brewster's colors seen with the right eye. The matching field was the same size, shape, and space-averaged hue as the test field. Matching stimuli containing chromatic splotches were generated with a PIXAR Image Computer. Each matching stimulus consisted of a 128x128 array of pixels in which the value of each pixel was chosen at random from a Gaussian distribution, centered on white and lying along a single direction in color space. The directions in color space for all the matching stimuli lay in the isoluminant plane, defined by pairwise flicker photometry of the monitor primaries. A total of 481 images were available to match Brewster's colors, representing 48 different color directions distributed around the color circle. For each color direction, 11 steps of chromatic contrast were available, each corresponding to a different standard deviation of the Gaussian distribution from which the pixel values were chosen. Both observers were satisfied that the matching images captured the general spatial structure of Brewster's colors, and determining the color direction that matched their hues was a relatively easy task.

Fig. 4a shows the individual matches for both observers (unfilled circles for observer NS and filled circles for DRW), plotted in the isoluminant color plane (MacLeod & Boynton, 1979). Chromatic contrast increases from white at the center of each circle. The horizontal axis represents stimuli that modulate the L and M cones but keep the S cone stimulation constant. The vertical axis represents stimuli that modulate the S cones while keeping the L and M cone signals constant. The unique hues (average of both observers) are indicated around each circle's perimeter, as are the major axes of the color plane. The color circle on the left shows the matches made when the eye was defocused by adding either -1.0, -1.25, or -1.5 diopters. In these refractive states, the appearance of Brewster's colors was always violet and greenish-yellow, corresponding roughly to the axis that modulates S cones only. The color circle on the right shows the matches made when the eye was within a quarter of a diopter of best focus (-0.25, 0.0, or +0.25 diopters). In this case, the splotches were always reddish and greenish, lying near the axis for which only M and L cones are modulated. At intermediate dioptric powers (-0.5 and -0.75) both observers reported a broad range of colors. This was revealed in their matches (not shown) which no longer lay close to any single direction in color space.

COMPUTATIONAL MODEL OF BREWSTER'S COLORS

We have implemented a computational model of Brewster's colors that accounts for the dependence of their hue on refractive state as well as their spatial structure. The stages of the model are shown in Fig. 5, and are described in turn below.

Optics of the Eye

The optics of the eye, the first stage of the model, were described as an idealized thin lens with a 3 mm pupil. For simplicity, we assumed that its monochromatic optical transfer function, $T_m(f, \lambda)$, at a particular wavelength, λ, was determined solely by diffraction at the pupil, the refractive state of

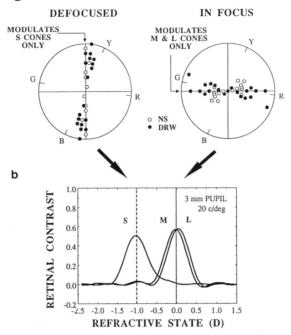

Fig. 4a. Coordinates in the isoluminant color plane of the chromatic noise images that matched Brewster's colors for observer NS (unfilled circles) and DRW (filled circles) for two ranges of refractive states of the eye (defocused = -1.0, -1.25, or -1.5 diopters, in focus = -0.25, 0.0, or +0.25 diopters).

b. The predicted effect of the refractive state of the eye on the contrast of a white 20 c/deg grating as seen by each of the three cone types. The grating is an effective stimulus for only the S cones when the eye is 1 diopter hyperopic, indicated by the dashed vertical line, and it modulates only the M and L cones when the eye is in focus, indicated by the vertical line at 0 diopters.

the eye, and axial chromatic aberration (Wyszecki & Stiles, 1982). These functions were computed as described by Hopkins (1955). The polychromatic optical transfer function, $T_p(f)$, was then computed for each cone type from the integral of the monochromatic transfer functions after weighting them in accordance with the emission spectrum of a typical fluorescent source, $E(\lambda)$, and the spectral sensitivity, $S(\lambda)$, of each cone photopigment (Smith & Pokorny, 1975). That is,

$$T_p(f) = \frac{\int_{\lambda_1}^{\lambda_2} T_m(f,\lambda) \, S(\lambda) E(\lambda) d\lambda}{\int_{\lambda_1}^{\lambda_2} S(\lambda) E(\lambda) d\lambda}$$

The predicted effect of the eye's optics on the 20 c/deg grating used in the experiments is shown in Fig. 4b, where the retinal image contrast seen by each of the three cone classes is plotted as a function of the refractive state of the eye. The dashed vertical line illustrates the situation when the eye is 1 diopter hyperopic. Due to chromatic aberration, the only wavelengths in focus on the retina are the short wavelengths to which the S cones are most sensitive. Thus S cones are exposed to a high contrast 20 c/deg grating, but the M and L cones see a practically uniform field because they are more sensitive to longer wavelengths that are strongly blurred by chromatic aberration. The violet and greenish-yellow Brewster's colors seen under these conditions are just those expected when only

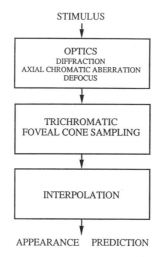

STIMULUS

OPTICS
DIFFRACTION
AXIAL CHROMATIC ABERRATION
DEFOCUS

TRICHROMATIC
FOVEAL CONE SAMPLING

INTERPOLATION

APPEARANCE   PREDICTION

Fig. 5. Three-Stage Model of Brewster's colors.

the S cones are modulated. But when the same grating is brought into best focus, as indicated by the vertical line at 0 diopters, the theory predicts that M and L cones should see a high contrast grating while the S cones see a uniform field. Correspondingly, in the psychophysical data, Brewster's colors shift to reds and greens, just what would be expected if only the M and L cones and not the S cones were involved in the production of these splotches.

Trichromatic Cone Sampling

The first stage of the model clarifies the dependence of Brewster's colors on refractive state, but it does not account for their splotchy spatial structure. Chromatic aberration changes the contrast seen by each cone class, but the retinal image still consists of orderly sinusoidal gratings, not irregular splotches. Sampling by a trichromatic cone mosaic, the second stage of the model, can account for the splotches. This is clearest for the case in which the eye is defocused by -1 diopter since that generates a grating seen only by S cones, with a spatial frequency in the aliasing range for that submosaic. Thus the violet and greenish yellow splotches seen under these conditions probably represent the same aliasing phenomenon reported by Williams, Collier, and Thompson (1983) under somewhat different stimulus conditions.

The mosaic we used is shown in Plate 1a. The S, M, and L cones are indicated in blue, green, and red respectively though the actual absorption spectra of the cone photopigments were implemented in the model. The coordinates of single cones in the central one deg of the primate fovea were taken from a photograph of a whole mount provided by Hugh Perry. There remains some uncertainty about the ratio of L to M cones, as well as their packing arrangement. We have assumed an L/M ratio of 2/1 here. We tried two packing arrangements that represent extremes along a continuum of possible degrees of regularity in cone assignment. The left half of the mosaic shows an assignment rule for L and M cones that is random, while the right half shows a more regular arrangement, assigned by hand, that is similar to that in Fig. 1. 10% of the cones were S cones. S cones were chosen so that their distribution was about as regular as the peripheral mosaic of cones as a whole. We also incorporated a 20 min diameter tritanopic area centered on the location of highest cone density, which is displaced slightly to the upper left from the center of the image.

Interpolation

The third stage of the model is an interpolation stage that generates the model output in the form of an appearance prediction. The model output is an interpolated image that contains three color signals at every point despite the absence of this information in the output of the cone mosaic. The interpolation rules used by the visual system are not well-understood and may involve the highest levels of object recognition in cortex. For our purposes, we chose a very simple interpolation scheme, which was two-dimensional linear interpolation (Akima, 1978) performed on the output of

17

each submosaic. This algorithm has the advantage (not shared by Gaussian convolution for example) that it does not distort the values at the sample points. Thus it generates true aliases in the sense that the model output is completely consistent with the data available at the photoreceptors. The three interpolated images were then recombined to produce an image with the three cone signals at each point. A linear transformation was then applied to convert these cone signals to the appropriate values for the red, green, and blue guns of the color monitor.

Model Results

Plate 1b-d shows the results for a spatial frequency of 20 c/deg which produces Brewster's colors experimentally. Due to the smaller eye of the monkey, the Nyquist frequency of the cone mosaic at the foveal center is only 91 c/deg, about 75% of the Nyquist frequency typical of the human fovea. This makes the model mosaic susceptible to aliasing effects at lower spatial frequencies than would the human retina. Thus the 20 c/deg grating used in the model based on a monkey would be expected to produce about the same sampling effects as a 27 c/deg grating in the human eye. Plate 1b shows that the model correctly produces violet and greenish-yellow splotches when the eye is hyperopic by 1 diopter. The input grating contrast was 100% in this case. Plate 1c shows that the model also correctly produces red-green splotches when the model eye is in focus. In this case, the input contrast was reduced to 25%, for reasons described later, and the assignment rule for M and L cones was the random rule.

For low spatial frequency gratings (less than about 5 c/deg), the model produces a more or less faithful rendition of the input, consistent with the psychophysical observation that Brewster's colors are not seen under these conditions. This is illustrated in Plate 1e which shows the model output for a 3 c/deg grating with 1 diopter of defocus. A small amount of distortion is present, mostly associated with errors in interpolating across the tritanopic area.

Thus the model is quite successful at capturing qualitatively the spatial structure of Brewster's colors and at capturing the quantitative dependence of their hue on refractive state supporting the view that Brewster's colors are explained by trichromatic cone sampling.

DISCUSSION

Alternatives to the Chromatic Aliasing Explanation?

In the past, Brewster's colors have usually been attributed to eye movements, under the assumption that the temporal modulation they produce generates Benham's colors (Luckiesh & Moss, 1933; Erb & Dallenbach, 1939). This explanation is incomplete in the sense that a satisfactory quantitative theory of the origin of Benham's colors is not yet available, nor is it clear that such a theory would predict the dependence on refractive state reported here. Moreover, our subjective observations suggest that steady fixation maximizes the visibility of the subjective colors, which does not favor an eye movement explanation. Finally, any theory based strictly on temporal modulation cannot account for the spatial structure of Brewster's colors. The effect of a given eye movement when a grating is imaged on the retina is to produce a temporal pattern of excitation that repeats itself across space with the same period as that of the grating. Thus a theory of Brewster's colors that does not incorporate spatial inhomogeneity (such as spatial sampling) predicts that the colors produced by eye movements should have the same spatial frequency as the test grating. It does not predict the generally low spatial frequency, irregular splotches of color that are observed.

Erb and Dallenbach (1939) reported that Brewster's colors are not observed immediately at the onset of a grating stimulus, requiring about several seconds delay to develop. They pointed out that flashed stimuli shorter than about three seconds did not produce Brewster's colors, an observation

---

Plate 1(a). The trichromatic cone mosaic used in the model of Brewster's colors. (b) Model output when the eye was defocused by -1 diopter, for a spatial frequency of 20 c/deg, with the random M and L cone assignment rule. Shows S cone aliasing. (c) Same as (b) but with the eye in focus. Shows M and L cone chromatic aliasing. (d) Same as (c) but showing chromatic aliasing with a regular assignment rule for M and L cones. (e) -1 diopter defocus, spatial frequency of 3 c/deg, random M and L cone assignment rule. (f) No defocus, a spatial frequency of 36 c/deg, random M and L cone assignment rule. (g) Same as (f) but with the regular M and L cone assignment rule.

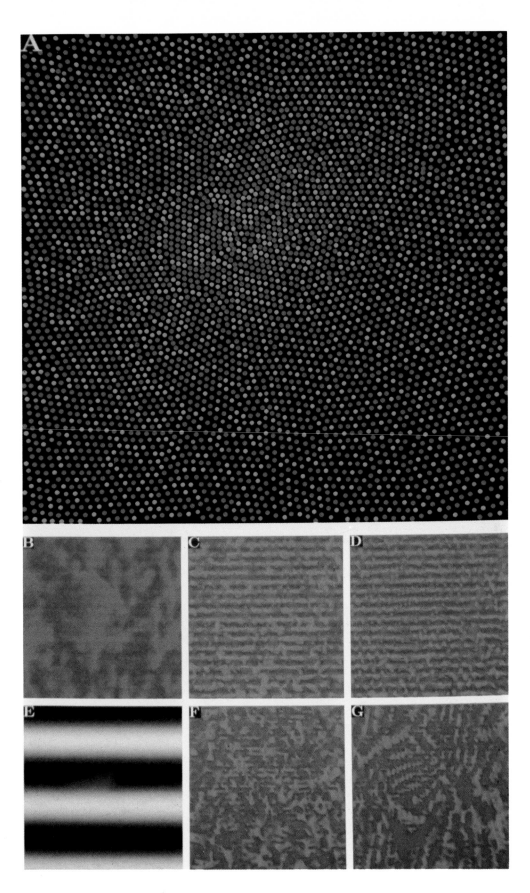

our observers generally corroborated. Erb and Dallenbach concluded from the existence of this delay that eye movements were necessary for the phenomenon. However, the probability or extent of eye movement during the first three seconds of an extended stimulus presentation is presumably little different than it would be at later times, so the logic that implicates eye movements is unclear. We do not view the delay as a repudiation of the sampling hypothesis because postreceptoral mechanisms could, depending on the stimulus conditions, bias the interpretation of a stimulus that sampling has made ambiguous. But the reason for the delay is a mystery, and we can only speculate that short duration flashes favor achromatic mechanisms over chromatic ones, biasing the visual system's interpretation toward a black and white percept and against a chromatic one.

Though we did not incorporate eye movements into our computational model, it is clear what effect they would have had. Brewster's colors should appear to flicker and their contrast should change rapidly over time, just as do the achromatic, primary zebra stripes seen with interference fringes (Williams, 1985a; Packer & Williams, 1990). Indeed, all observers reported this temporal aspect of Brewster's colors. Eye movements may also account for the variability in matches of the apparent contrast of Brewster's colors seen in Fig. 4a.

Another alternative to the sampling hypothesis for Brewster's colors is that they are some kind of optical distortion, perhaps related to the chromatic fringes generated by chromatic aberration. We have rejected this possibility with observations of monochromatic yellow gratings (570 nm), which could not produce chromatic fringes in the retinal image. All three observers who viewed such gratings insisted that reddish and greenish splotches could still be seen.

B.F. Skinner (1932), in a brief report, described color effects seen with a pattern of white disks on a black background. Skinner speculated that his effect was caused by variation in the ratio of cone types beneath each disk, which is the same explanation we propose here. However, the discs he used were 1.4 deg, large compared with the fine dots or gratings that we find are required to generate Brewster's colors, so that these phenomena may not be one and the same.

Finally, we cannot exclude the hypothesis that Brewster's colors are caused by spa: al aliasing at a site subsequent to the receptors. However, cone sampling is sufficient to account for the observations made so far, if anything predicting larger effects than we have observed psychophysically as discussed later.

Implications for the Assignment Rule for M and L Cones

The assignment rule for human M and L cones is not known. The packing arrangement of these two submosaics could be any of a number of regular arrangements, it could be random, or it could exhibit some intermediate degree of regularity. The packing of S cones shows that a mechanism exists for establishing regularity in a primate cone submosaic, though it is uncertain whether this mechanism extends to M and L cones. There is evidence in fish (Hashimoto & Shimoda, 1990) that random and regular assignment rules can coexist in the same retina. Williams (1988) used the dependence of primary zebra stripes on the spatial frequency and orientation of interference fringes to measure the packing geometry of cones in the living foveal mosaic. These moiré patterns revealed the spacing and triangular packing of receptors, though they did not distinguish between M and L cone submosaics. One of our primary motivations for studying Brewster's colors was that they could be used to reveal the submosaic packing geometry in the same way.

We have used our model to compare the effect of regular and random assignment rules of M and L cones. Plate 1c shows the effect of a random assignment rule, and Plate 1d shows the effect of a regular rule. The chromatic alias is somewhat higher contrast with the random than the regular packing arrangement. However, the contrasts predicted by the model are very different from the observed contrasts, a point we will return to, so this difference does not provide a clear way to determine the regularity of the assignment rule. The spatial structure of the predicted Brewster's colors with random and regular assignment rules provides a somewhat more optimistic approach. Though the predicted aliases at 20 c/deg do not differ very much in spatial structure for regular verses random assignment rules, larger differences are predicted at higher spatial frequencies (Ahumada, 1986; Sekiguchi et al., in press). Plate 1f and g show the model output at 36 c/deg for the random and regular assignment rules respectively. This frequency was chosen because it corresponds to the average sampling frequency of the M cone submosaic across the whole image, and this produces the coarsest chromatic alias. The regular assignment rule produces a more coherent alias that resembles zebra stripes, while the random rule produces a noisy chromatic alias. Because the mosaic as a whole is not perfectly crystalline, neither the M nor the L cone submosaic can be crystalline. This disrupts the regularity of the chromatic alias even when the assignment rule is regular, making the predictions from these two extreme cases less different.

In practice, it is difficult to use Brewster's colors to distinguish between assignment rules for several reasons. First, the spatial frequencies that would be expected to produce the clearest differences between regular and random rules are relatively high. For example, the spatial frequency used in Plate 1f and g would correspond to about 48 c/deg in the human, which would be strongly attenuated by the optics of the eye, making the alias more difficult to see. Though our subjective experience of Brewster's colors is that they are fairly disordered, we have not yet established conditions that allow a clear view of them at the very center of the fovea where, due to the overall regularity of the mosaic there, a regular assignment rule would have the best chance to reveal itself.

Extensive efforts to observe M and L cone chromatic aliasing with cone isolation provided by chromatic adaptation were unsuccessful (Packer et al., 1989; Williams, in press). It may be that the M and L cones are assigned with sufficient irregularity that no clear zebra stripes appear at any particular spatial frequency. Instead, chromatic aliasing noise may be generated over a very large range of high spatial frequencies, and this noise may be difficult to detect amidst laser speckle, which can act as a potent mask (Williams, 1985b). However, our models are not yet refined enough to exclude a regular assignment rule that produces submosaic zebra stripes with contrasts that are subthreshold. Green (1968) and Cavonius and Estevez (1975) showed that acuity with the M or L cones alone is no poorer than with both operating together. We confirmed this result with laser interference fringes (Packer et al., 1989; Williams, in press), showing that acuity with either cone class alone is close to the Nyquist frequency for the total M and L cone mosaic. One possible explanation for this would be that the M and L submosaics are irregular enough that there are occasional relatively large patches of foveal cones all of the same type and that these patches mediate high resolution. However, it may also be that the M and L cone mosaic is regular and that the postreceptoral visual system is clever in the way it combines information from the M and L cones, a possibility we address below. If our hypothesis is correct about the origin of Brewster's colors, their splotchy appearance indicates that the packing geometry of M and L cones is not perfectly crystalline, but the phenomenon is too fickle to allow us to reach quantitative conclusions about the amount of disorder that is actually present.

The Cost of Trichromatic Spatial Sampling in a Single Mosaic

The subtle and fleeting nature of Brewster's colors suggests that the visual system does not pay a substantial price for incorporating trichromacy. The similarity in the spectral sensitivities of the M and L cones plays some role in easing the chromatic aliasing problem. It means that signals from the two submosaics will be correlated. Thus, for example, the L cone value at a given point is a good predictor of what the M cone signal would have been had one existed at that point in the image. The model incorporated the overlapping spectral sensitivities of the M and L cones and, even so, large and highly visible chromatic aliases could be produced. The model output produced colors that were invariably much more visible than the psychophysical phenomenon. For example, when the contrast of the grating at the model input exceeded about 25% and the model eye was in focus, the predicted contrast of Brewster's colors was so high that they fell outside the gamut of our display. This is interesting since the model provides an indication of how severe chromatic aliasing could have been, had the visual system not had some additional protective mechanisms against it that we have yet to build into our model.

There are a number of ways such protection might be implemented, but perhaps the most interesting avenue to explore would be to incorporate a better description of the interpolation rules used by the visual system. The visual system is probably much better at guessing the external stimulus than our linear interpolation algorithm, and this could reduce the chromatic distortions that would otherwise arise from submosaic aliasing. When viewing stimuli for generating Brewster's colors, the visual system can presumably bring to bear a number of strategies for object recognition, more akin to hypothesis testing, that go well beyond passive interpolation. Brewster's colors can appear simultaneously with the original achromatic grating that generated them, and these two percepts wax and wane in relative strength over time. Though eye movements may also play a role here, this apparent competition between the two percepts is consistent with a multistable recognition process that settles briefly on one or another interpretation of the original stimulus. We saw additional evidence for this competition in preliminary observations of the effect of grating orientation on the visibility of Brewster's colors. The colors were most visible for oblique gratings rather than horizontal and vertical, possibly because the competitiveness of the percept of an oblique achromatic grating over a chromatic alias was reduced by the oblique effect (Emsley, 1925).

One way the post-receptoral visual system might reduce chromatic aliasing would be to bias the interpretation of very rapid changes in quantum catch over small distances toward changes in brightness rather than changes in hue. There may be some reason to believe that the visual system does this. First, Hayhoe and Williams (1987) showed that high spatial frequency, high contrast interference fringes appear desaturated compared with a uniform field of the same wavelength and average intensity. Second, contrast sensitivity functions for luminance modulations have a higher spatial bandwidth than those for isoluminant modulations (Mullen, 1985). Though such post-receptoral spatial filtering does not offer the protection against aliasing provided by pre-receptoral filtering, it may favor the interpretation of a high frequency pattern as a brightness instead of a chromatic variation.

The S cone mosaic does not usually alias under ordinary viewing conditions due to chromatic aberration, and aliasing in M and L cones produces only subtle distortion in the hues of tiny point sources and high spatial frequency achromatic patterns. More typical visual scenes are apparently unaffected by chromatic aliasing. The challenge for the future is to uncover the neural mechanisms that allow the visual system to make such clever guesses about the color everywhere in a scene when it only has one spectral sample at each point.

ACKNOWLEDGEMENTS

Supported by EY01319, AFOSR 85-0019, EY04367. We thank Hugh Perry for providing the photograph of the monkey fovea from which the trichromatic mosaic of Plate 1a was constructed.

REFERENCES

Ahumada, A., 1986, Models for the arrangement of long and medium wavelength cones in the central fovea, J. Opt. Soc. Am. (Suppl.), 3:92.

Akima, H., 1978, A method of bivariate interpolation and smooth surface fitting for irregularity distributed data points, ACM Trans. Math. Software, 4:148.

Brainard, D. H., Wandell, B. A., & Poirson, A. B., 1989, Discrete analysis of spatial and spectral aliasing, Invest. Ophthalmol. & Vis. Sci. (Suppl.), 30:53.

Brewster, D., 1832, On the undulations excited in the retina by the action of luminous points and lines, London and Edinburgh Philosoph. Mag. and J. Sci., 1:169.

Cavonius, C. R., & Estevez, O., 1975, Contrast sensitivity of individual colour mechanisms of human vision, J. Physiol., 248:649.

Cicerone, C. M., & Nerger, J. L., 1989, The relative numbers of long-wavelength-sensitive to middle-wavelength-sensitive cones in the human fovea centralis, Vis. Res., 29:115.

Coletta, N. J., & Williams, D. R., 1987, Psychophysical estimate of extrafoveal cone spacing, J. Opt. Soc. Am. A, 4:1503.

Coletta, N. J., Williams, D.R., & Tiana, C., Consequences of spatial sampling for human motion perception. Vis. Res., in press.

Curcio, C. A., Allen, K. A., Lerea, C., Hurley, J., Klock, I., Bunt-Milam, A., 1989, Distribution and morphology of human cones stained with anti-blue cone opsin, Soc. Neurosci. Abstr., 15:1206.

DeMonasterio, F. M., Schein, S. J., & McCrane, E. P., 1981, Staining of blue-sensitive cones of the macaque retina by a fluorescent dye. Science, 213:1278.

Emsley, H. H., 1925, Irregular astigmatism of the eye: Effect of correcting lenses, Trans. Opt. Soc., 27:28.

Erb, M. B., & Dallenbach, K. M., 1939, 'Subjective" colors from line patterns, Am. J. Psych., 52:227.

Green, D. G., 1968, The contrast sensitivity of the colour mechanisms of the human eye. J. Physiol., Lond., 196:415.

Hashimoto, Y., & Shimoda, Y., 1990, Identification of photoreceptor cell types and their distribution in the dace retina by NBT labeling (abstr.), Proc. Int. Soc. Eye Res., VI:42.

Hayhoe, M. M., & Williams, D. R., 1987, Spatial frequency dependence of the color of monochromatic light (abstr.), J. Opt. Soc. Am., A, 4:51.

Holmgren, F., 1884, Uber den Farbensinn, in: "Compt Rendu du Congres International de Science et Medecine," (Vol. 1. Physiology), Copenhagen, pp. 80-98.

Hopkins, H. H., 1955, The frequency response of a defocused optical system, Proc. R. Soc. Lond., A231:91.

Krauskopf, J., 1964, Color appearance of small stimuli and the spatial distribution of color receptors, J. Opt. Soc. Am., 54:445.

Krauskopf, J., 1978, On identifying detectors, in: "Visual Psychophysics and Physiology," J. C. Armington, J. Krauskopf, & B. R. Wooten, eds., Academic Press, Inc., New York, pp 283-295.

Krauskopf, J., & Srebro, R., 1965, Spectral sensitivity of color mechanisms: Derivation from fluctuations of color appearance near threshold, Science, 150:1477.

Luckiesh, M., & Moss, F. K., 1933, A demonstrational test of vision, Am. J. Psychol., 45:135.

MacLeod, D. I. A., & Boynton, R. M., 1979, Chromaticity diagram showing cone excitation by stimuli of equal luminance, J. Opt. Soc. Am., 69:1183.

Marc, R., & Sperling, H. G., 1977, Chromatic organization of primate cones, Science, 196:454.

Mullen, K. T., 1985, The contrast sensitivity of human colour vision to red-green and blue-yellow chromatic gratings, J. Physiol., 359:381.

Packer, O., & Williams, D. R., 1990, Eye movements and visual resolution, Invest. Ophthalmol. Vis. Sci. (Suppl.), 31:494.

Packer, R., Williams, D. R., Sekiguchi, N., Coletta, N. J., & Galvin, S., 1989, Effect of chromatic adaptation on foveal acuity and aliasing, Invest. Ophthalmol. Vis. Sci. (Suppl.), 30:53.

Sekiguchi, N., Williams, D. R., & Packer, O., Nonlinear distortion of gratings at the foveal resolution limit, Vis. Res., in press.

Skinner, B. F., 1932, A paradoxical color effect, J. Gen. Psychol., 7:481.

Smith, V. C., & Pokorny, J., 1975, Spectral sensitivity of the foveal cone pigments between 400 and 500mm, Vis. Res., 15:161.

Szél, A., Diamantstein, T., & Röhlich, P., 1988, Identification of the blue-sensitive cones in the mammalian retina by anti-visual pigment antibody, J. Comp. Neurol., 273:593.

Tiana, C., Williams, D. R., Coletta, N. J., & Haake, P. W., A model of aliasing in extrafoveal human vision, in: "Computational Models of Visual Processing", M. Landy, & A. Movshon, eds, MIT Press, Cambridge, in press.

Vimal, R. L. P., Pokorny, J., Smith, V. C., & Shevell, S. K., 1989, Foveal cone thresholds, Vis. Res., 29:61.

Williams, D. R., 1990, Photoreceptor sampling and aliasing in human vision, in: "Tutorials in Optics," D. T. Moore, ed, Optical Society of America.

Williams, D. R., 1985a, Aliasing in human foveal vision, Vis. Res., 25:195.

Williams, D. R., 1985b, Visibility of interference fringes near the resolution limit, J. Opt. Soc. Am. A, 2:1087.

Williams, D. R., 1986, Seeing through the photoreceptor mosaic, Trends Neurosci., 9:193.

Williams, D. R., 1988, Topography of the foveal cone mosaic in the living human eye, Vis. Res., 28:433.

Williams, D. R., The invisible cone mosaic, Proc. Natl. Res. Council Symp. Photorecept., in press.

Williams, D. R., & Coletta, N. J., 1987, Cone spacing and the visual resolution limit, J. Opt. Soc. Am. A, 4:1514.

Williams, D. R., & Collier, R. J., 1983, Consequences of spatial sampling by a human photoreceptor mosaic, Science, 221:385.

Williams, D. R., Collier, R. J., & Thompson, B. J., 1983, Spatial resolution of the short-wavelength mechanisms, in: "Colour Vision Physiology and Psychophysics," J. D. Mollon, & L. T. Sharpe, eds, Academic Press, London, pp 487-503.

Williams, D. R., MacLeod, D. I. A., & Hayhoe, M. M., 1981a, Foveal tritanopia, Vis. Res., 21:1341.

Williams, D. R., MacLeod, D. I. A., & Hayhoe, M. M., 1981b, Punctate sensitivity of the blue-sensitive mechanism, Vis. Res., 21:1357.

Wyszecki, G., & Stiles, W. S., 1982, "Color Science," Wiley, New York.

Yellott, Jr., J. I., 1982, Spectral Analysis of spatial sampling by photoreceptors: Topological disorder prevents aliasing, Vis. Res., 22:1205.

Yellott, Jr., J. I., 1983, Spectral consequences of photoreceptor sampling in the Rhesus monkey, J. Sci., 221:383.

Yellott, Jr., J. I., Wandell, B., & Cornsweet, T., 1984, The beginnings of visual perception: The retinal image and its initial encoding, in: "Handbook of Physiology: The Nervous System III," I. Darien-Smith, ed., American Physiological Society, New York, pp. 257-316.

VARIABILITY IN CONE POPULATIONS AND IMPLICATIONS

Joel Pokorny, Vivianne C. Smith and Michael F. Wesner

Visual Sciences Center, University of Chicago
939 East 57th St., Chicago, IL. 60637

INTRODUCTION

There have been a number of techniques directed toward estimating the relative numbers of the three cone types in the retinae of living humans. For the S-cones, psychophysical measurements which are dependent on the spatial distribution of S-cones (Stiles, 1949; Wald, 1967; Williams et al., 1981) give data which are consistent with the relative sparcity observed anatomically (Ahnelt et al., 1987; deMonasterio et al., 1985; Marc and Sperling, 1977). For typical color-normal observers, the L-and M-cones are both in relatively high spatial density, and the direct psychophysical techniques which are effective for characterizing the retinal distribution of S-cones do not yield interpretable information (Brindley, 1954; Green, 1968; Kelly, 1974). Additional techniques in the literature which have been suggested as capable of revealing L/M cone ratios include Weber fractions for mechanisms isolated by chromatic adaptation (Vos and Walraven, 1971) and retinal densitometry with chromatic bleaching lights (Rushton and Baker, 1964). In this paper we wish to describe our work on two techniques we have used to estimate the L/M cone ratio, spectral sensitivity from heterochromatic flicker photometry and point source detection for lights of varying wavelength. We then consider the perceptual consequences of individual variation in L/M cone populations.

HETEROCHROMATIC FLICKER PHOTOMETRY

For normal color vision, the luminosity function may be modeled as a linear sum of the L- and M-cone spectral sensitivities (Smith and Pokorny, 1975). A number of investigators (Adam, 1969; Crone, 1959; deVries, 1947; Lutze et al., 1990; Wallstein, 1981) have suggested that individual variation differences in the ratio of heterochromatic flicker sensitivity for "red" and "green" lights (red/green ratio) reflects the relative proportion of L to M cones. The evidence usually cited for this view is Rushton and Baker's (Rushton and Baker, 1964) comparison of red/green flicker ratios with retinal densitometry.

While there is evidence that L/M cone ratios are reflected in HFP spectral sensitivities, there are other biological variables such as possible differences in photopigment spectra among color normal observers (Alpern and Pugh, 1977; Eisner and MacLeod, 1981; Neitz and Jacobs, 1986; Smith et al., 1976; Webster and MacLeod, 1988), photopigment optical density (Burns and Elsner, 1985; Smith et al., 1976) and pre-receptoral filtering (principally the lens, (Pokorny et al., 1987)) which could

From Pigments to Perception, Edited by A. Valberg and
B.B. Lee, Plenum Press, New York, 1991

23

contribute to the measured spectral sensitivities. A straightforward way to evaluate the effect of these variables is to compare population variability of anomaloscopic color matching data with the population variability of HFP spectral sensitivity using wavelengths similar to or identical with the color matching primaries. The HFP logarithmic G/R flicker ratios have a standard deviation of 0.07-0.10 log unit (Adam, 1969; Lutze et al., 1990; Rushton and Baker, 1964; Wallstein, 1981), two to three times larger than the standard deviation of log G/R ratios of Rayleigh matches (Adam, 1969; Helve, 1972; Lutze et al., 1990; Marré and Marré, 1984; Rushton and Baker, 1964; Schmidt, 1955; Wallstein, 1981). The color matching data place constraints on the contributions of variables other than the L/M cone ratio to measured HFP spectral sensitivities; less than 19% of the HFP spectral sensitivity variance may arise from the pre-receptoral and receptoral factors cited above (Pokorny et al., 1988). We are left with the L/M cone ratio as a prime candidate for the large population variability in HFP spectral sensitivity.

POINT SOURCE DETECTION

Recently, two laboratories have used point source detection techniques to estimate the relative populations of L to M cones in the human retina (Cicerone and Nerger, 1989a, 1989b; Vimal et al., 1989; Wesner et al. in press; Wesner, et al., 1988). The technique and model we have employed yields estimates of the foveal L/M cone ratio, the minimum number of quanta per cone required for detection (C) and the effective number of cones (N) stimulated by the test.

Estimates of the L/M cone ratio are determined from the effect of test wavelength on the shapes of psychometric functions for the detection of point sources of light. We will outline the rationale of our approach by first developing a prototype retinal mosaic. Then we will explore the consequences of change in the major independent variable, wavelength. Figure 1 shows a mosaic of L-and M-cones arranged in a regular triangular array. The cone types have been assigned to locations in a random manner, with an L/M cone ratio of 2:1. The successive panels from left to right show how the relative probability of quantal catch would change as a function of wavelength. Here, probability of quantal catch is represented by a grey scale, with higher reflectances indicating higher probabilities. For the purpose of demonstration, the L-cone probability of quantal catch is represented by the same grey for all the wavelengths. For the shortest wavelength, 520 nm, the M-cones have a slightly higher probability of quantal absorption than the L-cones. For a wavelength where the two cone types have similar probability of absorption, (for example, 546 nm), all of the cones in the mosaic have equal probability of quantal catch and are depicted by the same reflectance grey. At longer wavelengths (600 and 650 nm), the M-cones locations form "probability troughs" (depicted in darker grey) where the depth of the troughs represents a decrease in quantal catch probability relative to the L-cone. Therefore it may be seen that the distribution of the probability of quantal catch by the cones which lie under the retinal image of a stimulus will vary with wavelength.

We use stimuli of small visual subtense, 1' diameter. With sufficiently small retinal stimuli, the change in the relative probability of quantal catch for the two cone types will be revealed by changes in the shapes of threshold psychometric functions, with shallower functions being associated with stimuli of longer wavelength. Mosaics composed of different L/M cone ratios will have different frequencies of the "probability troughs". Specifically, the greater the proportion of M-cones, the greater the expected change in the shape of the psychometric function with wavelength. The experimental paradigm we used involves measurements of psychometric functions at each of a series of wavelengths presented to the dark-adapted fovea. Since the adapting conditions are identical for all measurements, any changes in the shape of the psychometric functions may be attributed to wavelength specific effects.

At any given wavelength, each trial consists of simultaneous brief presentations of two point sources. The observer reports seeing 0, 1, or 2 stimuli. In the Vimal et al. (1989) study, we derived psychometric

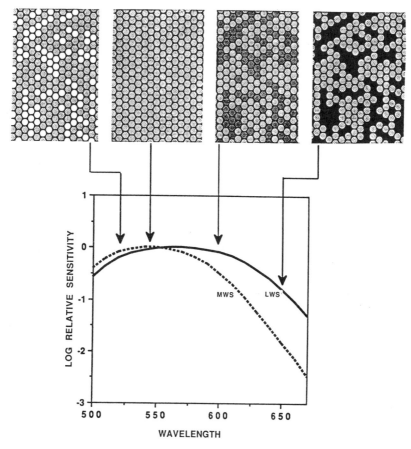

Figure 1. The lower panel shows the Smith and Pokorny (1975) L- and M-cone spectral sensitivities normalized to equal sensitivity at their respective peaks. The upper panels all represent the same mosaic of L- and M-cones with successive panels from left to right showing how the relative probabilities of quantal catch for M- and L-cones would change as a function of wavelength. The cone types have been assigned to locations in a random manner, with an L/M ratio of 2:1. The grey scale is used to represent probability of quantal catch with higher reflectances indicating higher probabilities. For the purpose of graphical presentation, the L-cone probability of quantal catch has been assigned the same grey for all the wavelengths (After Vimal et al., 1989).

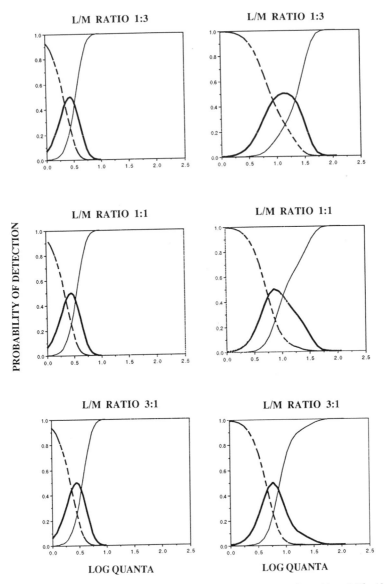

Figure 2. Predicted detection functions for wavelengths 546 (left) and 620 nm (right). The probability of detection for zero [P(0)], exactly one [P(1)], and exactly two points [P(2)] are plotted (dashed, thick, and thin lines respectively) as a function of log relative quanta (i.e. average number of quanta absorbed at the photoreceptor) to allow comparison of the slopes. The average number of quanta, C, is assumed to be 4; the number of cones attaining the average number, N, is 2; the parameter P_L is the proportion of L-cones. The upper panels show predictions for a P_L of 0.25 (L/M ratio of 1:3); middle panels for P_L of 0.5 (L/M ratio of 1:1) and the lower panels for a P_L of 0.75 (L/M ratio of 3:1).

functions for the detection of both stimuli, and either one or both
stimuli when the stimulus is two points. A least squares minimization
procedure yielded a single best estimate of the solution parameters.
These fits required estimation of 15 parameters, raising the question of
which alternative fits to the data provide nearly as good fits as the
single optimal solution.

Wesner et al. (in press) used a Chi-square fitting procedure which
allowed statistical confidence bounds to be established for the best
parameter estimates. The chi-square method required that the raw data be
expressed in terms of the proportion of neither point $[(1-P)^2]$, exactly
one point $[2(1-P)P]$, and exactly two points $[P^2]$ detected when two points
were presented. Figure 2 shows the predicted psychometric functions for
two of the wavelengths Wesner et al. (in press) used (546 nm in left
panels and 620 nm in right panels) as a function of the average number of
quanta absorbed when the stimulus consists of two points of light (other
modeling parameters are given in the figure caption). The upper left
graph is for a mosaic with a L/M cone ratio of 1:3, the middle graph is
for a ratio of 1:1 and the lower graph is for a ratio of 3:1. As can be
seen, with an increase in the density of "probability troughs" (created

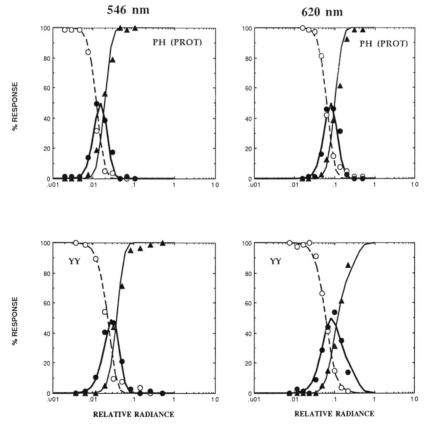

Figure 3. Data and predicted functions for a protanope (top panels)
and a color-normal observer (bottom panels). Details as for Figure
2. The model solutions for observer YY are: P_L is 0.61, C is 4 and
N is 2. For the protanope, P_L is 0.00, C is 6 and N is 1 (data and
theoretical solutions from Wesner et al., in press).

in the locations of the M-cones at longer wavelengths), the 620 nm
psychometric functions change shape and for the L/M cone ratio modeling
parameter, decreases lead to the $P(1)$ function broadening, and the $P(2)$
function shallowing. For the 1:3 L/M cone ratio, to achieve a
probability of detection approaching 1.0, stimuli need be sufficiently
intense so that the M-cones in the mosaic each approach a 1.0 probability
of reaching the detection criterion. The change in the shape of the
psychometric functions with wavelength is the property of the model that
establishes the value of P_L.

Vimal et al. (1989) obtained data from 2 normal trichromats. Wesner
et al., (in press) added three more normal trichromats and also obtained
interpretable data from three dichromatic observers, two protanopes and
one deuteranope. For the dichromats, the prediction of the model is such
that parameter P_L should converge to either 0.0 or 1.00 (i.e. only 1
active cone type). The only difference between fits with P_L of 0.00 and
1.00 is found in values associated with scaling factors which act on the
position of the psychometric function along the relative radiance axis.
We assign the appropriate P_L value based upon independent evidence from
spectral sensitivity. The minimum Chi-square residuals for each of the
protanopes occurred at P_L values of 0.00 and for the deuteranope the best
P_L value was 0.98.

Figure 3 shows data and best fit theoretical functions for two of
Wesner et al's (in press) observers, a protanope (PH) and a color-normal
observer (YY). As can be seen, the data are well characterized by the
best fitting theoretical functions. The psychometric functions for the
protanope at 546 and 620 nm are closely similar whereas these functions
differ systematically with wavelength for the color-normal.

Figure 4 shows estimates of the best P_L values derived from the
point source data and from the HFP data. Although the number of
observers is limited, there appears to be some congruence between the two

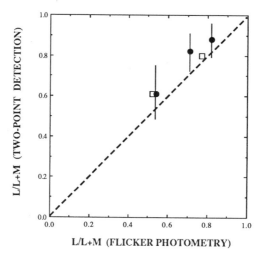

Figure 4. L/L+M estimated for color-normal observers from the point
detection experiment as a function of L/L+M estimated from flicker
photometry. The squares are the solutions from the Vimal et al.
(1989) study and the circles are the best solutions from Wesner et
al. (in press). The vertical bars represent the P_L ranges defined by
the 98% confidence interval.

TABLE 1a

MEAN AND STANDARD DEVIATIONS OF EQUILIBRIUM YELLOW WAVELENGTH

AUTHOR	# OBSERVERS	EQUILIBRIUM YELLOW	SD
Dimmick & Hubbard, 1939	10	583 nm	2.9
Rubin, 1961	278	577 nm	2.0
Hurvich & Jameson, 1964	40	584 nm	3.4
Wallstein, 1981	16	580 nm	5.3

TABLE 1b

STANDARD DEVIATIONS OF FLICKER PHOTOMETRIC MATCHES OF "RED" AND "GREEN" LIGHTS

AUTHOR	# OBSERVERS	SD (log units)
Rushton & Baker, 1964	197	0.100
Adam, 1969	175	0.077
Wallstein, 1981	16	0.056
Lutze et al., 1990	72	0.096

estimates. All observers show estimates of L/M cone ratio greater than 1:1 by both techniques, and the observers showing higher estimates by flicker photometry also show higher estimates by the point source detection paradigm.

PERCEPTUAL CONSEQUENCES OF INDIVIDUAL VARIATION IN L/M CONE POPULATIONS

The psychophysically defined chromatic channels are characterized by their equilibrium hues. equilibrium hues are colors which appear unitary or as psychologically equilibrium percepts. For spectral colors, four equilibrium hues may be identified; red, yellow, green and blue. The four equilibrium hues are used as parameters in color vision models based on the perceptual aspects of color appearance (eg. Boynton, 1979; Hurvich and Jameson, 1955; Ingling and Tsou, 1977). For example, equilibrium yellow is neither reddish nor greenish and is considered a null (or balance) point for a red/green color opponent mechanism. To achieve the correct spectral position for equilibrium yellow, the weighting of the L- and M-cone contribution must be approximately equal. Thus there is a difference in average weighting of the L- and M-cones when the luminance and chromatic channels are computed. Is it possible to estimate the spectral position of equilibrium yellow from the estimates of the L/M cone populations? To examine this question we wish to compare the individual differences in the cone weightings for the luminance system with individual differences in the weightings for the red/green chromatic system. The question we pose is: Can the locus of equilibrium yellow be predicted by the estimated L/M cone ratio derived from flicker photometry?

Two lines of evidence suggest not. The first is taken from a literature survey of population data concerning HFP spectral sensitivity and the spectral locus of equilibrium yellow, and the second from data for both measures on a more limited number of observers.

A number of investigators report data on the spectral locus for equilibrium yellow (Table 1a). The spectral location of equilibrium yellow ranged from 577 nm to 584 nm and the standard deviation ranged from 2.0-4.8. Additionally, Cicerone (1987) surveyed seven studies not included in Table 1a (total of 25 observers) and stated a range of 568-588 nm. If we assume a mean of 580 nm and a standard deviation of 4.0, then the expected values of equilibrium yellow in the population (+/- 3 SD) might range from 568 nm to 592 nm. If such variation were produced by variation in relative populations of the M- and L-cones, then (assuming the M-cone weight is twice as great for the red/green opponent channel than for the luminance channel) flicker photometric ratios would have a range of only 0.084 log unit. Literature data (Table 1b) show a far greater range of flicker photometric matches in the population, with standard deviations of 0.07-0.10 log unit. The range predicted from 3 standard deviations of the equilibrium yellow population data represents only about 1 standard deviation of the population flicker photometric data (Pokorny and Smith, 1987).

A second line of evidence that L/M ratios and equilibrium yellow are not linked can be found in Wallstein's (1981) data. He compared equilibrium yellow with red/green ratios from heterochromatic flicker photometry in 16 observers. Data were gathered at four field sizes, 0.5° 1.0°, 2.0° and 4.0°. Here we report only the 2° data though the same trends were found for all field sizes. Wallstein's first step was to obtain values for equilibrium yellow in the same metric as the heterochromatic flicker photometric ratios. He obtained estimates of the spectral position of equilibrium yellow from a device with continuous variation of wavelength and expressed the results in terms of a transformation to the (Judd, 1951) standard observer. Wallstein's observers also set a mixture equilibrium yellow using primaries of 545 nm and 670 nm (Figure 5). The correlation between the wavelength equilibrium yellow (transformed to an R/G mixture) and the mixture equilibrium yellow was 0.83, indicating that the observers were using the same perceptual criterion in both cases. Further, the high correlation indicates that prereceptoral filtering, which would alter mixture equilibrium yellow settings but not wavelength equilibrium yellow settings, was not a significant factor in this study. Finally the observers made an HFP match using the 545 nm and 670 nm primaries.

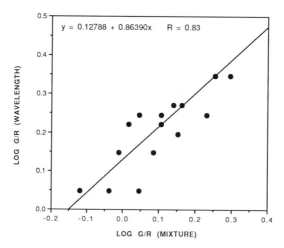

Figure 5. Equilibrium yellow obtained by two methods. Mixture equilibrium yellow was obtained by adjustment of the relative amount of two primaries, 545 nm and 670 nm. For the second method, the wavelength for equilibrium yellow was measured, and the result expressed in G/R by transformation using the (Judd, 1951) standard observer (Data from Wallstein, 1981).

The correlation between the log R/G mixture equilibrium yellow and the log
R/G for heterochromatic flicker photometry was −0.4; the correlation
between the spectral equilibrium yellow expressed in log R/G (from the
colorimetrically transformed data) and the log R/G for heterochromatic
flicker photometry was −0.27 (Figure 6). Thus Wallstein's data show a low
association between equilibrium yellow and HFP in the incorrect direction.
In a further step, we can convert the R/G ratios to L/M cone ratios using
the Smith and Pokorny L- and M-cone spectral sensitivities. The L/M
ratios for heterochromatic flicker photometry have a range greater than
20:1, while the L/M ratios for equilibrium yellow have a range of only
2:1. The correlation of these L/M ratios is −0.29. This calculation
emphasizes that the variation in L/M ratio from flicker photometry is
large compared with that predicted by the variation in equilibrium yellow.
We suggest that, while HFP spectral sensitivities reflect receptor
populations, the normalization for the red/green chromatic mechanism
depends on other factors, perhaps on a normalization to the "average
white" of the individual's environment (Pokorny and Smith, 1977).

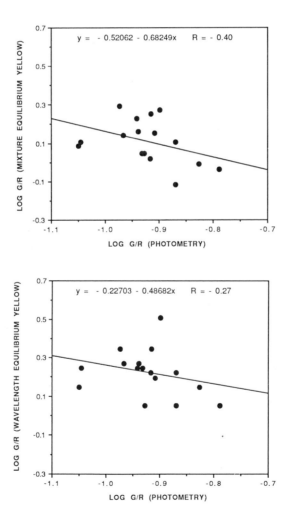

Figure 6. Equilibrium yellow log G/R plotted against log G/R from
flicker photometry. The top panel shows equilibrium yellow as measured
by color mixture, the bottom panel shows results of measurement by
wavelength variation (Data from Wallstein, 1981).

SUMMARY

Two techniques were used to estimate the L/M cone ratio, best-fit linear sum of the L- and M-cone spectral sensitivities to heterochromatic flicker photometric spectral sensitivity and psychometric function shape for point source detection of lights of varying wavelength. Data from five color-normal observers run on both paradigms all are consistent with a preponderance of L-cones relative to M-cones though there are individual differences in the estimates of L/M cone ratio. The observers showed congruence in their individual results across technique. In a separate study, the perceptual consequences of individual variation in L/M cone populations were evaluated by looking for a relation between flicker photometric spectral sensitivity and the spectral locus of equilibrium yellow. No significant relation was found, suggesting that receptor populations do not play a major role in the normalization of the perceptual red/green opponent channel.

ACKNOWLEDGEMENT

Supported in part by USPH NEI grant EY00901.

REFERENCES

Adam, A. (1969). Foveal red-green ratios of normals, colorblinds and heterozygotes. Proceedings Tel-Hashomer Hospital (Tel-Aviv). **8**: 2-6.

Ahnelt, P., H. Kolb and R. Pflug. (1987). Identification of a subtype of cone photoreceptor, likely to be blue sensitive, in the human retina. Journal of Comparative Neurology. **255**: 18-34.

Alpern, M. and E. Pugh. (1977). Variation in the action spectrum of erythrolabe among deuteranopes. Journal of Physiology (London). **266**: 613-646.

Boynton, R. (1979). <u>Human Color Vision</u>. New York, Holt, Rinehart and Winston.

Brindley, G. (1954). The summation areas of human colour-receptive mechanisms at increment threshold. Journal of Physiology (London). **124**: 400-408.

Burns, S. and A. Elsner. (1985). Color matching at high illuminances: the color-match-area-effect and photopigment bleaching. Journal of the Optical Society of America A,. **2**: 698-704.

Cicerone, C. (1987). Constraints placed on color vision models by the relative numbers of different cone classes in human fovea centralis. Die Farbe. **34**: 59-66.

Cicerone, C. and J. Nerger. (1989a). The relative numbers of long-wavelength-sensitive to middle-wavelength-sensitive cones in the human fovea centralis. Vision Research. **26**: 115-128.

Cicerone, C. and J. Nerger. (1989b). The density of cones in the fovea centralis of the human dichromat. Vision Research. **26**: 1587-1595.

Crone, R. (1959). Spectral sensitivity in color-defective subjects and heterozygous carriers. American Journal of Ophthalmology. **48**: 231-238.

deMonasterio, F., E. McCrane, J. Newlander and S. Schein. (1985). Density profile of blue-sensitive cones along the horizontal meridian of macaque retina. Investigative Ophthalmology and Visual Science. **26**: 289-302.

deVries, H. L. (1947). The heredity of the relative numbers of red and green receptors in the human eye. Genetica. **24**: 199-212.

Dimmick, F. and M. Hubbard. (1939). The spectral location of psychologically unique yellow, green, and blue. American Journal of Psychology. **52**: 242-254.

Eisner, A. and D. MacLeod. (1981). Flicker photometric study of chromatic adaptation: selective suppression of cone inputs by colored backgrounds. Journal of the Optical Society of America. **71**: 705-718.

Green, D. (1968). The contrast sensitivity of the colour mechanisms of the human eye. Journal of Physiology (London). **196**: 415-429.

Helve, J. (1972). A comparative study of several diagnostic tests of colour vision used for measuring types and degrees of congenital red-green defects. Acta Ophthalmologica Supplement. **115**: 1-64.

Hurvich, L. and D. Jameson. (1955). Some quantitative aspects of an opponent-colors theory. II. Brightness, saturation and hue in normal and dichromatic vision. Journal of the Optical Society of America. **45**: 602-616.

Hurvich, L. and D. Jameson. (1964). Does anomalous color vision imply color weakness? Psychonomic Science. **1**: 11-12

Ingling, C. and B. Tsou. (1977). Orthogonal combination of three visual channels. Vision Research. **17**: 1075-1082.

Judd, D. (1951). Colorimetry and artificial daylight, in Technical Committee No. 7 Report of Secretariat United States Commission, International Commission on Illumination, Twelfth Session, Stockholm, pp 1-60.

Kelly, D. (1974). Spatio-temporal frequency characteristics of color-vision mechanisms. Journal of the Optical Society of America. **64**: 983-990.

Lutze, M., N. Cox, V. Smith and J. P. J. (1990). Genetic studies of variation in Rayleigh and photometric matches in normal trichromats. Vision Research. **30**: 149-162.

Marc, R. and H. Sperling. (1977). Chromatic organization of primate cones. Science. **143**: 454-456.

Marré, M. and E. Marré. (1984). Rayleigh equation in acquired color vision defects. Documenta Ophthalmologica Proceedings Series. **39**: 165-170.

Neitz, J. and G. Jacobs. (1986). Polymorphism of the long-wavelength cone in normal human colour vision. Nature. **323**: 623-625.

Pokorny, J. and V. Smith. (1977). Evaluation of single pigment shift model of anomalous trichromacy. Journal of the Optical Society of America. **67**: 1196-1209.

Pokorny, J. and V. Smith. (1987). L/M cone ratios and the null point of the perceptual red/green opponent system. Die Farbe. **34**: 53-57

Pokorny, J., V. Smith and M. Lutze. (1987). Aging of the human lens.Applied Optics. **26**: 1437-1440.

Pokorny, J., V. Smith and M. Lutze. (1988). Sources of shared variance in Rayleigh and photometric matches. OSA Annual Meeting, 1988 Technical Digest Series, Washington, DC. **11**: 64.

Rubin, M. (1961). Spectral hue loci of normal and anomalous trichromates. American Journal of Ophthalmology. **52**: 166-172.

Rushton, W. and H. Baker. (1964). Red/green sensitivity in normal vision. Vision Research. **4**: 75-85.

Schmidt, I. (1955). Some problems related to testing color vision with the Nagel anomaloscope. Journal of the Optical Society of America. **45**: 514-522.

Smith, V. and J. Pokorny. (1975). Spectral sensitivity of the foveal cone photopigments between 400 and 500 nm. Vision Research. **15**: 161-171.

Smith, V., J. Pokorny and S. Starr. (1976). Variability of color mixture data - I. Interobserver variability in the unit coordinates. Vision Research. **16**: 1087-1094.

Stiles, W. (1949). Increment thresholds and the mechanisms of colour vision. Documenta Ophthalmologica. **3**: 138-163.

Vimal, R., J. Pokorny, V. Smith and S. Shevell. (1989). Foveal cone thresholds. Vision Research. **29**: 61-78.

Vos, J. and P. Walraven. (1971). On the derivation of the foveal receptor primaries. Vision Research. **11**: 799-818.

Wald, G. (1967). Blue-blindness in the human fovea. Journal of the Optical Society of America. **57**: 1289-1301.

Wallstein, R. (1981). Photopigment variation and the perception of equilibrium yellow. Doctoral Dissertation, University of Chicago.

Webster, M. and D. MacLeod. (1988). Factors underlying individual differences in the color matches of normal observers. Journal of the Optical Society of America A. **5**: 1722-1735.

Wesner, M., J. Pokorny, S. Shevell and V. Smith. (in press). Foveal cone detection statistics in color-normals and dichromats. Vision Research. **in press**:

Wesner, M., J. Pokorny, V. Smith and D. Feige. (1988). Effect of detection criterion on foveal cone thresholds. OSA Annual Meeting, 1988 Technical Digest Series, Washington, DC. **11**: 67.

Williams, D., D. MacLeod and M. Hayhoe. (1981). Punctate sensitivity of the blue-sensitive mechanism. Vision Research. **21**: 1357-1375.

DISCUSSION: BIOPHYSICS AND PSYCHOPHYSICS OF PHOTORECEPTORS

John Mollon

Dept. of Experimental Psychology
Downing Street,
Cambridge, England

Mollon: We have had this morning four masterly but heterogeneous papers, and I can identify only one recurrent theme, and that is the numerosity and packing of the receptors. We will take that as a general topic to discuss first.

Jim Bowmaker reported that in a large sample of cones from Old World monkeys the L- to M- cone ratio comes out at unity; this value is different from the ratio obtained for a somewhat smaller sample from human retinae, which was closer to the higher ratio required by some psychophysical procedures. If this species difference is real, then it reminds us that even macaques are not always going to be a suitable model for man. The organizers have set us the task of correlating physiology and psychophysics, but we should remember that in the time since our ancestors diverged from the catarrhine monkeys we have evolved all the changes in the vocal apparatus and in the left cerebral hemisphere that allow us to conduct a linguistic discourse; and therefore we should not assume there have been no substantial changes in the visual system.

So the difference in cone ratios may be a real species difference, as may the difference in the wavelength of peak sensitivity of the short-wave cones that Jim Bowmaker mentioned. However, there is one thing that Jim did not have time to mention If you give him a monkey retina, he can take his sample always within a couple of millimeters of the fovea, even if the fovea is difficult to see under the dim red dissection-light. But the human retinae that he and I get are invariably from cases of enucleation for melanoma, and we have to take the part of the retina that either the surgeon or pathology is willing to give us. Often the melanoma may have lifted away the foveola. So the ratios for human tissue are not to be taken as ratios for foveal vision, whereas the monkey ratios are all drawn from close to the anatomical fovea.

I should like to move on to the psychophysical estimates of Joel Pokorny. The first two methods that he discussed - flicker photometry and the ratio method - both give a predominance to the long-wave cones. I should like in a moment to ask Joel about the modelling that Chris Tyler was offering at the Alpern Symposium, where he simply models the absolute cone thresholds for small foveolar sources and finds he needs a ratio more like unity. But for now, let us pass on quickly to the discrepancy with measurements of unique yellow. The variance in unique yellow (the wavelength that looks neither reddish nor greenish) is much less than would be expected from flicker- photometric studies of populations. My own view is that unique yellow should behave more like a Rayleigh match than like flicker photometry. I put the view forward in 1982 in 'Annual Reviews of Psychology', but I suspect the idea goes back further. Unique yellow is that wavelength that produces in the long- and middle-wave cones the same ratio of absorptions as does the average illuminant of our world. That is to say, the tritanopic sub-system of our colour vision is in equilibrium for lights that produce the same ratio of quantum catches as does the mean chromaticity that we experience in our everyday life; its job is to signal departures from that equilibrium point. And therefore the variation in the population should have the same characteristics as the variation in Rayleigh matches: it should depend on the shape of the absorption curves and not at all on the relative numbers of long- and middle-wavelength cones.

Unique yellow should also correspond to what is called Sloan's notch. (In the case of increment thresholds, the notch should perhaps be called the Stiles- Crawford notch, since those authors were the first to report it for coloured increments on a white field). That notch, I know experimentally, does correspond to the tritanopic equilibrium point. At ARVO in 1987, Clemens Fach and I showed results for what happens to the Stiles-Crawford notch as you change the colour temperature of the white adaptation field: the bottom of the notch always occurs at the wavelength that lies on the tritanopic confusion line that passes through the chromaticity of the field. So the notch has nothing to do with the relative numbers of long- and middle-wave cones. In modelling it, we should, I think, take into account only the spectral sensitivity curves of the long- and middle-wave pigments: the notch will occur when the ratio of absorptions in those pigments is the same as the chromaticity of the background used in a given experiment.

That is one view of why unique yellow should not give the same result as flicker photometry. Donald MacLeod , at the end of his talk, came to slightly different reasons for a conflict between flicker photometry and unique yellow. When you change the chromatic adaptation of the eye you produce more shift in the flicker photometric results than in unique yellow. Now, Donald has a very ingenious way of accounting for this, by supposing that the adaptation is occurring not grossly in the cones, but at synaptic sites for different bipolars.

I have an alternative suggestion about this particular experimental discrepancy, and it goes back to an alternative view of what the subject is doing in flicker photometry. It's a view that I put forward in 'Die Farbe' for 1987. Flicker does not tap an additive channel, as everybody assumes. Rather, according to this view, the channel that we use in flicker photometry is a.c.-coupled to both long- and middle-wave cones; and what the observer is doing at the final setting is making a compromise. And of course this hypothesis has some physiological basis in the results from Goettingen, in that at high adaptation levels there is a frequency doubling of the response that is visible at the null setting when the response is minimal. If a person is a protanope and you ask him to make a flicker-photometric null, it is straightforward what he has to do: he goes for what Rushton calls the isolept, that is, the ratio of the two stimulus lights that produces the same quantum catch in the middle- wave cones. And conversely, if he were a deuteranope, he would go for the isolept for the long-wave cones. If he is normal, he has got to compromise, and what he does is go somewhere in the middle, between the two isolepts. And by putting his setting somewhere in the middle, he will give you something that looks extremely like the photopic sensitivity function.

Now, in flicker photometry, subjects find it easiest to do the experiment if you allow them to change the temporal frequency. If they can increase frequency so that only cone mechanism is above threshold in the region of the null, then the setting is as convenient for them as it is for a dichromat. This manoeuvre is particularly open to the subject if an asymmetry in the state of chromatic adaptation increases the separation of the sensitivities of the cone mechanisms at threshold. You may then find a bias totally in favour of one isolept or the other.

Lennie: Perhaps Jim Bowmaker would like to comment on the difference between his measurements and the electrophysiological measurements of the S-cones.

Bowmaker: The S-cones have only been measured in both ways in macaques, and both come out with a peak about 430 nm. Baylor's group has not measured human S-cones, we have measured very few but they always come out round 420 nm.

Dow: I want to ask about the ratio of the L- and M-cones at different eccentricities. You suggested that perhaps in the centre in the foveola the ratio is closer to one to one, and then as you proceed away from the foveola the ratio may drop to two to one with the long-wavelength cones predominating. Maybe this is a kind of relative protanopia in the foveola, and perhaps this may assist our visual acuity in the sense that the chromatic aberration will be least for the M-cones. There is a second point related to that, which has to do with the input to yellow cells. I think the traditional view has been that the cone input to yellow cells is a sum of the L- and M-cones, both acting in a positive way. But there is some evidence from single-cell recordings that the cone input to yellow cells may actually be entirely from the M-cones, and that the L-cones and that the L-cones may provide largely inhibitory input to yellow cells, and this may also make yellow cells have a higher acuity than red cells.

Shapley: There are actually some data about the possible dependence of L and M ratios on eccentricity. One comes from measurements of π mechanisms, which show no difference, and the other comes from the Stabells' measurements of the photopic luminosity function with eccentricity, showing no difference whatsoever at longer wavelengths than 500 nm. There is a difference in the S-cones' possible input to luminosity, but zero change in the L and M cones. So two very good quantitative psychophysical experiments indicate very little difference in the L- to M-cone ratio with eccentricity.

Mollon: I have a counter-comment, which is that there are some long- forgotten data of Stiles [Documenta Ophthalmologica, 1949] that suggest a rapid and striking variation in π_5 to π_4 ratios across the one degree of the central foveola.

Creutzfeldt: Most neurophysiologists agree that the ratio of red-cone excited cells far exceeds the number of M-cone excited cells all the way up to cortex. I wonder what that means when we hear that in sub-human primates the cone ratio is one-to-one, while if you take all the literature together, all come up with at least a two-to-one ratio in the parafoveal region.

Lennie: I want to dispute the observation. A couple of observations suggest that the ratio might be two-to-one, but in fact if you look at a large number of published observations, the ratios are very close to unity in the LGN. In our observations there are no doubts about it. I had thought, like you did, that the published observations favoured a two-to-one ratio, but that is not the case.

Mollon: I should like to ask John Krauskopf whether he has any comments on Professor Pokorny's technique, and then I should like to ask Joel himself whether he has any comments.

Krauskopf: I would support one of the points that was disputed earlier. I think our experiments in the Dark Ages do support the idea that detection is mediated by the signals that pass through individual cones. I don't mean to say that detection is determined in individual cones, but particularly the fact that the saturation judgements people made, saturated reds and saturated greens, were the only judgments that decreased in relative frequency with frequency of seeing, argued that single cones did mediate a very high percentage of detections. So that supports the methodology.

Pokorny: You asked me about Tyler's work. I am reluctant to talk about Chris's work when he is not here, but the experiments he reported on were data of Graham and Hsia, Guth, and Hurvich and Jameson, and none of those experiments used small, brief stimuli. The experiment that would be comparable with ours was Hood and Finkelstein's, and they obtained a Vl spectral sensitivity..The other question was the one of unique yellow, and our view is very similar to yours, and that is that there is some life-averaged illuminant, and our colour judgements are made relative to that.

Williams: Can you clarify your results on unique yellow in comparison with Carol Cicerone? Are you in conflict there?

Pokorny: We are in conflict and I don't understand it. David Wallstein in our lab in 1981 found a slight negative correlation, and Carol finds a strong positive correlation between unique yellow and flicker photometry. Her methodology looks good to us, our methodology looks good to her: it's not clear to me what's going on.

MacLeod: With the model relating detection to the cone mosaic, your model assumes that individual cones determine threshold by responding if and only if they absorb a large number of quanta, you showed 5 quanta. This feature seems implausible, for evidence suggests that cones are linear transducers and in view of the psychophysics of spatial summation, which does not seem to provide an advantage for small stimuli, as such an assumption would require. How important is this assumption?

Pokorny: We have been working on a model incorporating different criteria and areal summation, and we hope to get a crisp answer.

Shapley: This is a question about the species differences between man and other primates. Two comments, one from Peter Schiller, which is that the photopic luminosity function for macaques is rather similar to that of humans, so that if it were determined by cone numerosity, you might expect that in fact the cone numerosities were similar for macaques and humans. Peter tells me that the luminosity functions are not exactly the same, but they are really much closer than a two to one ratio would lead you to believe. The other point is that in Barry Lee's experiments and in ours on equal luminance points for macaque ganglion cells, we get very good agreement between macaque M-cells and human data. So I think we should be cautious about jumping to the conclusion from one set of data that humans are very different from the macaques.

Hayhoe: This has nothing to do with numerosity: I am just picking up on your comment on Donald's work. Can either you or Donald take your suggestion one step further as to how your idea of what is going on in flicker photometry relates to his results with different gains for unique yellow and flicker photometry?

Mollon: What I was suggesting was that in a judgement of hue you must use both classes of cone, but if you are making flicker photometric settings you may, depending how you are adapted in the experiments, go towards one isolept or towards the other isolept. You may may be able to place yourself in a region where you are getting modulation from only one class of cone.

MacLeod: I also missed the rationale for that in terms of your idea that the observer might be monitoring the L- and M-cone signals independently. It is not very clear to me why, if he does that, which I find very implausible to begin with, he should then vary all the way to one or the other as the adaptation conditions change. But I also want to register my scepticism about that basic view of flicker photometry, because I am one of those who adhere to the standard model where there is a single colour-blind channel that responds to rapid flicker, and adds together the signals from the L- and M-cones. And I wonder whether your model can explain such things as the very strong dependence of flicker sensitivity on the relative phase of the L- and M-cone stimuli, and on their relative amplitude. One finds a very pronounced trough in sensitivity as one varies relative amplitude or phase, and I cannot really see how that is explainable with a model that keeps the two cone signals separate at the seat of consciousness, the Helmholtz model as it were of the visual system. I also don't see why the observation of frequency doubling that you cited as support for your view, does support your view. It seems to me that it is exactly what would have been expected on the classical model that you are arguing against. When you have arranged for the L- and M-cone signals to cancel in that model, then the fundamental disappears, and all that is left is harmonic distortion, the frequency doubling.

Mollon: It's not a Helmholtzian model in that you don't consciously have access to the signals from the two classes of cones. The model says that, at a very early stage, the channel on which flicker photometry depends (perhaps the magnocellular pathway) is ac-coupled to both classes of cones, so that any modulation in the L- or M-cones is passed on. The great Instrument Maker wouldn't design a transient-detecting channel that allowed transients in different cones to cancel out. If that were the case, when you got yourself between the isolepts for the long- and middle-wave cones, then you would see the residual transients from the two classes of cones. But once you get outside, when you go beyond one isolept or the other isolept, then both classes of cones must be changing in the same direction. Then you will get your response that is dominated by the fundamental frequency

Lee: Physiologically, it would seem that when this frequency-doubled response becomes too large, the subjects do report difficulty with the task. Also, a marked degree of cancellation of the M- and L-cone signals does seem to occur in M-cells. It would seem that subjects can only perform the task when M- cells are performing predominantly in an additive and linear manner.

Mollon: It always seems to me dangerous that the flicker photometrists allow themselves to change the frequency, which may allow oneself to get oneself into that narrow region between the frequency at which neither class of cones is detecting the flicker and the lower frequency, at which both are. Anyway we have spent too long on what to me are private issues. And I should like to pass along to discuss Dave Williams' very fresh and novel account of the Brewster colours.

Walraven: I would ask of Dave Williams and Joel Pokorny that in their analyses they start with random M- and L-cones, but the the S-cones were very strictly arranged. On the other hand, you say it doesn't make much difference for the analysis whether cones are strictly arranged or not. Are there reasons to assume that we don't need ordering of the cones, is that a better hypothesis or is it better to have the cones organised? You do have receptive field organisation, which implies order in retinal connectivity.

Williams: One comment I would make is that if the two overall cone mosaics were quite regular to begin with, really crystalline, then it could make a substantial difference for visual performance, whether you had a regular assignment rule for L- and M-cones versus an irregular one. In fact, taking all the cones together, the array is reasonably distorted to begin with. No matter how well you try to assign M- or L-cones in a regular arrangement, you are stuck with the essential disorder in the mosaic, you can never produce a very regular packing scheme. So that tends to reduce possible discrepancies in performance between a perfectly crystalline assignment rule and a random one.

Walraven: You showed this picture of the foveal mosaic, and it looked like a regular array to me, only you had to fill in the red, green and blue spots.

Williams: I wouldn't want to argue there is a compelling case for a more random scheme, but there is some evidence for it. Most people describe Brewster's colours as quite irregular, and they don't appear as you might expect from regularity. We have done other experiments, where I attempted to isolate the M- or L-cone with chromatic adapting backgrounds, and in that case you would expect a kind of luminance aliasing for a cone sub-mosaic, like the luminance aliasing I described for the cone mosaic as a whole. We fail to find this under the best conditions we can devise. This tends to suggest that the arrangement of the sub-mosaics is random. Also, it is well known now in a number of studies that there is very little difference in visual acuity when L- or M-cones are isolated or with both together. This is not compelling evidence either, but it does favour a kind of random packing scheme.

Krauskopf: A question about the question of randomness. In your model calculations, if you shift the grating around, do the colours shift a great deal? With eye movements, this might relate to arguments for a random distribution.

Williams: We considered the possibility of eye movements as the source of the discrepancy, but the effect is a lot less than we expected. We have little direct evidence, but Orin Packer and I have done a number of studies on the visibility of luminance aliasing in very high spatial frequencies, around 120 cycles/degree, and we find almost no effect of eye movements. The temporal frequency bandwidth for luminance detection is higher than for chromatic detection, and it may be that in the chromatic case you can get temporal smearing through the eye movement more easily. But my hunch is that eye movements don't play a major role.

Mollon: We have not discussed at all MacLeod's paper, in which he suggested that cones locally adapt and rods don't seem to at all, persuasive because Donald gave us a functional reason why it is so.

Spillmann: Westheimer's results on lateral desensitisation and sensitisation suggest that there are adapting pools for rods as well as cones. Pools for the cones, however, are smaller than those for rods. Could the difference between 'local' adaptation (cones) and 'global' adaptation (rods) in your experiments be attributable to the difference in pool size? Furthermore, do you find 'local' adaptation for cones also in extra-foveal vision where pool size has been shown to be larger than in foveal vision?

MacLeod: Some results I showed were obtained for eccentricities up to 20°, so that is one comment. In our experiments, the line spread function for cone adaptation are proportional to the inner segment size of an individual cone. These spread functions are very much smaller than those of Westheimer. I think in the situation Westheimer used, with small steady backgrounds, additional factors come into play. Specifically, small adapting fields may saturate a bipolar cell which can then be relieved by horizontal cells on expanding the background. In our experiments with gratings local saturation of bipolar cells should not happen. In answer to the second question, the difference-frequency grating experiments do indicate strictly cone-system sensitivity regulation for parafoveal as well as for foveal vision, although the point-spread function for the parafoveal case is broader in proportion to the cone inner segment diameter.

Rodieck: In terms of rods, how does the rod summation pool compare with the dendritic field size of rod bipolars or the spatial acuity of rod monochromats.

MacLeod: We have made those comparisons as best we could, but tentatively the pool is a little larger than the dendritic field of the rod bipolar, and the pool is a little smaller than the spatial integration in rod vision.

Walraven: There is a lot of literature on spatial integration in cones differing when you have small and large test spots on a background and measure adaptation. Is there still room for the idea that when you have higher luminance you decrease the integrating area, or is there just one cone to work with all the time?

MacLeod: It is true that psychophysically, when measuring the dependence on sensitivity on background intensity for test stimuli of different sizes, the change of sensitivity is greater for large flashes at threshold than for small ones, and that is at first sight inconsistent with a strictly local adaptation process. On closer reflection, it turns out to be consistent with such a local process, because in those experiments when threshold are measured for large and small flashes, there is a purely local difference between the two test stimuli situations, and specifically the small flash at threshold is locally of much higher intensity that the large one. So one can in fact in principle explain the different threshold elevations for large and small flashes with a model in which local adaptation in the cones or at some other local stage of processing attenuates the responses to weak stimuli by a greater factor than it attenuates responses for strong stimuli. And we actually tested that conjecture and the results that we got suggested that the effect of adaptation is indeed only a function of the test stimulus intensity and not of its size. Specifically, the dependence of apparent brightness for large test flashes goes along with the dependence of threshold for small test stimuli of the same intensity on the different backgrounds. I trust that is totally obscure.

Hayhoe: I wondered if your results are quantitatively consistent with recent physiological results in the cat, showing some adaptation in the rods by some authors whose names I cannot remember.

MacLeod: That was Tomoto, Nakatani and Jal, though they have evidence that their results applied to other animals beside the cat, their results are not inconsistent with those I showed psychophysically, because in their paper the rod sensitivity drops to half with an illumination of 35 quanta/rod/sec. That is a level that is not very far below rod saturation, and quite a lot above cone threshold, so that if the rods were completely blind at that illumination level, the difference in vision will not be conspicuous. Down at the level where we depend on the rods to see, quantum fluxes are not more than about one quantum/rod/sec. That is where our experiments were done. I might mention that in some unpublished experiments that Mary Hayhoe and I did years ago, we were unable to find that bleaching the rods can ever protect them from saturation by steady backgrounds. That psychophysical result may indeed be somewhat difficult to reconcile with the physiological data cited above.

Mollon: There are many parts of the speaker's papers that we have not discussed; we must apologize to them.

TRANSITION FROM PHOTOPIC TO SCOTOPIC LIGHT ASSESSMENTS

AND POSSIBLE UNDERLYING PROCESSES

Françoise Viénot

Laboratoire de Physique Appliquée, CNRS UPR 257
Muséum National d'Histoire Naturelle
Paris, France

INTRODUCTION

Mesopic vision covers the intermediate range between photopic and scotopic vision, when both rods and cones are active. The transition from photopic to scotopic vision produces a change in the overall spectral sensitivity of the visual system, which is manifested as the Purkinje shift, and which is attributed to a progressive change from cone activity to rod activity. Assessment of mesopic vision is complex and what is actually happening in mesopic vision is not fully understood.

Many researchers have determined mesopic luminosity curves, mainly using the Direct Comparison Brightness matching method (DCB) (Walters and Wright, 1943, Kinney, 1958, Palmer, 1967, Ikeda and Shimozono, 1981, Sagawa and Takeichi, 1986). There is overwhelming evidence that mesopic luminous efficiency curves cannot be described using linear combinations of photopic and scotopic functions. Rather than a simple systematic change of sensitivity curves from photopic $V(\lambda)$ to scotopic $V'(\lambda)$, the transition curves display distortions in shape.

Measurements by heterochromatic flicker photometry (HFP) have also been reported. Walters and Wright (1943) preferred to avoid "uncertainty" in interpreting results obtained by the flicker method, so found it "not-satisfactory" for low-brightness photometry. Palmer (1966) also found this, when, on himself, he made both flicker and direct comparison 10° matches at 62 and 6.2 Td for the whole spectrum and at 0.62 Td for short wavelengths only. He found a significant departure from his DCB matches. Kinney (1964) concluded that it was "inadvisable" to use the flicker method for mesopic viewing conditions, because of the low speeds needed at these luminances. Nevertheless, Ikeda and Shimozono (1978) investigated the effect of temporal frequency on light assessments, on 1° fields and at 1.2 log unit higher than the absolute threshold for a white reference light. They showed that, even with low frequency alternation, provided that the criterion used is minimum flicker perception, an observer can obtain luminous efficiency values which agree closely with higher frequency photometry. This contradicts previous results from Sperling (1961) who has noted that the spectral sensitivity function obtained

From Pigments to Perception, Edited by A. Valberg and
B.B. Lee, Plenum Press, New York, 1991

by flicker photometry on a 45' field with a dark background, presents the same humps and dips as the curve obtained at dark-adapted threshold.

The CIE has documented 5 methods of computing mesopic brightnesses (CIE, 1989). All methods predict an equivalent brightness in approximate agreement with experimental visual direct comparison brightness matches performed by observers on 10° fields over the entire mesopic range. However, the formulation is strictly empirical and does not relate to any proposed mechanism of mesopic vision.

In the photopic domain, where colour vision is trichromatic, the roles of the spatial and temporal parameters have been shown to be important. High spatial frequency or high temporal frequency stimuli yield approximately additive light assessments based on the $V(\lambda)$ luminous efficiency function, while extended and steady stimuli yield non-additive assessments (Kaiser, 1971, CIE 1978). In the scotopic domain, where rods are the only contributing receptors, visual functioning is greatly simplified. In the mesopic domain, additional problems occur as sub-additivity and supra-additivity failures have been observed with brightness matches (Nakano, Ikeda, 1986, referenced in CIE, 1989) and as the state of adaptation of the whole retina is predominant (Yaguchi and Ikeda, 1984).

Until recently, a model built on cone inputs and 3 channels - an achromatic one and 2 chromatic opponent ones - was sufficient to describe a large amount of psychophysical data (Guth et al., 1980). Only the achromatic channel was assumed to contribute to flicker judgments, while the achromatic channel and the chromatic opponent channels were assumed to mediate brightness.

However, based upon electrophysiological data, no evidence exists of only three channels at photopic levels or of three neural substrates for these channels. Moreover, doubts have arisen over the assumption that a unique neural pathway would specifically code for luminance (Mollon, 1987, Zrenner, 1987) and that specialized pathways would sustain the two chromatic channels. Firstly, several physiological pathways are candidate for an additive achromatic channel in which the summation of cone contributions provides a $V(\lambda)$ shaped spectrally sensitive mechanism. Secondly, the same colour-opponent ganglion cells are able to change their behaviour from antagonistic to synergistic, depending upon the temporal frequency of the stimulus (Zrenner, 1983). Besides, at scotopic levels, psychophysical observations imply that rod signals travel through two retinal pathways (Sharpe et al., 1989) and physiological findings show that rods have several ways of connecting to ganglion cells (reviewed by Frumkes and Denny, 1987). Therefore, in the mesopic domain, where the rods and the three types of cones are operating, one can expect as many functioning mechanisms as there are in the scotopic and the photopic domains together, and there may be additional complexities originating from rod-cone interactions. Therefore, from a physiological perspective, only the receptor stage of the former scheme remains valid, whereas conceptualizations of post-receptoral stages have become increasingly complex.

An effective way to examine a multiple channel hypothesis is to test the same subjects with several experimental tasks. In this paper, a procedure is used which was designed to compare relative spectral sensitivities using HFP and DCB over the full mesopic range. The results reveal several aspects of mesopic processes.

EXPERIMENTS AND RESULTS

I shall report on three series of experiments in which the same observers were asked to assess HFP matches and DCB matches of monochromatic
lights at several mesopic levels. The field size was 10° and the reference was a 3800K white.

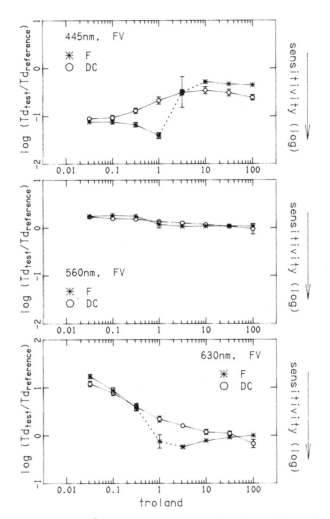

Fig. 1. Relative spectral sensitivities (right ordinates) at 445,
560 and 630 nm of the normal observer FV using HFP (✳)
and DCB techniques (○) at different illuminance levels.
Abscissae: retinal illuminance produced by the white
3800K reference. Ordinates: ratio of the retinal illuminance of test and reference. All photometric measurements
are in photopic trolands. Error bars show the standard
deviation of 30 matches recorded during 3 sessions. Field
size is 10°.

For each wavelength, matches were recorded during 3 sessions. Each session was divided in blocks of adaptation and measurement, including one minute or more of adaptation to the illuminance level and to the flicker configuration, 10 flicker matches, one minute of adaptation to the illuminance level and to the direct comparison configuration and 10 direct comparison matches. The luminance was then decreased by 0.5 log unit and the next block of adaptation and measurement was run. Timing of each session was strictly controlled in order to ensure adaptation to the illuminance level and to the spatial and temporal stimulus configuration. For HFP measurements, the observer was free to adjust the flicker frequency to assess the most reliable match (Chiron and Viénot, 1988).

In the first experiment, we concentrated on wavelengths which are known to reveal the characteristics of HFP and of DCB: 445 and 630 nm, where DCB sensitivity is high as compared to HFP sensitivity at photopic luminous levels, and 560 nm where they are almost equal (Wagner and Boynton, 1972). Three normal observers participated in the experiment.

Figure 1 shows that the changes in relative spectral sensitivity as a function of retinal illuminance given by HFP and by DCB are clearly different (fig. 1). Flicker changes are composed of two branches. The disruption appears in the range from 1 to 3 photopic Troland and is accompanied by a change in the flicker frequency chosen by the observer. Within each branch, the red scotopic branch excepted (this case will be examined in the third experiment), the spectral sensitivities show a reverse Purkinje shift, quite marked on the red photopic branch and moderate on the others. For 445nm, at 1Td, the match is greatly influenced by the flicker frequency. Left graph on figure 2 shows the results of one observer who has been able to assess two different matches, using two

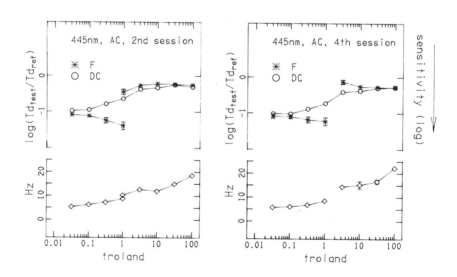

Fig. 2. Relative spectral sensitivities at 445 nm measured by the normal observer AC using HFP (✳) and DCB (◯) techniques at different illuminance levels, from two separate experimental sessions. Lower graphs show the temporal frequency chosen by the observer in order to assess the most reliable flicker match. Dispersion bars are standard deviation of 10 matches recorded during the session. Other details as for figure 1.

slightly different flicker frequencies. This possibility explains the large dispersion at 3 Td on figure 1, which results from averaging settings which were positioned on one or the other branch. With DCB, the transition from photopic levels to scotopic levels is smooth. Intra-observer variability of DCB assessments is small, provided that the test/reference illuminance ratio is slowly adjusted (around 0.5 Hz). Then, when adjusting his match, the observer is aware of following some "center of gravity of light" which gently moves from one side to the other as the balance of light is varied.

A second experiment was completed with dichromatic observers (Viénot and Chiron, 1990). Two protanopes and three deuteranopes participated using exactly the same conditions as in the first experiment. As far as red matches are concerned, the reverse Purkinje shift obtained with HFP at high mesopic levels was considerably reduced, compared to normals, by the same amount for protanopes and deuteranopes. At low mesopic levels, the pattern of HFP matches versus DCB matches depends upon the type of defect. With protanopes, the HFP curve and the DCB curve cross somewhere between 1 and 0.03 Td, with the flicker sensitivity becoming significantly lower than the DCB sensitivity at the lowest scotopic levels. With deuteranopes, the HFP curve joins the DCB curve at 0.1 Td, then running parallel to it (fig. 3).

In a third experiment, we extended the comparison of HFP and DCB matches to several long wavelengths and added two experimental illuminance levels toward the scotopic domain (fig. 4). The actual experiment was made with 3 normal observers. An overall examination of HFP curves obtained with observer JL who gave typical results, show that (i) the amplitude of the reverse Purkinje shift on the photopic branch continuously increases from a null value around 560 nm (not shown, similar to the results of observer FV in figure 1) to a maximum around 610 nm. (ii) The low mesopic branch resembles a plateau which is progressively shifted toward the scotopic end (0.003 Td) as the wavelength increases and is no longer observed beyond 610 nm. Consequently, between the two branches, the HFP sensitivity changes abruptly and passes the DCB sensitivity. (iii) The temporal frequency adjusted by the observers to assess HFP provides auxiliary information. On the photopic branch, it decreases from over 20 Hz to more than 10 Hz when illuminance decreases to 1 Td. Then it remains stable around 10 Hz until the sensitivity reaches the other branch and finally decreases to about 5 Hz when the matches settle on the scotopic branch. Intra-observer variability is small.

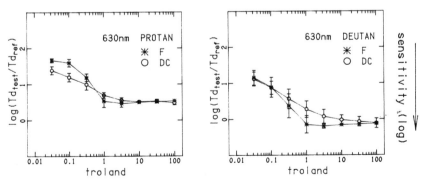

Fig. 3. Average relative spectral sensitivities at 630 nm of 2 protanopes (left) and 3 deuteranopes (right) using HFP (✳) and DCB (○) techniques at different illuminance levels. Other details as for figure 1.

To summarize the experimental results:
1) The HFP and the DCB techniques yield significantly different assessments over the whole mesopic range.
2) As far as the HFP assessments are concerned, the mesopic domain is split in two parts: a high mesopic section which is photopically related and a low mesopic section which is scotopically related. At long wavelengths, the flicker relative spectral sensitivity on the photopically related section increases sharply as illuminance decreases and is associated with a decrease in temporal frequency. However, dichromats do not show such a large sensitivity enhancement.
3) As far as the DCB assessments are concerned, the change in relative spectral sensitivity following a change in illuminance is smooth.

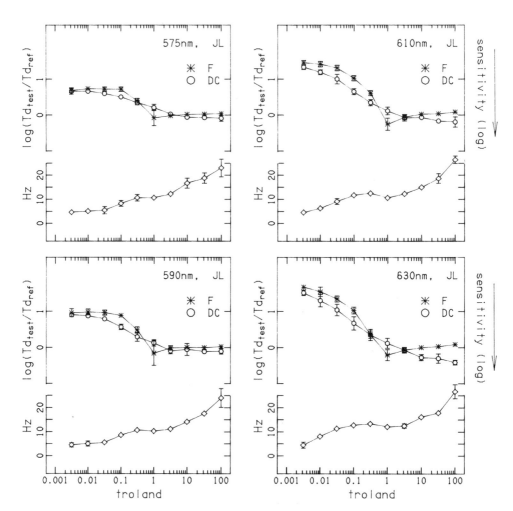

Fig. 4. Relative spectral sensitivities at 575, 590, 610 and 630 nm of normal observer JL using HFP (✻) and DCB (○) techniques at different illuminance levels. Lower graphs show the temporal frequency chosen by the observer in order to assess the most reliable flicker match. Other details as for figure 1.

DISCUSSION

The HFP and the DCB processes show considerable differences

"Mesopic vision" means "vision intermediate between photopic and scotopic vision", where both the rods and the cones are active. However, there should be different terms to distinguish between the two situations which we have encountered. In the DCB experiment, the wide mesopic range corresponds to a progressive transition from photopic process to scotopic process. In the flicker experiment, in terms of cooperation between photopic vision and scotopic vision, there is no mesopic range.

With two so different functions for assessing light in the mesopic range, mesopic operation cannot be defined as a simple reduction of a trivariant photopic operation to a univariant scotopic operation.

In terms of mechanisms, the considerable differences between HFP and DCB over the mesopic range makes it unlikely that the two mechanisms share the same pathway, although the receptor inputs are similar in both situations. Neither can the flicker spectral sensitivity, which is discontinuous in the middle of the mesopic range, be a constituent of the DCB spectral sensitivity which shows a continuous transformation over the whole mesopic range. As a consequence, even if the DCB matches were processed at a higher level than the HFP assessments, no hierarchy between the two processes should be assumed.

The questions addressed to the neurophysiologists is to find two distinct neural substrates for the HFP and the DCB processes, with an early separation based on temporal parameters.

The flicker processes

a. The presence of a photopically related section and of a scotopically related section which appear distinct all over the spectrum suggests two different flicker processes: photopic and scotopic. The transition between the two domains is represented as a discontinuity. When this is narrow, as with blue matches, there is almost no overlap of the two clearly separate domains. When it is wide, as with red matches, the discontinuity further indicates an absence of natural cooperation between the photopic and the scotopic activities. (In this case, the task is very hard to perform). Two underlying processes could account for these experimental results.

b. As far as the photopic flicker process is concerned, three characteristics stand out. (i) It is operative at retinal illuminances higher than 1 photopic troland. (ii) Its spectral sensitivity is changeable. (iii) It is sensitive at high temporal frequencies.

The lower bound at 1 photopic troland has appeared with all wavelengths and with all observers. It is linked with cone thresholds (Boynton and Whitten, 1970) and serves as a valid photopic limit for other functions, such as CFF.

The reverse Purkinje shift, which provides evidence for a change in spectral sensitivity of the photopic flicker process, was mentioned as early as 1912 by Ives who worked above illuminations where he could secure reliable flicker with 5.2° x 8.6° fields, both with the flicker method and the CFF method (Ives, 1912a, 1912b). It has also been reported by De Vries (1948) who found that the luminous efficiency of red lights vs green ones, using 1.6° field, is less when the illuminance increased to over 50 Td with normal and protanomalous observers, but not with dichromatic or deuteranomalous observers. It has been noted by Landis (1954) in a study of the effect of colour as a determinant of CFF. Non-linearity of flicker matches between a 680 nm light and a 550 nm light above 10 Td has been found by Ingling et al. (1978).

The presence of the reverse Purkinje shift means that the flicker channel is not an additive channel. Three additional observations give hints of antagonistic activity aroused with flicker in the high mesopic range. Firstly, the reverse Purkinje shift on the red photopic branch is considerably reduced with dichromats. Presumably, this reflects a loss in colour-opponency originating from the absence of one cone input in red-green dichromats. This result is consistent with the findings of Pokorny and Smith (1972) that luminosity measurements obtained on dichromats by HFP and DCB paralleled the absolute threshold spectral sensitivity over the luminance range 4 - 400 Td. Secondly, with normal observers, the reverse Purkinje shift increases from 560 to 610 nm. This is the wavelength where increment threshold spectral sensitivity, which is known to address antagonistic mechanisms, is at a maximum (Sperling and Harwerth, 1971) with normal observers but not with dichromats (Verriest and Uvijls, 1979). Thirdly, lower frequency may allow antagonistic mechanisms to mediate responses. The flicker frequency adjusted by the observer to assess the most reliable flicker matches decreases continuously from over 20 Hz at 100 Td to 10-12 Hz around 1 Td. Several authors have agreed about the existence of an achromatic channel with a maximum modulation sensitivity around 10 Hz, and an opponent-colour channel, approximately lowpass, with a maximum sensitivity around 2 Hz. After Kelly and Van Norren (1977), the flicker response is controlled by these two frequency-bands. Then a decrease in temporal frequency, which necessarily accompanies the lowering of illuminance is capable of favouring an antagonistic response. Conversely, an increase in temporal frequency would favour the achromatic response, as it does in the photopic range.
We suggest that the characteristics of HFP in the high mesopic range might indicate that flicker matches are not taken over by an achromatic channel *per se* but only by a channel whose spectral sensitivity happens to become the $V(\lambda)$ at high temporal frequencies.

Extra difficulties arise since retinal organization changes with eccentricity. Between 3 and 0.3 Td, all the observers have noticed the inhomogeneity of the flickering field. If the test radiance was adjusted in order to null the flicker sensation in the periphery of the field, but not in the center, then the long wavelength test looked too dark at 3 Td, compared to the reference, but too luminous at 0.3 Td. This resembles the phenomenon described as the yellow spot effect by Ives (1912a). He has reported that, on 6° field, the Purkinje shift appears with the DCB and a reverse Purkinje shift with the HFP, but both shifts are reduced on 2° fields. Clearly, there is a difference in flicker processing between the fovea and the parafovea. Reporting on flicker detection, Coletta and Adams (1986) have shown cone inhibition at nearly photopic levels (10-30 Td), which is cone shaped in the fovea, but resembles the R/G responses in the parafovea.

c. As far as the scotopic process is concerned, its slow temporal characteristics are obvious: The frequency curve for blue matches shows a sudden fall from 12-14 Hz to 8-9 Hz at 1 Td, where the assessments jump from the photopic branch to the scotopic branch. For all wavelengths, every time an observer settles his matches on the scotopic branch, he adjusts the rate lower than 10 Hz. At very low illumination, he can easily operate at a frequency as low as 5 Hz. However, the decrease in frequency may reflect a property of the rods themselves or of their immediate connections.

d. Transition from the photopic flicker process to the scotopic flicker process: when the observer switches from photopic flicker to scotopic flicker (below 1 Td), he sets the frequency slightly over 10 Hz and approximately maintains this rate for all the assessments which are on the transition branch between the photopic and the scotopic sections. But this is only a provisional expedient. Actually, the flicker sensation passes through a flat and wide minimum, around which very small imbalances of the test and the reference radiances make the flicker appear, and small increases in the flicker rate make it disappear. This is consistent with the phase shift between the cone signals and the rod signals assumed by MacLeod (1972, Sharpe et al., 1989). According to these authors, at mesopic levels and 7.5 Hz, neural signals are in phase when the stimuli are out-of-phase. The slight but significant rebound of the flicker frequency at illuminances just below 1 Td is a consequence of the observer's search for a best phase shift to optimize the detection of flicker.

The results concerning low mesopic assessments of red light by the dichromats are surprising. Whereas inter-observer variability among normals made unclear whether the HFP sensitivity joined the DCB sensitivity at scotopic illuminances, the relative positions of HFP and DCB assessments clearly differ for dichromats. With deuteranopes, HFP sensitivity gets closer to brightness sensitivity around 0.1 Td, and the two sensitivities develop in parallel toward the lower scotopic illuminance levels. Alternatively, with protanopes, HFP sensitivity falls under DCB sensitivity below 1 Td. This means that in the low mesopic domain at long wavelengths, if only red cones are involved, as in deuteranopes, the flicker assessments do not reach the scotopic plateau and a stable scotopic flicker process cannot be reached, although the brightness process develops normally.

To summarize the characteristics of the flicker processes, there seem to be two. The photopic flicker process originates from cones. It is sensitive to high temporal frequencies. Its spectral sensitivity, somehow additive at photopic illuminances and above 20 Hz, changes with the decrease of illumination and/or the lowering in temporal frequency and tends to reflect antagonism at 10 Hz. The scotopic flicker process originates from rods. It is sensitive to low temporal frequencies only. Its spectral sensitivity could reflect some red cone - green cone asymmetry. These two processes share a common section which is sensitive to temporal frequencies from 5 Hz to over 25 Hz. At temporal frequencies just above 10 Hz, the photopic flicker process and the scotopic flicker process outputs are nearly out-of-phase, rendering the cooperation between the two processes impossible.

The question addressed to the neurophysiologist is to find two neural pathways originating from cones and rods respectively, with appropriate temporal responses and adaptation characteristics. For the photopic flicker process, neural units should change their spectral sensitivity. For the scotopic process, the neural units should be fed with a red cone-green cone asymmetry.

The direct comparison brightness matching process

We noticed that some observers obtained the most reliable DCB assessments when the ratio of the test and the reference radiances were varied slowly, at about 0.5 Hz. This was very critical in the mid-mesopic and the low mesopic range, where a more rapid rate suddenly upset the apparent balance of brightness of the two half-fields. When adjusting the illuminance ratio, the observer could use the criterion of "center of gravity" to judge the brightness balance. With such a criterion, he was able to judge the appearance of light over the totality of each half-field and to search for a balance of luminous levels between the two, thus departing as far as possible from a MDB judgment.

In terms of processes, all the receptor inputs converge to the brightness matching process. It is a very slow process, its sensitivity being maximal around 0.5 Hz.

The question which I would like to address to the neurophysiologist is to find a pathway to the cortical level, where rod and cone inputs would be pooled at some stage. This pathway should contain slow neural components resulting in a maximum modulation sensitivity around 0.5 Hz. We suggest that the neural units involved at the last stage should receive non-localized brightness signals in order to ensure the brightness assessment is based on a centre of gravity of light.

CONCLUSION

We have outlined the characteristics of a photopic flicker process, of a scotopic flicker process and of a brightness matching process that describe the assessments of light that normal and dichromatic observers make over the full mesopic range. Considerable differences appear between these processes. Spectral sensitivity and temporal frequency data suggest that the processes are segregated and that the flicker processes do not contribute to the brightness process.

ACKNOWLEDGMENT

I thank Alain Chiron for his skilled contribution to the experiment, and Hans Brettel for fruitful discussion.

REFERENCES

Chiron A. and Viénot F., 1988, Direct comparison and flicker matches at high mesopic levels, Perception, 17:371.

CIE, 1978, "Light as a true visual quantity: principles of measurement," CIE Publ. n°41 (TC 1-4), Bureau central de la CIE, Paris.

CIE, 1989, "Mesopic photometry: history, special problems and practical solutions," CIE Publ. n°81, CIE Central bureau, Vienna, Austria.

Coletta N.J. and Adams A.J., 1986, Adaptation of a color-opponent mechanism increases parafoveal sensitivity to luminance flicker, Vision Res., 26:1241-1248.

De Vries H. L., 1948, The luminosity curve of the eye as determined by measurements with the flickerphotometer, Physica, 14:319-348.

Frumkes T.E. and Denny N., 1987, Types of rod-cone interactions, Die Farbe, 34:205-242.

Guth S.L., Massof R.W. and Benzchawel T., 1980, Vector model for normal and dichromatic color vision, J. Opt. Soc. Am., 70:197-212.

Ikeda M. and Shimozono H., 1978, Luminous efficiency functions determined by successive brightness matching, J.Opt. Soc. Am., 68:1767-1771.

Ingling C.R., Huong-Peng Tsou B., Gast T.J., Burns S.A., Emerick J.O. and Riesenberg L., 1978, Vision Res., 18:379-390.

Ives H.E., 1912a, Studies in the photometry of lights of different colours. I. Spectral luminositiy curves obtained by the equality of brightness photometer and the flicker photometer under similar conditions, Phil. Mag., 24:149-188

Ives H.E., 1912b, Studies in the photometry of lights of different colours. II. _Phil. Mag._, 24:352-370.

Ikeda M. and Shimozono H., 1981, Mesopic luminous-efficiency functions, _J. Opt. Soc. Am._, 71:280-284.

Kaiser P.K., 1971, Luminance and brightness, _Applied Optics_, 10:2768-2770.

Kelly D.H. and van Norren, 1977, Two-band model of heterochromatic flicker, _J. Opt. Soc. Am._, 67:1081-1091.

Kinney J.A.S., 1958, Comparison of scotopic, mesopic and photopic spectral sensitivity curves, _J. Opt. Soc. Am._, 48:185-190.

Kinney J.A.S., 1964, Effect of field size and position on mesopic spectral sensitivity, _J. Opt. Soc. Am._, 54:671-677.

Landis C., 1954, Determinants of the critical flicker-fusion threshold, _Physiol. Rev._, 34:259-286.

MacLeod D.I.A., 1972, Rods cancel cones in flicker, _Nature_, 235:173-174.

Mollon J.D., 1987, On the nature of models of colour vision, _Die Farbe_, 34:29-46

Nakano Y. and Ikeda M. 1986, A model for the brightness perception at mesopic levels, (in Japanese) _KOUGAKU (Jpn. Jr. Opt.)_, 15.

Palmer D.A., 1966, A system of mesopic photometry, _Nature_, 209:276-281

Palmer D.A., 1967, The definition of a standard observer for mesopic photometry, _Vision Res._, 7:619-628.

Pokorny, J. and Smith, V. C. Luminosity and CFF in deuteranopes and protanopes, _J. Opt. Soc. Am._ 62:111-117 (1972).

Sagawa K. and Takeichi K., 1986, Spectral luminous efficiency functions in the mesopic range, _J. Opt. Soc. Am._, A, 3:71-75.

Sharpe L.T., Stockman A. and Macleod D.I.A., 1989, Rod flicker perception: scotopic duality, phase lags and destructive interference, _Vision Res._, 29:1539-1559.

Verriest G. and Uvijls A., 1977, Central and peripheral increment thresholds for white and spectral lights on a white background in different kinds of congenitally defective colour vision, _Atti Fond. G. Ronchi_, 32:213-254.

Viénot F. and Chiron A., 1990, Mesopic luminous matches of protanopic and deuteranopic observers, _in_: "Colour Vision deficiencies X," Drum B., Moreland J.D. and Serra A., eds., Kluwer Academic Publishers, Dordrecht. (to be published)

Wagner G. and Boynton R.M., 1972, Comparison of four methods of heterochromatic photometry, _J. Opt. Soc. Am._, 62:1508-1515.

Walters H.V. and Wright W.D., 1943, The spectral sensitivity of the fovea and extrafovea in the Purkinje range, _Proc. Roy. Soc. (London)_, B, 131:340-361.

Yaguchi H. and Ikeda M., 1984, Mesopic luminous-efficiency functions for various adapting levels, _J. Opt. Soc. Am._, A, 1:120-123.

Zrenner E., 1987, Are there separate channels for luminance and color?, _Die Farbe_, 34:285-289.

DUAL ROD PATHWAYS

Lindsay T. Sharpe[1] and Andrew Stockman[2]

[1] Neurologische Universitätsklinik, Hansastraße 9, D-7800 Freiburg im Breisgau, West Germany

[2] Department of Psychology, University of California, San Diego, La Jolla, California 92093 U.S.A.

INTRODUCTION

Psychophysical and electroretinographic observations in normal and achromat observers suggest that human rod signals travel through two retinal pathways (Conner, 1982; Sharpe, Stockman & MacLeod, 1989; Stockman *et al.*, 1991). One pathway--slow and sensitive--is accessible in the dark; whereas the other--fast and insensitive--becomes prominent at higher intensities. These two pathways must diverge at or before the outer plexiform layer, since both the slow and the fast signals are evident in the b-wave of the electroretinogram (ERG). But, because the two signals can also be demonstrated to interact and extinguish the b-wave, they probably also reconverge at a very early stage in the retina. These findings accord to some extent with anatomical and electrophysiological observations of the mammalian retina.

PSYCHOPHYSICAL EVIDENCE FROM ROD FLICKER THRESHOLDS

The earliest psychophysical evidence for a duality in the transmission of rod signals was provided by Hecht *et al.* (1938, 1948), who observed that both the increment threshold-versus-intensity and flicker sensitivity-versus-intensity data of a totally colour-blind observer were distinctly double-branched, with the transition occurring at a background illuminance near 5 scotopic trolands (scot. td). This was puzzling since detection on both branches was mediated by the rod photopigment, the observer revealing no cone function under any test condition. Hecht *et al.*'s observations have since been replicated for other totally colour-blind observers (Alpern *et al.*, 1960; Blakemore & Rushton, 1965), most recently for typical, complete achromat observer K.N. Like Hecht *et al.*'s observer, K.N. displays lower and upper intensity branches in his critical flicker frequency (CFF) function (Hess & Nordby, 1986) and, under some conditions, in his incremental threshold function as well (Sharpe *et al.*, 1988b). Although the transition in his increment threshold data is somewhat equivocal (see Sharpe & Nordby, 1990), that in his flicker data is very robust indeed: it occurs near 5 scot. td for flickering stimuli of different modulation depth, angular subtense and retinal position (Hess *et al.*, 1987).

That a transition from a low intensity to a high intensity rod CFF function is not unique to the achromat, but is also displayed clearly by the normal trichromat observer has also been established. By exploiting the rod-isolation procedures of Aguilar and Stiles (1954), Conner and MacLeod (1977) were able to show not only that light-adapted rods in the normal eye detect flicker frequencies as high as 28 Hz, but also that a break in the function relating rod CFF to stimulus intensity occurs near 1 scot. td.

Such data enforce the conclusion that the rod visual system is inherently duplex. Current evidence (see also below) points to a system in which a slow rod signal predominates at low scotopic luminances, but is displaced by a faster rod signal at higher mesopic luminances. At intermediate luminances both types of signal can

From Pigments to Perception, Edited by A. Valberg and B.B. Lee, Plenum Press, New York, 1991

be shown to coexist in the retina. How is this to be explained? One possibility is that there are two types of rod photoreceptor with markedly different temporal properties. Hecht *et al.* (1938, 1948), for example, proposed that there were two types distinguished by different concentrations of rhodopsin (see also Lewis & Mandelbaum, 1943; Sloan, 1954, 1958), although different concentrations do not necessarily imply different speeds. However, the two receptor hypothesis seems incompatible with the physiological and anatomical evidence. Only in the all-rod retina of the skate (*Raja erinacea* and *Raja oscellata*) is there any indication of a duality at the receptor level (Green & Siegel, 1975). Other nocturnal species, notably *Tarentola mauritanica* and *Hemidactylus turcicus*, possess only one sort of rod and yet display duplex responses in the electroretinogram (Dodt & Jessen, 1961). So the presence of anatomically distinct rod photoreceptors cannot be regarded as a *sine qua non* for duplex responses.

The two rod signals could also originate from a single class of rod photoreceptor. Suppose, for example, that there is an abrupt speeding up of the rod response at higher scotopic luminances, but that across the population of rods there are small differences in the luminance at which the change occurs. As a result, there would be a transitional intensity region where both slow and fast rod signals are present in the retina. There is evidence from suction electrode recordings from single toad, *Bufo marinus*, outer segments that adapting lights do shorten the time course of the rod response to light flashes (Baylor *et al.*, 1979), but there is little support for it in the few suction electrode recordings that have been made from single macaque rod outer segments (Baylor *et al.*, 1984). Monkey rods show no evidence for an abrupt change in the rod temporal response at those intensities where the psychophysics indicates a transition from a slow to a fast rod signal. Furthermore, there is little or no decline in response amplitude and/or a significant (or abrupt) speeding up of the rod response with backgrounds, and the maximum reduction in light-integration time is only about 1.4 times.

If the slow and fast rod signals do not arise from different rods, then perhaps they arise from within the same rod. Membrane recordings from individual toad rods suggest that the rod response is faster at the base of the outer segment (the part nearest to the inner segment and to the source of light) than it is at the tip (Baylor et al., 1979). But, even if there are similar differences in mammalian rods, it seems unlikely that they can be the cause of the slow and fast signals that we find psychophysically. First, the change in time course with distance along the outer segment seems to be a continous function, so that discrete fast and slow signals are unlikely. Second, the transition from the slow to the fast rod signal that we find is intensity-dependent. For such a transition to occur within the rod, there would need to be some local adaptation of the rod outer segment. The best evidence, however, suggests that light adaptation does not take place within individual primate rods (Baylor *et al.*, 1986; but see Tamura *et al.*, 1989), but rather within postreceptoral networks. Parenthetically, we note that the transition from the slow to fast rod signal occurs when only a very small fraction of the rod photopigment has been bleached. In this intensity region, then, there is unlikely to be a change in the portion of the outer segment where flicker is most likely to be detected caused by changes in self-screening.

If we reject the duality as inherent in the rods themselves, we must seek for it elsewhere. The simplest alternative hypothesis compatible with the flicker data is that there are two independent (or semi-independent) pathways proceeding from the rod photoreceptors for the transmission of rod signals. Anatomical and physiological support for this type of hypothesis is very strong (see below). And, a second line of psychophysical evidence, involving the discovery of a perceptual null in rod flicker sensitivity (Conner, 1982; Sharpe, Stockman & MacLeod, 1989), is consistent with it. As shown in Fig. 1 (lower right panel), at 15 Hz, there is an abrupt discontinuity in the rod flicker threshold-versus-background intensity curve. The threshold (■) rises steeply until about 0.1 scot. td, where it levels off, rising again at rod saturating intensities. This discontinuity is reflected in the CFF versus intensity function: in the intensity range between 0.01 and 1 scot. td, the CFF does not exceed 15 Hz limit, whereas at intensities slightly above 1 scot. td it begins to rise sharply (see Conner & MacLeod, 1977; Hess & Nordby, 1986).

Though the evidence for two branches in the flicker threshold-versus-intensity profile is most apparent with a 15 Hz flickering stimulus, it can be seen in profiles measured with 1, 8 (Fig. 1, upper right panel), 12 and 21 Hz flickering stimuli as well (Conner, 1982; Sharpe, Stockman & MacLeod, 1989; Sharpe, Fach & Stockman, 1991). What is found only at or near 15 Hz, however, is the intensity region, demarcated by broken lines, adjoining the discontinuity, within which flicker vanishes, even though the stimulus amplitude is well above conventional rod flicker threshold (Conner, 1982; Sharpe, Stockman & MacLeod, 1989). That is, as the observer increases the target's flicker amplitude above its threshold intensity (■)--whether in the dark or against a dim background--flicker first becomes more pronounced, then suddenly disappears as the intensity approaches the level indicated by the lower broken line (●). Flicker only reappears, when the intensity is further increased and exceeds the level indicated by the upper broken line (♦).

A comparable perceptual null or loss of flicker perception is found for suprathreshold *mesopic* flicker at 7-8 Hz (though not under the conditions shown in Fig. 1). That perceptual null can be explained by assuming an interaction between rod and cone signals, which are close to out-of-phase near such frequencies (MacLeod, 1972; van den Berg & Spekreijse, 1977). The perceptual null near 15 Hz, however, can not be so explained; first, because it occurs well below cone flicker threshold (see Fig. 1, right panels, o) and second, because it is also found in the achromat observer K.N. who lacks functioning cone vision (Fig. 2, right panel).

Figure 3 presents a model of how the cancellation of 15 Hz flicker could come about. The rod signals produced by the flickering target are assumed to be transmitted through two visual pathways, one fast and the other slow. At 15 Hz the signal emerging from the slow pathway is delayed by half a cycle (i.e. 180°) relative to the signal emerging from the fast pathway. When the outputs from the two pathways are of the same magnitude, their recombination in a later common pathway produces a steady, non-flickering signal by destructive interference. Such destructive interference is not encountered at frequencies less than 15 Hz because the slow rod signals lag the fast ones by less than half a cycle. Nor is it encountered at higher frequencies because the phase lag is more than half a cycle. To explain the restricted range of intensities within which the null is found (see Figs 1 and 2), it is assumed that the intensity-dependences of the two signals differ, such that the slow signal predominates at intensities below the null, and the fast signal at intensities above the null, the two being approximately equal at intensities within the null. This interpretation gains support from measurements in the normal observer of the relative phase delay between rod and cone signals, which suggest a clear transition from a slow to a fast rod signal as the intensity level is increased (Sharpe, Stockman & MacLeod, 1989).

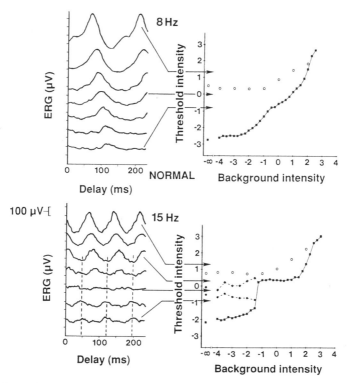

Fig. 1. Right panels: the 8 and 15 Hz rod flicker threshold-versus-background intensity curves for a normal observer (in $\log_{10}$ scot. td). The 500 nm, 6° square-wave flickering target was centred 13° temporally from the fovea and superimposed upon a steady 640 nm, 16° background. Its entry point in the fully dilated pupil was 3 mm eccentric to favour the rods. The (■)s represent the lowest amplitude at which flicker can just be seen. At both 8 and 15 Hz, there is an break in the curve at about 1 scot. td. At 15 Hz, the (●)s and (♦)s designate the lower and upper limits, respectively, of a nulled region, within which flicker cannot be seen. Strong flicker is seen both above and below this region. No nulled region is found for 8 Hz flicker. The (o)s are cone thresholds measured during the cone phase of recovery following a 7.7 $\log_{10}$ photopic td-s (3100K) bleach. Left panels: the 8 and 15 Hz Ganzfeld ERG flicker responses for the same observer. At 15 Hz, increasing flicker intensity first causes the b-wave response to grow in amplitude, without causing a significant change in the phase delay between the stimulus and response. (The flicker intensity increases upwards in steps of approximately 0.3 $\log_{10}$ units.) At a intensity near 1 scot. td, however, the b-wave suddenly diminishes in amplitude; and this diminuition corresponds directly with the perceptual null in flicker detection. At higher intensity levels, the b-wave reemerges, but with a phase shift of 180°. In contrast, at 8 Hz, there is neither a sudden loss of the b-wave nor an abrupt phase shift with increasing flicker intensity. (From Stockman et al., 1991.)

OTHER PSYCHOPHYSICAL EVIDENCE FOR TWO ROD PATHWAYS

The existence of two separate rod pathways, with distinct response properties is suggested not only by flicker sensitivity and incremental threshold measurements, but also, to a lesser extent, by the results of temporal-frequency adaptation (Nygaard & Frumkes, 1985), incremental threshold (Adelson, 1982), spatial summation (Bauer, Frumkes & Nygaard, 1983) and brightness matching (Whittle & MacLeod as reported by MacLeod, 1974) experiments. As has been observed by Bauer *et al.* (1985), many of these experiments concur in finding that an illuminance near 1 scot. td represents a transition point in rod visual processing. For instance, this value--which is at least 3 $\log_{10}$ units lower than usual estimates of rod saturation (Aguilar & Stiles, 1954)--demarcates, not only the point of abrupt transition in the incremental threshold-, critical flicker fusion- and flicker sensitivity-versus-intensity curves of normal and achromat observers, but also: (i) the upper limit of the rod influence on cone-mediated increment threshold in spatial summation effects (Bauer *et al.*, 1983); (ii) the saturating limit of the rod contribution to brightness sensations (MacLeod, 1974); and (iii) the saturation of the rod visual system when incremental threshold is measured on briefly flashed backgrounds (Adelson, 1982).

Given that the two types of rod flicker signals differ so radically in their temporal properties, one might expect that they should differ in their spatial response properties as well. This prediction can be tested by replacing the uniform flickering field with counterphase flickering gratings (in which alternate bars flicker out-of-phase with each other). If the grating is fine enough so that two or more adjacent bars fall within the individual excitation pools of either the slow or the fast rod pathways, there will be cancellation within each pathway (because adjacent bars produce out-of-phase signals). This cancellation wil be in addition to the cancellation found for uniform fields between the two pathways. As a result, the use of fine counterphase gratings may actually extend the flicker null to higher and lower intensities than for the uniform field. But, suppose that the excitation pools of the slow and the fast rod pathways differ in size. In that case, there will be some sizes of gratings for which the extent of cancellation within each pathway is different. Consequently, when the remaining uncancelled signals from the two pathways are subsequently recombined, they will no longer cancel each other as they did for a uniform field of the same time-average intensity. Under these conditions, then, a counter-phase flickering grating will be seen to flicker, at an intensity at which a uniform field appears steady. In fact, we could not find a grating for which this happened, suggesting that the two processes have excitation pools of similar spatial extent (Sharpe, Stockman & MacLeod, 1989). Further (indirect) evidence that the neural excitation pools of the two processes is the same comes from the observation that, for the achromat used in this study, a monotonic relation between grating acuity and retinal illuminance exists in the region of the null (Hess & Nordby, 1986). The matter, however, warrants closer examination; for there is other evidence supporting a duality in spatial organisation within rod vision (Hallett, 1962; Hofmann *et al.*, 1990).

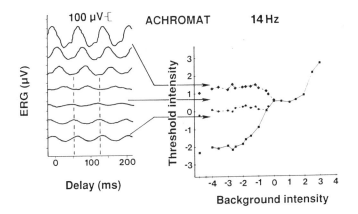

Fig. 2. The 14 Hz psychophysical flicker detectability (right panel) and ERG (left panel) data for typical, complete achromat observer K.N. Similar stimulus conditions as in Fig. 1. As for the normal observer, the achromat displays a break in his flicker function as well as a flicker null region (for K.N. 14 Hz is better than 15 Hz for eliciting the perceptual null). Likewise, increasing flicker intensity first causes the b-wave response to grow in amplitude, without causing a significant change in the phase delay between the stimulus and response. At a intensity near 1 scot. td, however, the b-wave suddenly diminishes in amplitude; and this diminuition corresponds directly with the perceptual null in flicker detection. At higher intensity levels, the b-wave reemerges, but with a phase shift of 180°. Both the psychophysical and the ERG null, however, are found at higher intensities for the achromat observer than for the normal. (From Stockman *et al.*, 1991.)

Evidence for two pathways for rod signals is further bolstered by human electroretinographic recordings. At (or near) 15 Hz, in accordance with the perceptual phenomenon, increasing flicker intensity causes the b-wave of the Ganzfeld electroretinogram (ERG) to decrease in amplitude until reaching a minimum at intensities associated with the perceptual null (Fig. 1, lower left panel; Fig. 2, left panel; Stockman et al., 1991). Above this narrow range of mesopic intensities, the ERG b-wave increases once again in amplitude. Moreover, the signal abruptly reverses in phase as the null is traversed, suggesting that interference is indeed the cause of the null. At lower and higher temporal frequencies, neither a minimum in the ERG b-wave nor a null in the psychophysics occurs above flicker threshold (for an example, see the 8 Hz data in Fig. 1, upper left panel). What is more, electrophysiological estimates of the phase difference between the slow (i.e. those below the null) and fast (i.e. those above the null) rod signals indicate that it grows monotonically with frequency and reaches about 180° at 15 Hz, corresponding to a 30-35 ms delay difference between the pathways. (The delay can be contrasted with that between the slow rod pathway and the cones, which is about 70-75 ms; MacLeod, 1972; Frumkes et al., 1973; van den Berg & Spekreijse, 1977; Denny, Frumkes & Goldberg, 1990.) Figure 4 (left panel) shows ERG recordings for the typical, complete achromat K.N. made below the null ('slow' response; left records) and above the null ('fast' response; right records). In each recording, the vertical line is an estimate of the time taken for the response to the flicker pulse at time zero to appear in the ERG. The difference in the delay or phase lag (in degrees) between the slow and fast responses is plotted as a function of flicker frequency in the righthand panel (■). The phase lags determined for K.N. agree well with those determined electroretinographically for the normal observer (Fig. 4, •) and with those determined psychophysically for the same normal observer using a cone reference standard (see Fig. 4, o). Hence, interference from cone signals cannot be responsible for the cancellation of flicker. (For the normal observer, the slow and fast signals reach out-of-phase at 15 Hz; whereas K.N.'s signals are out-of-phase at a slightly lower frequency; for a discussion, see Stockman et al., 1991.)

That the Ganzfeld ERG shows a diminuition of the b-wave coincident with the perceptual null at 15 Hz and that the phase relation of the b-wave abruptly reverses as the null is traversed suggests not only that rod flicker signals travel over two separate retinal pathways, but also that the pathways conveying the slow and fast rod signals converge in the outer plexiform layer, either at the horizontal cell level or at the level of input to the bipolar cells. If the nulling found in the ERG is not truly neural, but instead results from electrical averaging in the retina, then the site of the neural null could be anywhere. But it seems unlikely that the psychophysical and electrophysiological nulls would occur at the same flicker intensities. Rather it seems more likely that the coincident perceptual and ERG nulls result from a convergence of the two pathways before the ganglion cell layer; for it is possible to record a normal ERG from humans whose optic nerve has been severed and whose ganglion cells have degenerated (Noell, 1954; see Dowling, 1967, 1987). The best current evidence suggests that the b-wave has its origin in the glial (Müller) cells (Miller & Dowling, 1970; Dowling, 1987; Wen et al., 1990), and that its precursor is the activity of the depolarizing (ON) bipolar cells (Dick & Miller, 1978; Kline et al., 1978; Stockton & Slaughter, 1987; Dowling, 1987; Gurevich et al., 1990).

15 Hz SELF-CANCELLATION

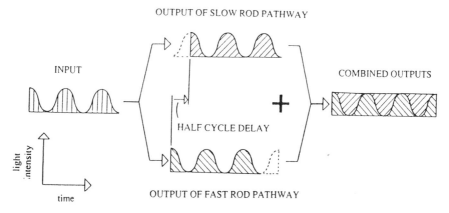

Fig. 3. An illustration of the self-cancellation of 15 Hz rod flicker. The flicker signal produced by a single stimulus is assumed to be transmitted by a slow and a fast rod pathway. At 15 Hz the signal emerging from the slow pathway is delayed by half a cycle relative to the signal emerging from the fast pathway. When the outputs from the two pathways are of the same amplitude, their recombination in a later common pathway produces a steady, non-flickering signal. Thus, the light will appear nulled to later stages of the visual system. (From Stockman et al., 1991.)

An attractive explanation of the adaptational and temporal differences between the two types of rod flicker signals is that the less sensitive, faster of the two travels in pathways intended for cones. Support for this conjecture comes from two types of observations. First, cone masking stimuli selectively suppress, albeit weakly, the salience of flicker signals of the faster rod process in the normal observer (Sharpe, Stockman & MacLeod, 1989). Second, cones influence the sensitivity of the upper branch of the double-branched rod-detected flicker threshold-versus-intensity curve in the normal observer (Knight, Sanocki & Buck, 1990; Sharpe, Stockman & Zrenner, 1990; Sharpe, Fach & Stockman, 1991); though not, of course, in the typical, complete achromat (Sharpe, Fach & Stockman, 1991). Field sensitivity measurements (Fig. 5) indicate that this cone intrusion is comparatively small and is only marked on long-wave backgrounds (see also Sharpe, Stockman &

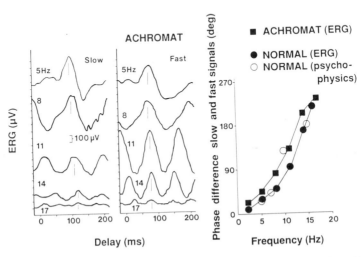

Fig. 4. The ERG measured as a function of stimulus frequency for two stimulus intensities, one directly below the perceptual null at 15 Hz, the other directly above it. The results are for the typical, complete achromat observer K.N. The left panel shows the b-wave responses for the 'slow' pathway (below the null) and for the 'fast' pathway (above the null) at various frequencies from 5 to 17 Hz (for scaling reasons, the data obtained at 2 Hz is not shown). The vertical line in each trace is an estimate of the peak in the ERG record corresponding to the flash that occurred at delay zero. For the slow pathway there is a delay of 90-115 ms between the flash and the ERG response, and for the fast pathway a delay of 70-80 ms. The right panel plots the difference in phase delay between the b-wave responses for the two different flicker intensities as a function of frequency (■). For comparison, the phase delays measured relative to a cone standard (o) and with the ERG (●) in the normal observer are shown. The psychophysical estimates were obtained by differencing the rod-cone phase lags measured at intensities just below the null from those measured just above the null. (From Stockman et al., 1991.)

MacLeod, 1989). (On the basis of field sensitivity data alone, we cannot determine if one or more of the cone classes are responsible for the influence at long-wavelengths; see Fach, Sharpe & Stockman, 1991). Thus, if the cones are sufficiently desensitized by backgrounds, they can alter the sensitivity of the upper branch to the flickering stimulus. However, the cones do not influence the position of the perceptual null, suggesting that the null itself does not depend on cone adaptation. And cones do not seem to influence the sensitivity of the lower branch to the flickering stimulus at any intensity. This finding, of course, does not prove that the slow rod pathway is better isolated from the influence of the cones than the fast pathway. The difference may arise simply because the slow rod pathway is active at intensities where the cones are themselves unadapted and therefore unlikely to have any measurable effect on rod sensitivity.

This result fits in with what we know about rod sensitivity control (e.g. Buck, 1985; Bauer *et al.*, 1983; Frumkes & Temme, 1977; Sharpe *et al.*, 1989). For incremental threshold detection, long-wave background lights (Sharpe *et al.*, 1989) and supplementary backgrounds (Fach *et al.*, 1991), which strongly stimulate the cones as well as the rods (Wald, 1945), affect the rod sensitivity to the test flash, steepening the incremental threshold function. It also accords with other psychophysical phenomena demonstrating that rod and cone flicker signals interact with one another (see Frumkes *et al.*, 1986). First, rod signals can sum together with cone signals to enhance flicker perception or cancel it (by destructive interference), depending upon the stimulus frequency and the relative phase lag of the two types of signal (MacLeod, 1972; van den Berg & Spekreijse, 1977). Second, surrounding rods can exert a suppressive or inhibitory influence on the sensitivity of neighbouring cones to a high temporal frequency stimulus (greater than 15 Hz), even though they are not directly stimulated by it (Coletta & Adams, 1984; Goldberg & Frumkes, 1983; Goldberg *et al.*, 1983; Arden & Hogg, 1985; Alexander & Fishman, 1984, 1985). In fact, the effect is drastically diminished by selectively bleaching or light-adapting the rods. Third, surrounding cones can exert an analogous influence upon neighbouring rods for flicker frequencies greater than 7 Hz (Frumkes *et al.*, 1986).

MAMMALIAN ROD PATHWAYS

Thus, the psychophysical evidence suggests that rod signals travel over two separate pathways, one of which (i.e. the faster) becomes prominent at levels where the cones are usually active. Is there any anatomical and/or physiological support for this conclusion? In the cat and rabbit, the mammalian species for which we have the most detailed information, there are at least two major pathways for conveying rod signals from the rod spherules to the ganglion cells (Kolb & Nelson, 1983; Mastronarde, 1983; Smith *et al.*, 1983; Sterling *et al.*, 1983; Daw *et al.*, 1990). One pathway proceeds through the rod bipolar cells, which contact the narrow-field AII amacrine cells as well as several other varieties of amacrine cell. The AII amacrines, in turn, contact hyperpolarising (type a1) cone bipolars and OFF-centre (both beta/X and alpha/Y) ganglion cells via inhibitory glycine synapses. They also contact depolarising (type b2) cone bipolar cells via large gap junctions, and then the bipolars excite ON-centre (beta/X and alpha/Y) ganglion cells via excitatory glutamate synapses (Kolb & Famiglietti, 1974; Famiglietti & Kolb, 1975; Kolb, 1979; Sterling, 1983; Nelson & Kolb, 1985).

The second pathway proceeds through the cones via gap junctions joining the basal processes of the cone pedicles and the neighbouring rod spherules (Raviola & Gilula, 1973; Kolb, 1977). The connexions must be functional because robust rod signals can be recorded in cones and in horizontal cells that have anatomical input entirely from cones (Steinberg, 1971; Kolb & Nelson, 1983; Nelson, 1977, 1982; Nelson & Kolb, 1983, 1985). From the cones, the rod signals travel over the various varieties of cone bipolar cell and merge with the signals from the rod bipolar pathway at the points where the AII amacrines make contact. What this means, in effect, is that in the cat (Barlow *et al.*, 1957; Daw & Pearlman, 1969; Andrew & Hammond, 1970)--and in the rhesus monkey as well (Gouras & Link, 1966)--the ON- and OFF-centre cells receiving a primary rod input (over the amacrine interneuron pathway) also receive a cone input and thereby a secondary rod input (over the cone gap junction pathway).

Besides the two major cellular pathways, there are several other subpathways and microcircuits in the cat retina available for transmitting rod signals. Rod signals from the rod bipolar infiltrate the cone bipolar pathway not only via the AII amacrine cell, but also via at least three other types of amacrine cell, A8, A6 and A13, all of which receive input directly from cone bipolar cells. These three types of amacrine, plus a fourth--the A17--synapse reciprocally with rod bipolar axon terminals. Unlike the others, the A17, however, never makes synapses with cone bipolar axon terminals.

In addition to contacting amacrine cells, rod bipolar cells in the cat retina also contact at least two other types of cell. (Rod bipolar cells only very rarely--if at all--make direct connections with ganglion cells; see McGuire *et al.*, 1984.) They reciprocally contact interplexiform cells (Kolb & West, 1977), which seem to form part of a feedback loop from the inner plexiform layer to the outer plexiform layer; and which also have synaptic contacts with both flat and invaginating cone bipolar types (Kolb & West, 1977; Nakamura *et al.*, 1980). They also receive input from the axon terminal system of H1 type horizontal cells (Kolb *et al.*, 1980; Boycott *et al.*, 1987), which may act as a feed-forward system to increase the field size over which rod bipolars function (Nelson *et al.*, 1975).

Much less is known about the cellular pathways of the primate retina, but recent evidence suggests that, for the primate as for the cat, there is always a direct feed between the cone bipolars and ganglion cells, whereas there is much more amacrine influence and/or intervention between the rod bipolars and ganglion cells (Chase & Dowling, 1990; Grünert & Martin, 1990). And, of the 25 amacrine cell types described so far in the primate (Rodieck, 1988; Mariani, 1988), one (A6) can be identified with the AII amacrine of the cat (Kolb & Famiglietti, 1974; Famiglietti & Kolb, 1975; Hendrickson *et al.*, 1988; Mariani, 1988). Nevertheless, there may be important differences between the microcircuitry of the cat/rabbit and primate retina. First, in the outer plexiform layer in the cat and rabbit retina, but not in the primate, there is a well-defined plexus of axonless horizontal cells, which spread activity laterally over a large retinal area (Nelson, 1977). Second, in the primate, but not in the cat or rabbit, a biplexiform cell (Mariani, 1982) has been identified at the ganglion cell level, which receives input directly from both rod and cone photoreceptors (Zrenner et al., 1983) as well as some additional input from bipolar cells. Third, the cat is largely a nocturnal animal with a much larger proportion of rods in its retina than man. Even in the area centralis the rod photoreceptors outnumber the cone photoreceptors by more than 10 to 1; and over much of the cat retina the rods outnumber the cones by 65 to 1 (Steinberg et al., 1973). In man, there are no rods in the central island of the foveola and, in the periphery, rods outnumber the cones by less than 20 to 1 (Østerberg, 1935; Curcio et al., 1987). Moreover, the situation is complicated by the synaptic wiring associated with our highly developed sense of colour vision.

DIFFERENTIAL SENSITIVITY OF THE ROD PATHWAYS

Which route rod signals take through the inner nuclear and innerplexiform layers may depend upon the temporal and spatial pattern of stimulation and upon the level of ambient illumination (Kolb & Nelson, 1983). It has been suggested that the rod bipolar pathway is optimised for transmission of signals at scotopic intensities, and that the cone bipolar pathways first become available for the conveyance of cone signals at mesopic intensities (Smith, Freed & Sterling, 1986; Sterling, Smith & Freed, 1986). In support of this hypothesis are measurements of the absolute thresholds for rod signals in the cat at various points along the cellular pathways. Although such estimates depend upon the stimulus conditions, the response measure and the animal preparation, they are useful for making relative comparisons.

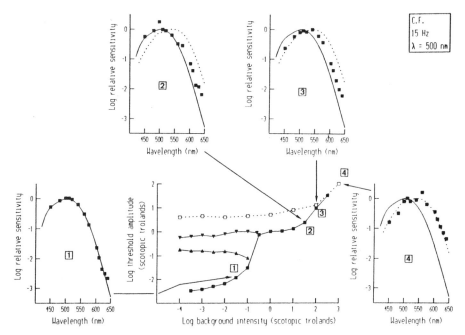

Fig. 5. The action spectrum of the receptors responsible for determining the lower and upper branches of the 15 Hz flicker threshold-versus-sensitivity function of observer C.F. (lower middle panel, ■). Same measurement conditions as in Fig. 1, except the background wavelength was varied. (The function shown was measured on a 640 nm background.) The (■)s in the four panels connected by arrows to the lower middle panel represent field sensitivities or the amount of field intensity as a function of background wavelength required to elevate the 15 Hz flickering target threshold to a criterion level. Functions are shown for both the lower (panel 1) and upper (panel 2) branches of the 15 Hz flicker threshold profile. The field sensitivities of the lower branch are well described by the the quantized CIE scotopic luminosity function (solid line), but those of the upper branch are not. They are better described by a combination of the scotopic luminosity function and the Smith and Pokorny (1975) middle-wave sensitive (M) cone function corrected for macular pigment absorption (dotted line). Above the cone bleach threshold (panel 3), where rods and cones are both mediating detection, the field sensitivities correspond very closely to the Smith and Pokorny M-cone function, as do the field sensitivities derived from the cone bleach thresholds themselves (panel 4). (From Sharpe, Fach & Stockman, 1991.)

In cat, the absolute threshold of dark-adapted ganglion cells has been estimated to be between 0.002 to 0.04 quanta $\mu m^{-2} s^{-1}$ or about 100 to 2000 quanta (500 nm) $deg^{-2} s^{-1}$ (Barlow & Levick, 1969; Barlow, Levick & Yoon, 1971; Harding & Enroth-Cugell, 1978; Shapley & Enroth-Cugell, 1986). These values can be compared with rod thresholds of 0.51 quanta (500 nm) $\mu m^{-2} s^{-1}$ or 24,500 quanta (500 nm) $deg^{-2} s^{-1}$ for cat A17 amacrine cells (Nelson & Kolb, 1985) and 0.26 quantum (500 nm) $\mu m^{-2} s^{-1}$ or 12,500 quanta (500 nm) $deg^{-2} s^{-1}$ for cat AII amacrine cells (Nelson, 1982; Nelson and Kolb, 1985). In contrast, the apparent threshold for the rod signal in cones (Nelson, 1977), cone-dominated horizontal cells (Steinberg, 1969, 1971) and cone bipolars after prolonged dark adaptation is between one and two log_{10} units higher, from 3 (Steinberg, 1969, 1971) to 10 (Nelson, 1977) quanta $\mu m^{-2} s^{-1}$ on the retina or 144,000 to 480,000 quanta (500 nm) $deg^{-2} s^{-1}$.

At 500 nm, 480,000 quanta $deg^{-2} s^{-1}$ is very close to a luminance of 1 scot. td. Thus, the threshold of the rod-cone bipolar pathway is found at an intensity that corresponds roughly to the 15 Hz null found in humans, and the threshold of the AII amacrines of the rod-rod bipolar pathway corresponds roughly to the absolute threshold intensity for 15 Hz rod flicker (see Fig. 1, bottom right). Given that ganglion cells, most of which summate signals from a group of cells, have lower thresholds than their input cells (cf. Nelson, 1982), it is tempting to conclude that the rod-rod bipolar and the rod-cone bipolar pathways found in the cat correspond to the slow and fast pathways, respectively, that we can demonstrate psychophysically and electrophysiologically in humans. Although the absolute thresholds for the rod-rod bipolar and rod-cone bipolar pathways do correspond reasonablly well with the psychophysical thresholds for the slow and fast pathways, it should be noted that the thresholds estimated neurophysiologically are for small spots of light--they are not thresholds for 15 Hz flicker. It is possible that the thresholds for 15 Hz flicker in the rod-cone bipolar pathway may be too high for that pathway to be plausibly identified with the fast pathway that we find psychophysically and electroretinographically. At the intensity level where the null occurs, the signals of the fast pathway must not only be above threshold, they must already be far enough above it to cancel the signals of the slow rod pathway.

THE SITE OF THE FLICKER NULL

The differential sensitivity of the rod-rod bipolar and rod-cone bipolar pathways provide some, albeit equivocal, support for their roles as the slow and fast rod pathways. What other evidence do we have? Lacking the requisite physiological recordings from the primate retina--or even from the cat retina--we cannot directly answer this question. But the answer hinges to a large extent on the answer to a second, related question: where does the interaction take place? From the human ERG evidence, we know that the two types of signal must converge at some retinal locus, perhaps as early as the bipolar cells.

Assuming that it is valid *at all* to compare human visual performance with cat electrophysiology, one attractive possibility is that the cancellation takes place at the depolarising (ON) cone bipolars, one of the points at which the two anatomical pathways converge. At the scotopic intensity levels at which the 15 Hz null occurs, the cone bipolars might be driven both by the rod signals travelling through the cone photoreceptors and the rod signals travelling through the rod bipolars and the AII amacrine cells (Nelson, 1982). The most obvious problem with this model is that the signal in the rod bipolar cell, although able to cancel the signal in the depolarising ON cone bipolar, remains itself uncancelled. And, despite being uncancelled, there is no trace of it in the b-wave of the ERG, even though its (assumed) signal is clearly evident below the null. Another problem with this type of model is that the rod signal entering the cones should appear in both ON- and OFF-cone bipolars, so that the signal in the OFF-cone bipolars remains uncancelled too. To some extent, these difficulties can be contended with by arguing that the null in the b-wave is the result of the *electrical* cancellation of signals in the separate rod and cone bipolars, whereas the perceptual null is due to neural cancellation at a later site, say at the ganglion cell layer. But then one must additionally assume that the electrical cancellation at the bipolar level coincides exactly with the physiological cancellation at the later stage (see Figs 1 and 2).

There is a second objection, however, to arguing that the faster rod signals travel over the cone gap-junction pathway and that the locus of interaction between the slower and faster signals is the depolarising cone bipolars. It concerns the typical, complete achromat observer; namely, how does the faster pathway originate in the achromat observer for whom there is no psychophysical evidence of post-receptoral cone vision and for whom the best anatomical evidence (Glickstein & Heath, 1975) suggests that cone photoreceptors are altogether missing or reduced to at most 5% of their normal population? By way of an answer, it could be argued that, in the peripheral retina, the achromat has morphologically intact cones--the histology supports this (see Sharpe & Nordby, 1990, for a discussion)--and that these residual cones, though insufficing to provide cone vision, suffice to allow rod signals via gap junctions to be transmitted over the cone bipolar pathway. Or, it could be argued that the rod signals infiltrating the cone bipolar pathway do so not only at the cone terminals but also at say the horizontal cells. Both answers, of course, assume that the post-receptoral cone connexions are intact in the typical, complete achromat. To some extent we know that they must be because in the ON-system of the mammalian retina all rod signals travel over the cone bipolars (via the AII amacrine-cone bipolar junction, or via the rod-cone gap junction); *ergo* the complete achromat must have intact cone bipolars (or at least their axonal trees must be intact). Still one would expect a striking difference between the perceptual null in the normal and the achromat, given the paucity or total absence of the latter's cones. This is, in fact, only partially borne out. In the achromat, the lower limit of flicker cancellation is more than 0.5 $\log_{10}$ unit higher than the limit for the normal observer, and the upper limit is more than 1.0 $\log_{10}$ unit higher, broadening the perceptual null and extending it into regions where cones predominate for the normal. Consistent with the displacement of the nulled region for the achromat, the change of slope in his rod flicker-versus-intensity threshold data--which like the nulled region is assumed to depend on interference between the slow and fast rod signals--is displaced to higher background intensities. These upward shifts suggest that the faster rod signals are being less effectively transmitted in the achromat eye.

Thus, attempting to correlate the psychophysically-isolated slow and fast pathways with the electro-physiologically-revealed rod bipolar and cone gap junction pathways, though initially attractive, may turn out to be misdirected. It seems more likely that the interaction leading to the cancellation is taking place either at the rod bipolars themselves or before the bipolar cell level, possibly within the rod spherules; and that horizontal cells are implicated somehow. If so, there are two possibilities. One is an inhibitory effect of the horizontal cells on the rod signal either at the receptor or at the bipolar cell layer. Rod modulation of horizontal cell feed-back onto cones (Baylor et al., 1971; Naka, 1972; Toyoda & Tonosaki, 1978) has been used to explain suppressive rod-cone interaction (e.g. Arden & Frumkes, 1986; Frumkes et al., 1986); and the explanation is supported by psychophysical studies in individuals with outer or inner retinal disorders (Alexander & Fishman, 1985; Arden & Hogg, 1985), by ERG data in normal humans and other mammals (Loew & Arden, 1985; Arden & Frumkes, 1986) and by electrophysiological and pharmacological evidence from various species including cat (Hassin & Witkovsky, 1983; Frumkes & Eysteinsson, 1987; Eysteinsson & Frumkes, 1989; Pflug & Nelson, 1986; Yang & Wu, 1989). Although horizontal cell feedback onto the rods has not yet been found in the mammalian retina (Nelson, 1977; Kolb & Nelson, 1984), it has already been demonstrated in the carp (Weiler, 1977). This type of sign-inverted feedback or feedforward could conceivably produce a null in the b-wave of the ERG, but it would be difficult to explain why such a null should be restricted to frequencies near 15 Hz as we find. Anyway, our phase delay measurements (see Fig. 4) suggest that the slow and fast rod signals have the *same* sign, not the opposite sign, since the phase delay between them falls to 0° as the frequency is reduced to 0 Hz.

A more reasonable possibility seems to be a sign-conserving feedback from the horizontal cells to the photoreceptors or a sign-conserving feedforward to the bipolar cells. A site where this might happen is at the axonal terminals of the short axon horizontal cells (B or H1 type) where they contact the rod photoreceptor and rod bipolar cells. In the cat (Nelson et al., 1975) and rabbit retinae (Bloomfield & Miller, 1982; Dacheux & Raviola, 1982), the axon terminals function as an independent dendritic system from the cell body (Nelson et al., 1975), integrating rod information over a large area (Wässle & Rieman, 1978). They may serve to enlarge the receptive fields of the rod bipolars (Nelson et al., 1975), either by contacting them directly or indirectly via the dopaminergic interplexiform cells, which are known to contact both horizontal cells and bipolar cells, as well as amacrine cells (Dowling & Ehinger, 1975; Dowling et al., 1980). For a sign-conserving signal from the horizontal cells to give rise to a null at 15 Hz, however, there must be a delay between the direct rod signals and those travelling through the horizontal cells. Given that the integration of signals by the horizontal cells should introduce some temporal delay, does the slower rod signal originate from the horizontal cells?

If feedback or feedforward from horizontal cells is the source of one of the two rod signals, then some differences would be expected in spatial properties above and below the null region, given the nature of the horizontal cell receptive fields (horizontal cells are argued to summate rod signals over large areas); but the psychophysical evidence for this is weak (see above). Interestingly, though, the perceptual null does seem to be spatially constrained: it is not found psychophysically with fields less than 3° in diameter (Sharpe, Fach & Stockman, unpublished observations).

CONCLUSIONS

We present human psychophysical and electroretinographic evidence for two types of rod signals: one type, slow and sensitive; the other, fast and less sensitive. And we relate it to principles of mammalian retinal organisation, largely derived from the cat, suggesting that the sensitive rod signals originate in the 'slow' pathway passing through the rod bipolar and AII amacrine cells; while the less sensitive rod signals originate either in the 'fast' pathway passing through the cone terminals and cone bipolar cells or in a pathway routed through the horizontal cells. Even if an exact correlation does not hold between the pathways identified psychophysically in man and those identified anatomically and electrophysiologically in cat, making comparisons between the two species can be useful in another regard; namely in laying to rest a strict--and false--interpretation of the Duplicity theory of vision. In its earliest versions, the Duplicity theory proposed that the mammalian visual system is divided into two morphologically and functionally distinct subsystems: the rod visual system mediating twilight and night vision and the cone visual system subserving day and colour vision (von Kries, 1894). And, indeed, initial psychophysical observations of the Purkinje phemonemon in normal, as well as colour-blind (von Kries, 1894) and night-blind observers (Parinaud, 1881), seemed to confirm histological observations in the retinae of nocturnal and diurnal vertebrates (Schultze, 1866; Ramon y Cajal, 1893) to this effect. However, our present knowledge of the anatomy (Kolb & Famiglietti, 1974), physiology (Nelson, 1977) and psychophysics of the mammalian visual system, building upon the earlier reservations of von Kries (1929) and Polyak (1941), renders such an extremist view impossible. Private channels do not exist in the mammalian retina for conducting the rod signals to the brain: all the rod pathways lead through those of the cones.

ACKNOWLEDGEMENTS

This research was supported by the Alexander von Humboldt-Foundation (Bonn-Bad Godesberg), the Deutsche Forschungsgemeinschaft, SFB 325, Tp B4 (Bonn), NSF grant BNS 88-12401, and NIH grant EY01711. We thank Clemens Fach, Donald MacLeod, Knut Nordby, Klaus Rüther and Eberhart Zrenner for helping to gather the psychophysical and electroretinographic data summarised here.

REFERENCES

Adelson, E.H., 1982, Saturation and adaptation in the rod system. *Vision Research* **22**:1299-1312.

Aguilar, M. and Stiles, W.S., 1954, Saturation of the rod mechanism of the retina at high levels of stimulation. *Optica Acta* **1**:59-65.

Alexander, K.R., and Fishman, G.A. 1984, Rod-cone interaction in flicker perimetry. *British Journal of Ophthalmolgy* **68**:303-309.

Alexander, K.R., and Fishman, G.A. 1985, Rod-cone interaction in flicker perimetry: evidence for a distal retinal locus. *Documenta Ophthalmologica* **60**:3-36.

Alpern. M., Falls, H.F., and Lee, G.B. 1960, The enigma of typical total monochromacy. *American Journal of Ophthalmolgy* **50**:996-1012.

Andrews, D.P. and Hammond, P., 1970, Suprathreshold spectral properties of single optic tract fibres in cat, under mesopic adaptation: Cone-rod interaction. *Journal of Physiology, London* **209**:83-103.

Arden, G.B. and Frumkes, T.E. 1986, Stimulation of rods can increase cone flicker ERGs in man. *Vision Research* **26**:711-721.

Arden, G.B. and Hogg, C.R., 1985, Rod-cone interaction and analysis of retinal disease. *British Journal of Ophthalmology* **69**:404-415.

Barlow, H.B., Fitzhugh, R., and Kuffler, S.W. 1957, Change of organization in the receptive fields of the cat's retina during dark adaptation. *Journal of Physiology, London* **137**:338-354.

Barlow, H.B. and Levick, W.R., 1969, Three factors limiting the reliable detection of light by retinal ganglion cells of the cat. *Journal of Physiology, London* **200**:1-24.

Barlow, H.B., Levick, W.R., and Yoon, M., 1971, Responses to single quanta of light in retinal ganglion cells of the cat. *Vision Research Supplement* **3**:87-101.

Bauer, G.M., Frumkes; T.E., and Nygaard, R.W., 1983, The signal-to-noise characteristics of rod-cone interaction. *Journal of Physiology, London,* **337**:101-119.

Baylor, D.A., Fuortes, M.G.F., and O'Bryan, P.M., 1971, Receptive fields of cones in the retina of the turtle. *Journal of Physiology, London,* **214**:265-294.

Baylor, D.A., Lamb, T.D., and Yau, K.-W., 1979, The membrane current of single rod outer segments. *Journal of Physiology, London,* **288**:589-611.

Baylor, D.A., Nunn, B.J., and Schnapf, J.L., 1984, The photocurrent, noise and spectral sensitivity of rods of the monkey Macaca Fascicularis. *Journal of Physiology, London* **357**:575-607.

van den Berg, T.J.T.P. and Spekreijse, H., 1977, Interaction between rod and cone signals studied with temporal sine wave stimulation. *Journal of the Optical Society of America* **67**:1210-1217.

Blakemore, C.B. and Rushton, W.A.H., 1965, Dark adaptation and increment threshold in a rod monochromat. *Journal of Physiology, London* **181**:612-628.

Bloomfield, S.A. and Miller, R.F., 1982, A physiological and morphological study of the horizontal cell types of the rabbit retina. *Journal of comparative Neurology* **208**:288-303.

Boycott, B.B., Hopkins, J.M., and Sperling, H.G., 1987, Cone connections of the horizontal cells of the rhesus monkey's retina. *Proceedings of the Royal Society* **B229**:345-379.

Buck, S.L., 1985, Cone-rod interaction over time and space. *Vision Research* **25**:907-916.

y Cajal, S.R., 1893, *The vertebrate retina* (translated by Maguire, D. and Rodieck, R.W.). In R.W. Rodieck, *The vertebrate retina: Principles of structure and function*, Appendix I 1973). San Francisco: W.H. Freeman.

Chase, L. and Dowling, J.E., 1990, A comparison of rod and cone pathways in the primate retina. *Investigative Ophthalmology & Visual Science Supplement* **31**:207.

Coletta, N.J. and Adams, A.J., 1984, Rod-cone interactions in flicker detection. *Vision Research* **24**:1333-1340.

Conner, J.D, 1982, The temporal properties of rod vision. *Journal of Physiology, London* **332**:139-155.

Conner, J.D. and MacLeod, D.I.A., 1977, Rod photoreceptors detect rapid flicker. *Science* **195**:689-699.

Curcio, C.A., Sloan, K.R., Packer, O., Hendrickson, A.E., and Kalina, R.E., 1987), Distribution of cones in human and monkey retina: individual variability and radial asymmetry. *Science* **236**:579-582.

Dacheux, R.F. and Raviola, E., 1982, Horizontal cells in the retina of the rabbit. *Journal of Neuroscience* **2**: 1486-1493.

Dacheux, R.F. and Raviola, E., 1986, The rod pathway in the rabbit retina: a depolarizing bipolar and amacrine cell. *Journal of Neuroscience* **6**:331-345.

Daw, N.W., Jensen, R.J., and Brunken, W.J, 1990, Rod pathways in mammalian retinae. *Trends in Neuro Sciences* **13**:110-115.

Daw, N.W. and Pearlman, A.L., 1969, Cat colour vision: one cone process or several? *Journal of Physiology, London* **201**:745-764.

Denny, N., Frumkes, T.E., and Goldberg, S.H., 1990, Comparison of summatory and suppressive rod-cone interaction. *Clinical Vision Sciences* **5**:27-36.

Dick, E. and Miller, R.F., 1978, Light-evoked potassium activity in mudpuppy retina: its relationship to the b-wave of the electroretinogram. *Brain Research* **154**:388-394.

Dodt, E. and Jessen, K.H., 1961, The duplex nature of the retina of the nocturnal gecko as reflected in the electroretinogram. *Journal of General Physiology* **44**:1143-1158.

Dowling, J.E., 1967, The site of visual adaptation. *Science* **155**:273-279.

Dowling, J.E., 1987, *The retina, an approachable part of the brain.* Belknap Press of Harvard University Press, Cambridge, Mass.

Dowling, J.E. and Ehinger, B., 1975, Synaptic organization of the amine-containing interplexiform cells of the goldfish and cebus monkey retina. *Science* **188**:270-273.

Dowling, J.E., Ehinger, B., and Floren, I., 1980, Fluorescence and electron microscopical observations of the amine-accumulating neurons of the cebus monkey retina. *Journal of comparative Neurology* **192**:655-685.

Enroth-Cugell, C., Hertz, B.G., and Lennie, P., 1977, Cone signals in the cat's retina. *Journal of Physiology, London* **269**:273-296.

Eysteinsson, T. and Frumkes, T.E., 1989, Physiological and pharmacological analysis of suppressive rod-cone interaction. *Journal of Neurophysiology* **61**:866-877.

Fach, C.C., Sharpe, L.T., and Stockman, A., 1991, The field adaptation of the human rod visual system (submitted).

Famiglietti, E.V. and Kolb, H., 1975, A bistratified amacrine cell and synaptic circuitry in the inner plexiform layer of the retina. *Brain Research* **84**:293-300.

Frumkes, T.E. and Eysteinsson, T., 1987, Suppressive rod-cone interaction in distal vertebrate retina: Intracellular records from xenopus and necturus. *Journal of Neurophysiology* **57**:1361-1382.

Frumkes, T.E., Naarendorp, F., and Goldberg, S.H., 1986, The influence of cone adaptation upon rod mediated flicker. *Vision Research* **26**:1167-1176.

Frumkes, T.E., Sekuler, M.D., Barris, M.C., Reiss, E.H., and Chalupa, L.M., 1973, Rod-cone interaction in human scotopic vision. I: temporal analysis. *Vision Research* **13**:269-1282.

Frumkes, T.E. and Temme, L.A., 1977, Rod-cone interaction in human scotopic vision--II. Cones influence rod increment thresholds. *Vision Research* **17**:673-679.

Glickstein, M. and Heath, G.G., 1975, Receptors in the monochromat eye. *Vision Research* **15**:633-636.

Goldberg, S.H., Frumkes, T.E., and Nygaard, R.W., 1983, Inhibitory influence of unstimulated rods in the human retina: evidence provided by examining cone flicker. *Science* **221**:180-182.

Gouras, P. and Link, K., 1966, Rod and cone interaction in dark-adapted monkey ganglion cells. *Journal of Physiology, London,* **184**:499-510.

Green, D.G. and Siegel, I.M., 1975, Double branched flicker fusion curves from the all-rod skate retina. *Science* **188**:1120-1122.

Grünert, U. and Martin, P.R., 1990, Rod bipolar cells in the macaque monkey retina: Light and electron microscopy. *Investigative Ophthalmology & Visual Science Supplement* **31**:536.

Gurevich, L., Stockton, R.A., and Slaughter, M.M., 1990, Comparisons of the waveforms of the B-wave of the ERG and ON bipolar cells. *Investigative Ophthalmology & Visual Science Supplement* **31**:114.

Hassin, G. and Witkovsky, P., 1983, Intracellular recordings from identified photoreceptors and horizontal cells of the xenopus retina. *Vision Research* **23**:921-932.

Hallett, P.E., 1962, Scotopic acuity and absolute threshold in brief flashes. *Journal of Physiology, London,* **163**:175-189.

Harding, T.H. and Enroth-Cugell, C., 1978, Absolute dark sensitivity and centre size in cat retinal ganglion cells. *Brain Research* **153**:157-162.

Hecht, S., Shlaer, S., Smith, E.L., Haig, C., and Peskin, J.C., 1938, The visual functions of a completely colorblind person. *American Journal of Physiology* **123**:94-95.

Hecht, S., Shlaer, S., Smith, E.L., Haig, C., and Peskin, J.C., 1948, The visual functions of the complete colorblind. *Journal of General Physiology* **31**:459-472.

Hendrickson, A., Koontz, M.A., Pourcho, R.G., Sarthy, P.V., and Goebel, D.J., 1988, Localization of glycine-containing neurons in the macaca monkey retina. *Journal of comparative Neurology* **273**:473-487.

Hess, R.F. and Nordby. K., 1986, Spatial and temporal limits of vision in the achromat. *Journal of Physiology, London* **371**:365-385.

Hofmann, M.I., Barnes, C.S., and Hallett, P.E., 1990, Detection of briefly flashed sine-gratings in dark-adapted vision. *Vision Research* **30**:1453-1466.

Kline, R.P., Ripps, H., and Dowling, J.E., 1978, Generation of b-wave currents in the skate retina. *Proceedings of the National Academy of Sciences, U.S.A.,* **75**:5727-5731.

Knight, R., Sanocki, E., and Buck, S.L., 1990, Field adaptation of dual rod mechanisms in the detection of 15 Hz flicker. *Investigative Ophthalmology & Visual Science* (supplement) **31**:494.

Kolb, H., 1977, The organization of the outer plexiform layer in the retina of the cat: electron microscopic observations. *Journal of Neurocytology* **6**:131-153.

Kolb, H., 1979, The inner plexiform layer in the retina of the cat: electron microscopic observations. *Journal of Neurocytology* **8**:295-329.

Kolb, H. and Famiglietti, E.V., 1974, Rod and cone pathways in the inner plexiform layer of cat retina. *Science* **186**:47-49.

Kolb, H., Mariani, A., and Gallego, A., 1980, A second type of horizontal cells in the monkey retina. *Journal of comparative Neurology* **189**:31-44.

Kolb, H. and Nelson, R., 1981, Amacrine cells of the cat retina. *Vision Research* **21**: 1625-1633.

Kolb, H. and Nelson, R., 1983, Rod pathways in the retina of the cat. *Vision Research* **23**:301-312

Kolb, H. and Nelson, R., 1984, Neural architecture of the cat retina. *Progress in Retinal Research* **3**:21-60.

Kolb, H., Nelson, R., and Mariani, A., 1981, Amacrine cells, bipolar cells, and ganglion cells of the cat retina: a Golgi study. *Vision Research* **21**:1081-1114.

Kolb, H. and West, R., 1977, Synaptic connections of the interplexiform cell in the retina of the cat. *Journal of Neurocytology* **6**:155-170.

von Kries, J., 1894, Über den Einfluß der Adaptation auf Licht- und Farbenempfindung und über die Funktion der Stäbchen. *Bericht der naturforschenden Gesellschaft zu Freiburg im Breisgau* **9** (2):61-70.

von Kries, J., 1929, Zur Theorie des Tages- und Dämmerungssehens. In A. Bethe, G. von Bergmann, G. Emden, and A. Ellinger (Eds), *Handbuch der normalen und pathologischen Physiologie*, Vol. XII (1), Receptionsorgane 2 (Photoreceptoren I) (pp. 679-713). Berlin: Springer-Verlag.

Lewis, S.D. and Mandelbaum, J., 1943, Achromatopsia: report of three cases. *Archiv Ophthalmologica* **30**:225-231.

Loew, E.R. and Arden, G.B., 1985, Inhibition of cones by rods in the mammalian eye as demonstrated electrophysiologically using flashing multipoint focal stimuli. *Investigative Ophthalmology & Visual Science Supplement* **26**:115.

MacLeod, D.I.A., 1972, Rods cancel cones in flicker. *Nature* **235**:173-174.

MacLeod, D.I.A., 1974, Psychophysical studies of signals from rods and cones. Unpublished doctoral dissertation, Cambridge University.

Mariani, A.P., 1982, Biplexiform cells: ganglion cells of the primate retina that contact photoreceptors. *Science* **216**:1134-1136.

Mariani, A.P., 1988, Amacrine cells of the rhesus monkey retina. *Investigative Ophthalmology & Visual Science Supplement* **29**:198.

Mastronarde, D.N., 1983, Correlated firing of cat retinal ganglion cells. II. Responses of X- and Y-cells to single quantal events. *Journal of Neurophysiology* **49**:325-349.

McGuire, B.A., Stevens, J.K., and Sterling, P., 1984, Microcircuitry of bipolar cells in cat retina. *Journal of Neurosciences* **4**:2920-2938.

Miller, R.F. and Dowling, J.E., 1970, Intracellular responses of the Müller (glial) cells of mudpuppy retina: Their relation to b-wave of the electroretinogram. *Journal of Neurophysiology* **33**:323-341.

Müller, F., Wässle, H., and Voigt, T., 1988, Pharmacological modulation of the rod pathway in the cat retina. *Journal of Neurophysiology* **59**:1657-1672.

Naka, K.-I., 1972, The horizontal cells. *Vision Research* **12**:573-588.

Nakamura, Y., McGuire, B.A., and Sterling, P., 1980, Interplexiform cell in cat retina: Identification of uptake of ∞-[3H]aminobutyric acid and serial reconstruction. *Proceedings of the National Academy of Sciences, Washington D.C.*, **77**:658-661.

Nelson, R., 1977, Cat cones have rod input: A comparison of the response properties of cones and horizontal cell bodies in the retina of the cat. *Journal of Comparative Neurology* **172**:107-135.

Nelson, R., 1982, AII amacrine cells quicken time course of rod signals in the cat retina. *Journal of Neurophysiology* **47**:928-947.

Nelson, R., Famiglietti, E.V., and Kolb, H., 1978, Intracellular staining reveals different levels of stratification for on- and off-center ganglion cells in cat retina. *Journal of Neurophysiology* **41**:472-483.

Nelson, R. and Kolb, H., 1983, Synaptic patterns and response properties of bipolar and ganglion cells in the cat retina. *Vision Research* **23**:1183-1195.

Nelson, R. and Kolb, H., 1984, Amacrine cells in scotopic vision. *Ophthalmological Research* **16**:21-26. *Neurophysiology* **54**:592-614.

Nelson, R., von Lützow, A., Kolb, H., and Gouras, P., 1975, Horizontal cells in cat retina with independent dendritic systems. *Science* **189**:137-139.

Noell, W.K., 1954, The origin of the electroretinogram. *American Journal of Ophthalmology* **28**:78-90.

Nygaard, R.W. and Frumkes, T.E., 1985, Frequency dependence in scotopic flicker sensitivity. *Vision Research* **25**:115-127.

Østerberg, G.A., 1935, Topography of the layer of rods and cones in the human retina. *Acta ophthalmologica, København*, supplement **6**:1-102.

Parinaud, H., 1881, Des modifications pathologiques de la perception de la lumiere, des couleurs, et des formes, et des differentes especes de sensibilite oculaire. *Comptes rendus hebdomadaires des seances et memoires de la Societe de biologie, Paris*, **33**:222.

Pflug, R. and Nelson, R., 1986, Enhancement of red cone flicker by rod selective backgrounds in cat horizontal cells. *Neuroscience Abstracts* **16**:402.

Polyak, S.L., 1941, *The Retina*. Chicago: Chicago University Press.

Raviola, E. and Gilula, N.B, 1973, Gap junctions between photoreceptor cells in the vertebrate retina. *Proceedings of the National Academy of Sciences, Washington D.C.*, **70**, 1677-1681.

Rodieck, R.W., 1988, The primate retina. *Comparative Primate Biology* **4**, Neurosciences:302-278.

Schultze, M., 1866, Zur Anatomie und Physiologie der Retina. *Archiv für mikroskopische Anatomie (und Entwicklungsmechanik)* **2**:175-286.

Shapley, R.M. and Enroth-Cugell, C., 1984, Visual adaptation and retinal gain controls. *Progress in Retinal Research* **3**:263-346.

Sharpe, L.T., Fach, C., Nordby, K.,, and Stockman, A., 1989, The incremental threshold of the rod visual system and Weber's law. *Science* **244**:354-356.

Sharpe, L.T., Fach, C.C., and Stockman, A., 1991, Rod flicker perception: scotopic duality and cone intrusion (submitted).

Sharpe, L.T. and Nordby, K., 1990, The photoreceptors in the achromat. In R. F. Hess, L.T. Sharpe, and K. Nordby (Eds), *Night Vision, basic, clinical and applied aspects*. Cambridge: Cambridge University Press.

Sharpe, L.T., van Norren, D., and Nordby, K., 1988, Pigment regeneration, visual adaptation and spectral sensitivity in the achromat. *Clinical Vision Sciences* **3**:9-17.

Sharpe, L.T., Stockman, A., and Macleod, D.I.A., 1989, Rod flicker perception: scotopic duality, phase lags and destructive interference. *Vision Research* **29**:1539-1559.

Sharpe, L.T., Stockman, A., and Zrenner, E., 1990, Dual rod pathways. *Perception* **19**: 350.

Sloan, L.L., 1954, Congenital achromatopsia: A report of 19 cases. *Journal of the Optical Society of America* **44**:117-128.

Sloan, L.L., 1958, The photopic retinal receptors of the typical achromat. *American Journal of Ophthalmology* **46**:81-86.

Smith, R.G., Freed, M.A., and Sterling, P., 1986, Microcircuitry of the dark-adapted cat retina: functional architecture of the rod-cone network. *The Journal of Neuroscience* **6**:3505-3517.

Smith, V. C. and Pokorny, J., 1975, Spectral sensitivity of the foveal cone photopigments between 400 and 500 nm. *Vision Research* **15**:161-171.

Steinberg, R.H., 1969, The rod after-effect in S-potentials from the cat retina. *Vision Research* **9**:1345-1355.

Steinberg, R.H., 1971, Incremental responses to light recorded from pigment epithelial cells and horizontal cells of the cat retina. *Journal of Physiology, London* **217**:93-110.

Steinberg, R.H., Reid, M., and Lacey, P.L., 1973, The distribution of rods and cones in the retina of the cat (Felis domesticus). *Journal of comparative Neurology* **148**:229-248.

Sterling, P., 1983, Microcircuitry of the cat retina. *Annual Review of Neuroscience* **6**:149-185.

Sterling, P., Freed, M.,, and Smith, R.G., 1986, Microcircuitry and functional architecture of the cat retina. *Trends in Neuro Sciences* **9**:186-192.

Stockman, A., Sharpe, L.T., and Zrenner, E., 1990, Scotopic duality in the ERG and in psychophysics. *Investigative Ophthalmology & Visual Science* (supplement) **31**:494.

Stockman, A., Sharpe, L.T., Zrenner, E., and Nordby, K., 1991, Slow and fast rod pathways in the human rod visual system (*submitted*).

Stockton, R.A., and Slaughter, M.M., 1987, ON bipolar cell potassium fluxes are uniquely associated with the ERG b-wave. *Investigative Ophthalmology & Visual Science Supplement* **28**:406.

Tamura, T., Nakatani, K., and Yau, K.-W., 1989, Light adaptation in cat retinal rods. *Science* **245**:755-758.

Toyoda, J.-I. and Tonosaki, K., 1978, Effect of polarization of horizontal cells on the on-center bipolar cells of the carp retina. *Nature* **276**:399-400.

Wald, G., 1945, Human color vision and the spectrum. *Science* **101**:653-658.

Wässle, H., Müller, F., Voigt, T., and Chun, M.H, 1989, Pharmacological modulation of the dark adapted cat retina. In R. Weiler, and N.N. Osborne (Eds), *Neurobiology of the Inner Retina* (pp. 247-259). Berlin: Springer-Verlag.

Wässle, H. and Riemann, H.J., 1978, The mosaic of nerve cells in the mammalian retina. *Proceedings of the Royal Society, London*, **B200**:441-461.

Weiler, R., 1977, Die Horizontalzellen der Karpenretina. Doctoral thesis, Universität München.

Wen, R., Tucker, J.L., and Oakley, B. I., 1990, Testing the K+/Müller cell hypothesis of the origin of the ERG B-wave. *Investigative Ophthalmology & Visual Science Supplement* **31**:114.

Wyszecki, G. and Stiles, W.S., 1982, *Color Science, concepts and methods, quantitative data and formulas* (2nd Edn). New York: John Wiley.

Yang, X.-L. and Wu, S.M., 1989, Effects of background illumination on the horizontal cell responses in the tiger salamander retina. *The Journal of Neuroscience* **9**:815-827.

Zrenner, E., Nelson, R., and Mariani, A., 1983, Intracellular recordings from a biplexiform ganglion cell in macaque retina, stained with horseradish peroxidase. *Brain Research* **262**:181-185.

WAVELENGTH-DISCRIMINATION WITH ONLY RODS AND BLUE CONES

Andreas Reitner[1], Lindsay T. Sharpe[2] and Eberhart Zrenner[3]

[1] Zweite Universitäts-Augenklinik, Alserstraße 2, 1090 Wien, Austria
[2] Neurologische Universitätsklinik, Hansastraße 9, D-7800 Freiburg im Breisgau, West Germany
[3] Department of Pathophysiology of Vision and Neuro-Ophthalmology, University Eye Hospital, Schleichstraße 12, 7400 Tübingen 1, West Germany

INTRODUCTION

At photopic levels, the ability to distinguish different wavelengths or different mixtures of wavelengths, independently of their luminance, depends upon the rate of change of quantum catches in the different classes of cones. Thus, the trichromat or normal observer--possessing all three cone classes--is very good at it, the dichromat or partially colourblind observer--lacking one cone class--is less so, and the monochromat or totally colourblind observer--lacking two or all three cone classes--is completely incapable.

At mesopic luminances, wavelength discrimination can depend upon the rate of change of quantum catch in the rods as well. This has (indirectly) been shown by comparing the red-green colour-mixtures (Nagel, 1905, 1907; McCann & Benton, 1969), metameric colour-matching (Smith & Pokorny, 1977) and colour-naming ability (Montag & Boynton, 1987) of dichromatic and normal trichromatic observers at mesopic and photopic levels. Montag and Boynton (1987), for instance, found that protanopes and deuteranopes have the ability to categorize colours along the red-green colour axis, but only when the rods are not excluded either by a bleach or by restricting the stimuli to the central fovea. The different conditions do not affect the performance of normal observers.

Why rod participation substantially benefits the dichromatic observer, but not the colour normal, can be explained in terms of the opponent-colour theory of colour vision. The three independent signals arising from the three classes of cones are transmitted along two chromatically-opponent and one non-opponent (luminance) retinal pathway. Since rods only have access to these same retinal pathways (Gouras, 1965; Gouras & Link, 1966; van den Berg, 1978; D'Zmura & Lennie, 1986), their signals can merely add to or subtract from the cone signals. Rod participation causes colour matches and colour naming made at one mesopic luminance level not to correspond to those made at other mesopic luminance levels (Trezona, 1970, 1973), but it does not provide a fourth independent sensation in the normal observer. This is confirmed by the findings that at any one level colour vision is never more than trichromatic and that the colour sensation associated with the rods does not occupy a unique locus in the CIE chromaticity diagram (Stabell & Stabell, 1974). For dichromatic observers missing one of the three cone classes, on the other hand, the rod signals at mesopic levels can supplement those of the two remaining cone classes, modulating activity in the opponent and luminance pathways which is otherwise unmodulated; thereby allowing discrimination in a missing dimension (i.e., the red-green colour axis in protanope and deuteranope observers), permitting large-field trichromacy and eliciting sensations that qualitatively differ from those attributable to the cones alone.

This leads to a question: if the rods can enhance colour discrimination in dichromats, can they also initiate it in monochromats? Obviously not in typical, complete achromats, who have only rod vision and no residual cone function (see Sharpe & Nordby, 1990a,b). But what about those monochromats who have one normally functioning cone class in addition to rods? Do they have dichromatic vision at mesopic light levels? This question can be answered by examining the wavelength discrimination of an extremely rare group of individuals, blue-cone monochromats (Blackwell & Blackwell, 1957, 1961; Alpern et al., 1965, 1971; Young & Price, 1985; Pokorny et al., 1970; Zrenner et al., 1988; Hess et al., 1989a,b), who have only rods and blue or short-wave sensitive (S) cones. We find that such individuals are, indeed, able to distinguish wavelengths between 440 and 520 nm, but only at mesopic levels where their rods and S-cones are simultaneously active.

From Pigments to Perception, Edited by A. Valberg and
B.B. Lee, Plenum Press, New York, 1991

METHODS

The observers were five blue-cone monochromats (X-chromosome linked incomplete achromats). Extensive tests were undertaken to exclude the possibility of residual middle-wave sensitive (M) and/or long-wave sensitive (L) cone function, which could provide the basis for wavelength discrimination. Such tests are necessary because, as Smith *et al.* (1983) have shown, many of the affected males in pedigrees of X-chromosomal incomplete achromatopsia display evidence of L-cone function in *large* field matches. In our observers, however, thorough spectral sensitivity and dark adaptation measurements confirmed that their vision could be fully described by the participation of normally functioning rods under scotopic conditions and normally functioning S-cones under photopic conditions (Zrenner, Magnussen & Lorenz, 1988; Hess, Mullen, Sharpe & Zrenner, 1989; Hess, Mullen & Zrenner, 1989). Under chromatic adaptation conditions, they failed to reveal the action spectra of the M- and/or L-cones and to display transient tritanopia, which in normal observers results from the interaction between the S- and the M- and L-cones (Mollon & Polden, 1977; Hansen *et al.*, 1978). Further, directional sensitivity measurements ruled out the presence in their fovea of anomalous cone photoreceptors, such as 'rhodopsin cones' (i.e. cone photoreceptors containing the rod photopigment), whose existence has been postulated in the eyes of blue-cone monochromats (Alpern *et al.*, 1971) and typical, complete achromats (Alpern *et al.*, 1960).

The psychophysical absence of the M- and L-cone sensitivities accords with the molecular genetics. Individual DNA blot hybridization patterns (Nathans, private communication) reveal that our five blue-cone monochromats have alterations in the M and L visual pigment gene cluster of their X-chromosomes. The pattern in four of the observers, F.B., S.B., M.P. and K.S., corresponds to one of the two recently reported by Nathans *et al.* (1989) for 14 blue-cone monochromats, in which only one of the tandem array of M and L visual pigment genes remains (either a L pigment gene or a M-L pigment hybrid) and is rendered nonfunctional by a point mutation. In the fifth observer, P.S., the pattern is of the second type: both the M and L visual pigment genes are rendered nonfunctional due to the deletion of DNA.

Wavelength-discriminability was measured by a computer-controlled procedure (for details, see Reitner, Sharpe & Zrenner, 1991a), The observers were required to regard a bipartite photometric field, presented in Maxwellian-view. One half-field (the standard field) was held fixed, while the other (the comparison field) was varied continously in wavelength, but kept constant in brightness. (To ensure that the standard and comparison fields were always equated in brightness, so that the observer could not discriminate the half-fields on any basis other than wavelength, the observer's spectral sensitivity between 420 and 650 nm was first measured in 5-nm or 10-nm steps by heterochromatic brightness matching to a 540 nm standard.) The standard and comparison wavelengths were selected by means of grating monochromators (Jobin Yvon H.10 Vis), whose 0.5 mm entrance and exit slits provided a bandpass of 4 mm (full-width at half-amplitude) and a resolution of 0.25 mm. Both halves of the field, which subtended 1° of visual angle in diameter, were presented simultaneously for 1 s. Between presentations, the observer adapted for 3 s to a homogeneous white field of the same intensity as the standard and comparison half-fields. The wavelength of the comparison half-field was changed in 1 nm steps from equivalence with that of the standard half-field. The discrimination threshold or just noticeable difference (JND) in wavelength was found where the observer signalled that the two halves of the field were no longer identical in appearance. This procedure was repeated five times for a series of standard wavelengths throughout the visible spectrum.

RESULTS

All five blue-cone monochromat observers were capable of discriminating wavelengths, equated on the basis of their individual heterochromatic brightness matches, at mesopic levels between 0.8 and 80 trolands (td). Figure 1A shows the average wavelength discrimination function obtained at the intermediate mesopic intensity level of 8 td. It can be compared with the average function of 8 normal observers (●) obtained under the same conditions. The average normal curve has the same general appearance as the classic curves reported in the literature (Wright & Pitt, 1935; Wyszecki & Stiles, 1982): a relative maxima (i.e. largest JND) is observed at approximately 530 nm, two relative minima (i.e. smallest JND) at approximately 490 and 590 nm. The average curve for the blue-cone monochromats, on the other hand, has only one minimum between 460 and 480 nm. Wavelength discrimination completely collapses for wavelengths longer than 520 nm, where, on average, they require a difference in wavelength of almost 70 nm before they can discriminate between the standard and comparison fields. However, within the narrow range between 440 and 520 nm, their discriminability (2-4 nm JNDs) compares favourably with that of the normal observers.

Figure 1B shows the individual wavelength discrimination functions of the blue-cone monochromat observers. In general, there is very good agreement between them. Only observer P.S. requires a larger JND in wavelength than the other observers at wavelengths between 440 and 490 nm. At wavelengths between 490 and 530 nm, however, his wavelength discrimination is slightly better than theirs. This difference may turn out to be important because P.S. has a different DNA hybridization pattern than the other four blue-cone monochromats (see above).

At 0.8 td and 80 td adapting levels, the observers were also able to discriminate wavelengths, but their performance tended to be poorer than that at the intermediate intensity of 8 td. In general, their wavelength discrimination extended to longer wavelengths as the field intensity increased, with the cut-off wavelength shifting rightwards from 480 to 520 nm. This is similar to what has been noted in the wavelength discrimination functions of deuteranopes and protanopes, which shift rightwards by about 20 nm from 470 to 490 nm when the troland value is increased. In the case of the blue-cone monochromat observers, the change with field intensity can be related to the change in their spectral sensitivity functions (Reitner, Sharpe & Zrenner, 1991b). At higher intensities the rod peak (near 520 nm) declines in magnitude relative to the S-cone peak (near 450 nm), so that the sensitivities of the two receptor types first converge then diverge and the spectral region where the sensitivities are most similar shifts from the short- to the middle-wavelengths. At light levels above 800 td, the spectral luminosity function of all five blue-cone monochromats becomes unimodal, with a peak near 450 nm. And, because their saturated rods no longer provide a differential signal which can be compared with that of the S-cones, their wavelength discrimination breaks down and their vision becomes monochromatic. A similar breakdown occurs at light levels below 0.08 td, at which the spectral luminosity function of the blue-cone monochromats has a single peak near 520 nm (the peak is near 520 nm rather than 510 nm because the curve is measured centrally and is subject to macular screening). These levels are too dim to excite strongly the S-cones and, consequently, too dim to provide the differential signals required for wavelength discrimination.

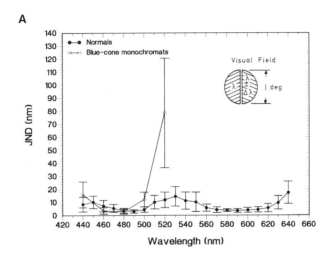

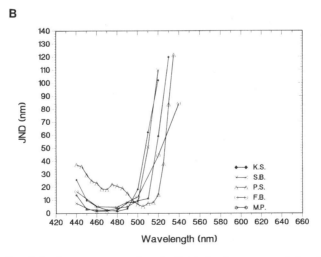

Fig. 1. Panel A: the average wavelength discrimination curves of five blue-cone monochromats (age 9-53; o) and of 8 normal trichromats (age 21-32; ●) measured at a retinal illuminance of 8 td. The inset shows the field of view, in which λ and λ+Δλ indicate, respectively, the standard and comparison wavelengths. JNDs in wavelength are plotted as a function of the standard wavelength. Panel B: The individual wavelength discrimination curves of the five blue-cone monochromats.

DISCUSSION

These results demonstrate that, within a limited intensity range, colour vision is possible with only rods and S-cones. This supports earlier reports that blue-cone monochromats can made consistent colour matches (Alpern *et al.*, 1971) or discriminate wavelengths (Young & Price, 1985). But the earlier reports relied on the assumption that blue-cone monochromats, have a third receptor type in their foveae--a rhodopsin cone--which is used in conjunction with the S-cones to discriminate colours. The existence of such receptors has not been verified psychophysically in the five blue-cone monochromats investigated here (Zrenner *et al.*, 1988; Hess *et al.*, 1989a), nor in another blue-cone monochromat observer investigated by others (Daw & Enoch, 1973). Moreover, such an hypothesis would have trouble explaining why our five blue-cone monochromats display no wavelength discrimination at photopic levels, where the rods saturate (Aguilar & Stiles, 1954), but where rhodopsin cones should not.

Since rods and S-cones are the only functioning photoreceptors in our observers, our results imply that some rod and cone signals travel by separate pathways to the visual processing stage where wavelength discrimination takes place. Normal wavelength discrimination is believed to be mediated by two chromatically-opponent mechanisms: the red-green pathway, comparing the signals arising from the M- and L-cones; and the yellow-blue pathway, comparing the signals arising from the S- and from the combined M- and L-cones (Walraven & Bouman, 1966; Judd & Yonemura, 1970; Bouman & Walraven, 1972). Because the blue-cone monochromats lack the function of the M- and L-cones, it seems reasonable to conclude that the rods must contribute their influence only through the second of these two pathways. That is, some rod signals could feed into the yellow-blue chromatic opponent pathway with the opposite sign as the S-cones. To demonstrate this conclusively, however, bichromatic field and/or test additivity experiments would need to be undertaken; for wavelength discrimination could, in principle, also take place without opponent mechanisms. If, on the other hand, the cellular connexions of the yellow-blue opponent pathway are intact, sustained by the signals of the rods and S-cones, then, in field additivity experiments, threshold for violet flashes on a blue field should fall if enough yellow light is added to the blue field to produce a composite field that is brighter but achromatic; and, in test additivity experiments, the detectability of a violet flash on a bright yellow background should be reduced if it is accompanied by a yellow flash. Although we have not yet conducted such tests in our blue-cone monochromat observers, we already have other indirect evidence for opponency from their spectral sensitivity curves. Curves measured at an intensity of 8.0 td, where their best wavelength discrimination is found, frequently have a deep notch between the S-cone (c. 450 nm) and rod (c. 520 nm) peaks. This feature, reminiscent of Sloan's notch (Fach & Mollon, 1987) which is a clear signature of detection by an opponent process (Mollon, 1982), suggests that the blue-cone monochromats' brightness matching in the wavelength region between 450 and 520 nm depends not on absorptions solely in the S-cones or solely in the rods, but rather on comparisons of absorptions in both.

ACKNOWLEDGEMENTS

This research was supported by the Alexander von Humboldt-Foundation (Bonn-Bad Godesberg), the Deutsche Forschungsgemeinschaft (Bonn), SFB 325, Tp B4, and Zr 1/7-1 and Hoechst AG. We thank Sonja Magnussen for her assistance throughout the project, Dr Jeremy Nathans (Baltimore, Maryland) for providing the DNA blot hybridization patterns and Professor Vivianne C. Smith (Chicago) for her helpful comments during the NATO workshop.

REFERENCES

Aguilar, M. and Stiles, W.S., 1954, Saturation of the rod mechanism of the retina at high levels of stimulation. *Optica Acta* **1**:59-65.

Alpern, M., Falls, H.F., and Lee, G.B., 1960, The enigma of typical total monochromacy. *American Journal of Opthalmology* **50**:996-1012.

Alpern, M., Lee, G.B., and Spivey, B.E., 1965, Π_1 cone monochromatism. *Archives of Ophthalmology* **74**:334-337.

Alpern, M., Lee, G.B., Maaseidvaag, F. and Miller, S., 1971, Colour vision in blue-cone 'monochromacy'. *Journal of Physiology, London,* **212**:211-233.

van den Berg, T.J.T.P., 1978, Rod-cone interaction with homogeneous field stimulation. *Modern Problems in Ophthalmology* **19**:341-343.

Blackwell, H.R. and Blackwell, O.M., 1957, Blue mono-cone monochromacy: A new color vision defect. *Journal of the Optical Society of America* **47**:338.

Blackwell, H.R. and Blackwell, O.M., 1961, Rod and cone receptor mechanisms in typical and atypical congenital achromatopsia. *Vision Research* **1**:62-107.

Bouman, M.A. and Walraven, P.L., 1972, Color discrimination data, in *Handbook of Sensory Physiology*, vol **VII/4**, ed. D. Jameson and L.M. Hurvich, pp. 484-516. Berlin, Springer.

Daw, N.W. and Enoch, J.M., 1973, Contrast sensitivity Westheimer function and Stiles-Crawford effect in a blue cone monochromat. *Vision Research* **13**:1669-1680.

D'Zmura, M. & Lennie, P., 1986, Shared pathways for rod and cone vision. *Vision Research* 26:1273-1280.

Fach, C.C. and Mollon, J.D., 1987, Predicting the position of Sloan's notch. *Investigative Ophthalmology & Visual Science* (supplement) 28:212.

Gouras, P., 1965, Primate retina: duplex function of dark adapted retinal ganglion cells. *Science* 147:1593-1594.

Gouras, P. and Link, K., 1966, Rod and cone interaction in dark-adapted monkey ganglion cells. *Journal of Physiology, London,* 184:499-510.

Hansen, E., Seim, T., and Olsen, B.T., 1978, Transient tritanopia experiments in blue cone monochromacy. *Nature* 276, 390-391.

Hess, R.F., Mullen, K.T., Sharpe, L.T., and Zrenner, 1989a, The photoreceptors in atypical achromatopsia. *Journal of Physiology, London,* 417:123-149.

Hess, R.F., Mullen, K.T., and Zrenner, E., 1989b, Human photopic vision with only short wavelength cones: Post-receptoral properties. *Journal of Physiology, London,* 417:150-169.

Judd, D.B. and Yonemura, G.T., 1970, CIE 1960 UCS diagram and the Müller theory of color vision. *Proceedings of the International Color Association, Stockholm, Sweden, 1969,* pp. 266-274, Göttingen: Munsterschmidt.

McCann, J.J. and Benton, J.L., 1969, Interaction of the longwave cones and rods to produce color sensations. *Journal of the Optical Society of America* 59:103-107.

Mollon, J.D., 1982, Color vision. *Annual Review of Psychology* 33:41-85.

Mollon, J.D. and Polden, P.G., 1977, An anomaly in the response of the eye to light of short wavelengths. *Philosophical Transactions of the Royal Society, London, B* 278:207-240.

Montag, E.D., and Boynton, R.M., 1987, Rod influence in dichromatic surface color perception. *Vision Research* 27:2153-2162.

Nagel, W.A., 1905, Dichromatische Fovea, trichromatische Peripherie. *Zeitschrift für Psychologie und Physiologie des Sinnesorganes,* 39:93-101.

Nagel, W.,A., 1907, Neue Erfahrungen über das Farbensehen der Dichromaten auf grossen Felde. *Zeitschrift für Sinnesphysiologie* 41:319-337.

Nathans, J., Davenport, C.M., Maumenee, I.H., Lewis, R.A., Hejtmancik, J.F., Litt, M., Lovrien, E., Weleber, R., Bachynski, B., Zwas, F., Klingaman, R., and Fishman, G., 1989, Molecular genetics of human blue cone monochromacy. *Science* 245:831-838.

Pokorny, J., Smith, V.C., and Swartley, R., 1970, Threshold measurements of spectral sensitivity in a blue monocone monochromat. *Investigative Ophthalmology* 9:807-813.

Reitner, A., Sharpe, L.T., and Zrenner, E. 1991a, Wavelength discrimination as a function of field intensity, duration and size (submitted).

Reitner, A., Sharpe, L.T., and Zrenner, E., 1991b, Is colour vision possible with only rods and blue cones? (submitted).

Sharpe, L.T. and Nordby, K., 1990a, Total colour-blindness: an introduction. Chapter 7 in R.F. Hess, L.T. Sharpe, and K. Nordby (Eds), *Night vision, basic, clinical and applied aspects,* pp. 253-289. Cambridge University Press, Cambridge.

Sharpe, L.T. and Nordby, K., 1990b, The photoreceptors in the achromat. Chapter 10 in R. F. Hess, L.T. Sharpe, and K. Nordby (Eds), *Night Vision, basic, clinical and applied aspects.* pp. 335-389. Cambridge: Cambridge University Press.

Smith, V.C. and Pokorny, J., 1977, Large-field trichromacy in protanopes and deuteranopes. *Journal of the Optical Society of America* 67:213-220.

Smith, V.C., Pokorny, J., Delleman, J.W., Cozijnsen, M., Houtman, W.A., and Went, L.N., 1983, X-linked achromatopisa with more than one class of functional cones. *Investigative Ophthalmology & Visual Science* 24:451-457.

Stabell, U. and Stabell, B., 1974, Chromatic rod-cone interaction. *Vision Research* 14:1389-1392.

Trezona, P.W., 1970, Rod participation in the 'blue' mechanism and its effect on colour matching. *Vision Research* 10:317-332.

Trezona, P.W., 1973, The tetrachromatic colour match as a colorimetric technique. *Vision Research* 13:9-25.

Young, R.S.L. and Price, J., 1985, Wavelength discrimination deteriorates with illumination in blue cone monochromats. *Investigative Ophthalmology & Visual Science* 26:1543-1549.

Walraven, P.L. and Bouman, M.A., 1966, Fluctuation theory of colour discrimination of normal trichromats. *Vision Research* 6:567-586.

Wright, W.D. & Pitt, F.H.G., 1935, Hue-discrimination in normal colour-vision. *Proceedings of the Physical Society* 46:459-473.

Wyszecki, G. and Stiles, W.S., 1982, *Color Science, concepts and methods, quantitative data and formulas* (2nd Edn). New York: John Wiley.

Zrenner, E., Magnussen, S., and Lorenz, B., 1988, Blauzapfenmonochromasie: Diagnose, genetische Beratung und optische Hilfsmittel. *Klinische Monatsblatter für Augenheilkunde* 193:510-517.

DENSITY OF BIPOLAR CELLS IN THE MACAQUE MONKEY RETINA

Paul R. Martin and Ulrike Grünert

Neuroanatomy Dept., Max-Planck-Institute for Brain Research
Deutschordenstrasse 46, D 6000 Frankfurt am Main, W. Germany

INTRODUCTION

The inner nuclear layer (INL) of the mammalian retina is composed of the cell bodies of bipolar, horizontal, amacrine, interplexiform and Müller cells. Of these, bipolar cells are responsible for the vertical flow of information from the photoreceptor matrix to the inner plexiform layer and hence to the ganglion cells. We have begun an analysis of the neuronal composition of the INL of the macaque monkey with emphasis on the different subpopulations of bipolar cells which are present there. In this chapter we describe the density and distribution of rod bipolar cells within the central 12 mm and provide an estimate of the total number of bipolar cells within one millimeter of the fovea. The results can be used to estimate limits to the spatial resolving power of rod and cone mediated vision in the central retina, and may thus have relevance for psychophysical estimates of these values.

A basic distinction in the bipolar cell population of the mammalian retina can be made between rod bipolar cells (which contact rod photoreceptors) and cone bipolar cells (which contact cones). Cajal (1893) recognized one morphological type of rod bipolar cell and several types of cone bipolar cell. This was confirmed for the primate retina by Polyak (1941), and by Boycott and Dowling (1969) but none of these authors could make any comment about the relative numerosity of bipolar cells or their distribution across the retina. The quantitative studies of Missotten (1974) and Krebs and Krebs (1987, 1989) suggest that there is a total of 2–3 bipolar cells per cone pedicle in the central primate retina. We have extended these studies by distinguishing rod bipolar cells with protein kinase C (PKC) immunocytochemistry, and by measuring cell densities over a range of eccentricities near the fovea to obtain cumulative estimates of the number of bipolar and ganglion cells within the central retina.

From Pigments to Perception, Edited by A. Valberg and
B.B. Lee, Plenum Press, New York, 1991

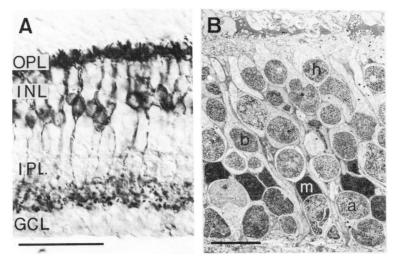

Figure 1. **A:** 12 μm vertical cryostat section through the macaque retina processed for PKC immunoreactivity. Abbreviations: OPL, outer plexiform layer; INL, inner nuclear layer; IPL, inner plexiform layer; GCL, ganglion cell layer. The labelled cell bodies of rod bipolar cells are seen in the middle to outer half of the INL; the dendrites terminate at the level of rod spherules and the axon terminals in the IPL are seen close to the GCL. Scale bar 50 μm. **B:** Low-power electron micrograph of a vertical section through the INL at about 2 mm eccentricity. The layer of electron dense Müller cells (m) separates amacrine cells (a) from bipolar (b) and putative horizontal cells (h). Other criteria used to distinguish these cell types are described in the text. Scale bar 10μm.

METHODS

All measurements were made from retinae of juvenile macaque monkeys (*M. mulatta* and *M. fascicularis*) which had been used for physiological recording of visual responses from the retina or lateral geniculate nucleus. These experiments were made under Halothane anaesthesia. Retinae were fixed immediately after administration of a lethal dose of pentobarbitone. For PKC immunocytochemistry, retinae were perfusion fixed as described in detail by Greferath *et al.* (1990). For determining cell density in Nissl stained semithin sections and electron micrographs, the posterior eyecup was immersion fixed as described by Grünert and Wässle (1990). To obtain vertical (radial) sections at known retinal positions, the retina was first cut into three pieces (Packer *et al.* 1989) then the piece containing the fovea was laid on a sheet of millimeter graph paper and viewed through a dissecting microscope. Blocks of defined size and at known eccentricity could then be cut with a razor blade broken to the required length. The remaining retina was either processed as a "wholemount" or cut horizontally (tangentially) at 70 μm on a freezing microtome. Cell densities could be determined directly from horizontal sections which included the entire INL, or calculated from the cell profile density in semithin vertical sections. The size of all nuclear profiles in each sample was measured and cell density was calculated using the recursive reconstruction method of Rose and Rohrlich (1988). Most samples were taken from 1 μm semithin sections, measurements made in the same way on low power electron micrographs or using the disector method on consecutive sections (Sterio, 1984) gave results which generally agreed to within 10%.

RESULTS AND DISCUSSION

Identification and distribution of rod bipolar cells

In the macaque, as in other mammalian species (Negishi *et al.*, 1988; Greferath *et al.*, 1990; Grünert and Martin, 1990) , rod bipolar cells can be identified by their PKC-like immunoreactivity (Figure 1A.) The density of rod bipolar cells along the ventral retinal axis of one retina is shown in Figure 2A. The peak density of rod bipolar cells occurs about 2 mm from the fovea; this is slighty central to the peak rod density which in this retina on the ventral axis was close to 3 mm eccentricity.

We have analysed rod bipolar cell density data from the horizontal retinal axis of two other retinae. Although there is some variation in the peak density, between 10,000–15,000 cells/mm^2, the peak density always occurs between 2–3 mm from the fovea, that is close to the rod density maximum. The inter-cell spacing for an hexagonal matrix of 16,000 cells/mm^2 is 8.5 μm; such an array could transmit spatial frequencies up to 12 cycles/deg without aliasing, assuming 200 μm per degree for central retina, which we calculated from the fovea–optic disk distance in the retinae we used. Since the total ganglion cell density at 2 mm eccentricity lies between 10,000–20,000 cells/mm^2 (Perry and Cowey, 1984; Wässle *et al.*, 1989), the potential resolving power of the rod bipolar cell matrix could only be utilised if the entire ganglion cell population formed a single channel for detection at this eccentricity.

The numerical convergence of rod photoreceptors onto rod bipolar cells is shown in Figure 2B; the rod density data were obtained from the same retinal positions as the rod bipolar cell density data of Figure 2A. Between one and 10 mm from the fovea the ratio rises gradually from 5:1 to over 20:1. We have not yet determined whether the approximately twofold higher rod density at equivalent eccentricities in dorsal compared to ventral retina (Packer *et al.*, 1989, Wikler *et al.*, 1990) is matched by a higher rod bipolar cell density, but some preliminary observations suggest that this might be the case.

Estimating total bipolar cell density

Figure 1B shows a low power electron micrograph of a vertical section through the macaque retina close to 2 mm eccentricity. Müller cells can be distinguished easily by their high electron density and angular profiles. We classify amacrine and interplexiform cells as those whose profiles lie below the layer of Müller cells, and bipolar and horizontal cells as those whose profiles lie above (distal to) those of Müller cells. We could not always distinguish horizontal from bipolar cells when counting from this or 1 μm semithin sections, so these cell types were always counted together and we relied on published estimates of horizontal cell density (Röhrenbeck *et al.*, 1989) to obtain final estimates of bipolar cell density. The values obtained within the first 2 mm on the temporal horizontal axis of one retina are shown in Figure 2C. Densities of both ganglion cells and the combined horizontal and bipolar cell population rise steeply to a maximum at 0.6–1 mm and 1–1.3 mm respectively. At higher eccentricities, ganglion cell density falls more steeply than bipolar and horizontal cell density. Krebs and Krebs (1989) measured INL cell densities at 500 μm eccentricity in two monkeys and obtained values which are slightly higher than ours if one assumes the same shrinkage factor (linear 10%).

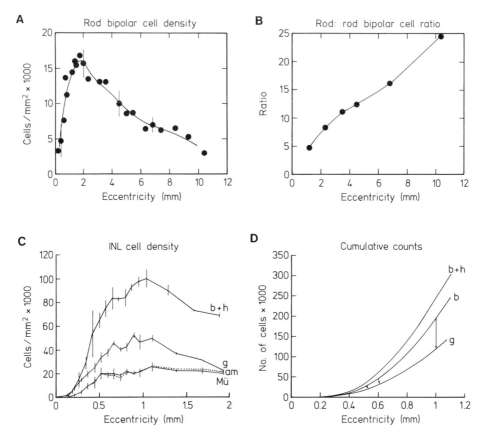

Figure 2. **A:** Distribution of rod bipolar cells along the ventral axis of one retina. Each point shows the mean of five samples with typical standard deviations. Retinal magnification 1 deg = 200 μm. Data corrected for shrinkage (linear factor 0.9, areal factor 0.81). **B:** Ratio of rod bipolar cell density to rod density measured at the same retinal locations as a subset of the sample points in A. **C:** Density of cell populations in the INL within the central 2 mm on the temporal horizontal axis of one retina. Data points show mean and standard deviations from at least three 1 μm semithin sections. Abbreviations: g, ganglion cells; b+h, bipolar and horizontal cells; am, amacrine cells; Mü, Müller cells. **D:** Cumulative ganglion and bipolar cell counts. Data shown are pooled values from the horizontal temporal axis of four retinae. Abbreviations as in C; the curve (b) shows an estimate for bipolar cells alone. The small arrows indicate the magnitude of the horizontal shift necessary to account for postreceptoral displacement.

The ratio of bipolar cells to ganglion cells in central retina

It can be seen from Figure 1B that the visible processes of bipolar cells do not follow a vertical course through the INL but run away from the fovea as they descend towards the IPL. One must take account of this postreceptoral displacement when the numbers of cells in different retinal layers are to be compared (Missotten, 1974; Schein, 1988; Wässle *et al.*, 1989). The simplest approach is to make cumulative estimates of the numbers of cells contained in volumes with increasing distance from the fovea; when cumulative cell number is then plotted against distance from the fovea the comparison between layers becomes equivalent to a shift on the horizontal axis. Such an analysis is illustrated in Figure 2D, which shows data pooled from four monkeys. It can be seen that the cumulative count for bipolar cells rises more steeply than that of ganglion cells.

For the retina shown in Figure 2C we obtain an average of 1.45 bipolar cells per ganglion cell between 0.4 and 1.4 mm eccentricity. Subtracting the cumulative count of rod bipolar cells (obtained from the density values of Figure 2A) produces an estimate of 1.3 cone bipolar cells per ganglion cell for the central 1.4 mm, a ratio which increases further at higher eccentricities. The most central (< 0.4 mm) eccentricities were not included in the estimate because small variations in the cumulative cell counts caused large fluctuations in the ratio; we are now making more exhaustive measurements to obtain reliable estimates of cell densities on the foveal slope. The total number of cells involved (< 5000) is relatively low, but as these cells almost certainly are associated with cone outer segments within the central degree, the task seems worthwhile.

Convergence of bipolar cells onto cones

We have not measured directly the density of cone outer segments or pedicles from our vertical sections, since we do not yet have a satisfactory description of cone pedicle shape that would allow an estimation of pedicle density from profile density. Estimates of the number of ganglion cells per cone in the central primate retina have varied between 1:1 to more than 3:1 but there are good reasons to believe that the ratio is at least 2:1 for the fovea (see Schein, 1988; Allen *et al.*, 1989; Wässle *et al.*, 1989)

The ratios 2:1 and 3:1 give lower and upper estimates of 2.6 and 3.9 cone bipolar cells per central cone. The former is probably more appropriate since the ganglion cell : cone ratio falls rapidly outside the foveal representation. (Schein, 1988; Wässle *et al.*, 1989) . Kolb and DeKorver (1988) analysed the synaptic connectivity of midget ganglion cells near the human fovea and found that that they received the vast majority of their direct bipolar input from a single midget bipolar cell. Our results show that there are enough cone bipolar cells to allow for specific input to the different ganglion cell populations present in the primate retina (see Rodieck, 1988; for review) and for each midget ganglion cell to possess a "private line" from a single cone. Whether the diffuse ganglion cell classes also receive predominate input from a single, presumably diffuse, bipolar cell is not known, but our results are compatible with this possibility.

Acknowledgements: We thank Barbara Tuschen, Walter Hofer and Felicitas Boij for technical assistance and Barry Lee for supplying the monkey retinas. We also thank Heinz Wässle for his generous support, practical assistance and encouragement during all stages of this project.

REFERENCES

Allen, K.A., C.A. Curcio, and R.E Kalina, 1990, Topography of cone-ganglion cell relations in human retina. *Investigative Ophthalmology and Visual Science*, 30(3):347.

Boycott, B.B. and J.E. Dowling, 1969, Organization of the primate retina: light microscopy. *Philosophical Transactions of the Royal Society of London B:*, 255:109–184.

Cajal, S. R., 1893, La rétine des vertébrés. *Cellule*, 9:121–255.

Greferath, U., U. Grünert, and H. Wässle, 1990, Rod bipolar cells in the mammalian retina show protein kinase C-like immunoreactivity. *Journal of Comparative Neurology. In press.*

Grünert, U. and P.R. Martin, 1990, Rod bipolar cells in the macaque monkey retina: light and electron microscopy. *Investigative Ophthalmology and Visual Science*, 31(4):536.

Grünert, U. and H. Wässle, 1990, GABA-like immunoreactivity in the macaque monkey retina : a light and electron microscopic study. *Journal of Comparative Neurology*, 297:509–524.

Kolb, H. and L. DeKorver, 1988, Synaptic input to midget ganglion cells of the human retina. *Investigative Ophthalmology and Visual Science*, 29:326.

Krebs, W. and I. P. Krebs, 1987, Quantitative morphology of the primate peripheral retina (*macaca irus*). *American Journal of Anatomy*, 179:198–208.

Krebs, W. and I. P. Krebs, 1989, Quantitative morphology of the central fovea of the primate retina. *American Journal of Anatomy*, 184:225–236.

Missotten, L., 1974, Estimation of the ratio of cones to neurons in the fovea in the human retina. *Investigative Ophthalmology and Visual Science*, 13:1045–1049.

Negishi, K., S. Kato, and T. Teranishi, 1988, Dopamine cells and rod bipolar cells contain protein kinase C–like immunoreactivity in some vertebrate retinas. *Neuroscience Letters*, 94:247–252.

Packer, O., A.E. Hendrickson, and C.A. Curcio, 1989, Receptor topography of the adult pigtail macaque (*macaca nemestrina*). *Journal of Comparative Neurology*, 288:165–183.

Perry, V.H. and A. Cowey, 1984, The ganglion cell and cone distributions in the monkey's retina: Implications for central magnification factors. *Vision Research*, 12:1795–1810.

Polyak, S. L., 1941, *The Retina*. University of Chicago Press, Chicago.

Rodieck, R.W., 1988, The primate retina. In *Comparative primate biology, volume 4:Neurosciences*, pages 203–278. Alan R. Liss, New York.

Röhrenbeck, J., H. Wässle, and B.B. Boycott, 1989, Horizontal cells in the monkey retina: Immunocytochemical staining with antibodies against calcium binding proteins. *European Journal of Neuroscience*, 1:407–420.

Rose, R.D. and D. Rohrlich, 1988, Counting sectioned cells via mathematical reconstruction. *Journal of Comparative Neurology*, 272:617.

Schein, S.J., 1988, Anatomy of macaque fovea and spatial densities of neurons in foveal representation. *Journal of Comparative Neurology*, 269:479–505.

Stereo, D.C., 1984, Estimating number, mean sizes and variations in size of particles in 3-d specimens using disectors. *Journal of Microscopy*, 134:127–136.

Wässle, H., U. Grünert, J. Röhrenbeck, and B.B. Boycott, 1989, Cortical magnification factor and the ganglion cell density of the primate retina. *Nature*, 341:643–646.

Wikler, K.C., R.W. Williams, and P. Rakic, 1990, Photoreceptor mosaic : Number and distribution of rods and cones in the rhesus monkey retina. *Journal of Comparative Neurology*, 297:499–508.

DISCUSSION: ROD VISION

William Swanson

Retina Foundation of the Southwest
Dallas,
Texas, U.S.A.

Swanson: Let me list a few questions to focus the general discussion.

The first question is "Why rods?" Except for Don MacLeod's talk, these are the only two talks in this workshop about rods. Most of the photoreceptors are rods, and certainly in primate evolution scotopic and mesopic vision have been important. However, Qasim Zaidi told me that nobody in New York City needs their rods, so we may not be used to thinking about scotopic vision.

The second question is "To what extent is flicker is an appropriate tool?" Both experiments used flicker, and Vienot also used brightness matching. I would like to use her work to distinguish between studying attributes and measuring thresholds. What I and other psychophysicists are used to doing is measuring thresholds, because thresholds are useful for trying to get at the most sensitive mechanism. It is common with thresholds to get a brisk transition from one mechanism (e.g., scotopic) to another (photopic), whereas with brightness matching there is a very slow and gradual systematic transition. The visual system is usually used not to detect thresholds, but to extract the attributes of stimuli. Van Essen may discuss attributes later, and he argues that when you are testing attributes it can be quite difficult to pin them down to any particular pathway. I think that this is exemplified with brightness matching. In normals, even at 0.01 troland you still don't have rods isolated, because above 600 nm you see red light, and cones contribute to brightness. Burns, Pokorny and Smith tried to model certain aspects of brightness by combining a range of different chromatic and achromatic pathways, but their systematic data set was difficult to fit with any of a variety of combinations. So I think that this shows the complexities we get into in dealing with attributes and this may be one reason people tend to focus more on thresholds. But in the long run if you want to understand the visual system we are going to have to be able to deal with attributes.

The third question is "What ganglion cells mediate rod vision?" This did not come up particularly in the talks, but some argue that rod signals are carried predominantly by the M-cells while some argue otherwise.

The last question is "What observers to test?" Traditionally in psychophysics you take a couple of young, healthy graduate students, and maybe the person whose lab it is, and run them through the experiments. If you are Don MacLeod, you have a deuteranomalous trichromat to use (deuteranomalous trichromats are useful for looking at the rapid rod response because it is easier to isolate rod responses in them). In both of these talks, people with unusual colour vision, dichromats or achromats, were used. The question is: "What is the appropriate use of these biological experiments". There are several other classes of unusual human. In addition to males with congenital colour defects, there are females who are carriers, with presumably unusual distributions of receptors in their retinae, there are incomplete achromats who primarily have S-cones and rods, and there are people with congenital nightblindness with normal rhodopsin but poor dark adaptation. Finally, there are ocular diseases, which are difficult to deal with because you have to have them carefully characterized. But if this is done, for example with an older patient who has had juvenile macular generation, it may be useful to test unusual observers.

MacLeod: Can Vienot comment on a straightforward explanation of the reverse Purkinje shift in the mesopic range, in which around 7 to 10 Hz the rod and cone signals are in opposite phase so that contamination of cones by rods above the discontinuity, and rods by cones below it, will drive the spectral sensitivity away from intermediate positions.

Vienot: I suggest that for the reverse Purkinje shift above 1 td it is only cone-cone interaction. Under 1 td there is rod-cone interaction. Above 1 td, the frequency was above 10 Hz, which would be too fast for rods.

Kulikowski: The preferential frequency for the brightness matching, 0.5 Hz, agrees with the preferential frequency of so-called sustained processes. Have you tried to vary spatial stimulus parameters to find out how invariant this figure is?

Vienot: We have not done this yet.

Walraven: Did you express your ratios for the match in photopic or scotopic trolands? Would using scotopic trolands affect the results?

Vienot: We used photopic trolands, always referred to the white. Using scotopic trolands will just shift the curves up and down.

Spekreijse: I am surprised by the long time difference between the pathways Ted Sharpe postulates; 33 msec is a lot of time. Is there physiological support for it?

Sharpe: The effect is represented in the b-wave of the ERG, and the bipolar cells seem involved in this. It would seem then that the two pathways are being combined at or before the bipolar cells, so any time difference in the ganglion cells is not involved. This is difficult however to relate to the underlying anatomy.

Lennie: Current electrophysiological evidence from ganglion cells is not of the sort to allow you to detect two pathways.

Mollon: On the upper branch of your curves, is there significant rod bleaching, which must, if present, cause a change in spectral sensitivity, and if it is present, is it in the right direction for the spectral sensitivity effect you see?

Sharpe: Rod saturation occurs well before significant bleaching occurs, so I don't think this can cause a large enough effect to explain the spectral sensitivity change we see between the upper and lower branches.

Smith: I don't understand how Ted Sharpe's model works in the achromat, are you telling us they have cone pedicles, or how is this model supposed to work?

Sharpe: I presented the simplest model of two pathways as a caricature to indicate that rod signals are not just being transmitted over one pathway, that it is not one monolithic system. Every pathway accessible to the rods is shared in part with the cones. But in the achromat, everything just does not work out very well. Histologically, complete achromats probably have no cones in the foveola, and 5 to 10 percent of the normal quantity in the periphery, with signals so small they are insignificant. So there is a possibility that they could be gap junction connections between rods and cones, but this is very unlikely. Other models than ours are possible.

Swanson: I would mention that it is easy to try this yourself. I did it last Thursday. The nulling of the ERG correlated well with the disappearance of the flicker.

Seim: Have you tried to activate the cones, which probably use one of your two channels, in antiphase so you cancel the signals from the rods in that channel?

Sharpe: We certainly could set up conditions psychophysically, but this is also more complicated in that you also can get cancellation between the rods and cones pathways at 7.5 Hz. You could however set up conditions that can get cancellation. At 15 Hz, we suspect, that both rod signals are 90 degrees out of phase with the cones. They are 180 degrees out of phase with one another.

Spillmann: It would be interesting if one could bias two spatially separated stimuli such that one is processed only by the fast system, and the other one by the slow system. These two stimuli would then be presented with the delay of 35 msec. They should appear to be simultaneous. Such an experiment might be a metacontrast experiment, for example with a disc and an annulus.

Sharpe: Yes, we would do that, but we have shown that when looked at globally, we didn't find any spatial organisation differences between the two mechanisms. That is surprising, if one pathway is intimately shared with the cones. It seems natural to expect that, since temporal characteristics are different, spatial ones should be as well.

Kaplan: You are talking about fast and slow, but your model could be handled simply by delay, and the two pathways could have identical speeds. If you go to higher and higher temporal frequency you should be able to distinguish the two.

Sharpe: We made this set of physical measurements at 8, 15 and 20 Hz; at 20 Hz the whole story was gone, probably because you get too much cone involvement. I tend to think in the direction you suggest, that there is probably something more to the story, but we have to analyze more electroretinogram results.

Mollon: Could you argue against the possibility that all the effects arise within a single rod? I think we have been treating the rod as rather a simple box but the electrophysiologists who work on amphibian rods, have shown very striking differences in the time constants between proximal and distal segments, a factor of two difference. And you would also expect a difference in the spectral sensitivity of the entire rod due to self-screening. Is it possible that the lower part of the curves represents the entire rod, eventually saturating, and that the upper branch represents the response of the proximal discs?

Sharpe: We can't exclude that possibility. We exclude the possibility that there are two types of rods, but the signals could be conveyed in two different parts of the rod. The light adaptation behaviour in the two branches is different. I don't know if I can argue that that would minimize the possibility that this is a phenomena within the rod itself.

Swanson: I'd like to address this question which ganglion cells are involved in Vienot's flicker matches. Bob Shapley, in terms of scotopic to photopic, do you think that all can be attributed to the M-pathway?

Shapley: Reid, Kaplan and I did record from M- and P-cells through the range of backgrounds that were studied in the psychophysical experiments. In our experiments the P-cells as a group were unresponsive to spatial patterns or anything except for diffuse fields at less than 1 td, whereas we did see responses from M-cells down to about 0.01 td.

Swanson: Did you have any information that might tell you if there is a abrupt transition?

Shapley: We didn't do that experiment but you might expect that from M-cells.

Kaplan: I would add that Wiesel and Hubel made a similar comment, in that they were surprised to find so many P-cells showing no sign of rod input.

Lennie: I want to discourage the inference that P-cells don't have rod input. Wiesel and Hubel were quite emphatic that many of their P-cells had rod input. We have found that by no means all of them have rod input, but clearly a good number of them do. Experiments that Mark Fairchild did looking at the scotopic acuity made it clear that at 6 to 10 degree eccentricity performance is much better than can be supported by the M-cells.

Swanson: Has anybody here any idea why there will be this sudden abrupt transition from photopic to scotopic.

Schiller: I would support what Lennie said on the basis of lesion experiments. When you ablate part of magnocellular geniculate, the monkey still has very good vision under dark-adapted conditions.

Creutzfeldt: Virsu, Lee and I tested this transition from photopic to scotopic. L-cone excited P cells seldom had rod input, but M-cone excited P cells did. Another question, if dark adaptation works so well, why does it look so dark in the dark?

WHICH CELLS CODE FOR COLOR?

R.W. Rodieck

Department of Ophthalmology, RJ-10
The University of Washington
Seattle, WA, 98195, U.S.A.

INTRODUCTION

In primates, information about color passes from the eye to the visual cortex via the parvocellular laminae of the Lateral Geniculate Nucleus (LGN). We know this from the work of Schiller, Logothetis and Charles (1990) and of Merigan (1989) who compared the visual performance of Rhesus monkeys, as tested behaviorally, before and after lesions were made to either a region that retinotopically included all the parvocellular laminae and none of the magnocellular laminae, or vice versa, or both. Any interruption to the pathways via the parvocellular laminae virtually abolished color sensitivity, whereas interruption of the pathways via the magnocellular pathways produced no detectable deficit. They also showed that fine pattern and texture information likewise required the parvocellular laminae, but not the magnocellular laminae.

Given that color information passes via the parvocellular laminae, which particular pathway conveys this information? To judge from recent reviews, there is general acceptance that color information is conveyed by the cells that compose the majority of ganglion cells that project to the parvocellular laminae (Shapley and Perry, 1986; Lennie and D'Zmura, 1988; Kaplan, Lee and Shapley, 1990). These are termed 'P cells' by those who record from the LGN (Shapley and Perry, 1986) and 'midget ganglion cells' by those who work on the retina (Polyak, 1941; Rodieck, Binmoeller and Dineen, 1985), and make up the great majority of retinal ganglion cells (ca 80%). Because not all ganglion cells project to the parvocellular laminae, the proportion of P cells that project to these laminae must be even greater.

P cells show both spatial and spectral opponency. Only they exist in sufficient number to account for fine pattern discrimination as measured psychophysically (e.g. Perry, Oehler and Cowey, 1984), and their preeminent role in this task is not in dispute. In the current paradigm, these same cells are presumed to convey color information as well, as noted above. Since it is accepted by all that information is conveyed via the rate of discharge of a cell (i.e. a scalar signal), this implies that the message sent by a single P cell is necessarily ambiguous, for it could be responding either to a spectral change, or to a spatial change, or to both. This leads to the notion that the visual cortex must engage in some form of multiplexing, combining the signals of a number of different inputs, so as to extract either the combined spatial signal or the combined spectral signal (e.g. Kaplan, Lee and Shapley, 1990). Since most cells in the striate cortex show no spectral antagonism, some mechanism of this type is presumably present as far as spatial coding is concerned. If the system can combine signals to remove spectral dependencies, thereby extracting spatial information, then there seems no reason why a similar mechanism might work the other way, extracting spectral information at the expense of nonspectral gradients. In principle, at least, there appears to be no added difficulty in this regard. Considerations of these sort appear to have lead to a hypothesis implicit in the current

From Pigments to Perception, Edited by A. Valberg and
B.B. Lee, Plenum Press, New York, 1991

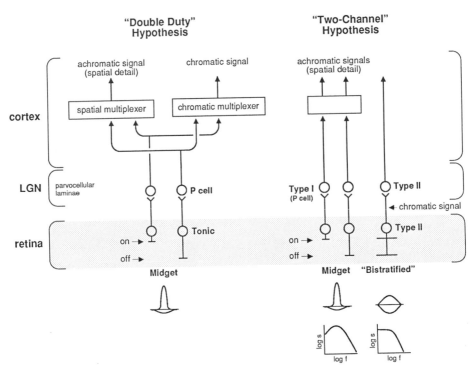

Fig. 1 Alternative hypotheses for the coding of color.

paradigm: *P cells convey essentially all the information that passes through the parvocellular laminae.* Indeed, it is sometimes claimed that P cells compose the *entire* projection to the parvocellular laminae, with the other cell types that have been reported considered to constitute normal parametric variations within the P-cell group, varying only in degree and without distinct functions (e.g. Lennie and D'Zmura, 1988). This notion is summarized schematically in the left half of Figure 1 as the 'Double Duty' hypothesis.

To me, all this seems too pat, and when it comes to the ganglion cell types that project to the parvocellular laminae, probably wrong. In order to sharpen the issue, I would like to put forward and defend an alternative hypothesis: *Color information is conveyed primarily or entirely by a group of ganglion cells that show a bistratified morphology, and synapse onto a minority population of cells in the parvocellular layers that have been termed Type II by Wiesel and Hubel (1966).* This 'Two Channel' hypothesis is shown in the right hand portion of Figure 1. In terms of this alternative hypothesis, what spectral information P cells do possess plays either a subsidiary role in color processing (weak form of hypothesis), or possibly no role whatsoever (strong form).

NOT ALL CELLS THAT PROJECT TO THE PARVOCELLULAR LAMINAE ARE P CELLS

On the basis of receptive-field analysis of parvocellular cells, Wiesel and Hubel (1966) described three distinct cells groups, their Types I, II and III. Briefly, and for point of reference, Type I cells show both spectral and spatial opponency, Type II cells show only spectral opponency, and Type III cells show only spatial opponency. Their findings have been confirmed by Dreher, Fukada and Rodieck (1976) in the LGN. Cells with identical properties have been described by DeMonasterio and Gouras (1975) and by DeMonasterio (1978) in retinal recordings; this indicates that these receptive fields are constructed within the retina, and relayed via the LGN. The Type I cells are equivalent to what are currently termed P cells; and, as noted above, correspond to what are known morphologically as midget ganglion cells (Leventhal, Rodieck and Dreher, 1981; Perry, Oehler and Cowey, 1984). For the purpose of this paper, we need not consider the properties of Type III cells in further detail.

Wiesel and Hubel (1966) characterized Type II cells as *...in many respects the most remarkable of the dorsal layer cells. Like type I they showed opponent-color responses, but their fields differed in having no trace of any center-surround arrangement.* Two types were observed, one with a neutral point at about 500 nm, the other at about 600 nm. The recorded incidence of these cells is as follows:

Table 1. Incidence of Type II cells in parvocellular laminae of macaque LGN

Reference	Region	500nm	600nm	%
Wiesel and Hubel, 1966	LGN	8	7	9
Dreher, Fukada and Rodieck, 1976	LGN	10	4	16
DeMonasterio and Gouras, 1975	Retina	6	3	
DeMonasterio, 1978	Retina	5	2	

There are two types of Type II cells, one with a neutral point near 500nm, the other near 600nm. The values in these columns are the number of cells of each type recorded. % is the ratio of Type II cells to Type I cells in each sample. Percentages for retinal studies are not given, because they were done near the fovea, where different ganglion-cell types stratify at different levels within the ganglion-cell layer (Polyak, 1941; Perry and Silveira, 1988) and direct recordings are strongly biased toward the most vitread somata.

TYPE II CELLS ARE DISTINCT FROM P CELLS

Figure 2 is a replot of receptive-field sizes of the color-opponent retinal ganglion cells as determined by DeMonasterio and Gouras (1975). The color-coding cells form three distinct clusters in this parametric space, indicating that they form at least three different types

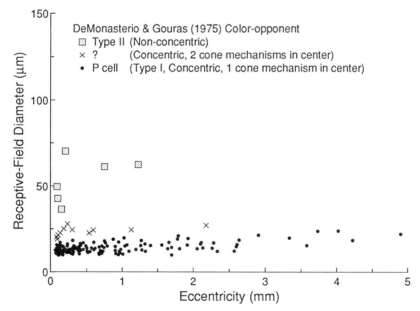

Fig. 2 Receptive-field size plotted against distance from the fovea for color-opponent retinal ganglion cells of the macaque retina, as determined by DeMonasterio and Gouras (1975). Replotted from their Fig. 16. The names they used are shown in parentheses, and the Discussion section of their paper indicates the correspondences shown. Conversion from visual angle to retinal distance assumed here and elsewhere to be 5°/mm (DeMonasterio et al., 1985).

(Rodieck and Brening, 1983). Ganglion cells that project to the magnocellular laminae of the LGN lack these properties, and none of the ganglion cells that project to the superior colliculus are color opponent (Schiller and Malpeli, 1977). Thus all three types may project to the parvocellular laminae of the LGN. In any case, Wiesel and Hubel (1966) described Type II cells as having larger receptive fields than Type I cells, and DeMonasterio and Gouras (1975) have demonstrated this quantitatively at the retinal level.

Figure 3 includes the data of Figure 2 and adds to it the receptive-field diameters of M cells, as determined by DeMonasterio and Gouras (1975), together with the dendritic-field diameters of midget and parasol cells, as determined by Watanabe and Rodieck (1989). There are two points to be made with regard to this figure. First, there is a reasonably close correspondence between the anatomical measure and the functional measure for the two types (M cell = parasol; P cell = midget). Second, the receptive field diameters of Type II cells lie within or slightly above the cluster of M cells.

There are more differences between Type I and Type II cells than simply their receptive field sizes. The papers listed in Table 1 are in general agreement as to these additional differences, as summarized in Table 2.

To summarize at this stage: Type I and Type II cells differ in a variety of ways, making it difficult to accept the notion that ... *there seem to be no distinct type I and type II groups.*

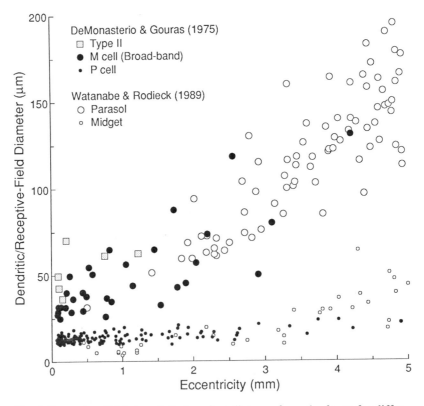

Fig. 3 Receptive-field size plotted against distance from the fovea for different retinal ganglion cells of the macaque retina, as determined by DeMonasterio and Gouras (1975), together with the dendritic-field diameters of midget and parasol cells, as determined by Watanabe and Rodieck (1989).

Table 2. Comparison of the properties of P cells and Type II cells.

Property	P cell (Type I)	Type II
Receptive-field size:	Small	Medium
Rod input:	Yes	No
Center/Surround:	Yes	No
Antagonistic response in center of receptive field:	No	Yes
Histogram of neutral points:	Continuous from about 450nm to about 690nm	Discrete at either about 500nm or about 600nm
Neutral point varies as a function of position within the receptive field:	Yes	No

The variation in neutral points for P cells was described by (De Valois, Abramov and Jacobs, 1966).

THE PARVOCELLULAR LAMINAE RECEIVE FROM MORPHOLOGICALLY DISTINCT GANGLION CELLS

In preliminary work aimed at determining the morphology of the different ganglion-cell types that project to the parvocellular laminae, Rodieck and Watanabe (1988) made a single central injection of a retrograde marker to the parvocellular laminae, which labelled a patch of ganglion cells in the nasal portion of the retina, that lay about 7-8 mm from the fovea. Histology showed that the injection site was confined to the upper portion of the parvocellular laminae. Most of the labelled cells appeared to be midget ganglion cells; however within the patch we attempted to inject the labelled cells with the largest somata.

Figure 4 compares the parameters of these cells with those of parasol and midget ganglion cells, taken from Watanabe and Rodieck (1989) and the receptive field sizes of the cells studied by DeMonasterio and Gouras (1975). Two morphologic forms were observed, which we have provisionally termed 'giant' and 'bistratified'. Each had its own size distribution. The 'giant' group consisted of cells with large and sparse dendritic fields that differed in their morphology from those that project to either the pretectum or the superior colliculus; they are not further considered here.

As their name implies, the 'bistratified' cells have dendritic fields that ramify at two different zones of the IPL. The lateral extent of the two ramifications were coextensive, and about the same or slightly greater than that of (unistratified) parasol cells at the same eccentricity. A drawing of a bistratified ganglion cell is shown in Figure 5. The ratio of their dendritic diameters to that of midget ganglion cells at the same eccentricity is about 4. Thus they could achieve the same dendritic coverage factor as that of the midget ganglion cells if their spatial density was about 6% of that of the midget ganglion cells.

Although we have not determined their levels of stratification within the inner plexiform layer (IPL), it seems probable that the two strata straddle the a/b border of the IPL. In primates this border lies at the middle of the IPL (Watanabe and Rodieck, 1989). They are thus candidates for coextensively receiving both 'on' and 'off' signals.

Given that there is a reasonable match between the receptive fields and dendritic fields of cells with known anatomical/functional correspondences (i.e. Fig. 3; M cell = parasol; P cell = midget), it is reasonable to expect the dendritic fields of Type II cells to be about the same or slightly larger than those of M cells (i.e. parasol cells). Given that Type II cells show two independent and coextensive center mechanisms of opposite sign (i.e. 'on' and 'off'), it is

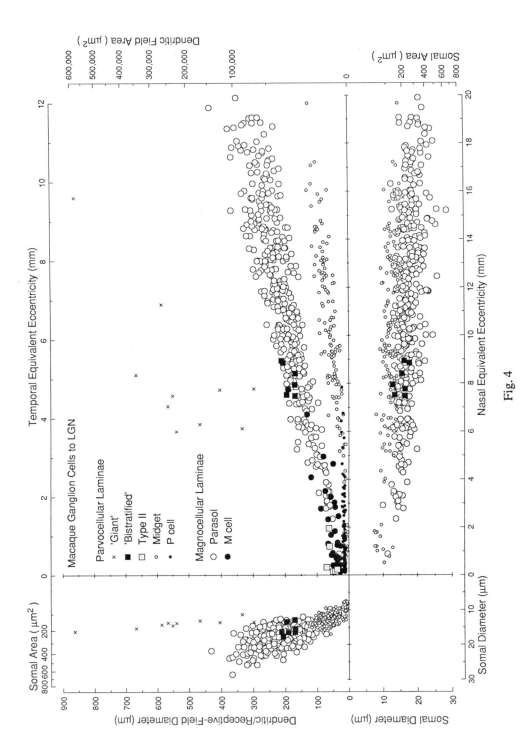

Fig. 4

88

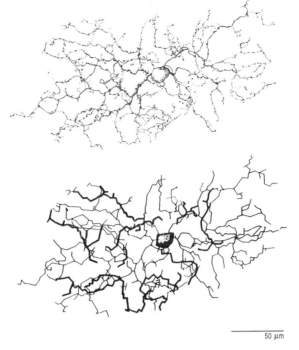

50 μm

Fig. 5 Upper: Bistratified ganglion cell labelled following an injection of a retrograde marker to the parvocellular laminae of the LGN, and intracellularly injected with HRP in an *in vitro* preparation. Lower: Cartoon of the above cell, with the inner dendritic arbor shown as thin lines, and the outer as thick lines.

reasonable to expect their dendritic field to extend coextensively into both sublaminae a and b of the IPL. Thus the 'bistratified' cells that project to the parvocellular laminae appear to have about the right properties one would expect Type II cells to have. Further work will be required to test this hypothesis. The relevant points here are: not all ganglion cells that project to the parvocellular laminae are midgets, and at least one group has anatomical properties compatible with those of Type II cells.

TYPE II CELLS APPEAR TO HAVE ALL THE PROPERTIES NECESSARY TO ACCOUNT FOR HUE DISCRIMINATION

The receptive-field properties of Type II cells (lack of center/surround antagonism; precise balance between opponent chromatic mechanisms, both within different locations of the receptive-field of each cell, as well as from cell to cell; lack of rod input) are just those needed to convey accurate color information. Indeed, DeMonasterio (1978) has remarked: *Type II cells may represent a "pure color" system whose opponent mechanisms have remarkably similar or perhaps identical retinal networks.*

About the only issue that remains is the question of whether Type II cells exist in sufficient number to account for the known psychophysical behavior of primates. The issue at stake is the minimum sampling period necessary to account for color discrimination. It is necessary to distinguish between studies in which the subject is asked to detect a color pattern (irrelevant here) and studies in that require hue perception (see De Valois and De Valois, 1988, p.214). All investigators agree that spatial acuity for color is significantly less than for luminance, the question is how much. Mullen (1985) found that, for either red/green and blue/yellow gratings, the maximum spatial frequency that could be resolved in the absence of luminance cues was down by a factor of about three compared to the maximum for an achromatic grating. Presuming a factor of about this magnitude, a minimum spatial density for color sampling of about 11% of that for the achromatic channel would be required. If the ratio of chromatic sampling to achromatic sampling corresponds to the ratio of Type II cells to P cells, then the values for this ratio thus far obtained (9% and 16%, Table 1) suggest that Type II cells are present in sufficient numbers to convey *all* the color information required by performance measures determined from psychophysical studies.

By comparison, P cells (Type I) convey a mixed signal of spatial contrast and a spectral antagonism. Lennie (1980) pointed out that the spectral opponency of P cells might be a trivial consequence of the receptive-field center being driven by a single cone, which is true within at least 4° of the fovea, where most recording is done. This hypothesis is consistent with the observation that, unlike Type II cells, P cells show a substantial variation from cell to cell in the relative strengths of the inputs from the R and G cones (e.g. Lennie and D'Zmura, 1988). This difficulty led Shapley and Perry (1986) to suppose that, although the aim of P cells was to distinguish color, the bipolar cells to which they were connected were unable during development to reliably distinguish the R and G cones. That may be so, but it is clear that the bipolars from which Type II cells receive their signals have no difficulty in this regard.

There are many more of P cells than Type II cells, so that, even with somewhat unreliable signals, it may be possible that the cortex somehow combines their messages so as to average out the variability in their chromatic properties and thereby extract reliable color information. But this has not been shown, and, given the properties of Type II cells, may not be required.

CHROMATIC GRATINGS, SPOTS, AND THE P/M PARADIGM

Chromatic gratings are elegant stimuli that provide one of the best means of relating the firing patterns of single cells to psychophysical measures. A good example of their worth is the correlation between the psychophysical measure of luminance and the spectral discharge properties of magnocellular cells demonstrated by Lee, Martin and Valberg (1988).

Given that the response properties of LGN cells are approximately linear, at least for low-contrast conditions, it might seem that the parametric domain encompassed by sinusoidal gratings would be sufficiently rich to reveal the full repertoire of coding of information by these cells. In highly linear systems, such as most optics and certain electronics, sinusoidal stimuli are sufficient to fully characterize the system. Unfortunately, this does not appear to be the case for P and M cells. For example, Derrington and Lennie (1984) demonstrated that there is little difference between the receptive field sizes of P and M cells at the same retinal eccentricity, when these sizes were calculated from the responses to a series of gratings. By contrast, responses to spot stimuli show large and nonoverlapping differences (e.g. Figure 2). The reasons why grating stimuli fail to find a sharp difference in this regard, even in the peripheral retina, is unclear.

In any case, the apparent uniformity of response properties among parvocellular cells, or among magnocellular cells, when tested with gratings, may have prompted the P/M paradigm, which implies that one need consider only a single cell group in the parvocellular laminae, and a single group in the magnocellular laminae. If P cells are a single group, and if both chromatic and luminance signals pass through the parvocellular laminae, then two inferences follow. First, P cells must somehow multiplex these two signals. But the converse also follows; since all P cells convey the chromatic signal, the response of *any* P cell to a chromatic stimulus becomes a measure of the manner by which color is coded.

The terms 'P cell' and 'M cell' have been given two somewhat incompatible definitions, the first as used above, and the second as designations of the zone of the LGN to which a ganglion cell projects (e.g. Kaplan, Lee and Shapley, 1990). This is a terminological issue that is without substance. By contrast, those who make use of the P/M paradigm accept its implications, as well as the model shown schematically in left portion of Figure 1, thereby allowing it to be falsified.

The finding that the receptive-field size distributions of retinal Type I and Type II cells do not overlap (Figure 2) constitutes a difficulty for the P/M paradigm and its implications. The finding that the parvocellular laminae receive from ganglion cells with dendritic-field size distributions that do not overlap (Figure 4), likewise constitutes a difficulty - independent of whether a correspondence exists between these functionally and morphologically characterized groups.

PROBLEMS WITH THE TWO-CHANNEL HYPOTHESIS

Finally, I would like to raise two difficulties associated with the strong form of the hypothesis presented here. Firstly, the locations of the pair of neutral points of the Type II cells (about 500nm and 600nm) define two unique colors within the spectrum. One might expect each to lie along one of the axes in the perceptual space formed by the two opponent color systems of Hering (1964). Now the perceptual change from yellow to blue occurs at about 500nm (reviewed by Hurvich, 1981); this matches the neutral point for one of the types of Type II cells (Table 1), and is thus consistent with the notion that these cells could subserve the postulated Y-B channel. But the perceptual change from red to green occurs at about 580nm (yellow) (Dimmick and Hubbard, 1939; Rubin, 1961; Hurvich and Jameson, 1964), over a total range of 568-588nm (Cicerone, 1987) or 568-598nm (Pokorny and Smith, 1987). By contrast, the neutral point for the other type of Type II cells in the LGN is reported to occur at about 600nm (orange) - and thus fails to match the postulated R-G channel. The set of interference filters used by both Wiesel and Hubel (1966) and by Dreher, Fukada and Rodieck (1975) were spaced at 20nm intervals and each included a 600nm filter, so the actual neutral point could have been as low as about 591nm. Macaques appear to categorize colors in a manner similar to that of humans (Sandell, Gross and Bornstein, 1979), and there is no evidence to support a trans-species difference in this regard. This discrepancy could possibly be explained in terms of the spectral composition of the backgrounds used in these studies. Some support for this notion comes from the fact that the Type II retinal ganglion cell illustrated by DeMonasterio (1978) had a neutral point between 570nm and 580nm. But it seems more useful at the moment to leave this as a problem to be resolved experimentally, rather than attempting to explain it away.

Secondly, the apparent lack of a rod input to Type II cells, although plausible from the point of view of reliable coding of spectral information, goes against the strong form of the hypothesis presented here, since Stabell and Stabell (1977) have convincingly shown by psychophysical methods that rod signals contribute directly to wavelength discrimination.

Supported in part by NIH grants EY02923, EY06098, and EY01730, by The E.K. Bishop Foundation, by Research to Prevent Blindness Inc., and by NIH grant RR00166 to the Regional Primate Research Center at the University of Washington. I would like to thank David Burr, Ehud Kaplan, Barry Lee, Peter Lennie, Donald MacLeod, Joel Pokorny, Robert Shapley, and Vivianne Smith for useful discussions on these issues, Masami Watanabe for the collaborative study that yielded the cells described, and Toni Haun for her many contributions to this study.

REFERENCES

Cicerone, C., 1987, Constraints placed on color vision models by the relative numbers of different cone classes in human fovea centralis. Die Farbe, 34:59.

DeMonasterio, F.M., 1978, Properties of ganglion cells with atypical receptive field organization in retina of macaques, J. Neurophysiol., 41:1435.

DeMonasterio, F.M., and Gouras, P., 1975, Functional properties of ganglion cells in the rhesus monkey retina, J. Physiol. (Lond), 251:167.

DeMonasterio, F.M., McCrane, E.P., Newlander, J.K., and Schein, S.J., 1985, Density profile of blue-sensitive cones along the horizontal meridian of macaque retina, Invest. Ophthalmol. Vis. Sci. 26:289-302.

Derrington, A.M., and Lennie, P., 1984, Spatial and temporal contrast sensitivities of neurones in lateral geniculate nucleus of macaque, J. Physiol. (Lond.), 357:219.

De Valois, R.L., Abramov, I., and Jacobs, G.H., 1966, Analysis of response patterns of LGN cells, J. Opt. Soc. Am. 56:966.

De Valois, R.L., and De Valois, K.K., 1988, "Spatial Vision", Oxford University Press, New York.

Dimmick, F., and Hubbard, M. 1939, The spectral location of psychologically unique yellow, green, and blue. Am. J. Psychol., 52:242.

Dreher, B., Fukada, Y., and Rodieck, R.W., 1976, Identification, classification, and anatomical segregation of cells with X-like and Y-like properties in the lateral geniculate nucleus of old world primates, J. Physiol. (Lond.), 258:433.

Hering, E., 1964, "Outlines of a Theory of the Light Sense", Hurvich, L.M., and Jameson, D. (transl.), Harvard University Press, Cambridge.

Hurvich, L.M., 1981, "Color Vision", Sinauer, Sunderland.

Hurvich, L.M., and Jameson, D., 1964, Does anomalous color vision imply color weakness? Psychonomic Sci. 1:11.

Kaplan, E., Lee, B.B., and Shapley, R.M., 1990, New views of primate retinal function, Progress in Retinal Research, 9:273.

Kelly, D.H., 1974, Spatio-temporal frequency characteristics of color-vision mechanisms. J. Opt. Soc. Am. 64:983.

Lee, B.B., Martin, P.R., and Valberg, A., 1988, The physiological basis of heterochromatic flicker photometry demonstrated in the ganglion cells of the macaque retina. J. Physiol. (Lond.), 404:323.

Lennie, P., 1980, Parallel visual pathways: a review, Vision Res., 20:561.

Lennie, P., and D'Zmura, M., 1988, Mechanisms of Color Vision, CRC Reviews in Neurobiology, 3:333.

Leventhal, A.G., Rodieck, R.W., and Dreher, B., 1981, Retinal ganglion cell classes in the Old World monkey: morphology and central projections, Science, 213:1139.

Merigan, W.H., 1989, Chromatic and achromatic vision of macaques: Role of the P pathway. J. Neurosci. 9:776.

Mullen, K.T., 1985, The contrast sensitivity of human colour vision to red-green and blue-yellow chromatic gratings, J. Physiol. (Lond.), 359:381.

Perry, V.H., Oehler, R., and Cowey, A., 1984, Retinal ganglion cells that project to the dorsal lateral geniculate nucleus in the macaque monkey, Neuroscience, 12:1101.

Perry, V.H., and Silveira, L.C.L., 1988, Functional lamination in the ganglion cell layer of the macaque's retina, Neuroscience, 25:217.

Polyak, S.L. 1941, "The Retina", The University of Chicago Press: Chicago.

Pokorny, J. and Smith, V., 1987, L/M cone ratios and the null point of the perceptual red/green opponent system. Die Farbe, 34:53.

Rodieck, R.W., Binmoeller, K.F., and Dineen, J.D., 1985, Parasol and midget ganglion cells of the human retina, J. Comp. Neurol. 233:115.

Rodieck, R.W., and Brening, R.K., 1983, Retinal ganglion cells: Properties, types, genera, pathways and trans-species comparisons, Brain Behav. Evol. 23:121-164.

Rodieck, R.W., and Watanabe, M., 1988, Morphology of ganglion cell types that project to the parvocellular laminae of the lateral geniculate nucleus, pretectum, and superior colliculus of primates, Soc. Neurosci. Abst., 14:1120.

Rubin, M., 1961, Spectral hue loci of normal and anomalous trichromates. Am. J. Ophthal. 52:166.

Sandell, J.H., Gross, C.G., and Bornstein, M.H. 1979, Color categories in macaques. J. Comp. Physiol. Psychol., 93:626.

Schiller, P.H., Logothetis, N.K., and Charles, E.R., 1990, Functions of the colour-opponent and broad-band channels of the visual system, Nature, 343:68.

Schiller, P.H., and Malpeli, J.G., 1977, Properties and tectal projections of monkey retinal ganglion cells, J. Neurophysiol. 40:428.

Shapley, R.M., and Perry, V.H., 1986, Cat and monkey retinal ganglion cells and their visual functional roles, Trends Neurosci., 9:229.

Stabell, U., and Stabell, B., 1977, Wavelength discrimination of peripheral cones and its change with rod intrusion, Vision Res, 17:423.

Watanabe, M., and Rodieck, R.W., 1989, Parasol and midget ganglion cells of the primate retina, J. Comp. Neurol., 289:434-454.

Wiesel, T.N., and Hubel, D.H., 1966, Spatial and chromatic interactions in the lateral geniculate body of the rhesus monkey, J. Neurophysiol., 29:1115-1156.

RECEPTIVE FIELD STRUCTURE OF P AND M CELLS IN THE MONKEY RETINA

Robert Shapley[1], R. Clay Reid[1,2] and Ehud Kaplan[3]

[1] Center for Neural Science
Departments of Psychology and Biology
New York University
New York, NY 10003

[2] Cornell University Medical College
New York NY 10021

[3] Laboratory of Biophysics
Rockefeller University
New York NY 10021

P (PARVOCELLULAR) AND M (MAGNOCELLULAR) PATHWAYS

The concept of P and M retinogeniculate channels has emerged from attempts to explain the layering of the primate Lateral Geniculate Nucleus (LGN). In the main body of the Old World primate's LGN there are six segregated layers of cells. The four dorsal layers are composed of small cells and are named the **Parvocellular** layers. The two more ventral layers, composed of larger neurons, are called **Magnocellular** layers. Recent work on functional connectivity and the visual function of single neurons has revealed that the different types of cell layers in the LGN receive afferent input from different types of retinal ganglion cells. The evidence on functional connectivity of retina to LGN came from Leventhal, Rodieck and Dreher (1981) and Perry, Oehler, and Cowey (1984) who labeled axon terminals in specific LGN layers of the macaque monkey with Horseradish Peroxidase (HRP) and looked back in the retina to see which ganglion cells were labeled retrogradely.

Direct electrophysiological evidence about retinogeniculate connectivity was provided by Kaplan and Shapley (1986) who recorded excitatory synaptic potentials (from retinal ganglion cells) extracellularly in different LGN layers and who found that different types of retinal ganglion cell drove different LGN layers. For example, LGN cells that are excited by red light but inhibited by green light (so-called red- green color opponent neurons) are only found in the Parvocellular layers. These "Red-Green Opponent" LGN cells receive excitatory synaptic input from "Red-Green Opponent" ganglion cells; "Red-Green Opponent" ganglion cells only provide direct excitatory input to Parvocellular LGN neurons of the "Red-Green Opponent" type. The specificity of ganglion cell types exactly matches that of their LGN targets (Kaplan and Shapley, 1986; Shapley and Kaplan, 1989). Therefore, Shapley and Perry (1986) proposed that the ganglion cells that drove Parvocellular LGN should be called P cells and the ganglion cells that drove Magnocellualr layers should be called M cells. This is the nomenclature we use here.

From Pigments to Perception, Edited by A. Valberg and
B.B. Lee, Plenum Press, New York, 1991

THREE PHOTORECEPTORS AND SPECTRAL SENSITIVITY

Discussion of the spectral sensitivities of the photoreceptors must precede consideration of the how the chromatic properties of P and M ganglion cells are constructed from their receptive field properties.

There are three cone photoreceptor types in human and macaque retinas. The spectral sensitivities of these photoreceptors have been determined for macaque retina by Baylor, Nunn and Schnapf (1987) and for human retina by Schnapf, Kraft, and Baylor (1987), using suction electrodes to measure cone photocurrent directly. These direct measurements of photoreceptor spectral sensitivities are in generally good agreement with microspectrophotometric measurements of cone absorption spectra (Bowmaker and Dartnall, 1980; Bowmaker, Dartnall, and Mollon, 1980). The photocurrent measurements agree even more closely with estimates of cone spectral sensitivity based on human psychophysics (Smith and Pokorny, 1975). The Smith and Pokorny fundamentals (estimated cone spectral sensitivities as measured at the retina after the light has been pre-filtered by the lens) are three smooth functions of wavelength peaking at 440 nm (S cones), 530 nm (M cones) and 560 nm (L cones).

Human sensitivity to light across the visible spectrum under photopic, daylight conditions is called the photopic luminosity function, denoted V_λ. The luminance of a light source is its effectiveness in stimulating the visual neural mechanism that has as its spectral sensitivity the photopic luminosity function. Thus, the luminance of any light may be computed by multiplying its spectral radiance distribution, wavelength by wavelength, by the photopic luminosity function, and summing all the products. The procedure known as heterochromatic flicker photometry has been employed to measure the luminosity function. Monochromatic light of a given wavelength is flickered against a white light at a frequency of 20 Hz or above, and the radiance of the monochromatic light is adjusted until the perception of flicker disappears or is minimized (Coblentz and Emerson, 1917). This technique exploits the fact that neural mechanisms that can respond to the color of the monochromatic light are not able to follow fast flicker. The photopic luminosity function has been measured more recently using contour distinctness (Wagner and Boynton, 1972) and minimal motion (Cavanagh, MacLeod, and Anstis, 1987) as response criteria. These measurements agree remarkably well with the luminosity function determined by flicker in the same subjects.

COLOR EXCHANGE AND EQUILUMINANCE

Color exchange, or silent substitution (Estevez and Spekreijse (1974; 1982) is a technique for identifying contributions from particular photoreceptors or spectral response mechanisms. For any spectral sensitivity function, and any two lights with different spectral distributions within the band of the sensitivity function, one can perform a color-exchange experiment that will provide a nulling color balance for that particular spectral sensitivity. For example, if one chooses two monochromatic lights with wavelengths such that they are equally effective at stimulating the L cone, then temporal alternation between these two lights at equal quantum flux should cause no variation in the response of the L cone. The same argument works for the photopic luminosity function which presumably is the spectral sensitivity of a neural mechanism that receives additive inputs from L and M cones. Two lights that, when exchanged, produce no response from the V_λ luminance mechanism are called equiluminant.

The results of a simulated color exchange experiment on cones and a macaque M retinal ganglion cell with a V_λ spectral sensitivity are illustrated in Figure 1. The calculations are based on the spectral sensitivities of the L and M cones and the photopic luminosity function (Smith and Pokorny, 1975). The spectral distributions of the light sources were those of the red and green phosphors on standard color television sets, designated P22 phosphors. The experiment that is simulated is color exchange between the red (denoted capital R) and green (denoted capital G) phosphors. We have scaled the x-axis so that when the G/R ratio is 1.0, the green phosphor is equiluminant with the red phosphor. In this experiment, the red phosphor's depth of modulation was kept constant while the green phosphor's modulation depth varied. When the modulation depth (in units of luminance) of the green phosphor is approximately 0.4 that of the red (G/R ratio 0.4), the response of the M cones is nulled. When the G/R ratio is 1.25, the L cone response is zero.

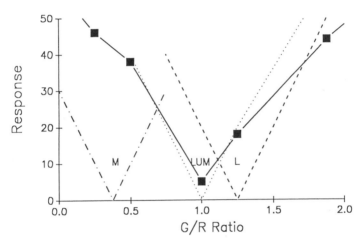

Figure 1. Color exchange response functions for L and M cones and luminance. The predicted response of the cones to different G/R ratios was calculated from the cross- product of the G and R phosphors with the spectral sensitivities of the L and M cones from Figure 1. Also shown are data from a single macaque M cell studied with heterochromatic drifting gratings at a spatial frequency of 1 c/deg and a temporal frequency of 4 Hz. Note the good agreement with the prediction from the LUM function.

A spectral mechanism that sums the responses of M and L cones will have a null in a color exchange experiment at a G/R ratio between the nulls of the two cones. If the spectral sensitivity of the summing mechanism is $K * L + M$, where K is a number between zero and infinity, then when K approaches zero, the color-exchange null approaches the M cone null from above. When K goes to infinity, the color-exchange null approaches the L cone null, from below. The null of the luminosity curve between the cone nulls in Figure 1 is a case in point. For that curve K is approximately 2. One must qualify the assertion to include the condition that the photoreceptor signals have the same time course, and that in the process of summation their time courses are unaffected. The existence of sharp V's in color exchange experiments on M ganglion cells and Magnocellular cells is reasonably good evidence that L and M cones have similar time courses under the conditions of those experiments (Lee, Martin, and Valberg, 1988; Kaplan et al, 1988; Shapley and Kaplan, 1989). An example is shown in Figure 1. The points plotted in that figure are from the amplitudes of response of an M ganglion cell in response to drifting red- green, heterochromatic gratings (see below). The response plummeted near a G/R ratio of unity, the equiluminant point.

Next, we consider what happens in a color-exchange experiment on a color-opponent neuron. In such a cell, L and M cone signals are not summed but subtracted. The results of Figure 2 would ensue. The luminosity color-exchange results are included for comparison with three different possible color-opponent cells: one in which the strength of L and M signals is equal but the sign is opposite $(M - L)$; one in which signals from M cones are twice as strong as those from L cones $(2M - L)$; and one in which signals from L cones are twice as strong as those from M cones $(2L - M)$. The curves would be unaffected if the signs of the cone inputs were reversed since only magnitude of response is plotted. What is striking about these simple calculations is that opponent neurons have no null response between the cone nulls along the G/R axis. The $M - L$ response is perfectly constant. The $2M - L$ and $2L - M$ cells show response variation but no null. This result is general for any neural mechanism with a spectral sensitivity equal to $K * L - M$ where K is a number greater than zero and less than infinity. As K goes to zero the null of the mechanism approaches the M cone null from below; as K goes to infinity, the null of the mechanism approaches the L cone null from above. As before, all these statements hinge on small-signal linearity, and identity of temporal response properties for M and L cones, which are reasonable assumptions for P cells at temporal frequencies of 16 Hz and below. As an example of what real P neurons do, data from a $M + L-$ neuron are plotted in Figure 2 also. The stimulus was full-field, red-green color exchange at 4 Hz. Notice that the response amplitudes were relatively constant at all G/R ratios, meaning that the cell responded to all color exchanges. Such a neuron must have approximately equal input from M and L cones, but the cone inputs must be antagonistic.

RESPONSES OF M AND P NEURONS TO EQUILUMINANT STIMULI

One particular color-exchange experiment has become a focus of interest because of speculations by Livingstone and Hubel (1987). This is the measurement of the responses of P and M neurons to equiluminant color-exchange. In their paper on perceptual effects of parallel processing in the visual cortex, Livingstone and Hubel (1987) assumed that because Magnocellular cells were broad-band, their responses would be nulled at equiluminance. This assumption is a non-sequitur. Based on Figure 1, and the discussion of that figure in the preceding paragraphs above, one knows that there could be a whole family of broad-band neurons in the visual pathway that summed signals from L and M cones with different weighting factors K_i, such that spectral sensitivity of the i-th mechanism was $K_i * L + M$. Each mechanism would have a null at a different point on the G/R axis. The striking thing about M cells and Magnocellular neurons is that, for stimuli that produce responses from the receptive field center mechanism, the position of the null on the color-exchange axis is close to that predicted from the human photopic luminosity function, V_λ (Lee et al, 1988; Shapley and Kaplan, 1989; Kaplan et al, 1990). Moreover, there is little variability in the position of the color-exchange null in the neurophysiological data on M/Magnocellular neurons; the variability is less than that measured in psychophysical experiments (Coblentz and Emerson, 1917; Crone, 1959).

98

Previously, we have inferred that psychophysical tasks that tap the M cell center will possess a spectral sensitivity like that of V_λ. Thus, we suppose that heterochromatic flicker photometry, minimally distinct border, and minimal motion are all psychophysical tasks that somehow select for the M-cell receptive field center because of its spatial and temporal filtering characteristics, relative to the different characteristics of the M cell receptive field surround.

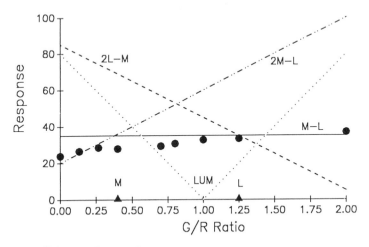

Figure 2. Color exchange functions for hypothetical chromatic opponent cells compared to luminance. Response magnitude as a function of G/R ratio is plotted for three different hypothetical chromatic opponent neurons, with cone balances as labelled. The color exchange response function for luminance is again shown and labelled LUM. Data plotted (as filled circles) are for an M+L-, M cone center, P ganglion cell, in response to full field (spatial frequency=0 c/deg) chromatic modulation (G/R color exchange) at 4 Hz temporal frequency.

There are other experiments that indicate that, under stimulus conditions where the center of the receptive field is not the only response mechanism contributing to the response, M and Magnocellular neurons do not have a color-exchange null at equiluminance. Lee et al. (1988) reported that large disks that stimulate center and surround have nulls away from equiluminance. Shapley and Kaplan (1989) used heterochromatic sine gratings to study chromatic properties of receptive field mechanisms. Heterochromatic sine gratings are formed by producing a sine grating on, say, the red phosphor of a color monitor, and producing an identical sine grating on the green phosphor except for an exact 180 degree phase shift. Thus where the red phosphor has a bright red bar the green phosphor has a dark green bar, and vice versa. The sum of these two grating patterns in antiphase yields as a spatial pattern a red-green, ergo heterochromatic, grating. Shapley and Kaplan (1989) reported that heterochromatic sine gratings of low spatial frequency may produce no color null in Magnocellular neurons. Derrington, Krauskopf, and Lennie (1984), using the technique of modulation in color space, found that many Magnocellular units exhibited properties expected of color-opponent cells when stimulated with spatially uniform fields. These results probably are related to the earlier work of Wiesel and Hubel (1966) who found that many Magnocellular neurons had a receptive field surround that was more red-sensitive than the receptive field center, the Type 4 cells. Such neurons could behave as color-opponent cells to stimuli which covered both center and surround if the spectral sensitivities of center and surround were different enough. Similar M ganglion cells were reported by DeMonasterio and Schein (1980). Thus, in psychophysical experiments, if the stimulus is designed to tap the receptive field center of cells in the M pathway, it will elicit a spectral sensitivity function like V_λ. Such a stimulus will be nulled in a color-exchange experiment at equiluminance. However, stimuli detected by the M-Magnocellular pathway that drive both the M-cell center and surround receptive field mechanisms will probably seem to be detected by a color-opponent pathway.

PHOTORECEPTOR INPUTS TO P AND M CELL RECEPTIVE FIELDS

The precise mapping of cone types to receptive field mechanisms is a problem not yet solved. Wiesel and Hubel (1966) postulated that color opponent cells received excitatory (or inhibitory) input from one cone type in the receptive field center and antagonistic inputs from a complementary cone type in the receptive field surround. However, the detailed quantitative evidence that would be needed to support or to reject this hypothesis was not available then. Though Wiesel and Hubel's (1966) proposal may be true, there are a number of other possibilities. One alternative hypothesis is that there is mixed receptor input to the receptive field surround, and only or predominantly one cone input to the center of the receptive field (Paulus and Kröger-Paulus, 1983; Kaplan et al, 1990). We have now made some progress on this problem by a combination of two methods: spatiotemporal mapping with m- sequences, and photoreceptor isolation by means of color exchange.

M-SEQUENCES AND THE MEASUREMENT OF SPATIOTEMPORAL PHOTORE-CEPTOR INFLUENCES ON RECEPTIVE FIELDS

In order to understand the photoreceptor inputs to receptive fields, we have embarked on a program of measurement of the two-dimensional spatiotemporal weighting functions of P and M neurons. In this paper we report some preliminary observations.

We have adapted an approach to spatiotemporal functional analysis pioneered by Erich Sutter (1987), and called maximal length shift register sequences, or m-sequences. An m-sequence of order n is a sequence of -1's and 1's of length 2n -1 within which no strings of length n are duplicated. Thus, an m sequence is the binary sequence of maximal diversity for its length. In our experiments, each picture element or pixel in a 16 by 16 array is modulated in time by a 16th order m-sequence. The sequences are derived by a computation to be described elsewhere (Reid, Victor, and Shapley, In Preparation), with a technique similar to that described by Sutter (1987). One important property of the m-sequence is that its power spectrum is white, and therefore the value of the m-sequence at any moment is uncorrelated with any time shift of itself. Such a signal can be used to recover the temporal weighting function of a linear system by cross-correlation. Following Sutter (1987), we have exploited the whiteness of the m-sequences by making each of the 256 pixels in the 16 by 16 array a time-shifted version of the same basic m-sequence. Therefore, the temporal stimulus at each spatial position is uncorrelated with all the others for all times less than the time-shift between pixels (usually 4 seconds). Then the temporal weighting function at each of the positions in the array can be recovered by computation of a single cross correlation between neuron output and m-sequence input. The first order response from each pixel occurs at a

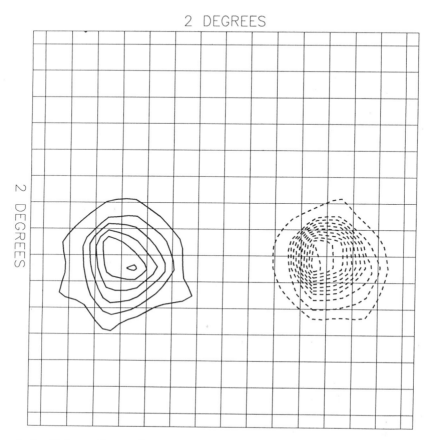

Figure 3. Spatial map from a first order spatiotemporal kernels of a macaque P cell (classified as M-L+ with conventional stimuli) in response to color-exchange m- sequences that only drove L cones (solid lines) and M cones (dashed lines). These are equivalent to two-dimensional distributions of sensitivity or contrast gain across the receptive field of the single cell. The kernel for each cone type was measured separately with a color-exchange m-sequence. The cone contrasts were both around 0.2. The kernels are plotted as contour plots in which each contour is equivalent to a response increment of 0.02 impulses/sec. The spatial kernels for the two different cones in fact overlapped almost completely. They have been separated by an arbitrary distance just to make clear that they have roughly the same shape, same spatial scale, and that they each are of one sign. The M cone input was much stronger as indicated by the higher density of contour lines for the M cone kernel. There is no Center- Surround organization within a cone class in P cells, as evidenced by the absence of a sign change for each separate cone kernel. The time of the response is 48 msec, near the initial peak of the first-order response.

separate, predetermined time in the cross correlation function. Another advantage of m sequences is that the correlation functions of all orders may be recovered with a single rapid computation referred to as the fast m-transform. Thus, first and second order spatiotemporal kernels may be computed very rapidly (Reid, Victor and Shapley, 1989).

The combination of spatiotemporal kernel measurement with m-sequences, and cone isolation by means of color-exchange (as in Figure 1), has produced experimental results that may settle some of the important issues concerning the functional connections of monkey photoreceptors to ganglion cells (Reid and Shapley, 1990). The most important results, in our opinion, are illustrated in Figure 3. There are shown the excitatory (with solid contours) and inhibitory (with dashed contours) spatial kernels, taken at the time of peak response, 48 msec, for the two different cone types (L and M) that converge onto a single P cell, an $M - L+$ opponent neuron. This cell was classified as an M-center neuron because of the stronger M cone input. The two contour plots have been displaced from each other for clarity's sake; in fact, they completely overlapped in space. The spatial overlap is one important point and the other is the absence of any sign change with position in either cone kernel. In other words, there was no center-surround interaction within one cone type in this representative P cell. The implication is that the standard "Center" and "Surround" mechanisms were cone-specific.

M ganglion cell fields were different in significant ways. For one thing, M and L cone inputs to the center of the receptive field were synergistic, and their relative strengths in the ratio of cone inputs to the photopic luminosity function; equiluminant red-green m-sequences elicited spatial kernels with small or absent central responses in M cells. In many M-cells the Surround received much stronger relative input from L cones than did the Center. Such neurons were Type 4-like, though some had "Off" Centers as mentioned above.

These new results on cone-specific response mechanisms in P cells and additive cone convergence in M cells will have to be explained by a fresh attack on the neuroanatomy, and new models of functional connections, within the primate retina. They may also provide new data that will help us understand the functional roles of these different cell types in color vision and pattern perception.

Acknowledgments: We would like to thank Norman Milkman and Michelangelo Rossetto for technical help. We are very grateful to Jonathan Victor for help on theory and practice in the m-sequence project. This work was supported by grants from the US National Eye Institute (EY-1472, EY-4888), from the US National Science Foundation (BNS 8708606), and by a grant from the Sloan Foundation.

REFERENCES

Baylor, D.A., Nunn, B.J. and Schnapf,J.L. (1987) Spectral sensitivity of cones of the monkey Macaca fascicularis . J..Physiol. 390, 145-160.

Bowmaker, J.K. and Dartnall, H.J.A. (1980) Visual pigments of rods and cones in a human retina. J. Physiol. 298:501-511.

Bowmaker, J.K., Dartnall, H.J.A. and Mollon, J.D. (1980) Microspectrophotometric demonstrations of four classes of photoreceptor in an Old World primate, Macaca fascicularis. J. Physiol. 298:131-143.

Cavanagh, P., Anstis, S.M. and MacLeod, D.I.A. (1987) Equiluminance: spatial and temporal factors and the contribution of blue-sensitive cones. J. Opt. Soc. Amer. A 4:1428-1438.

Coblentz, W.W. and Emerson, W.B. (1917) Relative sensibility of the average eye to light of different colors and some practical applications to radiation problems. Bull. of Bureau of Standards 14:167-236.

Crone, R. (1959) Spectral sensitivity in color defective subjects and heterozygous carriers. Am. J. Ophthalmol. 48:231-235.

DeMonasterio, F.M. and Schein, S.J. (1980) Protan-like spectral sensitivity of foveal Y ganglion cells of the retina of macaque monkeys. J. Physiol. 299:385-396.

Derrington, A.M., Krauskopf, J., and Lennie, P. (1984) Chromatic mechanisms in lateral geniculate nucleus of macaque. J. Physiol. 357, 241-265.

Estevez, O. and Spekreijse, H. (1974) A spectral compensation method for determining the flicker characteristics of the human colour mechanism. Vis. Res. 14:823-830.

Estevez, O. and Spekreijse, H. (1982) The "Silent Substitution" method in visual research. Vis. Res. 22:681-691.

Kaplan, E. and Shapley, R. (1982) X and Y cells in the lateral geniculate nucleus of macaque monkeys. J. Physiol. 330:125-143.

Kaplan, E. and Shapley, R. (1986) The primate retina contains two types of ganglion cells, with high and low contrast sensitivity. Proc. Nat. Acad. Sci. USA, 83:2755-2757.

Kaplan, E., Shapley, R., and Purpura, K. (1988) Color and luminance contrast as tools for probing the organization of the primate retina. Neurosci. Res. (suppl.) 2:s151-s166.

Kaplan, E., Lee, B.B., and Shapley, R. (1990) New views of primate retinal function. in Progress in Retinal Research, vol. 9, ed. Osborne and Chader, Pergamon, Oxford, In Press

Lee, B.B., Martin, P.R. and Valberg, A. (1988) The physiological basis of heterochromatic flicker photometry demonstrated in the ganglion cells of the macaque retina. J. Physiol. 404: 323-347.

Leventhal, A.G., Rodieck, R.W. and Dreher, B. (1981) Retinal ganglion cell classes in the old-world monkey: morphology and and central projections. Science 213:1139- 1142.

Livingstone, M.S. and Hubel, D.H. (1987) Psychophysical evidence for separate channels for the perception of form, color, motion, and depth. J. Neurosci. 7:3416- 3468.

Paulus, W. and Krger-Paulus, A. (1983) A new concept of retinal colour coding Vis. Res. 23:529-540

Perry, V.H., Oehler, R., and Cowey, A. (1984) Retinal ganglion cells that project to the dorsal lateral geniculate nucleus in the macaque monkey. Neuroscience, 12:1101-1123.

Reid, R.C. and Shapley, R. (1990) Spatial and temporal characteristics of cone inputs to Macaque LGN cells as mapped by pseudorandom stimuli. Inv. Ophthalmol. adn Vis. Sci. Suppl., 31, 2108

Reid, R.C. Victor, J.D. and Shapley, R. (1989) A new two-dimensional pseudorandom stimulus for the study of receptive fields in the LGN and striate cortex. Soc. Neurosci. Abstr. 15, 323.

Schnapf, J.L., Kraft, T.W. and Baylor, D.A. (1987) Spectral sensitivity of human cone photoreceptors. Nature 325:439-441.

Shapley, R. and Kaplan, E. (1989) Responses of magnocellular LGN neurons and M retinal ganglion cells to drifting heterochromatic gratings. Inv. Ophthalmol. and Vis. Sci. Supplement 30:323.

Shapley, R. and Perry, V.H. (1986) Cat and monkey retinal ganglion cells and their visual functional roles. Trends in Neurosci. 9: 229-235.

Shapley, R., Kaplan, E. and Soodak, R. (1981) Spatial summation and contrast sensitivity of X and Y cells in the lateral geniculate nucleus of the macaque. Nature 292:543-545.

Smith, V.C. and Pokorny, J. (1975) Spectral sensitivity of the foveal cone photopigments between 400 and 500 nm. Vis. Res. 15: 161-172.

Sutter,E. (1987) A practical non-stochastic approach to nonlinear time-domain analysis. Adv. Methods of Physiol. Systems Modelling, vol 1. Univ. Southern California

Wagner, G. and Boynton, R.M. (1972) Comparison of four methods of heterochromatic photometry. J. Opt. Soc. Amer. 62:1508-1515.

Wiesel, T.N. and Hubel, D.H. (1966) Spatial and chromatic interactions in the lateral geniculate body of the rhesus monkey. J. Neurophysiol. 29:1115-1156.

Zrenner, E. and Gouras, P. (1983) Cone opponency in tonic ganglion cells and its variation with eccentricity in rhesus monkey retina. in Colour Vision ed. J.D. Mollon and L.T.Sharpe,

ON THE RELATION BETWEEN CELLULAR SENSITIVITY

AND PSYCHOPHYSICAL DETECTION

Barry B. Lee

Department of Neurobiology
Max Planck Institute for Biophysical Chemistry
D-3400 Göttingen, FRG

INTRODUCTION

Recent interest in the parvocellular and magnocellular pathways within the primate visual system has a dual origin. Firstly, the physiology of these systems is of intrinsic interest. Some features appear unique to the primate, and are of relevance to wiring diagrams emerging from primate retinal anatomy. Secondly, there is abundant psychophysical evidence for the existence of chromatic and achromatic channels within the human visual system. The relationship of different cell types and systems to these channels has been controversial (see Shapley and Perry, 1986; Kaplan et al., 1990 for review). Although the parvocellular and magnocellular systems have been proposed as physiological substrates for the chromatic and achromatic channels (Crook et al, 1987; Lee et al, 1988; Lee et al., 1989a; Kaiser et al., 1990), there is evidence contrary to this viewpoint (e.g., Schiller et al., 1990). In this paper, I shall review physiological evidence relevant to certain standard psychophysical paradigms and consider some of the problems associated with specifying the substrates of performance from physiological data.

Linking physiological measurements with psychophysics is not straightforward. Assumptions are inherent in the linking hypothesis itself, and there are also practical considerations as to the mode of operation of central mechanisms processing afferent signals. Over the past three decades, there have been several attempts to provide a formal basis for such hypotheses. The discussion over the roles of the parvo- and magnocellular systems in perception provides an opportunity to test the applicability of some of these doctrines.

A primary assumption is the similarity of the visual systems of man and old-world monkeys such as the macaque. At least as far as the anatomy of the afferent visual systems is concerned, the evidence in favour is strong. It has yet to be shown that the similarity extends to all details of the psychophysics, although present evidence indicates close correspondence. However, more extensive monkey psychophysics seems desirable, for some of the linking hypotheses to be described rely on a precise psychophysical correspondence between the two species.

LINKING HYPOTHESES AND NEURON DOCTRINES

Brindley's (1962) distinction between class A and class B linking hypotheses is well known. In a class A hypothesis, if two stimuli evoke identical responses when physiological measurements are made, they must be indistinguishable psychophysically; it is implied that a change in one of the stimuli would lead to a response both from an observer and from some physiological system under study, further implying the physiological system is a necessary substrate for performance. Trichromatic colour matching provided an example of a class A hypothesis; Brindley also drew

From Pigments to Perception, Edited by A. Valberg and
B.B. Lee, Plenum Press, New York, 1991

threshold measurements into this category. Class B hypotheses involve some correlation between psychophysical and physiological data, which need not imply a causal relationship between the physiological response and behaviour. Although not without problems (Boynton and Olney, 1962), the distinction between class A and B hypotheses has generally been perceived as valuable. More recently, the implications of such hypotheses have been more fully explored (Teller and Pugh, 1983; Teller, 1984).

It has become apparent that comparison of physiological and psychophysical data is much less straightforward than might be thought on the basis of the distinction between class A and B hypotheses. In practice, the distinction is seldom clear-cut. For example, cells of the M-pathway are much more sensitive to luminance flicker than cells of the P-pathway. However, it could be argued that cells of the P-pathway also contribute to detection, since they are more numerous. Furthermore, since neural responses vary from trial to trial, it becomes necessary to formulate class A hypotheses in statistical terms (Teller, 1984). These factors tend to blur the distinction between class A and class B hypotheses. In fact, the difficulty in defining a neuronal 'threshold', has tended to make comparisons of cellular and psychophysical sensitivities a rather disreputable pastime.

A neuron doctrine relating signal units and sensation was formulated by Barlow (1972), who proposed five 'dogmas' intended to relate neural activity to experience, based on evidence partly derived from stimulus specificity in retinal ganglion cells and partly from responses of cortical neurones. Although the presence of 'trigger feature' neurones may not be in accord with more recent neurophysiological results, the arguments as to ganglion cell sensitivity remain valid. In what follows, I shall consider some of the problems involved in linking physiological and psychophysical data, taking examples from studies of the ganglion cells of the primate retina, and show some of the kinds of evidence required to give support to such hypotheses.

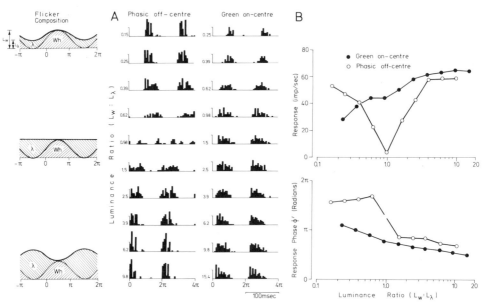

Fig. 1. Response histograms of a phasic off-centre cell and a green on-centre cell to heterochromatic flicker (10 Hz, 4^0 field), the composition of which is illustrated on the left. White was alternated with 506 nm. Each histogram represents response to two cycles of flicker. Luminance ratio ($L_\lambda : L_w$) is indicated for each histogram. The response of the phasic cell is minimal close to a luminance ratio of one, at which point an abrupt change in phase is seen. For the green on-centre cell there is a gradual change in response phase without any indication of a minimum. B. Amplitude of fundamental and response phase plotted against luminance ratio for the two cells. Reprinted from Lee et al. (1989c) with permission.

HETEROCHROMATIC FLICKER PHOTOMETRY; A CLASS *A* TASK?

Heterochromatic flicker photometry (HFP) is a psychophysical task of some practical importance, since it provided much data for specification of the human photopic luminous efficiency function. Two lights are alternated at ca. 10-40 Hz, and the subject must adjust their relative intensities until the sensation of flicker is minimised or abolished. By testing lights across the spectrum against a reference wavelength, a spectral sensitivity curve may be obtained. Recently, responses of ganglion cells have been studied under stimulus conditions made to match as closely as possible the psychophysical paradigm (Lee et al., 1988). Typical results are shown in Fig. 1. The stimulus paradigm is sketched at left, and in the centre panels are shown the responses of a phasic, off-centre, M-pathway cell and a green-on P-pathway cell to two cycles of modulation. White was alternated with 506 nm, and the luminance ratio for each histogram indicated. The activity of the phasic ganglion cell goes through a minimum close to equal luminance (a ratio of one), corresponding to the minimization of flicker sensation psychophysically. The responses of the green-on cell show a gradual change in phase without any indication of a minimum; this gradual phase drift can be attributed to a frequency-dependent centre-surround phase delay. At first sight, the phasic ganglion cell thus displays the properties which might support HFP. It was possible to show that all major photometric properties demonstrable psychophysically (e.g., additivity, transitivity, frequency independence) were directly demonstrable in cells of the M-pathway.

HFP appears to be a candidate for class *A* hypothesis status. If two lights are alternated at equal luminance, little or no M-pathway response is present and no flicker is visible; the alternated lights are indistinguishable from unmodulated light of the same mean chromaticity. However, this is not an example of a class *A* hypothesis, or an identity proposition (Teller, 1984), because the two physiological states are not indistinguishable; very vigorous responses are present within the P-pathway under such circumstances. As discussed by Teller (1984), if two visual stimuli are statistically indiscriminable psychophysically, this does not imply that the corresponding physiological states are also statistically indiscriminable. In fact, the paradigmatic example cited by Brindley also suffers from this defect. Two metameric lights will generally not produce equal quantal absorption in rods, and measures must be taken to minimize rod intrusion into the matches.

It is remarkable that observers can carry out HFP in the face of these vigorous responses in the P-pathway, and it is necessary to suppose that some low-pass filter operates on their signals at a cortical level. However, their remains the possibility that a central achromatic mechanism is generated from P-pathway cell activity, and that this mechanism is also nulled when performing HFP. We might therefore postulate two alternative linking hypotheses; one is that HFP is solely the responsibility of the M-pathway, and the other is that it is a collaborative effort between M- and P-pathways. A hypothesis of this latter sort was proposed by Gouras and Zrenner (1979).

To distinguish between these alternatives, several types of ancillary evidence may be utilised. Firstly, it can be shown that it is extremely difficult to generate a suitable achromatic channel for HFP from P-pathway activity. The centre-surround phase delay and cone weightings varies among P-pathway cells, so some achromatic channel generated by vector summation of their activities would be very idiosyncratic; changing modulation frequency or the colours used would necessitate a change in model parameters, which does not seem feasible neurophysiologically.

Secondly, the more completely the properties of the M-pathway can be used to account for psychophysical results with flicker photometry, the more probable it is that this pathway provides the sole substrate for the task. For example, an oddity of HFP is that at frequencies below 10 Hz, subjects can set a better flicker null if a small phase shift is introduced between the flickering lights (e.g., Lindsey et al., 1986). This puzzling psychophysical effect is directly reproduced in cells of the M-pathway, as described elsewhere in this volume (Smith et al., 1991). A further example of a psychophysical effect finding a direct explanation in responses of M-pathway cells is the effect of increasing retinal illuminance above about 200 td. With red and green lights, HFP becomes very difficult for subjects under these circumstances, since no clear flicker minimum is present; flicker is always seen, unless the frequency is increased to, say, 40 Hz. The effects of changing retinal illuminance on the behaviour of a phasic ganglion cell is shown in Fig. 2. Histograms for two cycles of flicker are shown at 200 and 2000 td. At the higher retinal illuminance, a much more substantial frequency-doubled response develops, apparently associated with a non-linearity of M- and L-cone summation (Lee et al., 1989b). The curves at right show the amplitudes of first and second harmonic Fourier components as a function of luminance ratio between the lights. It is plausible to suppose that this non-linearity within the M-pathway at high retinal illuminance underlies the psychophysical result; the non-linearity largely disappear at 40 Hz. Psychophysically, the residual flicker which interferes with performance appears of elevated frequency, although this is difficult to quantify.

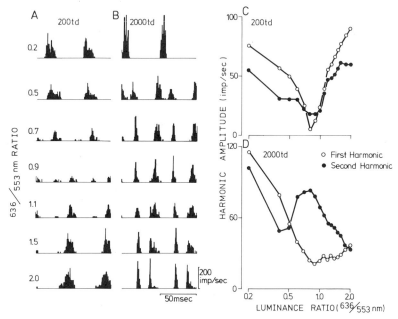

Fig. 2. Responses of phasic on-centre cell to red-green heterochromatic flicker (20 Hz), in a similar format to Fig. 1, at two levels of retinal illuminance (A,B). In the histograms, and in C, the development of a substantial frequency-doubled component at 2000td can be seen

It is thus possible to provide a convincing argument that cells of the M-pathway are the sole substrate for HFP. This argument rests on the implausibility an achromatic mechanism built from P-pathway cells and on a precise comparison of physiological and psychophysical results. If properties of the M-pathway had failed to correspond to psychophysical results, this would have substantially weakened the linking hypothesis.

DETECTION OF LUMINANCE AND CHROMATIC MODULATION

Comparison of cell sensitivity and psychophysical detection thresholds provides another comparison that might be formulated in terms of a class A hypothesis. Psychophysical sensitivity to luminance modulation is maximal around 10 Hz, and is attenuated at lower temporal frequencies. To chromatic modulation, sensitivity is maximal at low frequency, and falls off rapidly above 5 Hz. Evidence of this sort was used to support the notion that different post-receptoral mechanisms supported detection in the two cases (Kelly and van Norren, 1976). Cells of the M-pathway are much more sensitive to luminance modulation than those of the P-pathway, whereas cells of the P-pathway are the most sensitive to chromatic modulation (Lee et al., 1989a). It is plausible to suppose they provide distinct substrates for the post-receptoral mechanisms postulated psychophysically, but this linking hypothesis rests on several assumptions.

Firstly, when comparing psychophysical threshold with cell sensitivity, some measure of the latter must be employed. Due to the statistical nature of neuronal firing, a cellular 'threshold' can only be defined in probabilistic terms. Alternatively, cell sensitivity may be expressed as 'contrast gain' (Shapley and Kaplan, 1983). Nevertheless, in a variety of paradigms (Crook et al., 1987; Lee et al., 1989a) we have found that psychophysical thresholds appear to correspond to a modulation in firing rate of about 10-20 impulses per second in the most sensitive neurones encountered. For this generalization to be meaningful, it is necessary that the power spectral densities of individual neurones' maintained discharge are similar. This appears to be approximately the case, for M- and P-pathway cells and their various sub-classes (Troy and Lee, unpublished observations).

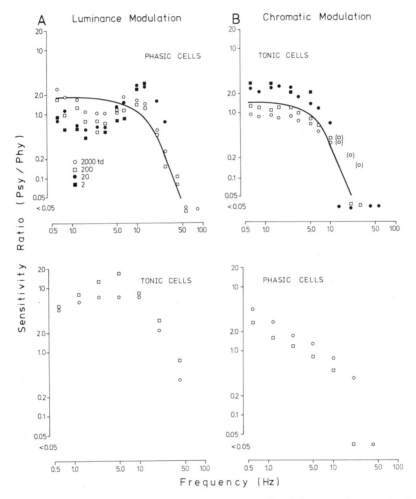

Fig. 3. Ratio of psychophysical to cell sensitivity for different cell types for luminance (A) and chromatic modulation (B) as a function of temporal frequency. For phasic cells and luminance modulation, sensitivity ratio is similar (close to unity) at all illuminances up to 20 Hz. Above 20 Hz, ratios fall steeply. Sensitivity ratio for tonic cells is much higher and dependent on retinal illuminance, few tonic cells responding at 20 or 2 td. For tonic cells and chromatic modulation, at low frequencies ratios are close to unity up to 5 Hz. Near 10 Hz, ratios for second-harmonic responses for phasic cells reach a level (about unity) consistent with their participation in detection. Curves shown indicate characteristics of hypothetical filters which may be present more centrally. Reproduced by permission (Lee et al., 1990).

Although it is plausible that the M-pathway thus underlies luminance flicker detection, it could be argued that cells of the P-pathway, through some kind of summation (probability or otherwise), are able to combine to provide higher sensitivity than individual neurones. Since the P-pathway contains more cells than the M-pathway (by a factor of 7-8), this suggestion is not unreasonable. Furthermore, vector-length models of psychophysical sensitivity implicitly assume that multiple mechanisms may contribute to detection in this way. It can be difficult distinguishing between a vector-length model and one in which separate, unique sets of visual mechanisms are involved in detection (e.g., Poirson et al., 1990).

One way to better define a correspondence between M- and P-pathways and post-receptoral channels is to show that a linking hypothesis holds along a variety of stimulus dimensions. We have attempted to do this be comparing physiological and psychophysical sensitivities at different adaptation levels.

As retinal illuminance is decreased, psychophysical sensitivity to luminance modulation remains similar at low frequency but there is progressive attenuation at higher frequencies. For chromatic modulation, sensitivity falls at all frequencies (Swanson et al., 1987). Measurements of sensitivity of M- and P-pathway ganglion cells reveal changes in modulation transfer function which closely resemble those seen psychophysically for luminance and chromatic modulation respectively. Results are summarized in Fig. 3. To compare physiological and psychophysical sensitivities, we have taken the ratios of psychophysical threshold to the modulation required to generate a 20 imp/sec modulation in cell firing; a more extensive description of these experiments may be found elsewhere (Lee et al., 1990). For luminance modulation and M-pathway cells, this ratio varies around unity at all contrast levels at frequencies below 20 Hz. Sensitivity ratios are much lower for P-pathway cells, and vary greatly as contrast level changes; it is difficult to evoke responses from many P-pathway cells below 20 td. With chromatic modulation, sensitivity ratios for P-pathway cells are close to one at lower temporal frequencies, implying that the decrease in psychophysical sensitivity to chromatic modulation with decreasing retinal illuminance is accompanied by a parallel change in P-pathway sensitivity.

At high temporal frequencies, P-pathway cells respond to chromatic modulation at frequencies well above fusion for human subjects. This implies operation of a cortical filter. Since M-pathway cells respond to luminance modulation up to 80 Hz at 2000 td, much higher than the flicker fusion frequency, a cortical filter is also implied for this system. A suggestion as to the properties of this filters is drawn into Fig. 3; their corner frequencies differ by a factor of two to three (Lee et al., 1990).

The parallel and distinctive changes in M- and P-pathway sensitivity on comparison with psychophysical sensitivity to luminance and chromatic modulation at different illuminances (and at different temporal frequencies) provides evidence that they provide unique and separable mechanisms for modulation detection, except perhaps for a 'luminance intrusion' in chromatic modulation detection around 10 Hz. Furthermore, the fact that sensitivity ratios are similar for the two comparisons indicate that the numerical advantage of P- over M-pathway cells does not bring about substantial improvement in sensitivity of the P-pathway as a whole.

THE MINIMALLY DISTINCT BORDER AND PATTERN PERCEPTION

The minimally distinct border method (MDB) shares with HFP all the characteristics of a photometric technique. This method, the parent of more recent psychophysical studies with isoluminant patterns, involves adjusting the relative intensities of two abutting coloured fields until the border between them is minimally distinct. It is parsimonious to suppose that M-pathway cells form the physiological substrate of both photometric techniques, and it may be shown that these cells display directly all the photometric properties observed psychophysically (Kaiser et al., 1990).

Residual distinctness after minimization depends on the colours either side of the border, and it can be shown that the residual responses of M-pathway cells (arising from the non-linearity of M- and L-cone summation mentioned above) correlates very well with subjective estimates.

Fig. 4 illustrates such estimates of residual distinctness for different wavelengths tested against a white reference and normalised to equivalent achromatic contrast. They are compared with averaged residual responses from M-pathway cells, normalized to their response to an achromatic edge.

MDB results provide a good example of how a linking hypothesis must postulate how a central mechanism handles afferent signals. With an edge moving across the retina, it would be difficult to conceive how on- or off-centre cells alone could form the substrate of the task because of the residual responses. It is necessary to postulate some mechanism which compares on- and off-centre signals. For example, a zero-crossing detector resembling an even-symmetric cortical receptive field would function in this way. Fig. 5 illustrates how this might work. A schematic receptive field is show above, together with a sketch of the stimulus situation. Responses of an on- and an off-centre M-pathway cell have been combined as indicated to generate an even-symmetric field. On the assumption that the distinctness of the border is a function of the output of such a detector, it can be seen that this output will be smeared spatially and of minimum amplitude at isoluminance. This would seem to provide a plausible explanation for the psychophysical result.

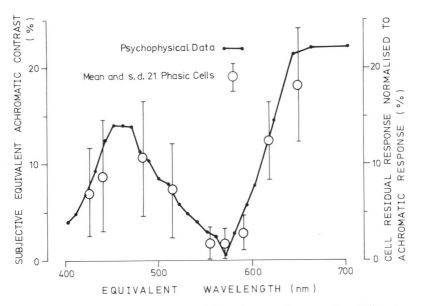

Fig. 4. Residual distinctness, estimated by a human observer when different monochromatic lights bordered onto an isoluminant white field, plotted as a function of wavelength, scaled relative to distinctness of an achromatic contrast (data of Tansley, taken from Boynton, 1978). In B, the data points show residual responses (mean, s.d., n=21) of phasic cells, scaled for each individual cell relative to its response to an achromatic contrast of 20%. Because our filters were not monochromatic, it was necessary to calculate the equivalent wavelength producing the same ratio of M/L cone excitations to make the comparison. Cell responses resemble psychophysical data in both shape of the function and its absolute magnitude.

Thus the M-pathway provides a signal adequate to account for border distinctness both in the presence and absence of a luminance difference. However, it has been proposed that an achromatic channel for pattern perception may be generated from the activity of P-pathway cells (e.g., Ingling and Martinez, 1983; Ingling, this volume). Although this viewpoint is supported by selective lesion experiments (Schiller, this volume), its origin lay in early physiological evidence that M-pathway cells were poorly represented foveally and did not respond to high spatial frequency gratings. Both these suppositions are now known to be incorrect (see Kaplan et al., for review). Nevertheless, the Nyquist limit of M-pathway cells is certainly lower than might be expected of a system supporting high resolution vision, probably by a factor of three to four.

If P-pathway cells played a significant role in pattern vision, it would be surprising if they did not contribute to border distinctness. It is thus necessary to consider the feasibility of a linking hypothesis for MDB in which an achromatic channel generated from some combination of P-pathway cells plays a role. A scheme is illustrated in Fig. 6A. A red on-centre and a green on-centre cell combine to provide input to some cortical neuron, which thus has a receptive field structure consisting of summed M- and L-cones to centre and surround, much as is required from an achromatic channel. There are several objections to such a model.Firstly, the diameter of the surround must be large in comparison with the centre. Although this is generally thought to be so, there seems to be substantial inter-cell variability, and some opponent surrounds seem to be little larger than the centre (Purpura et al., 1990). Secondly, it would break down on blurring the retinal image, but MDB is perfectly possible under those circumstances (Lindsey and Teller, 1989). Thirdly, actual responses of P-pathway cells are not consistent with the way such a model should function. For example, P-pathway cells respond very vigorously to isoluminant borders, and around equal luminance the response of the cell shows little change. This is illustrated in Fig. 6B; responses of a green on-centre and a red on-centre cell have been combined such that at equal luminance a minimum modulation of firing occurs. At luminance ratios of 0.4 and 2.5, the combined signal shows little change from the isoluminant condition; only at very high luminance contrasts do the predicted effects become apparent.

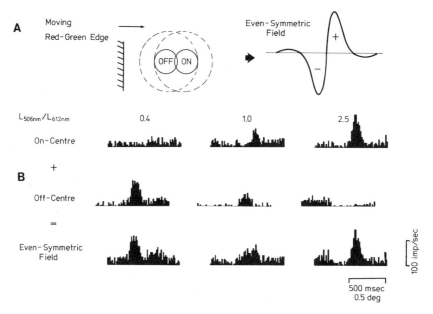

Fig. 5A. Sketch of a model in which an on- and an off-centre cell fields may be combined to give an even-symmetric receptive field. B. Responses of the cells have been added after a suitable shift in position to show how the output will vary as function of luminance contrast across the edge. Note that the output has changed spatial location on reversing contrast. At equal luminance a smeared, low amplitude response is present.

It thus seems unlikely that a signal produced by combining P-pathway cell activity can play a significant role in MDB. This is not entirely consistent with lesion experiments, where some MDB-like tasks are retained psychophysically after destruction of part of the magnocellular layers of the geniculate nucleus (Schiller, this volume; Merigan, this volume). These two sets of data appear in direct contradiction, and the resolution of this issue remains obscure.

CONCLUSIONS; VALIDATION OF LINKING HYPOTHESIS

I have tried to show some of the problems that occur in practice when trying to link physiological and psychophysical data. It seems that linking hypotheses can seldom be given a clean bill of health and assigned class A status; the physiologist (or psychophysicist) is usually faced with a set of alternative possibilities when seeking the physiological substrate of some particular task, in that if a particular cell system has properties capable of supporting a specific performance, this does not show that that system forms the sole substrate of the task; other systems may contribute. An example is the detection of small chromatic spots upon a background (Crook et al., 1987). One can nevertheless attempt some recommendations as to the criteria required to provide a secure link between physiological and psychophysical data.

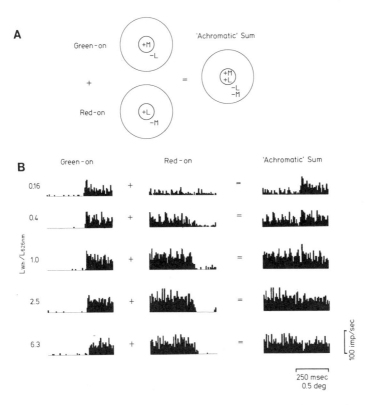

Fig. 6A. Possible scheme for generation of an achromatic cell from combining P-pathway cells. B. Responses of the cells have been added to show how output varies as a function of luminance contrast across the edge; a small shift was introduced to give minimum output at equal luminance.

Firstly, in order to convincingly demonstrate such a connection, it is necessary to compare in some detail results acquired under very similar stimulus conditions; in the past, physiological data have often been compared with psychophysical results in contexts much beyond those under which they were acquired. Secondly, linking propositions require detailed examination of underlying assumptions; Teller (1984) uses the oft-cited parallel between ganglion cell responses and Mach bands as an example of an unsatisfactory, *ad hoc* linking hypothesis. To show a simple parallel between cell response (or sensitivity) and psychophysical performance is not enough; it is necessary to show such a parallel remains valid when as many stimulus dimensions as possible are explored. If a previously puzzling psychophysical result then finds a direct physiological substrate (such as the phase shift effect with HFP), this provides particularly compelling evidence. Thirdly, there must be a plausible model of how central mechanisms make

113

use of afferent information. In the case of MDB, it is easy to provide a model of how cells of the M-pathway may support the task. Although the kind of model proposed whereby P-pathway cells provide an achromatic channel is plausible *a priori*, examination of real cell responses, as in Fig. 6, show it is difficult to implement. It is encumbent upon a cortical model to provide a plausible central mechanism which works with actual physiological data, does not require a change in parameters with changing stimulus conditions, and has reasonable assumptions as to signal-to-noise ratios in the nervous system.

THE NEURON DOCTRINE REVISITED

I have tried here to defend the identification of different cell types and systems within the primate visual pathway with different chromatic and achromatic channels postulated psychophysically. This involves the assumption that the most sensitive cells generally set the limits of detection. Such an assumption is not necessarily inconsistent with vector-length models of detection, since if one cell set is much more sensitive than another, such models predict detection primarily based on that set. Only when different elements have similar sensitivities are vector-length models distinguishable from those with no interaction between different mechanisms. A further complication is that individual cells within one pathway may vary in their properties (e.g., variability in spectral sensitivity between M-pathway cells). If a system is made up of non-homogeneous elements, it may not be univariant.

On the assumption that psychophysical detection threshold usually correspond to a modulation of firing of 10-20 imp/sec in the most sensitive ganglion cell, knowing variability of cell responses it is possible to calculate how many cells' activity need to be combined to achieve an adequate signal-to-noise ratio. In the case of flash responses, standard deviations are of the order of 20-30 imp/sec (Lee and Kremers, unpublished observations). This would indicate that only a few cells are necessary to give a 50% probability of detection.

A very similar argument was used by Barlow (1972) in developing a single neuron doctrine for perception. Referring to absolute threshold experiments, where absorption of only a few quanta are involved, he states 'quantitative knowledge of the noise level and reliability of single retinal ganglion cells enables one to see that the performance of the whole visual system can be attributed to a single cell'. The data presented here support the hypothesis that detection may depend on a change in firing of only a few ganglion cells. However, extrapolation of this notion to the cortical level would not generally be supported by recent evidence, which is generally in favour of different attributes of a stimulus being processed in different cortical areas (see, for example, van Essen, this volume), rather than specific 'trigger feature' neurones emerging further into the cortical hierarchy.

Acknowledgements: I thank my co-workers who have participated in the experiments described, and the several colleagues who provided comments on the manuscript.

REFERENCES

Boynton, R.M., 1978, Ten years of research with the minimally distinct border, in "Visual Psychophysics and Physiology" J.C. Armington, J. Krauskopf and B.R. Wooten, p. 193, Academic Press, London.

Gouras, P. and Zrenner, E., 1979, Enhancement of luminance flicker by color-opponent mechanisms. Science, 205:587.

Ingling, C.R. and Martinez, E., 1983, The spatio-chromatic signal of the r–g channel, in "Colour Vision; physiology and psychophysics" J. Mollon and L.T. Sharpe ed., Academic Press, London.

Kaiser, P.K., Lee, B.B., Martin P.R. and Valberg, A., 1990, The physiological basis of the minimally distinct border demonstrated in the ganglion cells of the macaque retina, J. Physiol., 422;153.

Kaplan, E., Lee, B.B. and Shapley, R.M., 1990, New Views of primate retinal function, Progress in Retinal Research, In Press.

Kaplan, E. and Shapley, R., 1986, The primate retina contains two types of ganglion cells with high and low contrast sensitivity, Proc. Natl. Acad. Sci., 83;2755.

Kelly, D.H. and van Norren, D., 1977, Two-band model of heterochromatic flicker, J. Opt. Soc. Am., 67;1081.

Lee, B.B., Martin, P.R. and Valberg, A., 1988, The physiological basis of heterochromatic flicker photometry demonstrated in the ganglion cells of the macaque retina, J. Physiol., 404;323.

Lee, B.B., Martin, P.R. and Valberg, A., 1989a, Sensitivity of macaque ganglion cells to luminance and chromatic flicker, J. Physiol., 414;223.

Lee, B.B., Martin, P.R. and Valberg, A., 1989c, A non-linearity summation of M– and L–cone inputs to phasic retinal ganglion cells of the macaque, J. Neuroscience, 9;1433.

Lee, B.B., Martin, P.R. and Valberg, A., 1989b, Amplitude and phase of responses of macaque ganglion cells to flickering stimuli, J. Physiol., 414;245.

Lee, B.B., Pokorny, J., Smith, V.C., Martin, P.R. and Valberg, A., 1990, Luminance and chromatic modulation sensitivity of macaque. ganglion cells and human observers, J. Opt. Soc. Am A, In press

Lindsey, D.T., Pokorny, J., and Smith, V.C., 1986, Phase-dependent sensitivity to heterochromatic flicker, J. Opt. Soc. Am. A, 3;921.

Lindsey, D.T. and Teller, D.Y., 1989, Influence of variations of edge blur on minimally distinct border judgements; a theoretical and empirical investigation, J. Opt. Soc. Am. A, 6; 446.

Poirson, A.B., Wandell, B.A., Varner, D.C. and Brainard, D.H., 1990, Surface characterizations of color thresholds, J. Opt. Soc. Am. A, 7;783.

Schiller, P.H., Logothetis, N.K. and Charles, E.R., 1990, Functions of the colour-opponent and broad-band channels of the visual system, Nature, 343;68.

Smith, V.C., Lee, B.B., Pokorny, J., Martin P.R. and Valberg, A., 1990, Responses of phasic ganglion cells of the macaque retina on changing the relative phase of two flickering lights. Submitted.

Swanson, W.H., Ueno, T., Smith, V.C. and Pokorny, J., 1987, Temporal modulation sensitivity and pulse-detection thresholds for chromatic and luminance perturbations, J. Opt. Soc. Am. A 4;1992.

P and M PATHWAY SPECIALIZATION IN THE MACAQUE

William H. Merigan

Department of Ophthalmology
and Center for Visual Science
University of Rochester
Rochester, NY 14627

The anatomical connections and physiological properties of the several parallel visual pathways in the Macaque offer tantalizing hints of their functional organization. One pathway is more robust than others, indicating dense sampling of the visual image. One has many neurons with color-opponent physiological responses, suggesting an important role in color vision. Some pathways provide the major inputs to cortical regions thought to be important for motion analysis, or the processing of visual shape.

Perhaps the best studied of these pathways are the two major retinogeniculate parallel pathways, the P pathway which projects from the retina through parvocellular geniculate, and the M pathway which projects through magnocellular geniculate (Shapley and Perry, 1986). These pathways could not be more different in anatomical and physiological characteristics. The P pathway includes many more retinal ganglion cells (about eight times as many as the M pathway) (Perry et al., 1984), and these cells inhabit the vitreal side of the ganglion cell layer (Perry and Silveira, 1988), they have small compact dendritic fields and somewhat smaller axons (Leventhal et al., 1981; Perry et al., 1984), and they show physiological color-opponency and rather low contrast sensitivity (Derrington et al., 1984; Derrington and Lennie, 1984; Kaplan and Shapley, 1982). The M pathway may have fewer cells, but each cell has a large, branched dendritic field and large axon, and while these cells are not color-opponent physiologically, they do show very high contrast sensitivity (Kaplan and Shapley, 1982; Leventhal et al., 1981; Perry et al., 1984).

The functional role of these two pathways has been extensively studied (Merigan and Eskin, 1986; Merigan et al., 1989; Schiller et al., 1990), and it is the purpose of this paper to summarize our current understanding of this research. It is not thought that these pathways play an important role in the processing of visual information, since basic visual descriptors, such as direction or orientation, are not seen physiologically before primary visual cortex. On the other hand, there is growing evidence from both macaque (Gross, 1972; Newsome and Pare, 1988) and human (Damasio et al., 1980; Hess et al., 1989) of the modular processing of particular dimensions of visual experience within localized cortical regions. The P and M subcortical pathways could thus contribute selectively to different aspects of visual perception if they provided almost exclusive input to specialized cortical modules. We will argue here that this possibility does not appear consistent with the results of lesion studies in the P and M pathways. A second possibility, which we support, and which does

From Pigments to Perception, Edited by A. Valberg and
B.B. Lee, Plenum Press, New York, 1991

117

appear consistent with lesion studies, is that, with the exception of color vision, which is dominated by the P pathway, the P and M pathways differ primarily in the range of spatio-temporal frequencies they provide to the visual cortex. This approach suggests that potentially modular functions of visual cortex, such as shape recognition, motion perception, or stereopsis can be done without either the P or the M pathway if the stimuli are chosen to fall within the spatio-temporal range of the remaining pathway.

SPATIO-TEMPORAL CONTRAST SENSITIVITY OF P AND M PATHWAYS

The first effort to separate the functions of P and M pathways followed the observation that acrylamide monomer selectively damaged P cells in the macaque retina (Merigan and Eskin, 1986). The major limitation of this model was the possibility that the M pathway might have suffered some subtle alteration. This remains a slight concern, although psychophysical and single unit physiological results indicate a normal M pathway, and converging evidence from studies of localized geniculate lesions (below) has confirmed the psychophysical observations made with this model.

Contrast sensitivity was measured in control and acrylamide exposed monkeys for several combinations of spatial and temporal frequency (Merigan and Eskin, 1986). Results for two typical monkeys, shown in Figure 1, indicate a very dramatic effect of degeneration of the P pathway. Spatial resolution was reduced, and contrast sensitivity was decreased at high spatial and lower temporal frequencies. On the other hand, contrast sensitivity at higher temporal and lower spatial frequencies appeared to be unaffected. In addition to the data shown here, sensitivity was also measured at 20 and 30 Hz (Merigan and Eskin, 1986), and, at least at the lower spatial frequencies, sensitivity was not affected by acrylamide exposure. These results suggest that the M pathway is specialized for transmitting higher temporal and lower spatial frequencies, while the P pathway is needed for vision at higher spatial and lower temporal frequencies.

In subsequent studies, e.g. (Merigan et al., 1989), we took advantage of the fact that the P and M pathways are spatially segregated in the lateral geniculate, making this the only location along these pathways in which localized lesions can be made. Lesions were centered in either magnocellular layer 1 (Merigan and Maunsell, (in press)) or in parvocellular layers 4 and 6 (Merigan et al., 1989), along the horizontal meridian about 6 deg from the fovea. Lesions

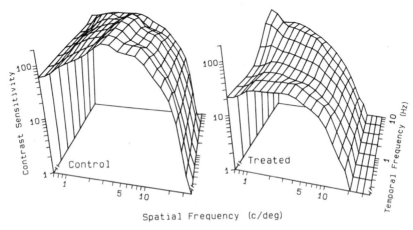

Fig. 1. Spatio-temporal contrast sensitivity of one control and one acrylamide-exposed monkey. From Merigan and Eskin, (1986).

were made by injecting ibotenic acid through a recording-injecting pipette
after establishing the correct location for the injection by physiological
recording. All psychophysical testing was done monocularly, using the eye contralateral to the injected geniculate. This was necessary because the ipsilateral eye projects to lateral geniculate layers 2 (magnocellular), and 3 and 5
(parvocellular), and it is not possible to completely lesion layer 2 without
damaging layer 3, or vice versa. After making the lesions, we carried out a
number of psychophysical studies in which the test stimulus was either
presented within the visual field location corresponding to the lesion, or as a
control, outside the lesion, usually at a comparable location in the opposite
visual field. The results to be described here were collected to determine the
spatio-temporal response profile of the P and M visual pathways.

Spatial Resolution

Spatial resolution was measured in two macaques after lesions of the P
pathway, and in one of these monkeys after a lesion of the M pathway (Merigan et
al., 1989). The results suggest a striking difference in the resolution of P and
M pathways that is consistent with the substantially greater sampling density
of the P pathway. The M pathway lesion did not decrease visual acuity, suggesting that the resolution of the P pathway was at least equal to or greater than
that of the M pathway. On the other hand, the P lesions caused a localized reduction in acuity of about a factor of four. This indicates that the acuity of the M
pathway is about four-fold worse than that of the P pathway, a result that is in
rough agreement with their relative retinal sampling densities.

Temporal Resolution

Temporal resolution was tested in two monkeys after lesions were made in
the M pathway (Merigan et al., 1989). Reconstruction of the lesion with physiological and anatomical measures has been completed in one of these monkeys, and
the representation of the M pathway in the tested part of the visual field was
completely destroyed. These same two monkeys were also used in measures of contrast sensitivity after M lesions described below. Thus, failure to find an
effect of M lesions is not due to incomplete lesions. The effects of the M pathway lesions on temporal resolution are shown in Figure 2. At the highest modulation depth there was virtually no effect of the lesion on temporal resolution. This
suggests that the spatio-temporal envelope for the P pathway extends to temporal

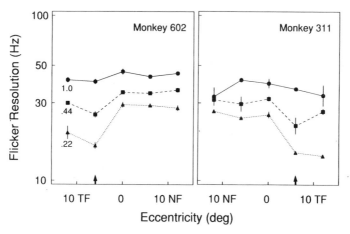

Fig. 2. Flicker resolution of two monkeys after M pathway lesions (arrows) for flicker modulation depths of
1.0 ((circles), 0.44 (squares), and 0.22 (triangles).
From Merigan and Maunsell (in press).

frequencies about as high as does that of the M pathway. On the other hand, at lower modulation depths, M lesions caused a reduction of temporal resolution. This suggests that while P and M pathways may have similar temporal resolution, the M pathway has considerable higher sensitivity to modulation at high temporal frequencies.

Contrast Sensitivity

Contrast sensitivity was measured at particular spatio-temporal frequencies chosen to indicate the shape of the contrast sensitivity envelope for P and M pathways (Merigan et al., 1989; Merigan et al., 1989). The first was a very low temporal frequency (slow onset, stationary grating) and moderate spatial frequency (2 c/deg). M lesions caused no effects, while P lesions severely reduced sensitivity. This finding, like that described above for acrylamide exposed monkeys, suggests that the M pathway makes little contribution to the detection of low temporal frequencies, and that such stimuli are detected primarily by the P pathway.

A second set of testing conditions involved a much higher temporal frequency of stimulation (10 Hz counterphase modulation), but about the same spatial frequency as used in the previous measurements (1 c/deg). M lesions in one study (Merigan et al., 1989) produced little or no effect on sensitivity, while in a second study, both P and M lesions caused slight decreases in sensitivity. Measurements in subsequent studies have shown that M lesions cause little loss at 2 Hz, inconsistent loss at 5 Hz, and substantial loss at 20 Hz. The implication of these findings for the contrast response of P and M pathways is that the sensitivity of the P pathway appears to decrease, while that of the M pathway is increasing, at temporal frequencies of 2 to 20 Hz. The crossover in relative sensitivity between the two is found somewhere in this temporal frequency range.

Contrast sensitivity was also tested at much lower spatial frequency (using an unpatterned gaussian blob as a stimulus) at a temporal frequency of 10 Hz (Merigan et al., 1989). Under these conditions (Figure 3), a lesion of the M pathway made it impossible for us to measure a contrast threshold. This final result explores the low spatial frequency portion of the spatio-temporal

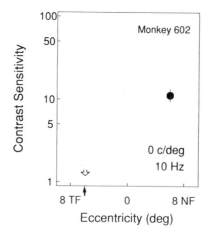

Fig. 3. The effect of an M pathway lesion (filled arrow) on contrast sensitivity for detecting a gaussian blob that was sinusoidally modulated at 10 Hz. From Merigan and Maunsell (in press).

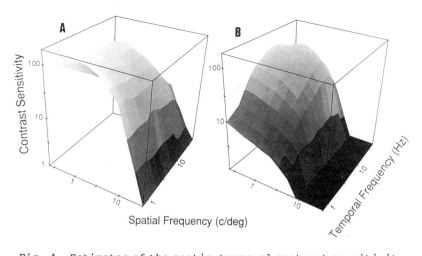

Fig. 4. Estimates of the spatio-temporal contrast sensitivity
of A) the P pathway alone and B) the M pathway alone.

space and indicates that the P pathway must have little or no sensitivity in this
region. It seems that sensitivity in this region is mediated solely by the M
pathway.

These results allow us to roughly sketch the spatio-temporal contrast sensitivity of the P and M pathways. Figure 4 shows an estimate of these sensitivity profiles derived from the above measurements. The primary implication
of these profiles is that the P and M pathways optimally transmit visual information in different portions of the spatio-temporal frequency space. The peak
responsivity of the two pathways differs in both spatial and temporal frequency, thus, it may be appropriate to describe the M pathway as responsive to
higher velocities (higher temporal and lower spatial frequencies), and the P
pathway as responsive to lower velocities. Two points should be kept in mind
when considering this analysis. The first is that both spatial and temporal
frequencies of best response vary with eccentricity across the visual field,
and thus, the locations shown in Figure 4 need to be scaled for different eccentricities. The second is that eye movements greatly enhance sensitivity at low
temporal frequencies (Kelly, 1979) and that these profiles would be different
if our measurements had been done with stabilized stimuli.

CHROMATIC SENSITIVITY

Chromatic contrast sensitivity has been assessed after both acrylamide-
induced lesions of retinal P ganglion cells and lesions in the lateral geniculate nucleus of the P or M pathways. In our earlier study (Merigan and Katz,
1989), we tested chromatic contrast sensitivity along four different directions of color space in acrylamide exposed monkeys. Figure 5 shows the results
of this measurement for one monkey (310), that had a very substantial loss of P
ganglion cells, as well as results for a normal control monkey. Color direction
0 deg represents a reddish-greenish stimulus and 90 deg represents a tritanopic
(roughly yellowish-bluish) stimulus. These two are the major or "cardinal"
directions of color space (Krauskopf et al., 1982) and the other two directions
(45 deg and 135 deg) are intermediate between them. One can see from this figure
that loss of P ganglion cells had a devastating effect on chromatic sensitivity, reducing it about 30 fold for all color directions. We are left with the
question of whether this result suggests a complete loss of color vision after
degeneration of the P pathway, (i.e. is the M pathway color blind?). In our

121

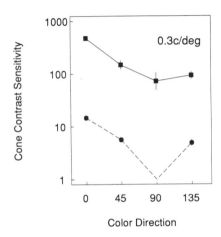

Fig. 5. Chromatic contrast sensitivity for
a control (squares) and an acrylamide-
exposed (circles) monkey. From Merigan
(1989).

view, it does not seem that the M pathway has any useful color vision, given that
it could be, at best, 30 fold less sensitive than the P pathway. Furthermore,
even this minimal residual sensitivity may overestimate that of the M pathway,
since it could reflect the function of a few remaining P cells, or luminance cues
introduced inadvertently in producing such high color contrasts.

We also tested chromatic sensitivity along one axis in color space after
lesions of P or M pathways in the lateral geniculate (Merigan et al., 1989).
These experiments resulted in a total inability to measure chromatic thresholds
after a complete P pathway lesion, and no effect on sensitivity after a lesion of
the M pathway. Together these and the above results suggest that the M pathway
has little or no sensitivity to chromatic variation, and thus, that all chromatically mediated discriminations rely only on the P pathway.

WHAT ROLE DO P AND M PATHWAYS PLAY IN MOTION
PERCEPTION, FORM DISCRIMINATION, ETC.

As noted in the introduction, the P and M subcortical pathways are not
thought to be important centers of visual processing. However, they could still
be quite specialized for certain visual functions if they provided the major
input to cortical areas identified with particular functions, e.g. the M pathway to areas MT and MST or the P pathway to area V4. This issue has not been
extensively explored, but what evidence there is to date suggests that particular visual capacities are not tied to particular retinogeniculate pathways.

We have measured three capacities relevant to this issue in acrylamide-
exposed and control monkeys. The first was contrast discrimination (contrast
increment thresholds) (Merigan, 1987). We used a two-alternative spatial
forced choice procedure, in which the monkey was required to choose whichever of
two grating stimuli were of higher contrast. A wide range of background contrasts were used and the two grating conditions were chosen to preferentially
stimulate the P or the M pathway. The stimulus that favored the P pathway was of
2 c/deg spatial frequency and of very low temporal frequency (stationary, with
slow onset). The stimulus chosen to favor the M pathway was of 1 c/deg spatial
frequency and 10 Hz temporal frequency. Contrast increment thresholds at 10 Hz
for acrylamide-exposed monkeys were virtually identical to those of controls,

which was consistent with the lack of any change in absolute contrast threshold for these conditions. On the other hand, absolute contrast threshold for the stationary grating was elevated by acrylamide exposure (Merigan and Eskin, 1986). With this stimulus, we found a uniform elevation of log contrast increment threshold of the same magnitude as the change in absolute contrast threshold. This result suggests that contrast discrimination thresholds are not especially dependent on either the P or M pathways, but rather that they can be mediated by either as long as the test stimulus is within the spatio-temporal range of the chosen pathway.

We subsequently measured form discrimination (Gross, 1972) in acrylamide-exposed and normal monkeys (Merigan and Polashenski, unpublished). Form perception has been studied previously in monkeys to examine the effects of inferotemporal cortex (IT) lesions (Gross, 1972). Initially we used stimuli to be discriminated that were derived from IT lesion studies, circles vs. squares, upright E vs a supine E, etc. We found no deficit in the performance of the acrylamide monkey relative to the control, but both monkeys reached perfect performance after only 5 to 10 trials. We then adopted more difficult stimuli which increased the number of trials to reach criterion performance (over 90% in 10 trials) to 100 to 200. Nonetheless, the performance of the acrylamide-exposed monkey was comparable to that of the control. This result suggests that function of the P pathway is not crucial for form discrimination, although the P pathway is thought to provide the major input to cortical areas important for form processing (Maunsell and Newsome, 1987).

Just as the P pathway provides the major input to the cortical "form and color" pathway, the M pathway provides the major input to the cortical "motion" pathway. We examined the role of the M pathway in motion processing by measuring opposite direction discrimination, and velocity difference thresholds after lesions of the M pathway (Merigan et al., 1990). We found that velocity difference thresholds were elevated by M lesions, but that this effect could be overcome by raising the contrast of the test stimuli. This suggested that M lesions decreased the visibility of the test stimuli, but did not affect analysis of their speed of motion. We also measured contrast thresholds for the detection of drifting Gabor stimuli, as well as for the discrimination of their direction of motion. We found that contrast thresholds for detection and direction discrimination decreased together as velocity was changed from 1 to 20 deg/sec velocity. This study suggests that while the M pathway is crucial for contrast sensitivity at higher velocities (see above), once the stimulus exceeds contrast threshold, velocity and direction can be processed normally.

A final example of the lack of dependence of specialized cortical visual capacities on the P and M pathways is from the work of Schiller and colleagues (Schiller et al., 1990). They were examining an earlier suggestion (Livingston and Hubel, 1987) that stereopsis depends on input from the M pathway. They tested stereo vision in monkeys after making lesions of either the P or M pathways (Figure 6), and found that disruption of the task by lesions depended on the spatial scale of the stereo elements. After lesions of the P pathway, stereo vision mediated by very small elements was disrupted, while performance mediated by coarser elements was not affected. This is another example of a particular visual capacity that could be mediated by the M or P pathway depending on the spatio-temporal content of the stimuli.

SUMMARY AND CONCLUSIONS

There is now substantial evidence from lesion studies that the P and M retinogeniculate pathways are specialized for different regions of spatio-temporal contrast space. The P pathway transmits information about the high spatial and lower temporal frequencies in the stimulus, while the M pathway transmits the higher temporal and lower spatial frequencies. In addition, it

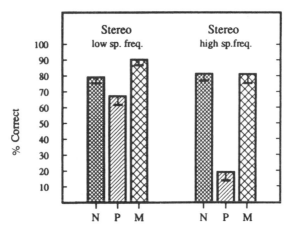

Fig. 6. Percent correct performance for detection of a displacement in depth of random dot stereograms made up of coarse (low sp. freq.) or fine (high sp. freq.) dots in normal (N), P pathway lesioned (P), or M pathway lesioned (M) portions of the visual field. From Schiller et. al. (1990).

appears that the P pathway is specialized for the transmission of chromatic information. Beyond this, there is, as yet, no indication that particular visual functions, such as motion or shape processing, depend exclusively on one of the retinogeniculate pathways. On the contrary, there is emerging evidence that certain specialized visual capacities, such as stereoscopic depth, can be mediated by either the P or M pathway, depending on the spatio-temporal characteristics of the test stimuli.

REFERENCES

Damasio, A., Yamada, Y., Damasio, H., Corbett, J. and McKee, J. (1980). Central achromatopsia: behavioral, anatomic, and physiological aspects. Neurology, 30, 1064 - 1071.

Derrington, A. M., Krauskopf, J. and Lennie, P. (1984). Chromatic mechanisms in lateral geniculate nucleus of macaque.. J. Physiol., 357, 241-265.

Derrington, A. M. and Lennie, P. (1984). Spatial and temporal contrast sensitivities of neurons in lateral geniculate nucleus of macaque. J. Physiol., 357, 219-240.

Gross, C. G. (1972). Visual functions of inferotemporal cortex. Handbook of Sensory Physiology, VIII/3B, 451-481.

Hess, R. H., Baker, C. L. and Zihl, J. (1989). The "motion blind" patient: low level spatial and temporal filters. J. Neurosci., 9, 1628-1640.

Kaplan, E. and Shapley, R. M. (1982). X and Y cells in the lateral geniculate nucleus of macaque monkeys. J. Physiol., 330, 125-143.

Kelly, D. H. (1979). Motion and vision. II Stabilized spatiotemporal threshold surface. J. Opt. Soc. Amer., 69, 1340-1349.

Krauskopf, J., Williams, D. and Heeley, D. W. (1982). Cardinal directions of color space. Vision Res., 22, 1123-1131.

Leventhal, A. G., Rodieck, R. W. and Dreher, B. (1981). Retinal ganglion cell classes in the old world monkey: morphology and central projections.. Science, 213, 1139-1142.

Livingston, M. S. and Hubel, D. H. (1987). Psychophysical evidence for separate channels for the perception of form, color, movement, and depth. J. Neurosci., 7, 3416-3468.

Maunsell, J. H. R. and Newsome, W. T. (1987). Visual processing in monkey extrastriate cortex. Ann. Rev. Neurosci., 10, 363-401.

Merigan, W. H. (1987). Role of P ganglion cells in vision: evidence from selective P cell degeneration. Society for Neuroscience 107.

Merigan, W. H., Byrne, C. and Maunsell, J. H. R. (1990). Does motion perception depend on the magnocellular pathway?. Soc. for Neurosci..

Merigan, W. H. and Eskin, T. A. (1986). Spatio-temporal vision of macaques with severe loss of Pb retinal ganglion cells. Vision Res., 26, 1751-1761.

Merigan, W. H. and Katz, L. M. (1989). Segregation of function between P and M pathways. Invest. Ophthalmol. Vis. Sci., 30,.

Merigan, W. H., Katz, L. M. and Maunsell, J. H. R. (1989). Contribution of the primate parvocellular pathway to acuity and contrast sensitivity. Invest. Ophthalmol. Vis. Sci., 30, 53.

Merigan, W. H. and Maunsell, J. H. R. ((in press)). Macaque vision after magnocellular lateral geniculate lesions. Visual Neurosci..

Newsome, W. T. and Pare, E. B. (1988). A selective impairment of motion perception following lesions of the middle temporal area.. J. Neurosci., 8, 2201-2211.

Perry, V. H., Oehler, R. and Cowey, A. (1984). Retinal ganglion cells which project to the dorsal lateral geniculate nucleus in the macaque monkey.. Neuroscience, 12, 1101-1123.

Perry, V. H. and Silveira, L. C. L. (1988). Functional lamination in the ganglion cell layer of the macaque retina. Neuroscience, 25, 217-223.

Schiller, P. H., Logothetis, N. K. and Charles, E. R. (1990). Functions of the colour-opponent and broad-band channels of the visual system. Nature, 343, 68-70.

Shapley, R. and Perry, V. H. (1986). Cat and monkey retinal ganglion cells and their visual functional roles. Trends in NeuroSciences, 9(5), 229-235.

THE COLOR-OPPONENT AND BROAD-BAND CHANNELS OF THE

PRIMATE VISUAL SYSTEM

Peter H. Schiller

Department of Brain and Cognitive Sciences
Massachusetts Institute of Technology
Cambridge, Massachusetts

ABSTRACT

To better understand the functions of the color-opponent and broad-band channels that originate in the primate retina, we examined the visual capacities of monkeys following their selective disruption. Color vision, fine but not coarse form vision and stereopsis are severely impaired in the absence of the color-opponent channel whereas motion and flicker perception are impaired at high but not low temporal frequencies in the absence of the broad-band channel. Much as the rods and cones of the retina can be thought of as extending the range of vision in the intensity domain, we propose that the color-opponent channel extends visual capacities in the spatial and wavelength domains whereas the broad-band channel extends them in the temporal domain.

INTRODUCTION

Physiological and anatomical studies have distinguished several distinct classes of retinal ganglion cells in the mammalian retina, each of which has been proposed to be involved in the analysis of a different aspect of the visual scene[1]. Two major classes of cells originating in the retina are the color-opponent and broad-band cells whose characteristics were first delineated in single-cell recording experiments[2-6]. The retinal ganglion cells of both classes have concentric, antagonistic center-surround organization. The color-opponent cells, as their name implies, receive input from different cone types in their center and surround regions, whereas the broad-band cells receive undifferentiated input from the cones throughout their receptive fields. The size of the color-opponent cell receptive field is small, the cells respond in a sustained fashion to visual stimulation, and they have medium conduction-velocity axons. By contrast, the broad-band cells have much larger receptive fields, are more sensitive to small changes in luminance, respond transiently, and have rapidly conducting axons[6,7].

Examination of the central connections of these two channels revealed

From Pigments to Perception, Edited by A. Valberg and
B.B. Lee, Plenum Press, New York, 1991

that they project to different laminae of the lateral geniculate nucleus of the thalamus: the color-opponent cells, which comprise about 90 percent of the neurons of the geniculo-striate system, terminate in the four parvocellular layers of the lateral geniculate nucleus; the broad-band cells on the other hand terminate in its two magnocellular layers[8-11].

METHODS

To assess the functions of the color-opponent and broad-band channels in vision we have undertaken to selectively block them in rhesus monkeys and to then test their visual capacities on a wide range of visual tasks [12-13]. We placed small lesions into either the parvocellular or the magnocellular portions of the lateral geniculate nucleus by injecting minute quantities of ibotenic acid subsequent to electrophysiological recordings which established the proper layers and visual field representations to be lesioned.

To confine the stimuli to the lesioned or the intact portions of the visual field, each trial was initiated by a small spot that appeared in the center of a color monitor. Following fixation of this stimulus, as assessed by eye-movement recordings, either a single stimulus or several stimuli appeared and the animal's task was to make a saccadic eye movement so as to direct his gaze at the proper target. A number of different visual capacities were assessed using either a detection or a discrimination paradigm, which included color, form, brightness, depth flicker and motion perception. In case of the detection paradigm a single target appeared following fixation in any one of several locations. Target acquisition with a single saccade was rewarded with a drop of apple juice. In case of the discrimination paradigm several stimuli appeared following fixation (typically eight), one of which was different from the other identical stimuli in either color, form, brightness, depth, etc. The animal had to saccade to the stimulus different from the others to be rewarded. Since the rhesus monkeys used in this work willingly perform several thousand trials per day and readily master all of the visual tasks, reliable psychophysical functions can be generated efficiently.

RESULTS

Summary data obtained in our lesion experiments are shown in Figure 1 for seven visual capacities. They are as follows:

Color vision: When color vision is tested by having monkeys discriminate stimuli only on the basis of color differences, severe deficits are incurred following parvocellular lesions: animals can no longer discern color differences at all. However, when a single, low spatial frequency color stimulus appears on an isoluminant background, they have no difficulty detecting it. These experiments establish then that parvocellular lesions abolish the capacity for discriminating color differences (to ascribe color values to various wavelength compositions) but do not interfere with the capacity to "see" stimuli on the basis of wavelength differences. Magnocellular lesions have no effect on color discrimination.

Form vision: At high spatial frequencies all three kinds of form vision studied, pattern, texture and shape, are severely compromised following parvocellular lateral geniculate nucleus lesions. The deficits are

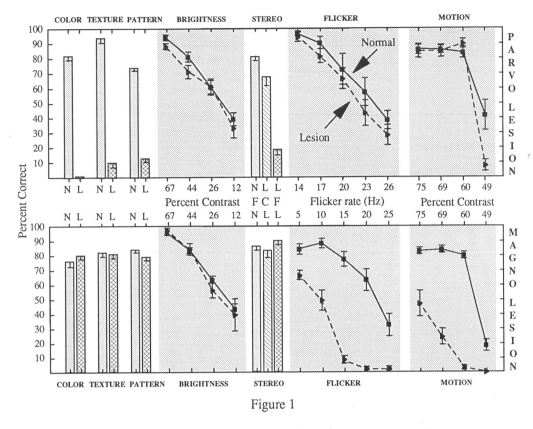

Figure 1

Percent correct performance is shown for color, texture, pattern
brightness, stereo, flicker and motion perception following
parvocellular (upper panel) and magnocellular (lower panel) lateral
geniculate nucleus lesions. In each case normal (N) and lesion (L)
data shown were collected concurrently at comparable retinal
eccentricities. Each point shown is based on a minimum of 100
trials. F = fine stereopsis, C = coarse stereopsis.

considerably less pronounced with low spatial frequency stimuli.
Magnocellular lesions have no discernible effect on any of these tasks.

Brightness perception: The detection or the discrimination of low
spatial frequency stimuli on the basis of luminance differences alone shows
no deficits with either lesion. This is true for both photopic vision
subserved by the cones and for scotopic or night vision, subserved by the
rods. Thus it appears that luminance information at low spatial
frequencies can be processed by both the color-opponent and the broad-
band systems. These findings also suggest that the broad-band channel
cannot be conceived of as a unique luminance channel.

Stereoscopic Depth perception: Studying stereopsis with random-dot
stereograms of various spatial frequencies and degrees of disparity shows
that parvocellular lateral geniculate nucleus lesions produce major deficits

in fine but not coarse static stereopsis; no deficits are obtained at all with magnocellular lesions. These findings dispute the claim that stereopsis is processed exclusively by the broad-band system as has been proposed by Livingstone and Hubel[14].

Motion and Flicker perception: The major deficits that arise following lesions of the broad-band system fall into the temporal domain. Most dramatically affected is the perception of flicker and motion, especially at low contrasts. Magnocellular lesions produce significant deficits in the detection of both monochromatic and heterochromatic flicker at high frequencies indicating that the temporal limits of flicker perception are set by the broad-band system. These and the color vision results fit with the observation that in the course of increasing red/green flicker rate, human observers experience three perceptions: at low rates the stimulus is seen as alternating between red and green; as rate is increased, the colors fuse and a yellow flickering stimulus is perceived; at high rates the sensation of flicker is lost and a steadily illuminated yellow stimulus is seen. The presumption is that the color sensations are mediated by the color-opponent system and rapid flicker by the broad-band system. It should be emphasized that while deficits in motion and flicker following magnocellular lesions are pronounced, neither of these capacities is eliminated; at low temporal frequencies and high contrasts the deficits are small, suggesting that the color-opponent system can make a significant contribution to the temporal aspects of perception[12,13,15].

DISCUSSION

The results of the lateral geniculate nucleus lesion studies lead to the conclusion that the color-opponent system is crucial for color discrimination, and is essential for form and depth perception at high but not at low spatial frequencies. The broad-band system plays an important role in motion and flicker perception; however, the color-opponent system can process these capacities at high contrasts and low temporal rates. Both systems can process brightness, shape and stereo information at low spatial frequencies. Thus while some of the perceptual losses incurred in these lateral geniculate nucleus lesion experiments are profound, it is important to note that except for color discrimination, the deficits for the various visual capacities are not all or none; they are graded: at low spatial and temporal frequencies, high contrasts and low disparities perceptual losses for the most part are small. Deficits become pronounced as spatial and temporal frequencies are increased and/or contrast and disparities are decreased.

What general statement might be made about why the color-opponent and the broad-band channels have evolved? In pondering this question it may be profitable to turn, as an analogy, to another system, the receptors of the retina, which were identified in 1866 by M. Schultze[16] to be dual in nature. He correctly hypothesized that the cones are for day and the rods for night vision. The hypothesis suggests that the evolution of rods and cones extended the range of vision in the domain of light intensity. In a similar vein, we propose that the evolution of the color-opponent and the broad-band channels also extended the range of vision, with the former extending it in the domain of wavelength and spatial frequency and the latter in the domain of temporal frequency as shown in Figure 2. Why could this not be accomplished within one system? Probably because of conflicting requirements. Conflicting, because for high resolution vision one needs small windows to the world, small receptive fields that is, that are available in relatively large numbers. One also needs cells that respond in a sustained fashion to optimize the extraction of static spatial information during each brief period of maintained gaze, the duration of which in diurnal primates is typically 200 to 500 milliseconds followed by a rapid saccadic eye movement to another part of the visual scene to repeat the process. In addition, for color vision one needs spatially segregated input from different cone types. The small degree of convergence of receptors in this kind of system and the sustained responses it gives, limit sensitivity and temporal resolution. For high temporal frequency selectivity and the concomitant sensitivity required for it, large receptive fields with non-specific convergence of receptors and with transient responses are required. What is given up in the broad-band system as a result is the processing of color and of pattern at high spatial frequencies.

In summary, we propose that the color-opponent and broad-band channels form two separate but overlapping systems the function of which is to extend the range of vision in the spatial, wavelength and temporal domains.

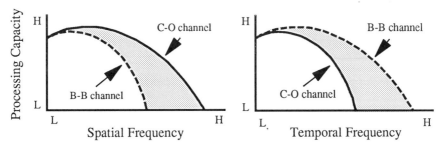

Figure 2

Schematic showing processing capacity for the color-opponent and broad-band channels along spatial and temporal frequency axes. In the spatial frequency domain the ability of the broad-band system for processing information falls off more rapidly with increasing spatial frequencies than the color-opponent system. In the frequency domain the opposite is the case: the broad-band system can process information to higher temporal frequencies than the color-opponent system. Further insights about the functions of these two channels could be gained in psychophysical experiments by testing subjects within the shaded areas of the figure. The Y axis shows information processing capacity where H stands for high and L for low. On the X axis L and H stand for low and high spatial or temporal frequencies.

ACKNOWLEDGMENTS

I thank Nikos K. Logothetis, Eliot R. Charles and Kyoungmin Lee for their participation in the research reported here from our laboratory which was supported in part by NIH grant EY00676.

REFERENCES

1. Schiller, P.H. (1986) The central visual system. *Vision Res.* 26, 1351-1386.
2. Wiesel, T.N. and Hubel, D.H. (1966) Spatial and chromatic interaction in the lateral geniculate body of the rhesus monkey. *J. Neurophysiol.* 29, 1115-1156.
3. De Valois, R.L. and Jacobs, G.H. (1968) Primate color vision. *Science* 162, 533-540.
4. Gouras, P. (1968) Identification of cone mechanisms in monkey ganglion cells. *J. Physiol. (London)* 199, 533-547.
5. De Monasterio, F.M. and Gouras, P. (1975) Functional properties of ganglion cells. *J. Physiology (London)* 251, 167-195.
6. Schiller, P.H. and Malpeli, J.G. (1978) Functional specificity of lateral geniculate nucleus laminae of the rhesus monkey. *J. Neurophysiol.* 41, 788-797.
7. Shaply, R.M., Kaplan, E. and Soodak, R. (1985) Spatial summation and contrast sensitivity of X and Y cells in the lateral geniculate nucleus of the macaque. *Nature* 292, 543-545.
8. Leventhal, A.G., Rodieck, R.W. and Dreher, B. (1981) Retinal ganglion cells in old world monkey: morphology and central projections. *Science* 213, 1139-1142.
9. Perry, V.H., Oehler, R. and Cowey, A. (1984) Retinal ganglion cells that project to the superior colliculus and pretectum in the macaque monkey. *Neuroscience* 12, 1101-1123.
10. Dreher, B., Fukada and Rodieck, R.W. (1976) Identification, classification and anatomical segregation of cells with X-like and Y-like properties in the LGN of old world primates. *J. Physiol. (London)* 258, 433-452.
11. Schiller, P.H. and Malpeli, J.G. (1978) Functional specificity of lateral geniculate nucleus laminae of the rhesus monkey. *J. Neurophysiol.* 41, 788-797.
12. Schiller, P.H., Logothetis, N.K., and Charles, E.R. (1990) Functions of the color-opponent and broad-band channels of the visual system. *Nature* 343, 68-70.
13. Schiller, P.H., Logothetis, N.K., and Charles, E.R. (1990) The role of the color-opponent and broad-band channels in vision. *Visual Neuroscience* (in press).
14. Livingstone, M.S. and Hubel, D.H. (1988) Segregation of form, color, movement and depth: anatomy, physiology, and perception. *Science* 240, 740-749.
15. Merigan, W.H. (1989) Chromatic and achromatic vision of macaques: role of the P pathway. *J. of Neuroscience* 9, 776-783.
16. Schultze, M. (1866) *Arch. Mikr. Anat.* 2, 175-286

DISCUSSION: P- AND M-PATHWAYS I

Eberhardt Zrenner

Universitäts Augenklinik
Tübingen
W. Germany

Zrenner: We have heard a number of conflicting views; it is natural to believe most strongly those things we have measured ourselves. I would like to start with a sketch of what is generally accepted. We have three different cones which input to both parvo- and magnocellular systems, although we have not discussed today if there is any or only minimal S-cone input to the magnocellular system. At the bipolar cell level, there are some data that S-cones have their own CCK- (cholycystokinin) bipolar. Certainly, we know we have rod bipolars and on- and off-bipolars of the midget type which presumably input to the parvocellular system. Coming to the magnocellular system, it is plausible that they have their own bipolars, perhaps the diffuse or brush bipolars, so they have their own, additive pathway. The role of different types of horizontal cells has not received much attention, and indeed recent data from Kolb suggests a differential input of S-cones to different horizontal cell types (H1,H2,H3), although this is not in accord with the analysis of Wässle's group.

What we have heard today concerns the tasks and structure of the parvocellular and magnocellular systems. Do the type I and type II parvocellular systems have different functions and what is the role of any non-opponent cells in the parvocellular system? As far as the magnocellular system is concerned, there seems to be an interesting failure of linearity. Also, whatever the role of the magnocellular system for luminance, the monkeys of Peter Schiller could still make brightness discriminations with the magnocellular system lesioned. There seem to be conflicting tasks; each cell can only transmit information in a limited range. Rather like having two cars, a small manœuvrable one and a large fast one, does the visual system just run with these two cars or are there more? This brings me to the question of how much variability there is within the two systems, so that tasks are shared within them as well as between them. Contrast can be processed in both systems, but with lower gain in the parvocellular system. At high temporal frequencies flicker is well signalled by the magnocellular system. In this slide of Peter Gouras and myself three magno cells follow very nicely the V_λ function. However, we should not disregard the parvocellular system, in which cells vary their spectral sensitivity with temporal frequency. We finally heard that lesion studies have created insight into function either at the ganglion cell level or at the geniculate level. We also have not spoken of opponent signals between rods and S-cones which can give good colour vision under some conditions, and there may be opponent interaction between the rods and the other cone systems as well. I would like to start the discussion by focussing on the differences that we have heard in the last talks; is the sharp distinction between parvo- and magno-systems valid or should we have a closer look at the variation in these systems and the tasks that can be handled by both systems.

Kaplan: To Bob Rodieck, from the fact that these cells are bistratified do you expect them to be on-off?.

From Pigments to Perception, Edited by A. Valberg and
B.B. Lee, Plenum Press, New York, 1991

Rodieck: Yes.

Kaplan: In recordings from hundred of cells in the P layers Bob and I have found one cell that gave such a response, and I think others find similar results. That is one point, The other is that you refer to type I as P and type II as something else, but I would keep P for all the cells that project to the parvocellular layers.

Rodieck: My comment is that you can't have it both ways. You can't both say that everything that projects there is a P-cell, and at the same time P-cells have certain receptive field sizes, and they have spectral this and they have spectral that. This is why I wanted to use the term type I, I wanted to deal with empirical issues and not terminological ones.

Spekreijse: In those responses you were showing for the M-cell, the impulse response for the M-cones was different from that of the L-cones, and does not that pose a problem for flicker photometry?

Shapley: I think that if you look at the center response mechanisms, the impulse responses go together very well. It is in the surround that they don't correspond. Vivianne Smith has a whole story about that. For the center mechanism, they would have to go together to explain the deep trough you get with the center isolating stimuli. So I don't think there is a problem.

Schiller: Since cones in central retinae are not stacked on top of another, and since we believe now that a single cone comprises the center mechanism of colour opponent cells, how can you have overlapping center and surround mechanisms.

Shapley: I don't think we can resolve single cones optically. I think we're measuring functionally how cell's look at natural stimuli. We're looking at spatial responses with the optics in series with the cell measurement.

Van Essen: Might not annuli evoke responses from the surround of P cells which you do not see with your spot technique?

Shapley: There might be weak inputs, but we think we can describe receptive field organisation on the basis of the measurements we have made. The contribution of the inhibitory cone to spatially displaced inhibition seems too weak too measure.

Spillmann: To Barry Lee, have you compared HFP and MDB with the same cells?

Lee: No, I'm afraid not.

Shapley: We have made comparisons with flicker and drifting gratings, and get very similar red-green ratios for a given cell.

Spillmann: If Bill Merigan's results can be related to those of Kelly in man, do the envelopes of P and M cells resemble the spatiotemporal and chromatic envelopes found in humans?

Merigan: Many of Kelly's recent measurements were done on stabilized images, but apart from that they look very similar.

Richter: I would like to bring up a question from psychophysical experiments. If you have a scene, you have a median luminance and you have different luminance ranges for one or the other task. By studying contrast effects over these ranges, I concluded there must be two different systems, one covering a small luminance range, about 1 log unit, compared to the median luminance and another system covering a large luminance range. The magno system may cover a small luminance range, and the other system - the parvo-may cover a larger range. Is this feasible?

Schiller: I would address this point about such a huge luminance range. Most of our vision is based on reflectances, which seldom exceed 70-80% contrast, so that most of our vision is in a range of less than a log unit. I do not see how these two systems could play significant differential roles in spanning luminance ranges.

Rodieck: I think that this makes the discussion too diffuse to explore boundaries that the speakers have not explored. There is enough interesting things on the differences that they have explored, and to try to push the results into frameworks of 'This does it' and 'That does it', is not that interesting. It's more interesting to see where the differences lie. As a comment, I think one of the terms that we have to address is the notion of P-cells, I am almost happy with any definition of that term, but I think that it has to be seen in the context of the different things that have been described. It looks like different types of cells in the parvocellular layers are not only functionally different, but that they are anatomically different, and we need a nomenclature to describe that. I would encourage either the type I and type II nomenclature at the geniculate level, at the retinal level the nomenclature of phasic, broad-band or tonic, and at the morphological level the terms midget and parasol. To call all of those cells either P-cells or M-cells fails to make a distinction that I think we need to incorporate into our thinking. In the same spirit, I am rather surprised that Peter Schiller calling the parvocellular layers the colour opponent channels, when he goes on so elegantly to show they do other things which have nothing to do with colour opponency.

Kaplan: I am not sure if my name is Kaplan or Kapley. I support the view of Rodieck. I have been impressed with your data showing the cells that project to the parvocellular layers that we may have not recorded from yet. So I completely agree that the term P-system as we use it, including all the cells which project to the parvocellular layers, is probably a heterogeneous group, and various sub-groups have different capacities, and it is because of that there is so much controversy. I would apply that too to the magnocellular layers. We have suggested that there are two kind of magnocellular cells, and I would be perfectly happy to meet new ones at the end of an electrode. Secondly, Bob Rodieck has so nicely described the classical view of which cells do what. I want to read from a recent paper by Hubel and Livingstone (J. Neuroscience, 1990). "This leads us to wonder whether the geniculate type I cells with their strange center-surround colour opponency play any role in colour perception. They may simply be used to build up cortical cells that are sensitive to both luminance and colour contrast. Type I geniculate cells were recognized at the start *(1966 is the start; E.K.)* as unlikely to play any part in colour constancy. The geniculate type II cells and cortical double opponent cells seem more likely candidates for colour perception."

Zrenner: This brings us back to the original question of the variability of the parvocellular system.

Shapley: I wanted to talk about the degree of heterogeneity in the magno system and emphasize a couple of points. One is that in studying the spectral characteristics of the M-cell system with heterochromatic red-green stimuli, either with gratings or with spots, at spatial frequencies high enough to isolate the central mechanism, we get very good agreement with what you would expect from a V_λ-like mechanism. However, if we include large spots or full-field stimulation, indeed, as others have found, and as we find, we do not get the same kind of colour-blind, V_λ-like responses that one observes with center isolating stimuli. The psychophysical tasks that tend to be associated with the luminance mechanism, high-frequency flicker, minimal distinct border or motion, are those which tend to isolate the center mechanism of the neurons, which seems to be very stereotyped in its spectral properties. And so I think that under neutral adaptation conditions, there isn't much variability. That is why I was surprised about Peter Schiller's results that these cells were not colour-blind with the red/green small spot flicker. Maybe in those experiments the background conditions were not such as to prevent light scatter on the surround, and the surround contributed to the response. But generally if you use conditions where you isolate the centers in the magno system, my belief is that you are working with the luminance system.

Zrenner: What you say then is that in a limited range of conditions there is very reduced variability, though there may be more under other conditions.

Schiller: In the 70s after Richard Gregory visited MIT, I recorded from the lateral geniculate but was disappointed to find under my conditions that the magno cells frequency-doubled and varied slightly, a tenth of a log unit, in their balance points. We then went on to record multiple units and were struck by magno responses at equal luminance. Quantitatively, with multiple units, we found there was no major difference in the degree to which responses in these two systems declined at isoluminance. When you lesion the magno system, you get a loss in flicker sensitivity, both for luminance and heterochromatic flicker. When you use heterochromatic flicker, at low temporal frequencies you see red and green alternating, which is probably the P-system. As you increase frequency, it becomes yellow but you still see flicker and it is my opinion that this is the magno system. In the spatial domain, with a red-green grating as you increase frequency you see yellow but you still see a grating, but this is different; what this is I cannot tell you.

MacLeod: Perhaps Merigan could comment on the discrepancy between his results and the physiological data, which show very low P-cell achromatic contrast sensitivity.

Merigan: It is clear from our studies that the sensitivity of the P-pathway as a whole is much higher than its individual cells. Going through the visual system, sensitivity doesn't seem to change very appreciably in V1. Where it occurs, or what is occurring, probability summation or whatever, is open.

Krauskopf: This intoxication with isoluminance causes us a lot of trouble, and I think it's time to once again point out some of the problems with it. Bob Rodieck and Peter Schiller both made mistakes in analyzing Cathy Mollon's data. You have to first decide how to measure contrast before you know where to place those curves on the vertical axis. Dave Williams and I studied temporal sensitivity to isoluminant stimuli and to luminous stimuli, and showed that if you equate cone contrast you get exactly the same temporal contrast sensitivity. And in the case of the Cathy Mollon data, if you plot in terms of cone contrast sensitivity, the high-frequency portions of those curves superimposed. Mark Farrell and I did similar experiments on vernier acuity with similar results. So the point is that you have to first have a theory about how you are going to use the signals, and maybe the correct one is that the cone contrasts are the fundamental element of the stimuli that you want to deal with. There are probably cases where once you have equated cone contrast, that you still get different answers; motion seems to be a case of this sort, When you get differences then you have something that is interesting to talk about.

Kulikowski: Perhaps we should remember that the visual system is not designed for our convenience as scientists but for seeing, and it consists of series of overlapping mechanisms, and depending of what kind of stimuli you use, you can get different results. I see absolutely no contradiction between the results of Schiller and Merigan because they used different stimuli. Schiller definitely biased his results towards the so-called parvo system. From working with the selective stimuli for some 20 years, we can say that with the sole exception of colour analysis, neither the parvo or magno systems are completely selective. It's been said that flicker is observed by the magno system, and I agree that this is so under normal circumstances for normal observers, but we have a patient after severe multiple sclerosis who did not show any evidence of the magno system working, but at 25 Hz showed a distinct peak of sensitivity in the blue, but with admittedly very low sensitivity. So 25 Hz is still within reach of the parvo system. The concept of selective stimuli should be treated with a pinch of salt.

Merigan: One of the major differences between Peter Schiller's work and ours is in the area of flicker. We have verified complete magnocellular lesions, and measured flicker resolution for small targets at 35 Hz although sensitivity is enormously reduced. In any case results indicate that the parvo system, as you would expect from the physiology, is not that slow, as indicated by Barry Lee's physiological measurements. Although flicker resolution may be subserved by the M-pathway, the P-pathway is close behind.

Zrenner: This also consistent with our physiological data.

Schiller: The degree of deficit one gets depends on the contrast that you use. When you use a very high contrast stimuli, the parvo system can follow to much higher rates. We were limited because we used LEDs which are fairly low contrast devices. Relative to what Kulikowski said of his patient, this might be correct, but there are several studies out that in those patients who supposedly lacked the magno system, there is a severe loss in flicker sensitivity.

Mollon: How important in Peter Schiller's experiments are the several non-geniculate visual pathways, and is this not especially important since the response is an eye movement.

Schiller: With lesions of the whole geniculate, moneys are virtually blind, so whatever the direct retino-tectal pathway subserves it doesn't help with the tasks we use here. With other evidence, I feel these pathways do not play a significant role.

Van Essen: I'd like to raise a puzzle in relation to yesterday, particularly David William's talk, and to some of the physiology we heard today from Bob Shapley. This stems from an issue Peter Schiller raised. If given cones do not physically overlap one another, how does one build opponent mechanisms that are spatially near by. David Williams showed a nice illustration of a randomly distributed collection of L- and M-cones in which large domains were dominated by one type or another, particularly by L-cones. It would seem to me difficult in those regions to find ganglion cells with strongly overlapping opponent mechanisms; one would expect to see much more spatial segregation in many parts of the retina. So either the cone mosaic is not so random, or there must be more spatial segregation than the impression I got from Shapley's description. Maybe one or both of them can comment on that.

Shapley: The question that you bring up is one that Dave Williams and I were talking about yesterday. We think in fact exactly as you do that models of the cone mosaic may be constrained by physiological data of the kind that we obtained, having to do with the degree of overlap or lack of overlap in cone mechanisms feeding into opponent cells. That is one constraint. Another constraint that one might use also is the degree of variability of the spectral sensitivity of the M-cell's centers. Again that would be something that might test these models of the cone mosaic. And I feel as you do, that a purely Poisson arrangement of cones with a 2 L- to 1 M-cone ratio seems to be inconsistent with the physiological data. But I think in this case more facts are needed before we can rule out one theory or another. But I think the direction things are going is that we think the cone mosaic may be more regular than random on the basis of these preliminary results.

Rodieck: As far as Dave Williams was concerned for his Brewster colours, he says that a systematic or a random array gives the same result, but Bob Shapley makes a very good point that the electrophysiology might clarify the situation. I would point out that de Monasterio in 1978 described profiles of two opponent mechanisms so much in balance throughout the entire extent of that receptive field that it seems difficult to believe that they could have resulted from a pure Poisson distribution. Bob Shapley's work is indeed a more powerful way of analyzing the receptive field.

Zaidi: I have a joint question to Bob Shapley and Dave Williams. Isn't there a difference in the size of the stimuli that you are talking about. Dave is talking at a level of the clusters, maybe a few cons together, but would that show up in the kind of experiment that you are doing?

Shapley: Judging from Dave Williams' picture, there were clusters that were on the order of 10 receptors wide in some places, and we have the resolution to see that. These predictions are so largely different from the physiology, I think it could put a constraint on the mosaic.

Williams: At this point we really don't know whether the physiology is inconsistent or not. Part of the reason is that the optics of the eye can in a physiological experiment do quite a lot to make these stimuli larger that they would otherwise be, and that could obscure subtle aspects in the organisation of receptive fields that might be revealed under other circumstances, namely with interferometry.

Dow: The genetics of the cones have been defined by Nathan's group. If the mosaic is really regular and not random, then the genetics are going to have to specify not only the cones but their positions in the retina.

Zrenner: If there are specificities in the horizontal cell, they may tap to a certain extent one or the other cone system.

Schiller: That claim is still under considerable debate, I don't think that we can take as a fact that there are three different horizontal cell types which differently sample cones. The opposite claim has been laid by Boycott and Wässle in the monkey retina.

Shapley: Something Peter Schiller said I just can't let go. He said that in his experiments they found that the parvocellular cells were compromised at equal luminance. We do not find this. We have done extensive experiments using heterochromatic gratings, and Read and I have done these spot experiments I described. We have also used spot that isolate the centers as well as larger spots that cover the whole receptive field. And in general our number of parvo cells that are compromised at equal luminance is vanishing small. Certainly under the conditions that the experiment that Peter reported on, I am certain that the parvocellular responses to red/green colour exchange were quite strong.

Spillmann: We shouldn't end the discussion without returning to the findings suggesting that there is selective impairment of vision under certain pathological conditions, showing that there maybe selective loss of colour or brightness vision or motion vision, and that seems to indicate that in man there are indeed highly specialized systems subserving these various functions.

Schiller: Just to be the perennial opponent here, I think that this kind of experimentation causes the biggest statistical biases that I ever encountered. There are hundred of thousands of patients with various cortical damage in the world. Most of them show little specific results of the damage, then you trot out one person which has this specific damage, and then you carry that analysis to the conclusion that you did. I think that that is a real fallacy.

Zrenner: I don't think it nice to end the session with such a statement!

Kaplan: With relation to Peter Schiller's idea that the magno system extends the spatial range relative to the parvo system, this comes from the idea that receptive fields of magno cells are larger, but these cells are also much more sensitive. The visual resolution of individual magno and parvo cells is similar, as found by Barry Lee and friends, and by Colin Blakemore and friends and by us. From the vantage point of someone with a microelectrode, I have a problem with that notion that the magno system is really a lower resolution system.

Merigan: To talk of the resolution of the individual cell makes about as much sense as talking about the resolution of one photo receptor. The important point is the matrix properties, and they suggest the P-system is a high resolution one, by more than a factor of three relative to the M-system.

Kaplan: For different tasks, for tasks that require identification rather than detection.

Merigan: Detection is a complicated problem because it is very likely that the M system aliases, and if aliased information can help in a detection task, then the M-system can do it perfectly well.

TEMPORAL CHARACTERISTICS OF COLOUR VISION:

VEP AND PSYCHOPHYSICAL MEASUREMENTS

Adriana Fiorentini, David C. Burr and Concetta M. Morrone

Istituto di Neurofisiologia del CNR
Via S. Zeno 51, Pisa
Italy

Recently there has been a great deal of interest by both physiologists and psychophysicists in the extent to which colour and luminance signals are processed independently by the visual system. The series of experiments reported in this chapter were designed to investigate the temporal properties of colour and luminance processing in human vision. The results of both the psychophysical and electrophysiological studies show that visual mechanisms responsible for processing colour patterns are more sustained than than those that process luminance patterns. The more sustained nature of the chromatic system was revealed by measures of temporal acuity, response latency and the form of the temporal frequency response curve, measured both at and above contrast threshold.

The stimuli for most experiments were plaid patterns modulated sinusoidally both horizontally and vertically, made by summing red and green sinusoidal plaids of equal but opposite contrast (see equation 1 in caption to figure 1). Following the procedure used by Mullen (1985), the ratio of the red-to-total mean-luminance (r in equation 1) could be varied from 0 to 1, where $r=0$ defined a green-black pattern, $r=1$ a red-black pattern and intermediate values a red-green pattern. For each observer there exists a value of r (near 0.5 for colour-normals) where the red and green luminances are exactly matched, making the pattern *iso-luminant*. To produce steady state VEPs, the patterns were reversed sinusoidally in contrast, and EEGs recorded with surface electrodes (OZ, CZ, with earth half way between). They were suitably filtered (1-100 Hz), amplified, and fed into a computer which performed on-line Fourier analysis.

The traces on the right of figure 1 show examples of VEP records at various colour ratios r. At all ratios, the records modulated smoothly at twice the counter-phase frequency (as previously shown by Regan, 1973; Regan and Spekreijse 1974), but modulation amplitude varied with colour ratio. On the left the second-harmonic amplitude and phase are plotted as a function of colour ratio. At this moderate contrast (30%) and high temporal frequency (7.5 Hz), amplitude varied with colour ratio symmetrically about a minimum value at 0.47. This ratio was very similar to the flicker-photometry estimate of iso-luminance, showing consistency between VEP and psychophysical techniques.

Having evaluated the iso-luminance colour-ratio for each observer,

From Pigments to Perception, Edited by A. Valberg and
B.B. Lee, Plenum Press, New York, 1991

that ratio was used for all further studies, both for chromatic and luminance stimuli. The chromatic stimuli were made as before by adding the red and green plaids of *opposite-phase* ($\theta=\pi$ in equation 1) and the luminance stimuli by adding them *in-phase* ($\theta=0$). This procedure ensured that both classes of stimuli had the same mean luminance and mean colour (a deep yellow). For all experiments spatial frequency was kept constant at 1 c/deg (minimizing possible chromatic aberrations: Flitcroft, 1988), while we varied temporal frequency to study the temporal properties of luminance and colour vision.

Figure 2 shows three representative VEP contrast response curves measured at three temporal frequencies (1, 4 and 10 Hz). One major difference between the luminance and chromatic responses was that at 1 Hz a reliable response could not be obtained from luminance stimuli of any contrast (although the stimuli were clearly visible to observers), while the chromatic responses were strong and reliable, and increased almost monotonically with contrast. Another difference is that the chromatic curves tend to be more linear than the luminance curves (on semi-logarithmic plot). The luminance curves have a shallow limb at low contrasts, followed by a steeper section at high contrasts (often separated by a local minimum). Curves of this form are quite characteristic for luminance stimuli, and have been interpreted as reflecting activity of separate visual mechanisms: M-neurones at low contrasts and P-neurones at high contrast (Nakayama and Mackeben, 1982). That the curves for chromatic stimuli (at all temporal frequencies) could be well fit by a single line is consistent with suggestions that only one class of neurones (presumably P-neurones) responds to these stimuli (see also Zeemon *et al*., 1990). But irrespective of speculations about M and P mechanisms, the fact that the curves are not simple translations of each other along the contrast axis is evidence for different neural mechanisms responsive to colour and luminance stimuli.

The arrows under the abscissa of figure 2 A-C show the psychophysical thresholds for detecting the stimuli, evaluated by a two-alternative forced-choice procedure. The psychophysical thresholds are close to those predicted by extrapolating to zero-amplitude the contrast response curves. The close correspondence between psychophysical and electrophysiological estimates of threshold gives us confidence that the VEP measurements reflect functional properties of luminance and colour vision.

Figure 2 D shows how contrast sensitivity for luminance and chromatic counter-phased stimuli varies as a function of temporal frequency. The squares depict psychophysical measurements and the circles estimates from extrapolation of VEP amplitude response curves to zero. The two curves are not simply translations of each other along the contrast axis, but have quite different form. The luminance thresholds are band-pass, peaking at around 6 Hz, while the chromatic thresholds are clearly low-pass, and start to decline at frequencies above 4 Hz. The difference in chromatic and luminance thresholds varies systematically with temporal frequency, from a factor-of-two at 1 Hz, to a factor-of-20 at 16 Hz.

One would expect chromatic contrast thresholds to be higher than luminance thresholds, for physical reasons. As the cone spectral sensitivity curves are broad and largely overlapping, the chromatic stimuli have a lower effective *cone contrast* than do the luminance stimuli. Our calculations suggest that the chromatic stimuli have an effective contrast 3-6 times lower than the luminance stimuli (depending on assumptions of how M and L cone output is combined). This factor does not vary with temporal frequency, and cannot explain the selective depression of chromatic contrast sensitivity at high temporal frequencies (cf Geisler, 1989). However, the ratio of best sensitivity for chromatic and luminance stimuli is within the range predicted by cone-contrast, in agreement with a physical limitation to both luminance and chromatic sensitivity.

The differences in temporal properties was even more evident in the supra-threshold VEP response. Figure 3A plots the VEP amplitude as a function of temporal frequency for luminance stimuli of 25% contrast, and chromatic stimuli of 90% contrast (equating for cone-contrast). At moderate to high temporal frequencies (above 6 Hz) the response to luminance stimuli was stronger than that to chromatic stimuli. However, at low temporal frequencies, the chromatic response was much stronger than the luminance response. At 1 Hz, the second-harmonic (and higher even-harmonic) response to luminance stimuli were barely reliable (although the contrast was 12 times detection threshold), while the amplitude to chromatic stimuli was near maximal at this frequency. Even at 0.25 Hz (one reversal every two seconds), the response to chromatic stimuli was of similar amplitude to the maximal response obtainable (at 8 Hz) from luminance stimuli. The response at very low temporal frequencies indicates that the chromatic responses are far more *sustained* than the luminance response.

Figure 3B plots the phase of the VEP response against temporal frequency on linear axes. Phase decreased linearly with temporal frequency for both luminance and chromatic stimuli, but at different rates. If one assumes that decrease in phase with temporal frequency results from response latency, then the slope of the linear regression of phase against temporal frequency gives an estimate of the latency (Regan, 1966). A weighted regression predicts delays of 88 ms for luminance and 128 ms for chromatic stimuli.

Figure 3 C and D show the response to an abrupt change of contrast of luminance (C) or chromatic (D) pattern (transient evoked potential). As previously observed (Murray et al., 1986), the responses to the two stimuli have different form. The primary component of the luminance VEP is a positive lobe at 92 ms, whereas the primary component for the chromatic response is a negative lobe at 139 ms. The latencies of these peaks and troughs are quite consistent with the estimates of response latency obtained from the phase response curves of the steady-state VEPs, and again point to the sustained nature of the chromatic VEP response.

Ohzawa and Freeman (1988) have recently devised a technique to investigate the response of binocular neurones with evoked potentials. In a dichoptic display, a drifting horizontal grating is displayed to each eye, drifting upwards for one eye and downwards for the other. Alone neither grating generates a VEP, as there is no systematic modulation of contrast. However, combination of the two drifting gratings (either physically or by binocular combination) produces a counter-phased grating, and this stimulus produces a phase-locked VEP. Thus any VEP produced by the dichoptic display must result from excitation of binocularly driven cortical neurones.

We used this technique with both chromatic and luminance stimuli. With both types of stimuli, the monocular display produced no measurable VEP while the binocular display produced strong and reliable VEPs at appropriate temporal frequencies. Figure 4A shows how VEP amplitude varied with temporal frequency for both chromatic stimuli (filled circles) and luminance stimuli (open circles). Again, the response to chromatic stimuli was confined to a lower frequency range than that to luminance stimuli.

We also applied the technique to investigate interactions between luminance and chromatic processes, by presenting a luminance grating to one eye and a chromatic grating to the other. This stimulus also produced a measurable VEP, but only over a very confined range of temporal frequencies. This suggests that there exist neurones that respond to both chromatic and luminance stimuli.

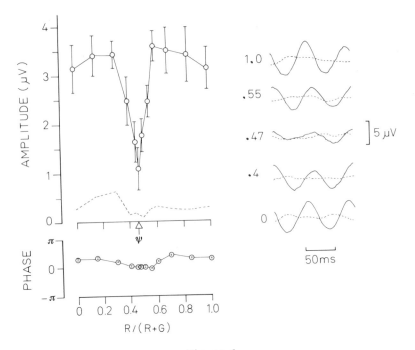

Figure 1

Examples of VEP recordings in response to coloured plaid patterns caused to
reverse in contrast over time. On the right are five examples of VEP
traces, for different colour ratios (r of equation 1 below). The records
were obtained by averaging the VEP response in synchrony with the rate of
contrast reversal (f_t in equation 1). The dotted lines show the response
averaged at 1.1 times this frequency, giving an indication of the noise
level. It is apparent that response amplitude varies with colour ratio,
being minimal at $r=0.47$. The effect of colour ratio is more apparent in the
curves at left, that plot second-harmonic amplitude and phase as a function
of colour ratio r. The dotted line shows the amplitude of asynchronous
noise, and the error bars the standard error of the amplitude and phase
distributions. At this contrast (30%) the minimum was quite clear, at 0.47.
There was also a slight but systematic phase change around the point of
iso-luminance.

The pattern was generated on a Barco oscilloscope by modulating the
luminance of the red and green phosphors (L_R and L_G) over space and time:

$$L_R(x,y,t)=r.L_o\{1+\tfrac{1}{2}m.cos(2\pi f_t t).[cos(2\pi f_s x)+cos(2\pi f_s y)]\}$$

$$L_G(x,y,t)=(1-r).L_o\{1+\tfrac{1}{2}m.cos(2\pi f_t t+\theta).[cos(2\pi f_s x)+cos(2\pi f_s y)]\}$$

$$(1)$$

r is the ratio of red-to-total mean-luminance, L_o the total mean-luminance,
m contrast, f_t temporal frequency and f_s spatial frequency. θ determines
the phase in which the red and green plaids are added together. For this
experiment $\theta=\pi$, so the red and green plaids were added out-of-phase to
produce red-green patterns. Spatial frequency was 1 c/deg, temporal
frequency 7.5 Hz and contrast 0.3. The pattern was viewed through an orange
filter (Kodak wratten number 16), to filter wavelengths less than 500 nm,
ensuring that the spectral range of the stimuli was within the Rayleigh
region.

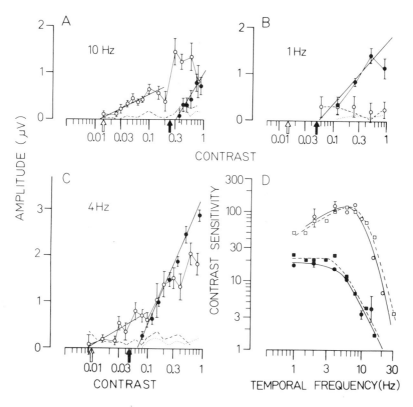

Figure 2

A B and C Examples of contrast response curves to chromatic and luminance
stimuli contrast reversed at 10, 4 and 1 Hz. The stimuli were given by
equation 1 of figure 1, with r=0.47 (judged as isoluminant by both flicker-
photometry and minimum VEP amplitude), and f_s 1 c/deg. Chromatic stimuli
were made by adding the red and green plaids of opposite-phase ($\theta=\pi$), and
luminance stimuli by adding them in-phase ($\theta=0$). In this and all following
graphs, open symbols represent results from luminance stimuli and closed
symbols chromatic stimuli. The curves for chromatic gratings tended to be
linear (on semi-logarithmic plot), and could be well fit by a single
regression line. The luminance curves tended to comprise to sections of
different slope, so the linear fit was applied only to the low-contrast
limb of the curve. The arrows under the abscissa indicate psychophysical
thresholds, measured by a two-alternate forced-choice technique. In all
cases (except the luminance curve at 1 Hz, where there was virtually no
response), extrapolation of the VEP curve to zero amplitude predicted
contrast thresholds close to those obtained by psychophysical means.

D Contrast thresholds as a function of temporal frequency. The circles
indicate estimates obtained form extrapolating VEP contrast response curves
(above), and squares psychophysical estimates (two-alternate forced-
choice). In all cases the two estimates correspond well, giving us
confidence in the VEP technique.

143

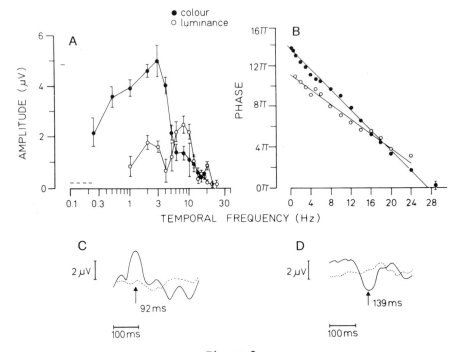

Figure 3

A Second-harmonic amplitude as a function of temporal frequency for luminance stimuli of 25% contrast (open-symbols) and chromatic stimuli of 90% contrast (closed-symbols). The contrasts were chosen to equate the effective *cone-contrasts* of the stimuli. However, we also measured responses for luminance stimuli of 90% contrast, and found results substantially similar to those shown here.

B Second-harmonic phase as a function of temporal frequency (on linear scale) for chromatic and luminance stimuli. The slope of the curves provide an estimate of response latency, 88 ms for the luminance stimuli and 128 ms for the chromatic stimuli. Measurements were also made for luminance stimuli of 90% contrast, yielding a latency estimate of 94 ms, indicating that precise choice of contrast is not crucial.

C and D Transient responses to an abrupt contrast reversal of a luminance pattern of 25% contrast (**C**) or chromatic pattern of 90% contrast (**D**). The primary response to the luminance stimulus was a positive peak around 92 ms, while the chromatic stimulus produced a negative trough at 139 ms.

144

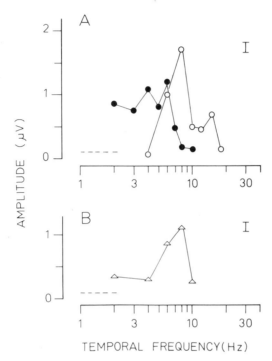

Figure 4

VEP response to dichoptic presentation. Each eye saw a horizontal grating of 1 c/deg. The direction of drift was upwards for the left eye, and downwards for the right. To compensate for the effects of cone contrast, luminance gratings were displayed at 25% contrast and chromatic gratings at 90% contrast. The open symbols of figure A show results for luminance gratings, and the closed symbols for chromatic gratings. For figure B, a luminance grating was presented to the left eye and a chromatic grating to the right eye.

In another series of experiments, we have applied psychophysical techniques to investigate the temporal properties and response latency of luminance and chromatic vision. We monitored *visual summation*, by measuring contrast sensitivity for detecting a pair of brief stimuli, separated in time by variable intervals. The manner in which sensitivity varies with presentation delay gives information about the temporal response properties of the visual mechanisms detecting the stimulus configuration (see for example, Ikeda 1965; Roufs and Blommaert, 1981).

Figure 5 A and B plots sensitivity for detecting the double flash (relative to that for a single flash) as a function of delay of the second presentation. In figure A the two stimuli had identical contrast (of the same "contrast-sign") while in figure B they were of equal but opposite contrast. For the same-contrast condition, there was an initial period of strong positive summation, occurring at different latencies for the luminance and chromatic stimuli. The difference in time-to-asymptote was about 40 ms, comparable with that observed with the VEPs. The opposite-contrast presentations have a different summation course, an initial stage of negative summation, followed by positive summation. Again there were differences in time course for the luminance and chromatic stimuli, with the time-to-peak differing by about 40 ms. Both these estimates of response latency agree reasonably well with the VEP latency estimates.

Finally, we asked what functional consequences may arise from the difference in temporal properties of the luminance and chromatic systems. An obvious candidate is motion perception, as all current models suggest that motion perception requires a band-pass temporal filter (e.g. Van Santen and Sperling, 1985; Burr, Ross and Morrone, 1986). We measured contrast thresholds for discriminating the direction of motion of a drifting grating (using a two-alternative forced choice procedure) and compared these thresholds with simple detection thresholds (see also Cavanagh, this volume). Figure 6 A and B reports the two classes of thresholds for both chromatic and luminance stimuli, as a function of temporal frequency. For luminance stimuli (A), discrimination thresholds were similar to detection thresholds at all temporal frequencies, showing that provided the pattern was visible, its direction of motion could be discerned. This was also true for chromatic stimuli of high temporal frequency. However, at frequencies of 2 Hz and below, the detection and discrimination curves diverge: the discrimination curve becomes band-pass, while the detection curve remains low-pass (as for detection of counter-phased plaids: figure 2). The extra chromatic sensitivity at low temporal frequencies clearly contributes nothing towards motion perception. These results agree with previous studies showing impairment of motion perception at iso-luminance (Ramachandran and Gregory, 1978; Cavanagh et al., 1984) and may explain why under some conditions (high temporal frequencies) colour may be an effective cue for movement (Gorea, 1989; Cavanagh, this volume).

All the experiments outlined above, both electro-physiological and psychophysical, point to major differences in the temporal properties of the mechanisms that process luminance and chromatic information. The chromatic mechanisms are more sustained, shown by the low-pass nature of the temporal contrast sensitivity curve (figure 2), by the strong VEP response at very low temporal frequencies (down to 0.25 Hz: figures 3 and 4), and by the longer latency predicted by both transient and steady-state VEPs (figure 3) and by two-shot summation studies (figure 5). The relatively poor performance of the chromatic system at motion discrimination at low temporal frequencies (figure 6) also reflects the lack of temporal inhibition in chromatic sensitive mechanisms.

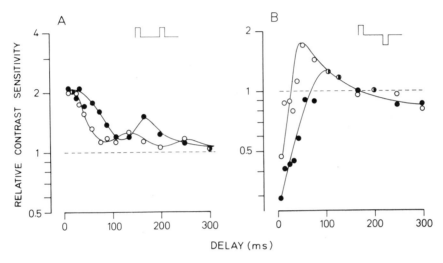

Figure 5

Summation between two pulses of the same contrast (A) and opposite contrast
(B), as a function of onset latency of the two presentations. The stimuli
were luminance (open symbols) or chromatic (closed symbols) gratings,
presented for 8 ms. The pattern of summation with time gives an indication
of the temporal response of luminance and colour mechanisms.

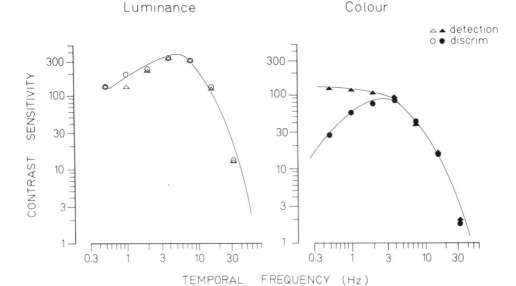

Figure 6

Contrast sensitivity for detecting drifting sinusoidal gratings (triangles) and for discriminating the direction of drift (circles). The gratings were presented within a Gaussian envelope of stand deviation 230 ms. Both discrimination and detection thresholds were measured by two-alternative forced-choice. For luminance gratings (left) the two types of thresholds were similar at all temporal frequencies. For chromatic gratings (right), however, they diverge at low temporal frequencies.

The differences in temporal properties are consistent with recent suggestions that colour information is processed primarily by the parvocellular stream, and luminance information by both parvo and magno streams (see chapters by Merigan, Rodieck, Schiller and Shapley as well as several discussion sections for a critical treatment of these claims). The fact that chromatic contrast response curves tend to have a single slope, while the luminance response curves have two distinct limbs also supports this suggestion. However, it is strange that there is no VEP response whatsoever to luminance gratings of temporal frequency below 1 Hz. If both magno- and parvo-neurones respond to luminance stimuli of 1 Hz, why is it not possible to register a response from the sustained parvo-neurones at this frequency, as it is with chromatic stimuli? Certainly some neurones respond to the stimulus, as it is clearly visible. To account for this result one would have to begin to speculate that the response of parvo-neurones to colour patterns is more sustained than that to luminance patterns, or for some reason the luminance response does not appear in the VEP at low temporal frequencies. While these possibilities are quite plausible, introducing caveats of this nature begin to erode the usefulness of the P-M classification for psychophysical data.

To conclude, the experiments presented in this chapter implicate separate neural mechanisms for colour and luminance processing, both at and above detection threshold, but whether these mechanisms can be usefully divided along the anatomically defined lines of M- and P-neurones remains to be proven.

REFERENCES

Burr, D.C., Ross, J. & Morrone, M.C. (1986) Seeing objects in motion Proc. Roy. Soc. (Lond) **B227** 249-265.

Cavanagh, P., Tyler, C.W. and Favreau, O.E. (1984) Perceived velocity of moving chromatic gratings J. opt. Soc. Amer. **1A** 893-899.

Flitcroft, D.I. (1989) The interactions between chromatic aberration, defocus and stimulus chromaticity: implications for visual physiology and colorimetry. Vision Res. **29** 349-360.

Geisler, W.S. (1989) Sequential ideal-observer analysis of visual discriminations. Psychological Review **96** 267-314.

Gorea, A., Papathomas, T.V. and Julesz, B. (1989) Colour against luminance in motion perception Perception **18** 536.

Ikeda, M. (1965) Temporal summation of positive and negative flashes in the visual system J. opt. Soc. Am. **55** 1527-1534.

Mullen, K.T. (1985) The contrast sensitivity of human colour vision to red-green and blue-yellow chromatic gratings J.Physiol.(Lond.) **359** 381-400.

Murray, I.J., Parry, N.R.A., Carden, D. and Kulikowski, J.J. (1986) Human visual evoked potentials to chromatic and achromatic gratings Clinical Vision Sci. **1** 231-244.

Nakayama, K. & Mackeben, M. (1982) Steady state visual evoked potentials in the alert primate. Vision Res. **22** 1261-1271.

Ohzawa, I. & Freeman, R.D. (1988) Cyclopean visual evoked potentials: a new test of binocular vision. Vision Res. **28** 1167-1170.

Ramachandran, V.S. and Gregory, R.L. (1978) Does colour provide an input to human motion perception? Nature **275** 55-56.

Regan, D. (1966) Some characteristics of average steady-state and transient responses evoked by modulated light EEG & Clin. Neurophysiol. **20** 238-248.

Regan, D. (1973) Evoked potentials specific to spatial patterns of luminance and colour. Vision Res. **13** 2381-2402.

Regan, D & Spekreijse, H. (1974) Evoked potential indications of colour blindness. Vision Res. **14** 89-95.

Roufs, J.A.J. and Blommaert, F.J.J. (1981) Temporal impulse and step responses of the human eye obtained psychophysically by means of a drift-correcting perturbation technique Vision Res. **21** 1203-1221.

Santen, J.P.H. van & Sperling, G (1984) Temporal covariance model of human motion perception. J. Opt. Soc. Am. **A1** 451-473.

Zemon, V., Gordon, J., Greenstein, V., Holopigian, K. and Seiple, W. (1990) Properties of chromatic and luminance channels measured electrophysiologically in humans Invest.Ophthal.Vis.Sci. **31** 263.

THE CONTRIBUTION OF COLOUR TO MOTION

Patrick Cavanagh

Department of Psychology
Harvard University
Cambridge, MA 02138
U. S. A.

INTRODUCTION

Anstis (1970) and Ramachandran and Gregory (1978) showed that the motion normally visible in random dot kinematograms could not be seen when the dots were presented in equiluminous colours. Based on this evidence, Ramachandran and Gregory suggested that colour and motion analyses were functionally independent and that the motion pathway responded only to luminance information. However, more recent studies (Cavanagh, Tyler & Favreau, 1984; Cavanagh, Boeglin & Favreau, 1985; Cavanagh & Favreau, 1985; Derrington & Badcock, 1985; Gorea & Pappathomas, 1989; Mullen & Baker, 1985) have shown that this may not be the case. There is a motion response, although somewhat degraded, to equiluminous coloured stimuli. In this chapter, I shall review several experiments that have examined the contribution of colour to motion in an attempt to identify whether the contribution is mediated through the magnocellular (M) or parvocellular (P) stream.

First, despite the dramatic examples of loss in motion perception at equiluminance (Moreland, 1982; Cavanagh, Tyler & Favreau, 1984), the visual system actually shows a high degree of sensitivity to the motion of chromatic stimuli when the stimulus strength is expressed in terms of cone contrast (Stromeyer, Eskew, & Kronhauer, 1990). A major factor in the losses at equiluminance may be due to the restricted contrast range available for chromatic stimuli.

Second, in collaboration with Stuart Anstis, I have measured the relative contributions of colour and luminance to motion (Cavanagh & Anstis, 1990). These results demonstrate that the contribution is based on opponent-colour mechanisms.

Third, we also tested the phase lag in the relative contributions of red and green to motion at equiluminance (Cavanagh & Anstis, 1990). These phase lags were similar to those for P units in the retinal ganglia and very different from those for M units.

Finally, motion can be seen for drifting patterns defined by attributes other than colour which do not stimulate luminance-based motion detectors (Cavanagh & Mather, 1989). In particular, stereo-defined (random dot stereograms) and texture-defined stimuli are detected only by the parvocellular stream (Schiller, Logothetis, & Charles, 1990), so that the perception of the motion for these stimuli must rely on signals carried by the parvocellular stream. Since the parvocellular stream can contribute to the perception of motion for these stimuli, it would be reasonable to assume that it also contributes to motion in the case of colour stimuli. In fact, it would be unreasonable to assume the opposite.

PATHWAYS

The luminance pathway takes its input from the sum of the long- and medium-wavelength sensitive cones (R- and G-cones, respectively) although some psychophysical studies indicate that the short-wavelength senstive cones (B-cones) may also contribute to some extent (Drum, 1983; Lee & Stromeyer, 1989; Stockman, MacLeod, & DePriest, 1987). The chromatic signals arise from the differences between cone signals: R-G and B-(R+G) for the red/green and blue/yellow opponent-colour pathways, respectively. An equiluminous stimulus is one that varies in colour but not in luminance. The purpose of such a stimulus is to provide information to the chromatic pathway but not the luminance pathway in order to test the capacities of the chromatic pathways in isolation.

The notion of equiluminance presumes one luminance pathway that has a single null, that is, one relative luminance between any two colours for which there is no response of the luminance pathway. In truth, however, there may many luminance pathways. Image information from separate spatial frequency bands pass through separate units in the visual system — each of these could be considered a separate luminance pathway (low-pass, high-pass, etc.) with potentially a different equiluminance point. Many physiological studies have revealed a variation in the equiluminance points of M units that project to the directionally selective cells in the cortex. In particular, Shapley and Kaplan (1989) and Lee, Martin, and Valberg (1988) have shown that individual M units do show a null activity point at a particular luminance ratio between the two colours of the stimulus and that this null ratio varies somewhat from unit to unit. Given that many pathways may be contributing to a final luminance stream, it is remarkable that there is a fairly well defined equiluminance point around which performance is degraded.

There have been many attempts to link the luminance and chromatic pathways identified psychophysically to the magnocellular and parvocellular streams (DeYoe & van Essen, 1988; Livingstone & Hubel, 1987; Maunsell & Newsome, 1987) of the primate visual system. The units in the magnocellular stream have little colour sensitivity and respond best to low spatial and high temporal frequencies whereas those in the parvocellular stream generally have colour-opponent responses and prefer high spatial and low temporal frequencies (see Schiller, Logothetis, & Charles, 1990; and Shapley, 1990, for reviews). However, there is no simple relationship between these properties and those of the luminance and chromatic pathways. For example, it has been argued (De Valois & De Valois, 1975; Ingling, & Martinez-Uriegas, 1985; Schiller et al, 1990) that both magnocellular and parvocellular streams are involved in carrying luminance information: The magnocellular stream is principally non-opponent but the parvocellular stream, although carrying colour-opponent information for low spatial and low temporal frequencies, also carries non-opponent (luminance) information at high spatial and temporal frequencies. Moreover, units in the magnocellular stream exhibit colour-opponency for large stimuli (Type IV units, Wiesel & Hubel, 1966). Several studies have revealed residual, opponent-colour responses in the magnocellular stream both in the retina (Gouras & Eggers, 1982) and in the magnocellular layers of the LGN (Krueger, 1979; Schiller & Colby, 1983; Derrington, Krauskopf & Lennie, 1984). No physiologically distinct structures have yet been identified whose properties correspond in a straightforward way to those of the luminance and chromatic pathways (see for example, Lennie, Krauskopf, & Sclar, 1990).

Schiller et al (1990) have directly tested the role of the magnocellular and parvocellular streams of macaque monkeys in various visual tasks. They made small lesions in the lateral geniculate in the magnoclelluar and parvocellular layers destroying cells that responded to different spatial areas in the two systems. They then presented tests to areas subserved by both magnocellular and parvocellular units, by only magnocellular or by only parvocellular. Two of their findings are of particular interest for this review. First, within areas subserved only by parvocellular units, motion detection was very poor if the temporal frequency was greater than 6 Hz but was little affected at lower rates. This indicates that the parvocellular stream can mediate motion responses although only for lower temporal frequencies. Second, areas subserved only by magnocellular units were unable to support the perception of either texture or random-dot stereograms. This will be important when we consider the substrate responsible for the perception of motion of stimuli defined by texture and by stereo.

CONTRAST THRESHOLD FOR MOTION

In collaboration with Stuart Anstis (Cavanagh & Anstis, 1990) I measured thresholds for one normal observer (colour-deficient observers were also tested but their results are not discussed here) on two threshold tasks. We collected the contrast thresholds for detection and for direction discrimination with chromatic stimuli (red/green), and luminance stimuli (yellow/black).

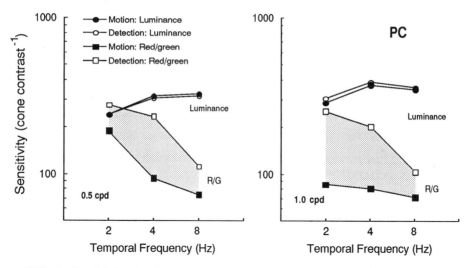

FIG. 1. Sensitivity (reciprocal of threshold cone contrast) for detection and direction discrimination as a function of spatial and temporal frequency. (Adapted from Cavanagh & Anstis, 1990.)

As has been reported previously for comparable stimuli, there is no difference between detection and motion thresholds for the achromatic stimuli and sensitivity increases with temporal frequency between 2.0 and 8.0 Hz (Kelly, 1979). The red/green detection sensitivity drops with increasing temporal frequency, as expected (Kelly, 1983), and the motion sensitivity drops with temporal frequency as well suggesting that the motion threshold is based on chromatic mechanisms. The sensitivity for direction discrimination is lower than that for detection and between these two thresholds, in the shaded area on the graphs, the colour bars could be seen but did not appear to move.

When expressed in terms of cone contrasts, the sensitivity to chromatic stimuli is quite high, comparable to that for luminance stimuli at the lowest spatial and temporal frequencies. Stromeyer et al (1990) measured even higher sensitivity to chromatic stimuli (as much as four times higher than to luminance stimuli) in a direction discrimination experiment. The difference between their results and those of our one observer may simply be that our observer (myself) falls on the low end of normal colour vision. Among the four normal observers in the next experiment to be described, my results were next to lowest in strength of contribution of colour to motion (Cavanagh & Anstis, 1990).

How can we reconcile this high sensitivity to chromatic stimuli with the many reports of degraded motion response at equiluminance? The explanation lies in the contrast available in chromatic and luminance stimuli. Luminance stimuli can modulate each cone class by as much as 100%, whereas the maximum differential modulation of the R- and G-cone classes attainable on a colour monitor is between 15 and 25%. In other words, the biggest loss for chromatic stimuli may be a stimulus factor and not a processing factor: luminance modulation drives the cones more effectively than chromatic modulation.

Our data (Cavanagh & Anstis, 1990) and the data of Stromeyer et al (1990) demonstrate that the visual system is in fact quite sensitive to chromatic stimuli but these data do not identify definitively the pathway that mediates the response. In the next experiment that I shall describe, the characteristics of the response help to identify the source.

OPPOSING MOTION

The contrast of an unknown grating can be measured by varying the contrast of an otherwise identical grating moving in the opposite direction. The direction of perceived motion of the combined gratings is determined by the grating with the higher contrast. When the two gratings have equal contrast, a motion null—counterphase flicker—is obtained. This same technique can be used to compare the relative contributions of luminance and colour to motion.

To do so, a red/green grating drifting in one direction was superimposed on a light yellow/dark yellow grating drifting in the opposite direction (Cavanagh & Anstis, 1986). Four normal and nine colour-deficient observers were tested. Observers nulled the motion of a drifting luminance grating of fixed contrast by varying the luminance contrast of a colour grating (the contrast between the luminances of the red and green components of the colour grating) that drifted in the opposite direction. At motion null, the total effective contrast of the colour grating is equal to that of the luminance grating whose motion it has nulled. In Fig. 2, the total effective contrast of the colour grating is shown as a function of its luminance contrast. Results for two observers — a normal observer, SA, and a deutan observer, BA, — are given for tests at 0.5 cpd and 2 Hz. The V-shaped dotted lines of unit slope rising from the origin indicate the effective contrast that the colour grating would have if colour made no contribution to the perception of motion. In this case, the effective contrast of the colour grating would be equal to its luminance contrast. The lines plotted through the data points indicate the observed effective contrasts in these two examples. The shift of these lines from the dotted lines indicates an increase in the effective contrast and the amount of the shift, shown as a vertical arrow, is the contribution of the colour: We labeled this the equivalent luminance contrast of the colour in the colour grating. Note that the equivalent luminance contrast is substantial, more than 10%, for the normal observer (note also that there is only one data point per line in this example!), but very small, less than 1%, for the colour-deficient observer.

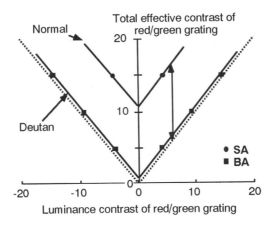

FIG. 2. Total effective contrast as a function of the luminance contrast between the red and green in the colour grating. Data from two observers, SA and BA, for 2 Hz, 0.5 cpd, red/green stimuli. The equivalent contrast of the colour in the grating is the difference (shown as a vertical arrow) between the luminance contrast of the colour grating (the dotted V rising from the origin) and the total effective contrast. (Adapted from Cavanagh & Anstis, 1990.)

What should we expect to see if the response to the colour grating is a residual response of a luminance pathway, due to the variation of equiluminance points among the units in the pathway? The left panel of Fig. 3 shows the response of a single unit in a luminance pathway to a drifting red/green grating as a function of its luminance contrast. The response function is given by the grating's absolute luminance contrast. At the centre of the horizontal scale, the red and green have equal luminance as measured by a photometer and the unit "sees" a uniform field with no contrast. At the right end of the horizontal scale, the grating is bright red and dark green while at the other end it is is bright green and dark red. In both cases, however, the unit merely detects an identical light and dark grating.

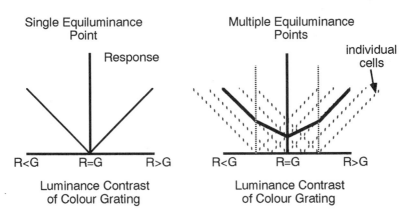

FIG. 3. The response of luminance-based units to a colour grating as a function of the luminance contrast between the two colours in the grating. On the left, the response for a single unit is shown and the response function follows the absolute value of the luminance contrast. On the right, different units have different equiluminance points. The total response is shown for graphical convenience as the average of the individual responses and indicated by the heavy curve. The curve has a minimum but the response is greater than zero at the minimum value.

We assumed that the contrasts sensed by individual units are summed to produce the net contrast. If all units had the same null point, there would be a single, true response null in the luminance pathway and the overall function would look like that of the single unit shown on the left in Fig. 3. On the other hand, in righthand panel of Fig. 3, we have the more likely situation of variable null points for individual units and overlapping functions. If we take the response to be the sum of the activity of the individual units (thick, curved line in right panel of Fig. 3 shows the average value for graphical convenience), we see that there is no longer a true null. The mean function dips to a minimum at photometric equiluminance but this minimum response is not zero. The shape of the actual function depends on the response characteristic of the individual units (shown as linear in Fig. 3) and the distribution of equiluminance points. We do not know either of these factors exactly, but we do know that M units have a moderate degree of scatter (probably not more than ±10% around the population average, Shapley & Kaplan, 1989; Lee et al, 1988). This allows us to predict that the V-shaped functions should be rounded off at the bottom within ±10% of equiluminance but should follow the V-shaped curve of the luminance contrast outside that range. That is, the net effectiveness of the colour grating in a *luminance pathway* only deviates from its luminance contrast within the range of scatter where the responses from some units are rising while others are falling.

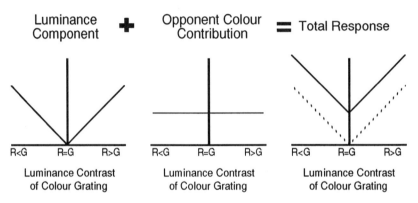

FIG. 4. The total response to a colour grating as the sum of luminance and opponent-colour contributions. Since the chromatic contrast and, therefore, the opponent-colour response, is fairly constant over the moderate range of luminance contrasts of the colour grating that were tested, the total effective contrast will be an elevated, V-shaped curve.

What would the total effective contrast look like if colour made a contribution to motion through an opponent-colour pathway and not through residual activation of a luminance pathway (i.e. if there were no interunit variation in equiluminance points)? We can assume that a contribution from an opponent-colour pathway would be a function of the chromatic contrast of the stimulus and this is fairly independent of its luminance contrast, at least over the range we looked at (maximum of ±30%). Assuming a simple linear model in which the colour contribution sums with the luminance contrast to produce the total effective contrast, the function would just be a V-shaped curve that is raised everywhere by the same amount (Fig. 4).

The data of Figure 2 are sufficient to state that, as expected, the colour-deficient observer gets no contribution from colour so that the response is purely determined by a luminance pathway. However, there are insufficient data in Figure 2 to draw any conclusions about the normal observer. In order to see the pattern of response for a normal observer across a wider range, we extended the luminance contrasts tested to ±25% and included a test at the equiluminance point. We used only one spatial and one temporal frequency (1 cpd and 8 Hz). Fig. 5 shows the total effective contrast of the red/green grating as a function of its luminance contrast. The data again appear to be V-shaped over this larger range.

There appears to be a fairly constant contribution of colour to motion at all the physical contrast values measured producing an elevated, V-shape curve that does not rejoin the dotted luminance contrast function. This behavior is more like that of a chromatic contribution than that of a luminance pathway with scattered equiluminance points.

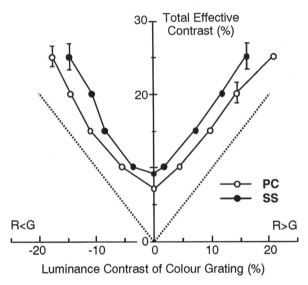

FIG. 5. Total effective contrast of a red/green grating at 1.0 cpd and 8 Hz as a function of the luminance contrast of the grating for two observers. The dotted lines show the total effective contrast the colour grating would have if its contribution to motion were determined solely by its luminance contrast. Vertical bars show standard errors (±1 S. E.) where they are larger than the data symbols. (Adapted from Cavanagh & Anstis, 1990.)

Could the contribution have resulted from some luminance artifact due to display alignment or calibration or to the optics of the eye? Any luminance artifact in the colour stimulus will add directly to its effective contrast, increasing it uniformly within a fairly wide range of luminance contrasts around equiluminance — a pattern very similar to that which we had measured. However, the luminance artifact sets a lowest possible value for the contribution of colour that will be present in all the measurements and, for colour-deficient observers who have little or no colour-opponent response, it must be the main or only contributor to the measured equivalent contrast of the colour. Figure 6 shows the equivalent contrast of a red/green grating (the difference between its total effective contrast and its luminance contrast) for normal and colour-deficient observers as a function of spatial frequency (averaged over the two temporal frequencies tested, 0.5 and 1.0 Hz). Note that the colour-defective observers had little or no equivalent contrast for the colour stimuli at 0.5 and 1.0 cpd. We can conclude that the stimuli at these low spatial frequencies produce luminance artifacts of less than 1% and that the readings for the normal viewers are true readings of the contribution of colour to motion.

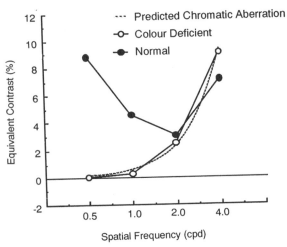

FIG. 6. Equivalent contrast of colour for red/green gratings as a function of spatial frequency for normal and colour-deficient observers. The curved line is the predicted luminance artifact produced by chromatic aberration.

The theoretical value of the luminance contrast generated by chromatic aberration increases with the square of the spatial frequency (Cavanagh & Anstis, 1990) and the data of colour-deficient observers for spatial frequencies covering the range of 0.5 to 4 cpd (averaged over the three temporal frequencies) showed this squared increase. The normal observers show a U-shaped curve resulting from two factors: 1) the colour contribution to motion that decreases with spatial frequency; and 2) the chromatic aberration artifact that increases with spatial frequency.

To summarize, the data showed a motion response to colour that did not arise from display or optical artifacts and was not due to interunit variability in equiluminance points in a luminance pathway. The evidence supported an opponent-colour source for the contribution of colour to motion.

PHASE LAG

Although the magnocellular pathway is characterized as broad-band or non-opponent (see Schiller et al, 1990), it is capable of colour-opponent responses. In particular, Type IV retinal ganglion cells (Wiesel & Hubel, 1966) show colour-opponency between centre and surround. Could this colour-opponency in the magnocellular stream account for the results described in previous sections? I shall use the phase characteristics of responses to chromatic stimuli to examine this point.

Cushman and Levinson (1983), deLange (1958), von Grünau (1977), Lindsey, Pokorny, and Smith (1986) and Swanson, Pokorny and Smith (1987) have reported that the relative phase between red and green producing minimum flicker sensitivity can deviate from the expected 180° by as much as 180°. We measured phase lags in the motion response using quadrature motion techniques (Cavanagh & Anstis, 1990; Cavanagh, Anstis & MacLeod, 1987; Shadlen & Carney, 1986) where a counterphasing luminance stimulus is positioned as a lure so that it will generate a moving stimulus when combined with the luminance artifact produced by the phase lag.

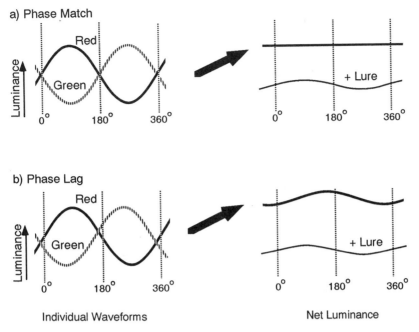

a) Phase Match

b) Phase Lag

Individual Waveforms

Net Luminance

FIG. 7. a) The sum of red and green sine-waves, 180° out of phase, produces no net luminance modulation if the amplitudes of the two sine-waves are equal and the phase offset is exactly 180°. b) The sum of red and green sine-waves having the same amplitude produces a net luminance modulation if their phase difference is not exactly 180°. The luminance waveform is a sine-wave with the same frequency as the colour waveforms and its phase is the mean of the phases of the two colour waveforms, about 90° as shown here. A luminance lure combined in quadrature phase will produce motion only when the luminance artifact is present.

A stationary red/green stimulus is set in counterphase temporal modulation and a stationary luminance lure is introduced having the same spatial and temporal frequencies but 90° out of phase with the chromatic stimulus in space *only* (Fig. 7). Since it is not in quadrature phase with the red/green stimulus, it does not produce any motion. However, any phase lag between the red and green waveforms in the red/green stimulus will produce a luminance component at 90° from the peaks and troughs of the red and green waveforms (Fig. 7). This artifact will then be in quadrature phase with the luminance lure that we have introduced in the stimulus and the combination of the two will produce motion. If there is a phase lag between the two colours in the neural response at some point, it can be canceled by introducing the opposite phase lag in the stimulus. When it has been exactly canceled, no motion will be visible and when it has been overcompensated, the motion will reverse direction. The motion reversal point can therefore be used to accurately measure phase lag in the pathways responding to counterphasing gratings.

The measured phase lags (Fig. 8) were much smaller than those reported for minimum flicker settings by Cushman and Levinson (1983), deLange (1958), von Grünau (1977), Lindsey et al (1986) and Swanson et al (1987). This may be due to the direct measurement technique that involves only the response of the motion pathway. There may be additional phase lags in a form pathway that contribute to flicker judgements especially at low temporal frequencies.

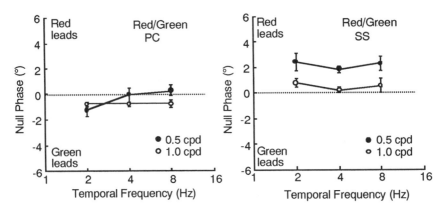

FIG. 8. Temporal phase lag for red/green stimuli as a function of spatial and temporal frequency for two observers. Vertical bars show standard errors.

The lags are comparable to those for parvocellular stream units reported by Smith, Lee, Pokorny, Martin, and Valberg (1989). M-stream units generate much larger phase lags at low temporal frequencies (Smith et al, 1989). These data argue strongly for the participation of the parvocellular stream in the contribution of colour to motion.

MOTION PERCEPTION FOR OTHER EQUILUMINOUS STIMULI

The typical motion detector in area V1 of visual cortex responds principally to drifting luminance contours. However, there are several motion phenomena that cannot be explained by these motion detectors. Ramachandran, Vidyasagar and Rao (1973) reported the perception of apparent motion in a two-frame display consisting of two completely uncorrelated dot patterns. The perception of motion in the absence of correlated luminance information has also been reported for stimuli defined by relative motion (Petersik et al., 1978; Anstis, 1980; Prazdny, 1986a, 1986b, 1987), random-dot stereograms (Julesz, 1971; Prazdny, 1986a, 1986b, Papathomas, Gorea, Julesz & Chang, 1988), flicker (Lelkens & Koenderinck, 1984; Mather, Cavanagh & Anstis, 1985; Prazdny, 1986a, 1986b, 1987; Chubb & Sperling, 1988) and texture (Cavanagh & Mather, 1989; Pantle, 1978; Turano & Pantle, 1989).

In collaboration with Martin Arguin and Michael von Grünau, I have replicated and extended these observations (Cavanagh, Arguin & von Grünau, 1989). The stimulus consisted of two disks which alternated at 2.0 Hz. Each disk could be defined by a difference in luminance, colour, binocular disparity, texture, or motion with respect to the random dot background. The observer's task was to decrease the separation between the two disks until motion was just visible. This separation was taken as an indication of motion strength. When the two disks that alternated were defined by the same attribute, the motion strength was comparable for all attributes, varying by at most a factor of 2. Motion could also be seen between disks defined by any two different stimulus attributes and the motion strength showed no systematic variation as a function of the attributes involved.

According to Schiller et al (1990), at least two of these stimulus types — texture and random dot stereograms — are detected entirely by the parvocellular stream (a lesion of the parvocellular layers drastically reduces texture and random dot stereo detection while a magnocellular lesion does not affect either). Since moving texture or random dot stereo borders produce impressions of motion, the parvocellular stream must be capable of mediating the motion responses for at least these stimuli. It seems illogical to propose that in the case of colour stimuli, but not texture or stereo, the parvocellular stream somehow becomes incapable of mediating motion responses.

CONCLUSIONS

The data reviewed here indicate that the visual system is quite sensitive to motion of chromatic stimuli. On the other hand, the restricted contrast range available in physical chromatic stimuli reduces their effective strength compared to luminance stimuli. The results of the opposing motion experiment showed that the contribution of colour is mediated by opponent-colour mechanisms and the phase lag data linked the mechanisms in question to the parvocellular stream. The evidence of motion perception for texture and random dot stereogram contours — stimuli detected by the parvocellular stream — also supported the motion capabilities of the parvocellular stream. Since the directionally selective units of MT have been shown to respond to drifting equiluminous stimuli defined by colour or texture (Albright, 1987; Charles & Logothetis, 1989; Saito, Tanaka Isono, Yasuda, Mikami, 1989) these responses may be mediated by signals projecting to MT from structures in the parvocellular stream such as area V4.

ACKNOWLEDGEMENTS

The preparation of this chapter was supported by grant A8606 from the Natural Science and Engineering Research Council of Canada and by a Harvard University Seed Grant.

REFERENCES

Albright, T. D. (1987). Isoluminant motion processing in macaque visual area MT. *Society for Neurosciences Abstracts* **13**, 1626.

Anstis, S. M. (1970). Phi movement as a subtraction process. *Vision Research* **10**, 1411-1430.

Anstis, S. M. (1980) The perception of apparent movement. *Philosophical Transactions of the Royal Society, London* **B290**, 153-168.

Cavanagh, P. & Anstis, S.M. (1990)The contribution of color to motion in normal and color-deficient observers. *Vision Research,* in review.

Cavanagh, P. & Favreau, O. E. (1985) Color and luminance share a common motion pathway. *Vision Research* **25**, 1595-1601.

Cavanagh, P., Anstis, S.M., & MacLeod, D.I.A. (1987) Equiluminance: Spatial and temporal factors and the contribution of blue-sensitive cones. *Journal of the Optical Society of America A* **4**, 1428-1438.

Cavanagh, P., Arguin, M. & von Grünau, M., (1989) Interattribute apparent motion. *Vision Research,* **29**, 1197-1204.

Cavanagh, P., Boeglin, J., & Favreau, O. E. (1985). Perception of motion in equiluminous kinematograms. *Perception* **14**, 151-162.

Cavanagh, P. & Mather, G. (1989) Motion: the long and short of it. *Spatial Vision* **4**, 103-129.

Cavanagh, P., Tyler, C. W., & Favreau, O. E. (1984) Perceived velocity of moving chromatic gratings. *Journal of the Optical Society of America A* **1**, 893-899.

Charles, E. R. & Logothetis, N. K. (1989). The reponses of middle temporal (MT) neurons to isoluminant stimuli. *Investigative Ophthalmology and Visual Science Suppl.* **30,** 427.

Chubb, C. & Sperling G. (1988). Drift-balanced random stimuli: a general basis for studying non-Fourier motion perception. *Journal of the Optical Society of America A* **5**, 1986-2007.

Cushman, W. B. & Levinson, J. Z. (1983) Phase shifts in red and green counterphase flicker at high temporal frequencies. *Journal of the Optical Society of America* **73**, 1557-1561.

De Valois, R. L. & De Valois, K. K. (1975) Neural coding of color. In *Handbook of Perception*, Vol 5, *Seeing* (Edited by Carterette E. C. and Friedman M. P.), pp. 117-166. Academic Press, New York.

de Lange, H. (1958) Research into the dynamic nature of the human fovea-cortex systems with intermittent and modulated light. II. Phase shift in brightness and delay in color perception. *Journal of the Optical Society of America* **48**, 784-789.

Derrington, A. M. & Badcock, D. R. (1985). The low level motion system has both chromatic and luminance inputs. *Vision Research* **25**, 1874-1884.

Derrington, A. M., Krauskopf, J., & Lennie, P. (1984) Chromatic mechanisms in lateral geniculate nucleus of macaque. *Journal of Physiology* **357**, 241-265.

Drum, B. (1983) Short-wavelength cones contribute to achromatic sensitivity. *Vision Research* **23**, 1433-1439 .

Gorea, A. & Papathomas, T. V. (1989). Motion processing by chromatic and achromatic pathways. *Journal of the Optical Society of America* **A6**, 590-602.

Gouras, P. & Eggers, H. M. (1982) Retinal responses to color contrast. *Investigative Ophthalmology and Visual Science Suppl.* **22**, 176.

Ingling, C. R. & Martinez-Uriegas, E. (1985) The spatio-temporal properties of the r-g X-cell channel. *Vision Research* **25**, 33-38.

Julesz, B. (1971). *Foundations of Cyclopean Perception.* University of Chicago Press, Chicago, IL.

Kelly, D. H. (1979) Motion and vision II: Stabilized spatiotemporal threshold surface. *Journal of the Optical Society of America* **69**, 1340-1349.

Kelly, D. H. (1983). Spatiotemporal variation of chromatic and achromatic contrast thresholds. *Journal of the Optical Society of America* **73**, 742-750.

Krueger, J. (1979) Responses to wavelength contrast in the afferent visual system of the cat and the rhesus monkey. *Vision Research* **19**, 1351-1358.

Lee, B. B., Martin, P. R., & Valberg, A. (1988) The physiological basis of heterochromatic flicker photometry demonstrated in the ganglion cells of the macaque retina. *Journal of Physiology* **404**, 323-347.

Lee, J. & Stromeyer III, C. F. (1989). Contribution of human short wave cones to luminance and motion detection. *Journal of Physiology* **413**, 563-593.

Lelkins, A. M. M. & Koenderink, J. J. (1984). Illusory motion in visual displays. *Vision Research* **24**, 1083-1090.

Lennie, P., Krauskopf, J., & Sclar, G. (1990). Chromatic mechanisms in striate cortex of macaque. *The Journal of Neuroscience* **10**, 649-669.

Lindsey, D. T., Pokorny, J., & Smith, V. C. (1986) Phase-dependent sensitivity to heterochromatic flicker. *Journal of the Optical Society of America* **A3**, 921-927.

Livingstone, M. S. & Hubel, D. H. (1987). Psychophysical evidence for separate channels for perception of form, color, movement and depth. *Journal of the Neurosciences* **7**, 3416-3468.

Mather, G., Cavanagh, P., & Anstis, S. M. (1985) A moving display which opposes short-range and long-range signals. *Perception* **14**, 163-166.

Maunsell, J. H. R. & Newsome, W. T. (1987) Visual processing in monkey extrastriate cortex. *Annual Review of Neuroscience* **10**, 363-401.

Moreland, J. D. (1982) Spectral sensitivity measured by motion photometry. In *Colour deficiencies VI.* (Edited by Verriest G.), pp. 61-66. Junk, The Hague.

Mullen, K. T. & Baker, C. L. (1985). A motion aftereffect from an isoluminant stimulus. *Vision Research* **25**, 685-688.

Pantle, A. J. (1978) On the capacity of directionally selective mechanisms to encode different dimensions of moving stimuli. *Perception* **7**, 261-267.

Papathomas, T. V., Gorea, A., Julesz, B., & Chang, J.-J. (1988). The relative strength of depth and orientation in motion perception. *Investigative Ophthalmology and Visual Science Suppl.* **29**, 449.

Petersik, J. T., Hicks, K. I., & Pantle, A. J. (1978). Apparent movement of successively generated subjective figures. *Perception* **7**, 371-383.

Prazdny, K. (1986a). What variables control (long range) apparent motion? *Perception* **15**, 37-40.

Prazdny, K. (1986b). Three-dimensional structure from long-range apparent motion. *Perception* **15**, 619-625.

Prazdny, K. (1987). An asymmetry in apparent motion of kinetic objects. *Bulletin of the Psychonomic Society* **25**, 251-252.

Ramachandran, V.S , Rao, V.M., & Vidyasagar, T.R. (1973). Apparent movement with subjective contours. *Vision Research* **13**, 1399-1401.

Ramachandran, V.S. & Gregory, R. (1978) Does colour provide an input to human motion perception? *Nature* **275**, 55-56.

Saito, H., Tanaka, K., Isono, H. Yasuda, M., & Mikami, A. (1989). Directionally selective response of cells in the middle temporal area (MT) of the macaque monkey to the movement of equiluminous opponent color stimuli. *Experimental Brain Research*, in press.

Schiller, P. H. & Colby, C. L. (1983). The responses of single cells in the lateral geniculate nucleus of the rhesus monkey to color and luminance contrast. *Vision Research* **23**, 1631-1641.

Schiller, P. H., Logothetis, N., & Charles, E. (1990). The role of the color-opponent and broadband channels in vision. *Visual Neuroscience*, in press.

Shadlen, M. & Carney, T. (1986). Mechanisms of human motion perrception revealed by a new cyclopean illusion. *Science* **232**, 95-97.

Shapley, R. & Kaplan, E. (1989) Responses of magnocellular LGN neurons and M retinal ganglion cells to drifting heterochromatic gratings. *Investigative Ophthalmology and Visual Science Suppl.* **30**, 323.

Shapley, R. (1990). Visual sensitivity and parallel retinocortical channels. *Annual Review of Psychology* **41**, 635-658.

Smith, V. C., Lee, B. B., Pokorny, J., Martin, P. R., & Valberg, A. (1989). Response of macaque ganglion cells to changes in the phase of two flickering lights. *Investigative Ophthalmology and Visual Science* **30**, 323.

Stockman, A., MacLeod, D. I. A., & DePriest, D. D. (1987). An inverted S-cone input to the luminance channel: Evidence for two processes in S-cone flicker detection. *Investigative Ophthalmology and Visual Science Suppl.* **28**, 92.

Stromeyer III, C. F., Eskew, R. T., & Kronauer, R. E. (1990). The most sensitive motion detectors in humans are spectrally-opponent. *Investigative Ophthalmology and Visual Science Suppl.* **31**, 240.

Swanson, W. H., Pokorny, J. & Smith, V. C. (1987) Effects of temporal frequency on phase-dependent sensitivity to heterochromatic flicker. *Journal of the Optical Society of America A* **4**, 2266-2273.

Turnano, K. & Pantle, A. J. (1989). On the mechanism that encodes the movement of contrast variations: velocity discrimination. *Vision Research* **29**, 207-221.

von Grünau, M. (1977) Lateral interactions and rod intrusion in color flicker. *Vision Research* **17**, 911-916.

Wiesel, T. N. & Hubel, D. H. (1966). Spatial and chromatic interactions in the lateral geniculate body of the rhesus monkey. *Journal of Neurophysiology* **29**, 1115-1156.

FUNCTIONAL CLASSIFICATION OF PARALLEL VISUAL PATHWAYS

Karen K. De Valois and Frank Kooi*

Group in Physiological Optics
University of California at Berkeley
Berkeley, CA 94720 USA
and
*TNO - Institute for Perception
Soesterberg, The Netherlands

The last decade has seen an explosion of renewed interest in the presence and function of multiple parallel pathways in the visual system (e.g., Ungerleider and Mishkin, 1982). Most of this has centered around analyses of the so-called M and P pathways which originate in the retina and remain largely separated through several visual cortical regions. The excitement results in large part from new evidence for remarkable differences in the sensitivity of the M and P systems to different visual dimensions, in association with psychophysical evidence for similar and potentially related behavioral differences.

The anatomical separation between the two pathways or processing streams is most obvious in the lateral geniculate nucleus (LGN) of the thalamus, where the dorsal four layers (numbered 3-6) (which contain cells with small somata) are collectively called parvocellular layers, and the ventral two layers (1 and 2) (with fewer cells and larger cell bodies) are referred to in turn as magnocellular layers. It is from the magnocellular and parvocellular LGN layers that the M and P nomenclature is derived, but the distinction begins in the retina, where morphologically different cell types provide separate inputs for the M and P pathways. The separation continues at least through several prestriate visual cortical regions. The characterization of the M and P streams on anatomical and physiological grounds has been extensively reviewed (see, for example, Shapley, 1990, or Maunsell and Newsome, 1987 for relevant references) and need not be belabored here. There are a few distinctions, however, which should be reviewed briefly.

The P system in its early stages contains neurons with small receptive fields (RFs) and high sensitivity to chromatic contrast. The cells in V1 that show fine orientation tuning are generally assumed to belong to this processing stream, as well. Although the P system is sometimes further subdivided into two additional streams corresponding to the striate cortex regions rich in cytochrome oxidase (cytox)--the so-called blob regions--and those intervening regions with less cytox (Livingstone and Hubel, 1984), agreement about this distinction is less widespread. We shall consider all the P system to form a single processing stream for purposes of this discussion. In part because of the concentration of neurons with high selectivity for color and for spatial structure, it is suggested that the P system may be responsible for the processing of information about color and form (e.g., Van Essen and Maunsell, 1983).

Most neurons of the M system, to the contrary, show little selectivity for or sensitivity to chromatic contrast, but many of them are highly selective for direction of motion (Blasdel and Fitzpatrick, 1984; Livingstone and Hubel, 1984; Movshon and Newsome, 1984). Partly because of the high incidence of direction selectivity in M pathway neurons, this system is assumed to be responsible for the analysis and encoding of visual motion.

From Pigments to Perception, Edited by A. Valberg and
B.B. Lee, Plenum Press, New York, 1991

It is not merely the presence and high incidence of neurons responsive to chromatic and spatial parameters (in the case of the P system) or direction of movement (for the M system) that has led to current models of separate processing streams for these parameters. There is other evidence, as well. For example, there are many demonstrations that lesions restricted to one system or the other selectively affect the perception or the ability to respond differentially to a particular visual dimension. In some cases this comes from the clinical literature (e.g. Zihl et al., 1983; King-Smith, 1987); in others, from experimental behavioral research using restricted lesions (e.g., Newsome et al., 1985; Wild et al., 1985).

This model is also supported by much psychophysical evidence. When the tasks and stimuli are appropriately chosen, pronounced perceptual differences rivaling those seen physiologically can be demonstrated. It is presumed that these reflect the differences between the M and P systems. The usual tactic is to choose a stimulus that is believed to be processed exclusively (or almost so) by one of the two streams and then examine the behavior of that system in isolation. One example of this is the use of isoluminant stimuli to eliminate substantial input from the M stream (see Livingstone and Hubel, 1987). For example, the perceived speed of a moving isoluminant heterochromatic grating can be markedly lower than the perceived speed of a similar luminance grating moving at the same velocity (Cavanagh, Tyler and Favreau, 1984). Motion segregation of a random dot kinematogram is severely compromised at isoluminance (Ramachandran and Gregory, 1978).*

Current models suggest that each stream, while operating essentially in parallel with the other stream(s), displays serial processing internally. Any given dimension (at least those of color and motion) is analyzed primarily within one or the other of these two processing streams. Evidence for serial processing of movement within the M stream and color within the P stream comes from studies in which neurons with new, emergent properties have been demonstrated.

One example of this is cells that show color constancy. Under normal viewing conditions objects generally appear to have the same color, even though the spectral content of the light they reflect can change drastically as the illuminant changes. This property does not appear in neurons in area V1, but Zeki (1983) reports that it can be found in some neurons in area V4 (part of the P system). These cells show strong effects from stimulation outside the classical RF similar to the spatially extended interactions seen in color constancy.

In the motion domain Movshon et al. (1986) provided evidence for a physiological analogue of motion transparency and rigidity. They examined the responses of cells to a moving plaid pattern comprising two orthogonally oriented sine wave gratings. If the two component gratings are identical in all respects save orientation and are seen through a round aperture, the pattern will appear to move coherently in a single direction. The two individual components will meld to produce a cohesive whole that appears to move as a rigid object. No cells in striate cortex respond to this coherent pattern movement direction, but a significant subset of neurons within area MT (part of the M system) do, suggesting that the latter area is functionally specialized for the analysis of pattern motion. The existence of such neurons with emergent properties reflecting the serial processing within each stream is perhaps the most compelling evidence for the functional specialization of these parallel pathways.

Eventually, the visual system must integrate across all the incoming information, building a complete representation of the seen object. Ultimately, information about each dimension is attributed to the visual object of interest. How can this integration be achieved? It has been suggested that not only are color and motion independently analyzed, but the

---

*Despite these observations, all of which support largely independent analysis of color/form and motion, it is clear that not all M system neurons are silent at isoluminance (Derrington et al., 1984; Lee et al., 1988; Shapley and Kaplan, 1989). There is also psychophysical evidence that color has a measurable input into the motion system (Cavanagh et al., 1984; Troscianko and Fahle, 1988). The relationship between the residual M system neural responses at isoluminance and the psychophysical influence of color on motion perception is not clear, however.

motion signal derived from luminance is then attributed without modification to the colored object. When several black dots are scattered on top of a yellow blob that is isoluminant with its background, coherent movement of the dots can make the blob appear to move in concert, even though it is actually stationary (Ramachandran, 1987). Carney et al. (1987) also demonstrated that the color information could be captured by the motion properties of the luminance component.

It is the nature and implications of separate processing followed by recombination that we wish to discuss. We shall argue that although there is a considerable degree of separateness in the processing of different dimensions at early stages, eventually the system both should and does allow the activity within one stream to influence either the activity or the interpretation of the responses within the other.

We have studied the relationship between the color and motion systems using two different paradigms. Kooi (1990), noting the fact that the apparent speed of stimuli is not perfectly correlated with their real speeds, predicted and confirmed the existence of a novel illusion. The direction in which a coherent sinewave plaid pattern appears to move is not always veridical. For example, the apparent direction of motion of a plaid with components that differ in contrast will appear to be biased in the direction of the higher contrast component. If the component-to-pattern transform (Adelson and Movshon, 1982) were to be based on a signal related to perceived speed rather than real speed, then the apparent direction of movement of the pattern is expected to deviate from the real direction in a predictable fashion. We have tested and confirmed this prediction with several different luminance-varying plaid patterns (Kooi, 1990; Kooi et al., 1990). It is possible to predict the apparent direction of plaid motion quite accurately, given knowledge of the spatial frequency, contrast and velocity of each of the components.

This paradigm now gives a new method by which to examine the possible input of color to the motion system (Kooi, 1990; Kooi and De Valois, 1990). The relative motion strength of a given chromatic and luminance grating can be measured using a plaid made of one luminance grating and one isoluminant chromatic grating. The chromatic component is fixed in contrast while the luminance component varies. The luminance contrast at which the apparent and real plaid directions are equal (zero direction bias) is defined to be the "equivalent luminance contrast" of the chromatic grating.

Figure 1 shows data from three subjects on this task. Direction bias, the measure of perceived direction error in degrees, is plotted against the contrast of the luminance-varying component. The contrast of the chromatic component was about 32 times detection threshold. When the direction bias is positive, the luminance component is dominant; when the bias is negative, the chromatic component dominates. It can be seen that for each of the subjects only 1-2% luminance contrast was required to balance (zero direction bias) the high-contrast chromatic component. Thus a luminance contrast of only 3 to 4 times detection threshold is equivalent to a chromatic contrast of 32 times threshold on this task, making the luminance a roughly 8 times more powerful motion stimulus. These results confirm the parallel processing hypothesis and support the notion of Troscianko and Fahle (1988) that a moving high-contrast chromatic pattern can be considered to be equivalent to a low-contrast luminance pattern.

In a related experiment we examined the effect of adding a chromatic component to a low-contrast luminance-luminance plaid (Kooi and De Valois, 1990). If the chromatic pattern behaves simply like a low-contrast luminance signal, then its equivalent luminance contrast should add linearly to the component to which it is added. This does not occur. The interactions between color and luminance in this case are in general quite small, but they do not reflect a simple linear addition of the luminance and chromatic signals. Under most conditions the presence of chromatic contrast added to one of the luminance components reduces the effective strength of that component rather than increasing it. It is as though the color inhibits or reduces the effectiveness of the luminance input. These results are analogous to the slowdown of a luminance grating produced by the addition of color (Cavanagh et al., 1984) and are very reminiscent of the effect of a chromatically varying mask on a simultaneously presented luminance test pattern. The color mask, once it exceeds its own threshold, significantly reduces the detectability of a luminance test (De Valois and Switkes, 1983; Switkes et al., 1988).

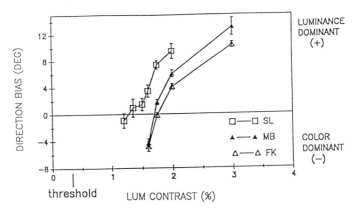

Fig. 1. Direction bias for plaids made of one luminance and one
red-green color component. The color contrast is fixed at
47%, the variable luminance contrast is plotted on the x-
axis. A positive direction bias indicates that the
luminance component dominates the direction of motion; a
negative direction bias indicates that the color
component dominates the motion direction. For quite small
luminance contrasts (1 to 2.0%) the direction bias is
zero. We define this to be the "luminance equivalent
contrast" of the color component.

To this point, then, color and motion appear to be processed largely
separately at early levels, with color providing only a weak input to the
motion system, although the situation is complex, since the interaction of
luminance and chromatic inputs to the motion system is not a simple linear
one. This is acceptably consistent with the general model of parallel
processing within multiple visual streams, after which the output of each
processing stream is attributed unchanged to the final object percept.

From the results of Ramachandran (1987) and Carney et al. (1987) the
integration of attributes across multiple subsystems appears to occur
without interactions between the subsystems. (Note that we use the term
"interaction" in a different sense than Ramachandran; 1987). Although our
results, like those of several other laboratories (see above), suggest that
there is a weak color motion signal, Ramachandran's conclusion could be
accepted in principle with minor modifications. Specifically, one could
argue that one system (perhaps the M system) determines everything there is
to know about the motion attribute without any influence from the system
that determines the color attribute (perhaps the P system). We need only
assume that there is a minor degree of chromatic sensitivity within the M
system. Each system could act independently, sending its information to
some eventual synthesizer. The final object construction could simply adopt
unchanged each attribute characteristic as it is relayed by the lower-level
analyzers.

But logical considerations force us to question this model. Although
such a strictly non-interactive, feed-forward system might be easy to design
and build, it would be not only simple, but simplistic. For any one object
there are many visual attributes such as color and motion. The ultimate aim
of the visual system must be to identify each object of interest and to
correctly determine all of its attributes. For a single object the non-
interactive, feed-forward system will work fine, but it is rare indeed that
only a single object is present. Only in a vision laboratory does such an
unnatural stimulus exist. In natural scenes there is a profusion of
objects, each with its own attribute signature. The first problem faced by
the visual system is to determine which parts of the scene go together.

Suppose there are n different motion signals within a restricted region
of the scene. They could arise from a complex single object which is
actually moving coherently. In that case, the appropriate response would be
to integrate across all the individual signals to determine a single
coherent direction of motion. On the other hand, the n motion signals could
be derived from n separate objects, each moving independently. In that
case, the appropriate response would be to keep each signal separate and

attribute each to its individual object. In essence, the component-to-pattern transformation should not occur at a low level before object segregation is accomplished. First, the visual system should determine what parts of the scene go together--segregate the objects--and only then should it begin to attribute particular features to the objects.

It is well known that object segregation can be based upon a variety of lower-level attributes such as similarity of color, brightness, depth or direction and speed of movement. (We use dimension to refer to an axis along which a visual stimulus can vary--e.g., color, direction of motion, orientation. Attribute refers to a particular value on a dimension--for example, 45° orientation.) Surely the visual system should combine information from as many of these as possible to arrive at an optimal decision about object segregation. Otherwise it is more likely that an error would be made. (For example, partially overlapping and transparent red and yellow objects could be perceived as orange.) But if object segregation based on information from all dimensions influences the final determination of each attribute, the "parallel" streams do not operate completely in parallel. The influence could be of three types. Either the processing streams must interact at a relatively low level, or object segregation must occur prior to and feed into the level at which attributes are determined, or there could be an iterative process with repeated cycles of feed forward and feed back until an optimal solution is achieved.

Thus a more complex scheme than a parallel feed-forward system should be employed. In the face of mounting evidence for independent, dimension-specific processing, however, it is reasonable to ask whether the visual system may have sacrificed versatility in the interests of simplicity. It has proven to be remarkably difficult to demonstrate any profound influence of some dimensions (such as color) on the decision made about others (such as motion or stereo). Recall the very impoverished contribution of color to the motion direction task described above.

The key to the problem was the realization that the tasks in which little cross-dimensional interaction is seen are ones in which objects have already been segregated (see Kooi, 1990). In order to determine the extent of possible interactions, one must devise a measure in which object segregation can vary and its influence can be observed. Accordingly, we (Kooi et al., 1989) examined the coherence versus sliding decision for a variety of colored plaid patterns.

When two orthogonally-oriented, moving sinewave gratings which are otherwise identical are superimposed, the result will most often appear to move in a coherent fashion. If the gratings are vertical and horizontal, the plaid will move on a diagonal. If the component gratings differ significantly on any dimension, the pattern will be less coherent and the two gratings may appear to separate and move independently. Visual transparency occurs. In this case, the directions of movement--clearly a motion attribute--will be quite different. Each grating will appear to move perpendicularly to its stripes. Using a 5-point scale in which 1 is completely coherent and 5 is completely sliding, we have examined the influence of color on the sliding versus coherence judgment.

With two luminance gratings that are identical in spatial frequency, contrast, speed and space-averaged chromaticity, the resulting plaid is coherent when viewed through a round aperture. (With other kinds of apertures, however, it is possible to produce a fair amount of sliding.) Changing either the spatial frequency or the contrast of one component will increase the degree of pattern sliding, as Adelson and Movshon (1982) reported. The most striking effects occur, however, when color is added. If the two components are identical isoluminant chromatic gratings, the plaids become virtually completely coherent. As Kooi (1990) notes, it is as though the color acts as a "glue", binding the two component gratings into a single, coherently moving object. If color is added asymmetrically to the components, however, the plaid will split and the two gratings will appear to move more independently. For example, the two gratings could lie along different axes in color space, or color variation could be added to one component but not the other, or identical color variation could be added to the two luminance gratings but with opposite spatial phase. Thus color symmetry reinforces apparent rigidity of motion; color asymmetry reinforces apparent transparency. Figure 2 illustrates averaged ratings showing the degree of coherence for 6 patterns. It can be seen that the effect of color on this motion decision is profound indeed.

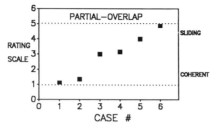

Fig. 2. Results of the rating experiment of six plaids. The
median rating results of 8 subjects are plotted
vertically (1 is complete coherence and 5 complete
sliding). The six plaid cases are in order of
increasing sliding to the right, their essential
component characteristics being described below:
1: equiluminant, equal contrast and spatial
   frequency.
2: equiluminant, unequal spatial frequency.
3: luminance (yellow-black), equal contrast and
   spatial frequency
4: luminance, unequal contrast
5: luminance, unequal spatial frequency
6: luminance, unequal hue (red-black and green
   black).

These observations lead us to the disquieting conclusion that whether
the multiple streams of visual analysis seem to operate independently and in
parallel depends largely on the experiment. That a supposedly high-level
task like object segregation can have strong effects on a supposedly low-
level judgment like direction of motion should not be surprising. It
reflects a visual system designed for versatility and accuracy in the face
of a complex natural world. It may also help explain the puzzling, massive
back projections seen at virtually every level in visual cortex. An
iterative system in which processing at each level can be modified by
feedback from higher centers (and probably extravisual input, as well) could
appear in simple tests to show independent, serial processing. But when
appropriate questions are asked, it becomes clear that the visual system is
much more complex, and thus much more interesting.

ACKNOWLEDGEMENTS

The research reported in this paper was supported by grants from the
National Eye Institute (EY00014) and the National Science Foundation (BNS
8819867).

REFERENCES

Adelson, E.H. and Movshon, J.A., 1982, Phenomenal coherence of moving visual
    patterns. Nature 300: 523-525.
Blasdel, G.G. and Fitzpatrick, D., 1984, Physiological organization of layer
    4 in macaque striate cortex. J. Neurosci. 4: 880-95.
Carney, T., Shadlen, M. and Switkes, E., 1987, Parallel processing of motion
    and color information, Nature 328: 647-649.
Cavanagh, P., Tyler, C.W. and Favreau, O.E., 1984, Perceived velocity of
    moving chromatic gratings. J. Opt. Soc. Am. A 1: 893-99.
Derrington, A.M., Krauskopf, J. and Lennie, P., 1984, Chromatic mechanisms
    in lateral geniculate nucleus of macaque. J. Physiol. 357: 241-65.
De Valois, K.K. and Switkes, E., 1983, Simultaneous masking interactions
    between chromatic and luminance gratings. J. Opt. Soc. Am. 73: 11-18.
King-Smith, E. (1987) Cortical colour defects. In Drum, B. and Verriest,
    G. (Eds.), Colour Vision Deficiencies IX, Proceedings of the
    International Symposium, Annapolis, 1987, Doc. Ophthal. Proc. Ser. 52,
    Dordrecht, The Netherlands: Kluwer Academic Publishers.

Kooi, F.L., 1990, <u>The Analysis of Two-Dimensional Luminance and Color Motion Properties</u>. Unpublished doctoral dissertation, The University of California at Berkeley, Berkeley, CA.

Kooi, F.L. and De Valois, K.K., 1990, The role of color in the motion system. Submitted to <u>Vision Res</u>.

Kooi, F.L., De Valois, K.K., Grosof, D. H. and De Valois, R.L., 1990, Properties of the recombination of one-dimensional motion signals into a pattern motion signal. Submitted to <u>Vision Res</u>.

THE RESPONSE OF MACAQUE RETINAL GANGLION CELLS TO

COMPLEX TEMPORAL WAVEFORMS

Jan Kremers[1], Barry B. Lee[1], Joel Pokorny[2] and Vivianne C.Smith[2]

1: Department of Neurobiology
Max Planck Institute for Biophysical Chemistry
D-3400 Göttingen, FRG

2: The Visual Sciences Center
The University of Chicago
Chicago, USA

INTRODUCTION

The responses of retinal ganglion cells depend on the temporal characteristics of the visual stimulus, and on the way visual information is processed in the retina. A powerful mathematical tool in relating the in- and output of the retinal system is linear systems analysis. If a system is linear, its output to a sinusoidal input is again a sine wave with the same frequency, although amplitude and phase may be modified. The modulation transfer function (MTF), in which amplitudes and phases are given as function of sine wave frequency, completely determines the system's properties. Thus, when the temporal MTF of a linear system is known, one can calculate the system's response to every temporal stimulus. We discuss here how far the responses of macaque retinal ganglion cells to complex temporal waveform can be predicted from the cell's temporal MTF. We further relate the cells' responses to human detection thresholds for complex waveforms.

METHODS OF ANALYSIS

Extracellular responses of macaque retinal ganglion cells were recorded as described elsewhere (e.g. Lee et al., 1989). Visual stimuli were generated by computer controlled modulation of a pair of red and green diodes. It was possible to apply different stimuli of any desired frequency, contrast, and type of temporal waveform e.g. sinusoidal modulation, sawteeth, squarewaves. The first harmonics of the response to sinusoidal modulation as a function of contrast were fitted by a Naka-Rushton equation (Lee et al., 1990). Sensitivity (contrast gain) is given by the initial slopes of these functions. The sensitivities and response phases were measured for different frequencies. Intermediate values were approximated by linear interpolation between these points.

A more complex periodic waveform, can be decribed by a Fourier expansion containing the first harmonic, and some combination of higher harmonics. The Fourier spectrum of a linear system's response to such a complex waveform can then be predicted by multiplying the factors of the Fourier expansion with the sensitivities and introducing a response phase term for each frequency. The predicted responses were compared with the measured responses to the appropriate stimuli.

From Pigments to Perception, Edited by A. Valberg and
B.B. Lee, Plenum Press, New York, 1991

A peak detector algorithm was applied to responses of tonic and phasic cells to luminance sawteeth stimuli, providing as output the peak firing rate to the stimulus within a 32 msec window. Sensitivity of this peak detector was calculated by fitting the responses at different contrasts with the Naka-Rushton equation and taking the inverse of the contrast at the threshold response of 20 imp/s.

RESULTS

Fig. 1 shows the amplitude and phase of the temporal MTF's of a tonic red on-centre cell to chromatic modulation and of a phasic on-centre cell to luminance modulation. Typically, the responses of the phasic cell to luminance modulation is more band-pass and has a higher cut off frequency than the tonic cell to chromatic modulation.

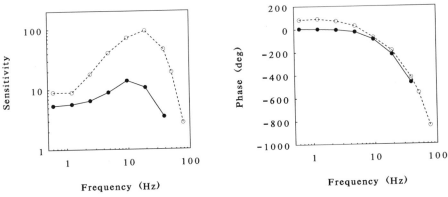

Fig. 1. Amplitude (left) and phase (right) plot of temporal MTF's of a phasic-on cell to luminance modulation (open symbols) and a red-on tonic cell to chromatic modulation (closed symbols).

Predicted and measured responses of the two cells, with the temporal MTF presented in Fig. 1, to chromatic sawteeth, squarewaves and pulses are compared in Fig. 2. For the tonic cell, there is a close resemblance, implying substantial linearity of the processes leading to this cell's response. Such results were common for all tonic P-cells, and for nearly all stimulus conditions. Near threshold, the predicted and measured responses of the phasic cell resemble each other closely. Suprathreshold responses were more spiky, larger, and earlier than the predictions. Often, at frequencies between 5 and 20 Hz, two or more bursts of spikes can be observed in an extremely rapid sequence, which could not be predicted.

The sensitivities of 6 red- and green-on centre tonic cells, 9 phasic-on, and 6 phasic-off cells to luminance sawteeth were analysed using the peak detector algorithm. The results are shown in Fig. 3. Clearly, sensitivities to rapid-on and rapid-off sawteeth of all cells are different for frequencies below about 40 Hz. Above 40 Hz, however, the cells respond equally to rapid-on and rapid-off sawteeth. Further, phasic-on cells showed the highest sensitivity to rapid-on sawteeth, whereas phasic-off cells were the most sensitive to rapid off sawteeth. Bowen et al. (1989) measured sensitivities of human subjects to luminance rapid-on and -off sawteeth. The mean sensitivities are presented in a seperate plot. The physiological and psychophysical data are only qualitatively comparable, because of differences in mean retinal illuminance, and in colour and retinal eccentricity of the stimulus.

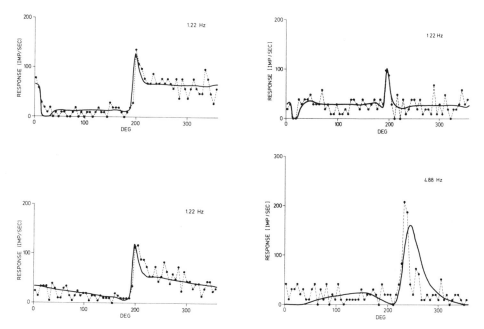

Fig. 2. Predicted and measured responses of the cell with the MTF's shown in Fig. 1. The two panels on the left present the red-on tonic cell's responses to 1.22 Hz, 50 % contrast, chromatic square waves (upper left) and sawteeth (lower left). The two panels on the right present the phasic-on cell's responses to 1.22 Hz, 6.25 % contrast, luminance square wave (upper right), and to 4.88 Hz, 12.5 % contrast, luminance sawteeth (lower right).

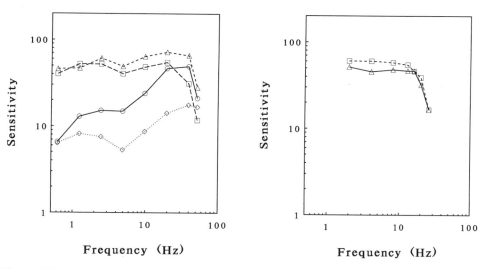

Fig. 3. Mean sensitivities of cells to luminance sawteeth using a peak detecting algorithm (left panel). Triangles: mean of 9 phasic-on cells to rapid-on sawteeth; squares: mean of 6 phasic-off cells to rapid-off sawteeth; circles: mean of phasic-on cells to rapid-off and of phasic-off cells to rapid-on sawteeth; diamonds: mean of tonic-on cells to rapid-on sawteeth; other conditions resulted in much lower sensitivities. The right panel shows mean human sensitivity to rapid-on (triangles) and to rapid-off (squares) luminance sawteeth as presented by Bowen et al. (1989).

DISCUSSION

Comparing measured and predicted responses for tonic cells leads to the conclusion that the processing of the visual information between photoreceptors and these cells is very linear. Lee et al. (1991) modelled the tonic cell's response as a simple subtraction of centre and surround signals with a latency difference. This linear model provided an excellent account of a tonic cell's behaviour. Phasic cells show less linearity. An accelerating non-linearity is present: the response is always larger than the predicted response for supra-threshold stimuli. Further, the responses are very 'spiky' and rapidly alternate with more or less silent periods, which could not be predicted. Possible explanations might be a short term (50 msec) adaption in the spike generating mechanism, as described by Lankheet et al. (1989) in cat ganglion cells, or non-linear cone interactions in the surround of the receptive field (Lee et al., 1989).

Using a peak detector algorithm, off-centre cells are the most sensitive for rapid-off saw-teeth, and on-centre cells for rapid-on sawteeth. The resulting sensitivity curves as function of frequency are low pass, and resemble those found in psychophysical studies (Bowen et al., 1989). This suggests the existence of some kind of central peak detector, and further that detection of luminance sawteeth is mediated by the magnocellular pathway. The suggestion that two different parts of the magnocellular pathway (the "on-" and "off-" parts) are involved in detection of the two types of sawteeth, is in agreement with the notion that adaptation to each type selectively affects sensitivity to incremental or decremental steps or pulses (Krauskopf, 1980). Bowen et al. found that human subjects are more sensitive to rapid-off than to rapid-on saw-teeth. However, our cell data did not reveal such a difference. This might reflect an asymmetry between the on- and off- elements of the magnocellular pathway at a later stage. Physiological and psychophysical sensitivity curves differ in cut off frequency, which is about 80 Hz for the physiological sensitivity and 40 Hz for the psychophysical sensitivity. These results are an indication for the existence of a cortical low pass filter in the luminance pathway, with corner frequency of about 20 Hz, as suggested by Lee et al. (1990) on the basis of similar discrepancies between physiological and psychophysical sensitivities to luminance sine wave modulation.

ACKNOWLEDGEMENTS

This work was partially supported by NIH USPH Research Grant EY 09001 and by NATO Collaborative Research Grant 0909/87.

REFERENCES

Bowen, R.W., Pokorny, J., and Smith, V.C., 1989, Sawtooth contrast sensitivity: decrements have the edge, Vision Res., 29:1501.

Krauskopf, J., 1980, Discrimination and detection of changes in luminance, Vision Res., 20:671.

Lankheet, M.J.M., Molenaar, J., and van de Grind, W.A., 1989, The spike generating mechanism of cat retinal ganglion cells, Vision Res., 29:505.

Lee, B.B., Martin, P.R., and Valberg, A., 1989, Nonlinear summation of M- and L-cone inputs to phasic retinal ganglion cells of macaque, J. Neurosc., 9:1433.

Lee, B.B., Pokorny, J., Smith, V.C., Martin, P.R., and Valberg, A., 1990, Luminance and chromatic modulation sensitivity of macaque ganglion cells and human observers, J. Opt. Soc. Am. in press.

Lee, B.B., Smith, V.C., Pokorny, J., Martin, P.R., and Valberg, A., 1991, Responses of tonic ganglion cells of the macaque retina on changing the relative phase of two flickering lights, to be submitted.

REMOTE SURROUNDS AND THE SENSITIVITY OF PRIMATE P-CELLS

Arne Valberg[1], Barry B. Lee[2], and Otto D. Creutzfeldt[2]

[1]Department of Physics, Section of Biophysics
University of Oslo, Box 1048 Blindern
0316 Oslo 3, Norway

[1]Department of Neurobiology
Max Planck Institute for Biophysical Chemistry
3400 Göttingen, F.R.G.

Colour constancy and the appearance of surfaces in natural scenes is dependent on the presence of extended surrounds. Surface colours in general, and dark colours like black, grey, brown, and olive, are only perceived with brighter surrounds, and are therefore often called 'related' colours.

Neurophysiologically, colour-coded cells in V4 of macaque monkeys may exhibit partial colour constancy (Zeki, 1983), because their responses are influenced by light stimuli outside their 'classical' receptive fields. Here, we report that remote surrounds also influence the responses of colour opponent parvocellular cells (P-cells) in the lateral geniculate nucleus (LGN) of macaques. Steady white annuli, well outside the classical receptive field set the contrast gain and provide for a spectral tuning of these cells that varies much less with the size of the stimulus area than when such stimuli are presented without a surround.

Previously we have shown that, in the presence of remote white surround annuli, the intensity-response curves for flashed, 300 ms, 4 or 8 deg. stimuli, large enough to cover centre and surround of receptive fields, are parallel shifted towards higher intensity by the addition of white surrounds (Valberg, Lee, Tigwell, and Creutzfeldt, 1985). An effect can be elicited with inner radii 20-30 deg. from the receptive field centre. This is an adaptive gain control that implies a von Kries sensitivity reduction by the same factor for both excitatory and inhibitory cone mechanisms having input to the opponent cell.

For small 0.3 deg. spots, flashed on the receptive field centre, spectral sensitivity in the appearance of a surround is broad due to isolation of the excitatory mechanism (Wiesel and Hubel, 1966). In this case, the addition of a white surround annulus, well outside the 'classical' receptive field, brings about both a sensitivity reduction and a narrowing of centre spectral selectivity through opponent inhibition (Figs. 1 and 2). Opponency is thus reintroduced by the peripheral white annulus. Effects were measured for annuli with up to 50 deg. inner diameter.

Consequently, with fields set in a context, the result is a more constant spectral tuning for small and large spots. This behaviour, combined with the parallel shift for

From Pigments to Perception, Edited by A. Valberg and
B.B. Lee, Plenum Press, New York, 1991

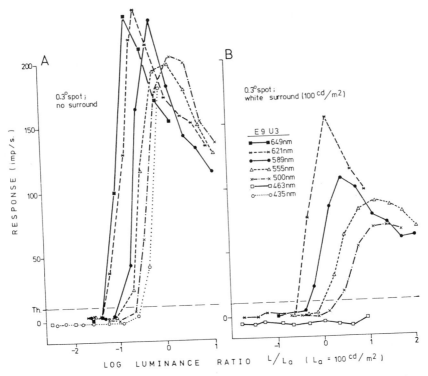

Fig. 1. The luminance-response curves for different wavelengths without (A) and with (B) the white surround. The cell is a L-M geniculate P-cell stimulated with a 0.3 deg. spot in the receptive field centre, without and with a 120 cd/m² steady white surround. The surround causes a shift of the curves toward higher luminance ratios, smaller amplitudes, and a change in spectral selectivity (see Fig. 2).

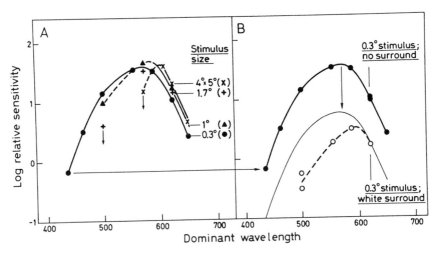

Fig. 2. A: The spectral sensitivity of the cell of Fig. 1 with spots of different sizes presented in a dark surround. Opponency increases as spot size increases. B: The spectral sensitivity for a small spot of 0.3 deg without and with a white surround annulus. With a surround, absolute sensitivity decreases and opponency increases, as for a medium size spot in A.

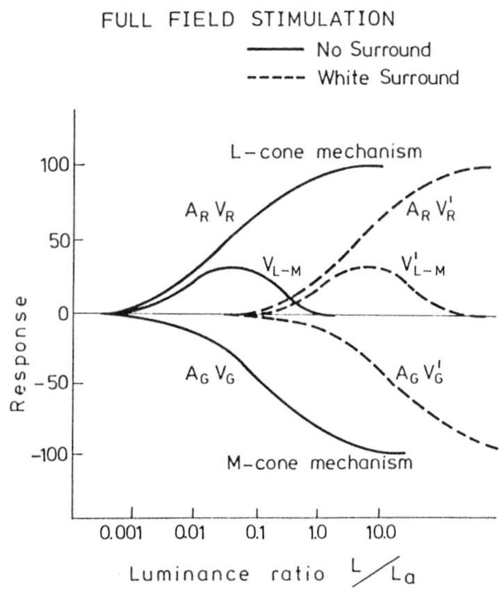

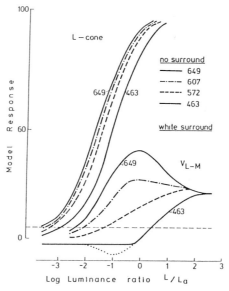

Fig. 3. Parallel shifts of intensity-response curves of a L-M cell with an excitatory L-centre and a suppressive M-surround cone pool. The von Kries adaptation (reduction of cone sensitivities by a constant factor) is the same for centre and surround mechanisms, provided the stimulus is large and a white steady surround, situated well outside the 'classical' receptive field, adapts the cell (Valberg et al., 1985).

Fig. 4. Model of the change in the response to a small spot of the same L-M cell as in Figs. 1 and 2 when a white peripheral annulus is added. The change in the response is brought about by two processes: A: A von Kries coefficient adaptation (multiplicative sensitivity reduction) with a factor of 0.1 for both opponent cone mechanisms (causing a parallel shift of response curves as in Fig. 3). B: Subtraction by the inhibitory M-cone mechanism of the surround reducing the response amplitude. In this example the weight of the surround mechanism is 0.8 of that of the centre. Subtraction causes opponency and a reduced maximum reponse. This model accounts well for the results of Figs. 1 and 2.

large stimuli mentioned above, seems to imply a two-process model of multiplicative and subtractive adaptation controlling the excitability of cells through remote surrounding fields (Figs. 3 and 4).

In conclusion, retinal illuminance far from the receptive field centre exerts some kind of 'gain control' over cell responses. Surrounds are most important in determining the perceived colour of objects; they provide for colour constancy, simultaneous contrast and light adaptation (Valberg and Lange- Malecki, 1990; Creutzfeldt et al., 1990). Surround phenomena at a retinal and geniculate level must therefore be considered before one searches for a physiological substrate of more global effects higher up in the visual pathway, and their relation to psychophysics.

REFERENCES

Creutzfeldt, O.D., Lang-Malecki, B. and Dreyer, E., 1990, Chromatic induction and brightness contrast; a relativistic color model. *J. Opt. Soc. Am. A*, 7:1644-1653.

Valberg, A., Lee, B.B., Tigwell, D.A., and Creutzfeldt, O.D., 1985, A simultaneous contrast effect of steady remote surrounds on the responses of cells in macaque lateral geniculate nucleus. *Exp. Brain Res.*, 58:604-608.

Valberg, A. and Lange-Malecki, B., 1990, "Colour constancy" in Mondrian patterns: A partial cancellation of physical chromaticitry shifts by simultaneous contrast. *Vision Res.*, 30:371-380.

Wiesel, T. N. and Hubel, D. H., 1966, Spatial and chromatic interactions in the lateral geniculate body of the rhesus monkey *J. Neurophysiol.*, 29:1115-1156.

Zeki, S., 1983, Colour coding in the cerebral cortex: The reactions of wavelength-selective and colour-coded cells in monkey cortex to wavelengths and colours. *Neurosci.*, 9:741-765.

ON NEUROPHYSIOLOGICAL CORRELATES OF SIMULTANEOUS COLOUR AND

BRIGHTNESS CONTRAST AS DEMONSTRATED IN P-LGN-CELLS OF THE MACAQUE

Sabine Kastner, Otto D. Creutzfeldt,
Chao-yi Li, John M. Crook, and Pei Xing

Department of Neurobiology
Max-Planck-Institute for Biophysical Chemistry
D-3400 Göttingen, FRG

INTRODUCTION

The colour of objects depends not only on the spectral composition and intensity of light reflected from them, but also on the spectral composition and intensity of objects surrounding them. Set in a monochromatic environment, the colour of an object generally appears as to be shifted in a direction complementary to that of the environment and its brightness also changes (colour and brightness contrast). Although simultaneous colour and brightness contrast phenomena have been investigated extensively, little is known about its neuronal correlate and the level at which it arises within the visual pathway. Some follow Helmholtz (1866) and place it in the primary visual cortex or beyond (Zeki, 1983; Land et al., 1983; Livingston and Hubel, 1984), while others find evidence supporting Mach's location of the mechanism in the retina (Pöppel, 1986) or even at a prereceptoral level (Walraven, 1973).

We present here some evidence that simultaneous colour and brightness contrast is encoded already at a subcortical level, in colour-opponent P-LGN- and retinal ganglion cells of macaques.

SIMULTANEOUS COLOUR CONTRAST

We have investigated in 281 colour-opponent P-LGN-neurones of anaesthetized macaques the effects of constant remote surrounds of achromatic light stimuli. The responses were determined for spots (0.2°-5°) flashed on the receptive field center for 3 sec. presented either alone or in the presence of an annular surround (inner dia. 5°; outer dia. 15°-20°). Center spots and surround illuminations were presented isoluminant to each other. Surround colours were usually red (664 nm) or blue (452 nm). Cells were classified for their spectral sensitivity (short-, middle-, longwavelength) and bandwidth (wide-, narrowband) (Creutzfeldt et al., 1979).

During blue surround illumination, neurones with a weak or strong excitatory input from S- or M-cones (WS- or NS-cells, respectively) showed a strong attenuation of responses to blue and green center spots and an increase of their maintained discharge rate (MDR). During red surround illumination the responses of NS- and WS-cells showed a clear increment. Fig. 1 shows a typical example of these effects for a WS-cell (responding over a wideband of shortwavelength). Conversely, L-cone excited WL- and NL-cells showed a decrement of spectral responses to red,

From Pigments to Perception, Edited by A. Valberg and
B.B. Lee, Plenum Press, New York, 1991

yellow and green center spots during red surround illumination and in the majority also an increment of MDR. Blue surround illumination affected WL-cell responses little and less consistently M-cone excited and S-cone suppressed WM-cells (3% of the P-LGN-cell population) were exceptional, because their responses were unaffected by non-opponent red surround illumination, but strongly suppressed by opponent blue surround illumination.

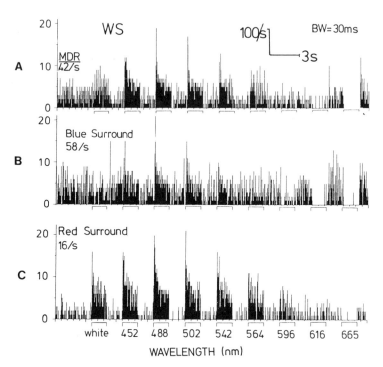

Fig. 1. Effects of red and blue surround illumination on spectral responses of a WS-cell (M-cone excited).

A: Responses to equiluminant 1.5° diameter stimuli of different wavelengths (as indicated on the abscissa) without surround illumination (control response). B, C: Same center stimulus as in A, but now with continuous surround illumination with blue (B) or red (C) light (inner dia. 5°; outer dia. 20°), respectively. Sum of 3 stimulus runs. MDR = maintained discharge rate. BW = binwidth.

BRIGHTNESS INDUCTION

To analyze the brightness induction effect we calculated the population response of P-LGN-neurones by summing up the activity of all weak opponent wideband-cells and subtracting that of all strongly opponent narrowband-cells. The resulting function deviates slightly from the equal luminosity function, but closely resembles the spectral brightness function. This corresponds to previous results that the activity of W-cells is positively, that of N-cells negatively correlated with the luminance of a spectral stimulus in relation to a white background (Creutzfeldt et al., 1986).

With red or blue surround illumination, the [WS+WL+WM-NS-NL]-population response was lowered nearly parallel to about 0.7 of the amplitude of the control (i.e. without surround presentation). Fig. 2 (hatched columns) demonstrates this for some selected colour stimuli.

In a psychophysical test on 4 observers we estimated the brightness induction of an equiluminous surround in a stimulus arrangement identical to the neurophysoplogical experiment, and found a brightness reduction for white, blue, green and red center stimuli to 0.5-0.7 of the brightness values without surround. A comparison of the neurophysiological and psychophysical results is given in Fig. 2.

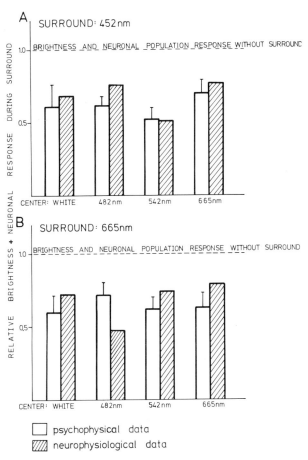

Fig. 2. Comparison between changes of neuronal population responses (hatched columns) and of psychophysical brightness perception (open columns) during surround illumination. Dashed horizontal line: Control values without surround. The brightness values of center spots of different colours (as indicated on the abscissa) during blue (A) and red (B) surround illuminations are the means from 4 observers with standard deviations from all measurements. The relative amplitudes of the neuronal population responses are the activity sums of all colour-opponent P-LGN-neurones with the following weighting: [WS+WL+WM-NS-NL]. For further informations see text.

DISCUSSION

We have demonstrated that the responses of P-LGN-neurones to light stimuli are altered by continuous surround illumination. The spectral bandwidths of colour-opponent cells were narrowed and their maximum responses were decreased by surround illuminations of a non-opponent wavelength (i.e. blue for blue-green sensitive cells and red for red-sensitive cells). In contrast, the spectral bandwidths were broadened and the maximum responses increased by surround illumination of the opponent colour. The direction of these changes is consistent with the psychophysical experience that perceived colours are shifted by a coloured surround in a direction complementary to the surround colour.

These surround effects are present in neurones of the P-LGN. They can also be demonstrated in prepotentials from retinal ganglion cells from which we recorded occasionally. They are therefore not due to corticofugal feedback, but must be present already at the retinal level.

The effects of continuous illumination of a surround far outside the 'classical' receptive field can be described as a steady excitation and gain reduction of the excitatory or inhibitory (opponent) cone inputs by the appropriate wavelengths. Such an effect is in the same direction as that exerted by direct illumination of the receptive field center with light of the same spectral composition but of lower intensity. As possible mechanisms lateral interaction through horizontal connections or physiological intraocular scattered light or a combination of both is possible. Lateral horizontal cell interaction meets with the difficulty that there is no evidence for spectral selectivity of retinal horizontal cells. Therefore, intraocular scattered light must be seriously considered as an early mechanism leading to the complex sensation of colour contrast phenomena, and data in support of this are presented.

The population response of P-LGN-cells was decreased by surrounds of any colour. This decrease comes close to the brightness reduction which a human observer recognizes during surround illumination. This indicates that the results may be directly related to brightness perception, and that P-LGN-cells not only signal for chroma but also for brightness, by combining their signals in a different way.

REFERENCES

Creutzfeldt, O.D., Lee, B.B., Elepfandt, A., 1979, A quantitative study of chromatic organisation and receptive fields of cells in the lateral geniculate body of the rhesus monkey, Exp.Brain Res., 35: 527-545.

Creutzfeldt, O.D., Lee, B.B., Valberg, A., 1986, Colour and brightness signals of parvocellular lateral geniculate nucleus neurones, Exp.Brain Res., 63: 21-34.

Helmholtz, H. von, 1866, Handbuch der physiologischen Optik, L. Voss, Hamburg, Leipzig.

Land, E.H., Hubel, D.H., Livingstone, M.S., Perry, S.H.,Burns, M.M., 1983, Colour-generating interactions across the corpus callosum, Nature, 303: 616-618.

Livingstone, M.S., Hubel, D.H., 1984, Anatomy and physiology of a colour system in the primate visual cortex, J. Neurosci., 1: 309-356.

Pöppel, E., 1986, Long-range colour generating interactions across the retina, Nature, 320: 1739-1753.

Walraven, J., 1973, Spatial characteristics of chromatic induction; the segregation of lateral effects from straylight artefacts, Vision Res., 13: 1739-1753.

Zeki, S., 1983, Colour coding in the cerebral cortex: The responses of wavelength-selective and colour-coded cells in monkey visual cortex to changes in wavelength composition, Neurosci., 4: 767-781.

DEVELOPMENT OF INFANT CONTRAST SENSITIVITY

AND ACUITY FOR COLOURED PATTERNS

David C. Burr, M. Concetta Morrone and Adriana Fiorentini

Istituto di Neurofisiologia del CNR
Via S. Zeno 51, Pisa
Italy

In this study we apply the technique of recording visual evoked potentials (VEPs) to investigate how the spatial characteristics of the infant colour system develops over the first six months. The stimuli for our experiments were plaid patterns modulated sinusoidally both horizontally and vertically, made by summing red and green sinusoidal plaids of equal but opposite contrast (see Fiorentini *et al.*, this volume, for details). The patterns were reversed in contrast at frequencies from 2 to 5 Hz, and VEPs were recorded from the infants in synchrony with the patttern reversal.

We first measured VEPs as a function of the colour ratio to establish the point of iso-luminance for each infant. Whereas for adults, there was a strong response for all colour ratios (see figure 1 of Fiorentini *et al.*, this volume), for very young infants (less than about 6 weeks of age) no reliable VEPs could be recorded near the colour ratio 0.5 (the isoluminant point for most adults). The existence of a point (presumably the iso-luminant point for the infant) where no VEP could be elicited indicates that the infant had no response to purely chromatic stimuli at this age. By 8 or 9 weeks of age responses began to be recorded from iso-luminant stimuli, but only at low spatial frequencies.

To quantify the development of colour vision, we measured both contrast sensitivity and acuity of eight young infants, using the extrapolation technique of Campbell and Maffei[1]. VEPs for both chromatic and luminance stimuli were recorded as a function of contrast and of spatial frequency, and the VEP amplitude curves extrapolated to zero response, to give estimates of both contrast sensitivity and spatial acuity.

Contrast sensitivity was measured at 0.1 c/deg, to favour the colour system which prefers lower spatial frequencies[3], and to minimize chromatic aberrations[2]. Before 4-7 weeks (varying from infant to infant) there was no response to chromatic stimuli, even at 100% contrast (figure 1A). At the same time, however, contrast sensitivity to luminance stimuli was 5-8 (threshold of 12-20% contrast). Both chromatic and luminance sensitivity increased with age, first rapidly then more slowly. Over the first 20 weeks, chromatic sensitivity increased 0.073 log-units per week, whereas luminance sensitivity increased at 0.060 log-units per week. As the standard errors of measurement were about 0.001, the difference in slope is

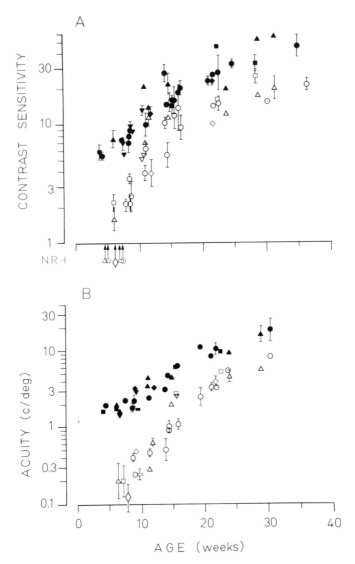

Figure 1

A Development of contrast sensitivity (the inverse of contrast threshold) for chromatic stimuli (open symbols) and luminance stimuli (filled symbols). The different symbol types refer to longitudinal measurements of the eight infants. The arrows below the abscissa indicate the latest recording session at which no-response could be elicited by chromatic stimuli at 100% contrast. Spatial frequency was always 0.1 c/deg, and temporal frequency varied with age, chosen to yield maximal response to chromatic stimuli: 2 Hz from 4 to 8 weeks, 3 Hz from 8 till 20 weeks, and 5 Hz thereafter. Over the first 20 weeks, chromatic sensitivity increased at the rate of 0.073 +- 0.001 LUs/week, and luminance sensitivity at 0.060 +- 0.001 LUs/week.

B Development of acuity for chromatic and luminance stimuli (open and closed symbols respectively). Contrast was always 0.8, and temporal frequency varied with age (see above). Over the first 20 weeks, chromatic sensitivity increased at the rate of 0.138 +- 0.002 LUs/week, and luminance sensitivity at 0.045 +- 0.001 LUs/week.

186

clearly significant. By about six months, both chromatic and luminance contrast sensitivity levels approached those of adults.

At the earliest age at which chromatic VEPs could be recorded, only very low spatial frequencies (around 0.1 c/deg) would elicit a response (figure 1B). During this period, the acuity for luminance stimuli was around 2 c/deg, 20 times higher. Like contrast sensitivity, chromatic and luminance acuity developed at a different rate: chromatic acuity at 0.138 log-units per week and luminance acuity at 0.045 log-units per week (SEs around 0.001 log-units/week). The rate of improvement of luminance acuity is consistent with previous studies[4]. By six months chromatic acuity was 1/3 luminance acuity, a ratio similar to that observed with adults[3].

As infants are tri-chromatic by 3 months, with adult-like spectral sensitivity curves[5], the different rate of development of chromatic compared with luminance acuity and sensitivity probably reflects maturation in the organization of chromatically opponent receptive-fields. Although we cannot speculate on what level of processing limits infant colour thresholds, we can be reasonably certain that the development of post-receptoral neural mechanisms are implicated in the differential maturation of chromatic spatial processing.

REFERENCES

1. Campbell, F.W.& Maffei, L. (1970) *J. Physiol. (Lond)* **207** 635-652.
2. Flitcroft, D.I. (1989) *Vision Res.* **29** 349-360.
3. Mullen, K.T. (1985) *J. Physiol (Lond.)* **359** 381-400.
4. Pirchio, M., Spinelli,D., Fiorentini, A. & Maffei, L. (1978) *Brain Res.* **141** 179-184.
5. Teller and Bornstein (1987) In *Handbook of Perception* (Eds. P. Salapatek & L.B. Cohen) Academic Press, NY.

PSYCHOPHYSICAL EVIDENCE OF TWO GRADIENTS OF NEURAL SAMPLING IN PERIPHERAL VISION

N Drasdo, C M Thompson and RJ Deeley

Department of Vision Sciences
Aston University
Birmingham B4 7ET

INTRODUCTION

Parvocellular (p) and magnocellular (m) processing systems have been studied extensively in primates. The p system processes fine detail, high contrast and colour, whereas the m system processes rapidly changing achromatic images of low contrast and lower spatial frequency (De Yeo and Van Essen, 1988; Livingstone and Hubel, 1988; Tootel et al., 1988). These may relate to transient/movement and sustained/pattern systems identified in psychophysical experiments, though the temporal frequency responses are slightly reduced (Kulikowski and Tolhurst, 1973; Anderson and Burr, 1985). Some evidence suggests that the p mechanism and sustained systems tend to be dominant in central vision, whereas the m and transient systems are increasingly evident in the periphery (Connolly and Van Essen, 1984; Schein and De Monasterio, 1987; Harwerth and Levi, 1978; Drasdo and Thompson, 1989). However, some contrary histological evidence has recently emerged. Perry and Silveira (1988) and Livingstone and Hubel (1988) have suggested that the numbers of m and p neurons do not vary in their relative proportions across the retina and striate cortex. We therefore applied stimuli based on these theoretical models to determine spatial threshold gradients in peripheral vision, which are believed to relate to receptive field centre separation (Weymouth, 1958; Klein and Levi, 1987).

Methods and materials

According to the physiological model, the threshold for a low contrast briefly presented grating would depend on m neurons, whereas p neurones would be dominant for a prolonged presentation at high contrast (Kaplan and Shapley, 1986; Tootel et al., 1988; Crook et al., 1988). We therefore determined the distribution of resolution threshold along the principal meridians by a perimetric staircase technique using appropriate stimuli (Fig 1). Square wave gratings with a circular field of five cycles diameter were projected in a Zeiss Kugel Perimeter fitted with a high speed magnesium alloy shutter, on a background luminance of 31.5 asb. Monocular stimulation was applied to the preferred eye of six normal subjects, two female and four male, aged 23 to 32 years.

According to psychophysical studies the sustained channel has a low pass response and the transient channel responds maximally at 5 to 9 Hz (Kulikowski and Tolhurst, 1973) or 7 to 13 Hz (Anderson and Burr, 1985). In another experiment therefore we aimed to determine the distribution of spatial selectivity of these channels along the horizontal meridian, for five normal volunteer subjects. Static, and 8 Hz counterphasing grating patterns with sinusoidal temporal and spatial luminance profiles were generated on a video monitor. A logarithmic series of spatial frequencies were presented in random order, using the increasing contrast paradigm (Ginsburg and Cannon, 1983) with an increment rate of .02 log units per second.

From Pigments to Perception, Edited by A. Valberg and
B.B. Lee, Plenum Press, New York, 1991

The stimulus field measured 2.5 cycles horizontally and 5 cycles vertically. The response criteria were detection of a static or moving unidirectional grating orientated vertically. The average pupil size was 5 mm for which the ocular modulation transfer function (MTF) can be approximated by the expression $EXP-((F/8.4)^{0.8})$. A correction was therefore applied to counteract the effects of optical degradation. The observed sensitivity was divided by this MTF, to yield a neural contrast sensitivity function (CSF).

Results

The group averaged data from the first experiment were fitted with polynomial functions (Fig 1a). The eccentricities (denoted E2) at which the threshold of resolution for the m and p stimuli had twice the foveal value (Klein and Levi, 1987) were 6.4° and 1.77° respectively. A marked difference in acuity was noted for central vision (Fig 1b). Similarly, in the second experiment, the spatial period for peak sensitivity, $SFmax^{-1}$ had an E2 value of 6° at 8Hz and 2.1° for the static grating (Fig 2b). The spatial dimensions for the p/sustained stimuli therefore varied more markedly with eccentricity from the fovea than did those for the m/transient stimuli.

Discussion

The results in Fig 1 seemed paradoxical at first acquaintance, causing us to recheck the data; because the thresholds for two paradigms were almost equal in the periphery even though there was a twenty-fold difference in the integrated contrast. Two possible explanations might be offered for this phenomenon and the correct one could only be identified by further experiments.

The first explanation is that envisaged at the outset of the experiment, which is that the spatial sampling of the two systems falls off at different rates with eccentric visual angle. However, in addition, it is necessary to suppose that there is a relatively higher efficiency in resolution for the m system compared to the p system for a given receptive field centre separation. This is conceivable for gratings, though not for small acuity targets, because of the spatial filtering properties of the m neurones (Crook et al., 1988) in an irregular array. When sampling an extended grating stimulus, these cells might exceed the predictions of the sampling theorem because this relates to a simple one dimensional array of linear detectors. The fact that the foveal resolution for the p stimulus was 2.4 times greater than for the m stimulus, despite the effects of optical degradation, is assumed to reflect the local difference in sampling density.

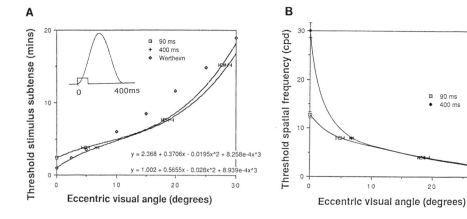

Fig 1 a) Group averaged resolution thresholds for gratings presented for 90 ms at 10% contrast, and for 400 ms with a filtered temporal wave-form at 80% contrast. b) The same data plotted in reciprocal form, emphasises the surprising similarity of grating acuity for the above m and p paradigms outside an eccentric visual angle of 5°.

However, due to the nature of the experiment it is difficult to exclude a second alternative possibility. This is that aliasing of spatial frequency may have occurred (Thibos et al 1987). Whether assuming disproportionate or proportionate m and p sampling densities across the visual field, a recent quantitive model (Drasdo 1990) predicts undersampling for the m but not for the p neurones for the gratings which were detected in the periphery (Fig 1). The previous literature shows that aliasing with moving gratings may occur in the periphery without optical bypass techniques, though these are usually necessary to reveal the effect with static gratings. This suggests that movement aliasing may relate to m neurones which correspond to a lower Nyquist limit in spatial frequency. However, an explanation of our data as an aliasing phenomenon is not totally convincing because no atypical appearances were reported by the observers, and the threshold eccentricity was quite similar when approaching from either direction.

Both of the above explanations are based on the assumption that the strategy used to isolate the m response was successful. The data on spatial selectivity from the second experiment could not be affected by aliasing and therefore supports the concept of differences in the gradients of spatial sampling for the two systems (Fig 2). We might expect this to correlate with receptive field centre diameter, but due to the broad bandpass, extensive overlap of receptive fields and the effects of spatial summation this also may tend to reflect receptive field centre separation.

Perry and Silveira (1988) and Livingstone and Hubel (1988) considered that the ratio of densities of the m and p neurons may be uniform across the ganglion cell layer of the retina and in the striate cortex. However, as we have seen, our experimental results, like some earlier studies, suggest a disproportion in spatial sampling by different sets of (m and p) neurones. This appears to be most marked in the central field up to 5° angular radius, which represents only 1 mm of the retina in the cynomolgus monkey. It coincides with the area of excavation, where the different types of ganglion cell are displaced laterally making it difficult to evaluate their relative densities. Furthermore according to Wassle et al. (1989), 5% of the neurons in the ganglion cell layer at the fovea and >50% in the periphery are displaced amacrines. Due to their size and laminar location these would tend to be counted as p cells, and thus to reduce the difference in m and p gradients of cell density.

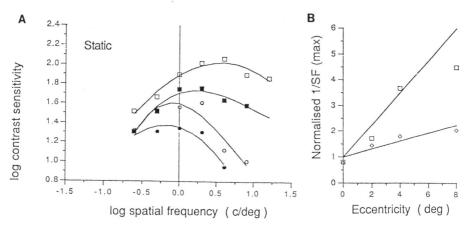

Fig 2 a) Contrast sensitivity functions (CSFs) for static and 8 Hz counterphasing gratings were divided by the estimated MTF of the eye, to yield neural CSFs. Group averaged data are shown for the static stimulus. Sensitivity declined with eccentricity (0,2,4 and 8°). b) Plots of the spatial frequency $^{-1}$ for the CSF peak value, against eccentric visual angle for counterphasing and static stimuli were normalised with respect to the foveal value to demonstrate the different gradients which may relate to spatial sampling by transient and sustained systems.

The observations of Perry and Silveira (1988) are not therefore very conclusive evidence in favour of proportional m and p sampling. The findings of Livingstone and Hubel (1988) are more persuasive, though in their discussion they did not press their argument with total conviction and in the light of the above evidence we conclude that it would be premature to reject the concept of different distributions of sampling by m and p neurones across the visual field.

REFERENCES

Anderson, S. J.and Burr, D. C., 1985, Spatial and temporal selectivity of the human motion detection system, Vision Res., 25:1147.

Connolly, M. and Van Essen, D., 1984, The representation of the visual field in the parvocellular and magnocellular layers of lateral geniculate nucleus of the macaque monkey., J.Comp. Neurol. 226: 544.

Crook, J. M., Lange-Malecki, B., Lee, B.B. and Valberg, A., 1988, Visual resolution of macaque retinal ganglion cells. J. Physiol. 396: 205.

DeYeo, E. A. and Van Essen, D. C., 1988, Concurrent processing streams in monkey visual cortex. Trends in Neuroscience, 11: 219.

Drasdo, N. and Thompson, C. M., 1989, Do visibility and colour recognition isopters relate to the distribution of Pa and Pb ganglion cells of the human retina? Ophthal. Physiol. Opt., 9: 447.

Drasdo, N., 1990, Neural substrates and spatial thresholds in peripheral vision, in : Kulikowski, J. J., Ed, "Limits of vision", MacMillan London (In Press)

Ginsburg, A. P. and Cannon, M.W., 1983, Comparison of three methods of rapid determination of threshold contrast sensitivity. Invest. Ophthalmol. Vis.Sci., 24: 793.

Harwerth, R. S. and Levi, D. M., 1978, Reaction time as a measure of suprathreshold grating detection., Vision Res., 18: 1579.

Kaplan, E. and Shapley, R. M. 1986 The primate retina contains two types of ganglion cells, with high and low contrast sensitivity. Proc. Natl. Acad. Sci., 83: 2755.

Klein, S. A. and Levi, D. M. 1987 Position sense of the peripheral retina. J. Opt. Soc. Am, 4, 1543.

Kulikowski, J. J. and Tolhurst, D. J. 1973, Psychophysical evidence for sustained and transient detectors in human vision. J Physiol , 232:149.

Livingstone, M. S. and Hubel, D. H. 1988 Do the relative mapping densities of the magno- and parvocellular systems vary with eccentricity? J. Neurosci. 8: 4334.

Perry, V. H. and Silveira, L. C. L., 1988, Functional lamination in the ganglion cell layer of the macaque's retina. Neurosci., 25: 217.

Schiller, P. H., Logothetis, N. K. and Charles, E. R., 1990, Functions of the colour-opponent and broad-band channels of the visual system. Nature, 343: 68.

Schein, S. J. de Monasterio F. M., 1987, Mapping of retinal and geniculate neurons onto striate cortex of macaque. J. Neuroscience, 7:996.

Thibos, L. N., Walsh, D. J. and Cheney, F. E., 1987, Vision beyond the resolution limit : aliasingin the periphery, Vision Res, 27:2193.

Tootel, R. B. H., Hamilton, S. L. and Switkes, E., 1988, Functional anatomy of macaque striate cortex. IV. contrast and magno-parvo streams. J. Neuroscience, 8: 1594.

Wässle, H., Grünert, U., Röhrenbeck, J. and Boycott, B.B., 1989, Cortical magnification factor and the ganglion cell density of the primate retina. Nature, 341: 643.

Weymouth, F. W., 1958, Visual sensory units and the minimal angle of resolution. Am. J. Ophthal., 46: 102.

DISCUSSION: P- AND M-PATHWAYS II

Ehud Kaplan

Rockefeller University
New York
N.Y., U.S.A.

Kaplan: We have witnessed a remarkable afternoon – three prominent scientists, working in related areas, agree among themselves. All three presented evidence of interactions between the M and P pathways; I wonder if we can find disagreement on this question in this group. I was glad to see these presentations, since it has become very fashionable recently to describe these pathways as very different and separable. I think that it is possible that such separation exists at the earliest levels of the visual system, in the retina and the Lateral Geniculate Nucleus, but once the information reaches the the cortical network, where the various elements are allowed to interact, we have ample opportunity for crosstalk between the systems, and reverberations and feedback within them. I feel that at least some of the phenomena that were discussed this afternoon can be attributed to what happens at a cortical level. Perhaps we can hear from van Essen something about the possible pathways for such interactions in V1 and beyond.

Another issue I hope will be discussed is relating the quantitative physiological knowledge we have from single cell studies to Visual Evoked Potentials (VEPs). Creutzfeldt once said that these potentials are very easy to measure and very hard to interpret, and that accounts for the huge literature about VEPs. In this context, I would like to relate one of the major differences between M and P cells to what we just heard from Dr. Fiorentini and her colleagues. M cells are eight times more sensitive to luminance contrast than are P cells, as Shapley and I have shown in 1986. It is tantalizing to interpret Fiorentini's results in relation to this difference in contrast sensitivity, but the relationship is not simple. She showed response vs contrast functions for luminance and for chromatic stimuli, and they had different slopes and different thresholds. The difficulty is that the responses to chromatic stimuli had a higher threshold, as you would expect, but they had a steeper slope than would be expected from the single cell responses. The luminance responses had a low threshold, again as expected, but with surprisingly shallow slopes. This is reminiscent of the results published by Nakayama and Mackeben (Vis. Res., 1982) on VEPs in the alert monkey. I was puzzled by their results, as I am by the ones we heard today.

Cavanagh: For the VEP, multiply the P cell data by 10, for that is the relative number of cells.

Kaplan: There are problems with this idea, because it requires that P cells should respond with little variation, if you assume algebraic summation of inputs to the cortex. There are great variations in the phase relationship of P-cells to the kind of luminance stimuli that we use. The incoherent of the response makes the summation much less effective. If you propose probability summation then you need many more P cells than are available.

Cavanagh: I didn't want to be so complicated. On Adriana's slide her axis was logarithmic, so that I don't think there are any problems with her data. The other question I had was

From Pigments to Perception, Edited by A. Valberg and
B.B. Lee, Plenum Press, New York, 1991

relating to the link between psychophysics and the physiology. The more physiologists tackle the properties of cortical neurons with respect to colour and luminance, the less it seems likely that they will ever be a luminance and a chromatic pathway found anywhere in the cortex. So I wanted to ask physiologists if we should really give up the ideas of luminance and chrominance.

Kulikowski: In 1983 we published a comparison of field potentials and spikes of multiunit and single unit recordings, and we found that there is a completely different transformation between the number of spikes against contrast and the amplitude of field potentials. Field potentials are definitely logarithmically related to contrast, but the spikes are a power function of contrast. This may be heretical, but the power function looks as if it was a mixture of P and M. Secondly, if you want really to analyse the pure colour response you have to be tremendously restrictive to a certain subtype of cells, showing linear antagonism.

Lee: Pat Cavanagh showed residual movement with equal luminance patterns was not due to phase, or inter-M-cell variability in spectral sensitivity, but what about the residual frequency-doubled response, because that is equivalent to about 10% achromatic contrast in the M-pathway.

Cavanagh: We tried to measure second-harmonic distortion components with our stimuli by a cancellation technique in the same way as we did for phase, and they were less than one percent, which is not to say that they could not have been much larger had we increased chromatic modulation amplitude or temporal frequency.

Lee: I don't think you can cancel out signals from both on- and off-centre cells by those means, by a linear compensating signal.

Valberg: Did you test the yellow-blue or tritanopic directions?

Cavanagh: In the opposing motion test we did test with tritanopic stimuli, there was much less contribution to motion than the red-green direction. The best red-green grating required about 12% luminance contrast to null its motion and the best tritanopic grating required only about 3-4%, which could be due to chromatic aberration. The contribution of tritanopic stimuli to motion was not strong enough to discriminate between luminance and truly chromatic motion.

Rodieck: Going on to Pat Cavanagh's and Karen DeValois' talks, one other thing which I feel strongly about is the fundamental difference between a percept and a pathway, and I think that these two things are being confused. It also came up with the emphasis of Bill Merigan versus Peter Schiller. Bill Merigan said motion is basically carried by both systems, and Peter was saying, well, look under these conditions of motion you just wipe out this system, or you just wipe out that system. Those two things reflect a difference of attitude. I think perhaps the easiest way to explain this is in terms the evolution of the visual system. Evolutionary pressures drive the organisation of this system, and not the notions of colour, motion etc. We can easily fall into a categorisation scheme which we think relates to the construction of the system. To give a simple example, take two critical issues in vision. First, you are looking off into the horizon and saying if something is there or isn't there. The other is that you are running for your life, either to catch something because you are very hungry, or to make sure that your not caught yourself. In the first situation movement has to do with the little motions of your eye as you gaze, in the second situation movement is because your visual world is jumping all over the place. Both of these situation we have to live with, and the convenient notion that all of this is contained within the concept of motion sensitivity, sort of misses the point - the pathways are results of evolution, they are not a result of a sort of theoretically analyses of what vision might be all about.

Cavanagh: To respond to Bob Rodieck's point about evolution. I can imagine that evolution does tend toward developing different streams, that's why the same cells don't control digestion and perception; it would be sloppy. I assume that the same thing must happen in vision, and I assume it's to capture independent attributes of the visual world. e.g. an object is an object, no matter what colour you present it. The visual system has an advantage in separating different attributes for the same reason that would separate digestion and perception. Each can then evolve optimally without having to rewire the entire system when one changes slightly. Anyone who does computer programming knows the enormous advantages of independent subroutines, which I see as equivalent to perceptual streams.

Rodieck: That's a good point. Let me give a specific example to clarify this. This is the direction selective system in the rabbit that is geared for image motion. There are four different types of cell, each pointing in different direction, and each is aligned to one of the four muscles of the eye and is concerned with a optokinetic nystagmus. There is another system also concerned with retinal image motion which is aligned with the three semicircular canals, and is concerned with the vestibular-ocular reflex. This may be a trivial example, but here are two examples of motion perception each of which is involved with a different pathway and a specific function. It was this sort of notion as to the distinction between percepts and pathways which I was trying to clarify. Here is a good example of two distinct pathways with no percept whatsoever.

Dow: Sometimes when we are busy looking at attributes we forget that in the real world they are attributes of stimuli, and even if there are these pathways at some point information has to combine to produce a final percept. We need somehow to tag the object in each pathway, and allow the common object in all these pathways to get back together.

Shapley: I just want to respond to a couple of things that Patrick Cavanagh said - one in the discussion and one in his talk. In his question about whether we should give up luminance pathways or chromatic pathways as a concept because we are having difficulty finding them in the visual cortex, I would just say that these concepts are very much test dependent and related to very precise psychophysical tasks. I think following the kind of approach that Lee and Valberg have done of trying to measure neurophysiological responses in situations which are very close to the psychophysical tasks, may actually illuminate why we are having difficulty and what the resolutions may be.

Schiller: I would like to turn this discussion to some considerations of what various streams might be doing in the system. I would like to use two analogies, and I would like to see if people could relate either or both of them to the M- and P-pathways as we discussed all day. The first one is the idea that two systems extract two different aspects of the visual environment, and as result, they can combine these two different impressions, and come up yet with another one, as yellow derives from an interactive process of M- and L-cones. Another analogy was to the rods and cones which interact very little. Rather what they do is to look at different aspects of the environment and feed into the same cells. By virtue of their existence, you extend the range you can see; they support two modes of operation. Is there any evidence in any of the physiological or human work which suggests that if the two systems extract different aspects of the environment, do they somehow combine to give you a kind of impression that neither of them have alone?

Richter: I think this may have to do with what I said this morning; combining the systems extends the luminance range.

van Essen: The idea of iterative processes Karen put forward is an attractive hypothesis, but I think we do not yet have a convincing example for her suggestion or for others in which the mechanisms can be tied down to feedback pathways between visual areas.

Creutzfeldt: Karen, could you comment more on what you think these reciprocal connections do. Is it more than saying that the brain has something to do with these perceptions and this is the the anatomy, or do you actually have a model. I can't imagine any model an engineer might design.

Karen DeValois: What strikes me thinking about it as a engineering problem is that you do not want to make either of these decisions, a decision about characteristics of motion colour or whatever, or a decision about segregation of objects in isolation from the other kind of information. There are a variety of ways and theories than one could build something to do that. I certainly have no compelling evidence for a strong model, wish I did.

Gilbert: An alternative mechanism to back connections is cross-talk between compartments at each stage, and so the iterative progress could be a progressive one.

Karen DeValois: It could indeed, and there are other suggestions as well. One reason I don't favour crosstalk is that with other tasks we find much better separation between the systems. With massive crosstalk, this might be difficult to get.

Krauskopf: Bart Farrel and I did very similar work. According to Edelson and Movshon's view, coherence was more likely to occur if the two gratings were similar, in contrast or spatial frequency, and we studied in particular the effect on coherence of chromatic gratings with different cardinal directions in the MacLeod-Boynton colour space. With different cardinal directions, the gratings always slid, were incoherent, but with the slightest bit of luminance contrast, coherence occurred. The striking result was that at 45° angles in the colour space, perfect coherence occurred. The similarity rule thus extends to colour space.

Kaplan: I would like to ask whether anyone thinks that the results that we heard this afternoon about interactions between M and P systems shed any light on the perplexing results that we heard before lunch about the lesion experiments. Can one reinterpret the lesion experiments in view of this demonstration of interaction between the supposedly separate streams.

Cavanagh: I didn't see any difference between this morning and this afternoon, one seemed to support the other. I guess we didn't have the right people here.

Schiller: A brief comment to follow up on what Bruce Dow said. I think it's the opposite of what Patrick was saying, that different systems evolve to analyse different aspects of the environment. Bruce said the opposite, and the example I would like to use is to talk about motion and colour. Colour vision may have evolved to defeat camouflage by colour. Now you add motion to this. When a perfectly camouflaged animal begins to move, we can again extract from the motion the shape of the object. So I think that both of these kind of processes have evolved in the course of time, the one that Patrick has talked about as well as the one that Bruce has.

van Essen: I want to elaborate on the point which Rodieck raised, in a somewhat different context. He drew the distinction between pathways and perception. The related point I have in mind is that we as neurophysiologists find certain properties in cells which is often a certain kind of selectivity or tuning, and our natural inclination is to assume not only that this information is used, but that it is used explicitly for certain kinds of perception, such as wavelength selectivity leads to colour perception. That need not be the case, however, and I think we also have seen examples where spectral information can serve as a cue, perhaps a concealed cue, which can help in a different kind of task, not in the perception of particular colours, but in tasks of figure segregation or motion analysis. Perhaps certain cell properties or certain pathways can be used in both direct and indirect fashion.

ON THE NATURE OF VISUAL EVOKED POTENTIALS, UNIT RESPONSES

AND PSYCHOPHYSICS

J.J. Kulikowski

Visual Sciences Laboratory, University of Manchester
Institute of Science and Technology
P.O. Box 88, Manchester (UK) M60 1QD

INTRODUCTION

There are two main electrophysiological measures of visual cortical responses: frequency of action potentials (spikes) and amplitude of slow potentials. Spikes are recorded with a microelectrode from single units, or groups of units. Slow field potentials (FP) can be recorded by the same microelectrode, or with another (local) macroelectrode. Another type of recording, occipital visual evoked potential (VEP), reflects global cortical activity and can be used in human as well as animal experiments. Although each of these electrophysiological responses is a different measure of visual cortical activity, and may reflect different neural processes (thereby providing complementary information about the visual system), all of them can be shown to have a general link with psychophysical thresholds, provided that visual stimuli are <u>suitably</u> chosen.

The present study illustrates how the neural basis of selected psychophysical phenomena can be elucidated by using global recordings. The choice of visual stimulus and its mode of presentation is particularly critical since this determines the range of mechanisms activated. If the range is too broad suprathreshold global responses (FP, VEP) reflect summed responses of various neural systems, each having a different threshold. Single unit recordings depend less on responses of other systems. With appropriately chosen visual stimuli, however, FPs and VEPs, have the advantage that they continuously sample neuronal responses.

The stimuli of choice used here to link psychophysical and electro-physiological experiments (see APPENDIX) are gratings switched on and off or, for comparison, reversed in contrast. The on-off presentation reveals linear components of responses (those which follow the fundamental frequency of presentation), characteristic of early stages of processing in pattern and colour vision (see separate on- and off- systems, Schiller 1982); reversal reveals even harmonics. Further emphasis may be put on the temporal aspects of responses, i.e. on testing such fundamental psychophysical functions as spatio-temporal sensitivity and resolution of linear elements of pattern and colour vision mechanisms. Specifically, with the aid of VEPs it may be concluded that spatial and temporal resolution of linear colour vision is low and determined by the cortical mechanisms, which provide the neural basis for clasical photometric phenomena.

BACKGROUND PSYCHOPHYSICAL AND VEP EXPERIMENTS

The electrophysiological studies summarised here have been selected for their correspondence with psychophysical phenomena in order to establish possible neurophysiological correlates with psychophysics. The following key psychophysical findings are relevant (see also APPENDIX: TERMINOLOGY):

1. Transient and sustained detection of luminance-modulated gratings

a. Coarse sinusoidal gratings (below 1 c/deg) are detected at threshold by a purely transient mechanism when presented abruptly. The threshold is then solely determined by a change in contrast irrespective of whether the grating is turned on or off, reversed in contrast (Kulikowski 1971; Kulikowski and Tolhurst 1973), incremented in contrast (Bain 1977; Kulikowski and Gorea 1978), or shifted abruptly sideways (Murray and Kulikowski 1984).
All these presentations produce a similar illusion of motion and also elicit similar occipital VEPs (Bain 1977; Kulikowski 1974; 1978).
b. Fine gratings (above 30 c/deg) are detected only by sustained mechanisms. The threshold is determined by standing contrast. Neither contrast reversal, nor abrupt sideways shifts are detectable (Murray an Kulikowski 1984).
No contrast reversal VEPs (which reveal transient components associated with apparent movement) are recordable (Kulikowski 1978; Russell et al. 1988).
c. Both mechanisms process intermediate spatial frequencies (see also 3).
d. Detection of apparent movement, connected with a transient mechanism, has a band-pass temporal frequency characteristic with a peak around 5-9 Hz for all spatial frequencies (Kulikowski and Tolhurst 1973).
e. Pattern detection (Kulikowski and King-Smith 1973; King-Smith and Kulikowski 1981) is based on sustained mechanisms (Tolhurst 1975) which are low-pass in the presence of eye movements; stabilised images have a decay time constant of about 2.5 s (King-Smith 1978).

2. Detection of chromatic gratings by sustained, low-pass mechanisms

Spatial and temporal characteristics of detection of isoluminant gratings depend critically on elimination of chromatic aberration effects (otherwise the eye focusses on one colour - Mullen 1985; Granger and Heurtley 1973). Chromatic aberration can be minimised by using gratings with few spatial cycles (8-10 for red-green). Spatial resolution limit at high chromatic contrast is around 11-12 c/deg, but only 8-9 c/deg at a contrast of 0.1. Beyond 13 c/deg it is not possible to identify colours of both chromatic components of chromatic patterns and textures, although such patterns may be visible, mainly due to chromatic aberation. At low contrasts (0.1), chromatic VEPs are obtained between 1 and 7 c/deg; above 7 c/deg chromatic gratings generate only residual achromatic-like VEPs (Murray et al. 1987).

3. VEPs and thresholds

Grating stimuli near psychophysical threshold elicit just resolvable VEPs in man (Kulikowski & Kozak 1967). Contrast thresholds can be predicted by extrapolation from the VEPs against a logarithmic scale of contrast (Campbell & Maffei 1970). Analysis of the VEP-contrast relationship can be further developed for low-rate presentations (< 3Hz), at which individual components (with different latencies) are separable. These different components predict different thresholds (note that stimuli which generate apparently simple, nearly sinusoidal waveforms, e.g. 8 Hz reversal, confound these components). Two modes of presentation, ON-OFF and REVERSAL, can be used.
(a) ONSET DURATION: TEMPORAL SUMMATION. For a brief onset (<60ms) the total VEP size is proportional to log suprathreshold contrast. Longer onset times reduce contrast thresholds, but the early components (~100ms) are hardly affected since they appear before the grating offset (Kulikowski 1972). The late components, not the early complex representing the initial responses of the visual cortex, account for temporal summation.

(b) REVERSAL reveals only transient components of global responses; hence the reversal VEPs correlate with motion detection thresholds (Kulikowski 1978).

METHODS

Stimuli. Vertical gratings were the main stimuli, common to all acute single- and multi-unit experiments on anaesthetised and paralysed monkeys, as well as in VEP experiments on sedated rhesus monkeys and humans. In single-unit (acute) experiments, bars, segments (and edges) were used to map the receptive fields; they were generated on a Tektronix display (phosphor P4). When necessary, hue was adjusted by using broad-band coloured filters combined with an additional front-projected neutral background to create coloured stimuli. This was particularly important when testing chromatic cells whose responses to luminance-modulated stimuli are generally weak, unless their wavelength composition is adjusted to elicit excitatory responses (Fig.1A).

In VEP studies, chromatic and achromatic gratings were generated on a Grundig colour TV monitor as described previously (Murray et al. 1987; Kulikowski et al. 1989a) and presented on and off, or in contrast reversal mode. The colour (hue) and luminance of the grating stripes were adjustable. The colour components of the chromatic gratings R (red) and G (green) are described by their luminance ratio $R/(R+G)$. This ratio for monochrome-luminance modulated gratings has a value of either 0 (green-dark green), or 1 (red-dark red). Intermediate values of $R/(R+G)$ then provide a numerical value for the relative intensity of the red and green bars.

For isoluminant, hue-modulated gratings, $R/(R+G)$ was measured for human subjects, using heterochromatic flicker photometry (HFP), and was found to be around 0.5 for normal observers. In monkeys, the electrophysiological equivalent of HFP can be performed by recording fast reversal VEPs at different values of $R/(R+G)$. The response is reduced to noise levels at around 0.5, and this is interpreted as indicating that macaque and human isoluminance are comparable in our subjects.

The achromatic stimulus consisted of yellow, luminance-modulated gratings (red and green gratings superimposed in phase). In all cases, the mean luminance and mean hue of the screen were kept constant throughout the presentation. Luminance contrast was defined as $(Lmin-Lmax)/(Lmin+Lmax)$ for luminance-modulated gratings. The stimulus subtended 10 or 20 deg, i.e. 5 or 10 spatial cycles for the 0.5 c/deg grating (note that transverse chromatic aberation for up to 10 cycles of a red-green gratings is minimal).

Recordings. In acute animal experiments, under anaesthesia (nitrous oxide plus Nembutal) and paralysis (Flaxedil and Tubarine, details in Kulikowski and Vidyasagar, 1986) a tungsten microelectrode was used to record 3 types of neuronal responses which were directed to 3 separate channels: single-units, multi-units and slow field potentials (filter: 1-30 Hz).
VEP recordings were from occipital disc electrodes (placed 10% of inion-nasion distance and 10% either side of the midline, or 15% in macaques), with a mid-frontal or linked-ears reference using a Medelec averager ER94 (filters with corner frequencies 10-30 Hz for 12.5 Hz reversal and 1-30, or 3-30 Hz for 2 Hz presentations). Typically 128 or 256 sweeps of the averager were used and each average took 1 or 2 minutes respectively to obtain.

Subjects. Overall, 10 Rhesus macaques were used in these experiments. VEPs were obtained in 6 animals; recordings were made whilst the animal was seated in a primate chair. Sedation with Ketamine (initial dose: i.m. 10-20 mg/Kg), augmented with Diazepam if necessary, was used.

ELECTROPHYSIOLOGICAL EXPERIMENTS

The experiments reported here are designed to show the response characteristics of units belonging to different sub-systems, with particular attention paid to linear mechanisms. To isolate mechanisms, visual stimuli were often chosen to activate channels described as achromatic-transient and chromatic-sustained (Kulikowski et al. 1989a). The results are grouped in 3 parts: (1) single unit recordings, (2) recordings of multi-unit responses and field potentials (1 and 2 under anaesthesia and paralysis), and (3) VEP recordings without anaesthesia in man (under slight sedation in macaques).

PROPERTIES OF SINGLE UNITS RECORDED IN THE STRIATE CORTEX

Single units in the macaque striate cortex were classified as concentric (chromatic and achromatic), simple (end-stopped included), complex or hypercomplex, according to standard criteria (Hubel & Wiesel 1968), modified by Kulikowski & Vidyasagar (1984, 1986) to select for further analysis cells (concentric and simple) with a substantial degree of linearity (Enroth-Cugell and Robson 1966; Kaplan and Shapley 1982).

Chromatic opponent cells were those which responded much less to white spots flashed on their receptive field centres than to spots whose colour was optimally adjusted (see also Hubel and Wiesel 1968; Dow and Gouras 1973). Chromatic opponency could not be easily verified in most receptive field surrounds, partly because of low-levels of spontaneous activity. Some chromatic opponent cells resembled the "modified type II" cells described by Ts'o and Gilbert (1988) in having conspicuous chromatic opponency in their receptive field centres, but differed by responding to long bars (which were used as search stimuli). Only 2 out of 14 cells examined could be unambiguously identified as double opponent, i.e showing opponent arrangement both in the receptive field centre and surround (Michael 1989; Thorell et al. 1984; Kulikowski and Vidyasagar 1987a); three cells gave small reponses to the complementary colour bar from the surround (see Kulikowski et al. 1989b, Fig.1.1) and the responses of two others were also consistent with double opponency. Such cells may project to colour areas and form the basis of colour perception (Zeki 1980). Cells not showing specific spectral biases were classified as achromatic.

Example of spatial and temporal properties of chromatic and achromatic units

The recordings obtained in the same electrode penetration allow the comparison of units representing the same location in the visual field. An important landmark is recording from fibres in lamina 4, whose properties resembled those of LGN cells (Wiesel and Hubel 1966; Hicks et al. 1983). An example is shown below. Figure 1A,B shows spatial and spatial frequency properties of units which were located between lower layer 3 and layer 4C and represented the visual field about 4 deg below the visual axis. There is a conspicuous difference in receptive field sizes of the input fibres (which are similar to those encountered in the parvo-cellular layers of the LGN) and of both types of cortical cells. Typically, chromatic-opponent cells in the visual cortex have larger receptive fields than parvo neurons in the LGN and this is illustrated in Fig 1A; the cell is chromatic-opponent only in the centre since a response to a light green bar was not obtained. If such LGN afferents were the main inputs to chromatic units, directly or via another neuron (see Livingstone and Hubel 1984; 1987; Ts'o and Gilbert 1988; Michael 1989), this would suggest extensive spatial summation with several LGN afferents contributing to the receptive field centre of a cortical unit. An alternative is that completely different LGN units form the inputs to linear chromatic cells (see below), whereas the type I LGN signals may be inputs to texture processing cells (Logothetis et al. 1989).

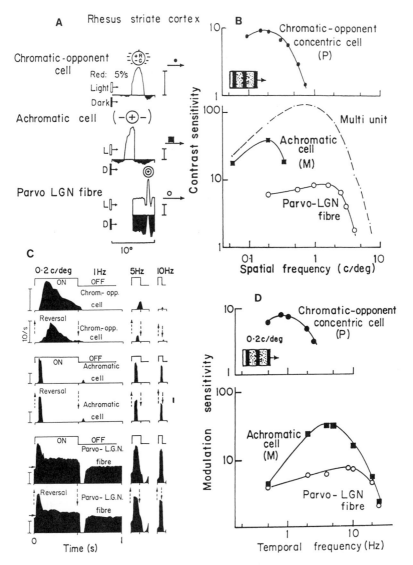

Fig. 1. The response characteristics of 3 types of units in V1 (one track,
from above lamina 4 to the site of LGN afferents): chromatic-opponent
cell, achromatic cell and a fibre with opponency type I.
(A) Responses to bars (light - upper histograms and dark lower
histograms), moving across the receptive fields. Calibration bars: 10
spikes/s; note a weak response of the chromatic opponent cell in spite of
high luminance-contrast (0.5) and red hue of bars.
(B) Contrast sensitivity against spatial frequency for all 3 units,
obtained with luminance-modulated (red) gratings drifting at constant
temporal frequecies (1 Hz for the chromatic cell, 5 Hz for the other
units). Note low sensitivity of both chromatic units.
(C) Responses of units to a 0.2 c/deg luminance-modulated grating,
presented on and off and contrast reversing. Contrast change is 0.5.
(D) Modulation sensitivity against temporal frequency of a drifting
grating (0.2 c/deg) for all units. Note similar temporal resolution limits
of the achromatic cell and the fibre.

Figure 1B shows contrast sensitivity as a function of spatial frequency of drifting luminance-modulated gratings for the same units. The poor spatial resolution of both cells is consistent with the sizes of their receptive field centres. However, the achromatic cell has much higher luminance contrast sensitivity and narrower spatial frequency tuning than the chromatic cell. There are many other achromatic units, having higher resolution limits, each covering a section of a broad range of spatial frequencies (see Fig.2, and Kulikowski and Vidyasagar 1986). The overall contrast sensitivity for this cortical recording site is best visualised by multi-unit responses (dash-dotted line). The achromatic cell in Fig 1 was specifically selected since (a) it represents a similar spatial frequency preference to the chromatic cell and can, therefore, be driven by the same grating stimulus); (b) it exhibits extremely transient responses to an onset or reversal of a grating (Fig.1C). Note that only fast, transient bursts of spikes are elicited and the temporal frequency increase up to 10 Hz hardly reduces the response. Moreover, onset of a grating of the same spatial frequency elicit the same transient responses provided that the step change of contrast is also the same (Fig.1C).

Responses of the chromatic cell to the same luminance-modulated grating stimulus have very different properties. These responses are rather weak, increase slowly, do not follow high temporal frequencies and also show a sustained component, which, although small (see below), is best visualised when comparing the response to the grating onset (Fig.1C, top histograms).
The parvo-fibre's responses do not exhibit the sluggish properties which are characteristic of cortical chromatic cells (although specific tests do reveal such properties in the receptive field surround, see - Dreher et al. 1976). The fibre's response shows both transient and sustained components. The transient response can follow high rates of presentation. These results are consistent with the responses of the parvo-LGN cells and their retinal inputs (Gouras and Zrenner 1979; Hicks et al. 1983; Lee et al. 1989; Zrenner 1983). Modulation sensitivity curves for the units in Fig.1D are consistent with their response properties to reversal (C) in that they show high temporal resolution limits. Thus it is obvious that temporal properties of chromatic opponent cortical cells are different from most P-LGN cells. This suggests that either some temporal integration mechanism, which slows the responses of chromatic cells must be sited in the striate cortex, a conclusion which has been reached independently by several researchers (Hicks et al. 1983; Lee et al.1988; Kulikowski and Russell 1989), or that completely different, as yet unidentified, inputs feed cortical cells.

Spatio-temporal properties of chromatic and achromatic cortical cells: Optimal velocities of cortical cells

Fig.1 showed an example of a chromatic cell in V1 combined with an achromatic cell of similar receptive field size which responded transiently. How typical are they? Fig. 2 shows a scatter plot of combined spatial and temporal properties of cortical cells. The choice of drifting stimuli to examine temporal properties is important for paralysed preparations since flashing stimuli reveal strong nonlinear relations in the temporal domain and also because moving (but not flashing) images can be perceived in the absence of eye movements (Sharp 1972).

All cells in this sample which showed linear chromatic opponency in their receptive field centres (Fig.2A cirles with dots) had both low spatial and temporal resolution and low achromatic contrast sensitivity. The other chromatic cells show similar properties. Conversely, the achromatic cells have a broad range of properties. The cells tuned to low spatial frequencies (with optimal spatial frequencies close to the maximal sensitivity for a site) responded transiently and had high contrast sensitivity with a peak at about 5 Hz (Fig. 2A); this temporal preference is comparable to that of the

motion detection (transient) channel (Kulikowski and Tolhurst 1973).
Transient achromatic cells tuned to higher spatial frequencies (5 Hz line in
Fig.2A) have lower peak sensitivity than the cells with slower velocity
preferences and sustained properties (see Kulikowski et al. 1989b), which is
in accord with psychophysical findings (Kulikowski & Tolhurst 1973). Fig. 2A
does not include cells with the finest receptive fields (with the highest
spatial resolution and most sustained) whose linearity could not be verified.
The cat visual cortex has no comparable chromatic cells; achromatic units in
V2 are very fast and transient, whereas the simple cells tuned to high
spatial frequencies are slow (Fig. 2B).

In summary, recording from individual cells in the visual cortex (V1)
helps to define the different properties of chromatic and achromatic cells.
The linear chromatic cells are generally sluggish, whereas those achromatic
cells which respond at low contrasts have high temporal frequency responses
and exhibit transient characteristics. In addition, by comparing responses
obtained from fibres in the input layer of V1 (considered to be inputs to
colour cells - see Michael 1989) with those of individual cells, it is
plausible to suggest that a substantial reorganisation of signal processing
occurs in the visual (striate) cortex, since the same stimuli elicit
different responses in the retina/LGN and V1 (see also responses in Fig.4).

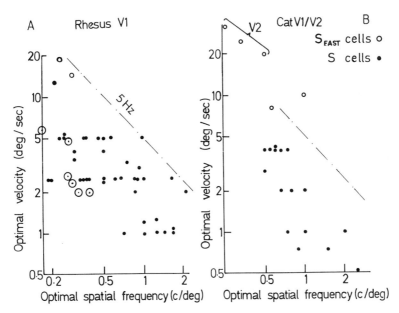

Fig. 2. Optimal velocities for "linear" visual cortical cells tuned to
various spatial frequencies, in macaque (A) and cat (B). The cells
described in: A: Kulikowski and Vidyasagar, 1984, 1986, 1987a,
B: Kulikowski and Bishop, 1981. The optimal velocity determined as giving
the maximal peak responses to drifting bars (or similar responses profiles
in both directions). The cells tuned to low spatial frequencies and high
velocities (optimal temporal frequency about 5 Hz) have highest contrast
sensitivities. Note similar optimal velocities are shared by simple cells
tuned to different spatial frequencies (as would be required by Gabor-like
analysis of moving retinal images); concentric chromatic double opponent
cells (marked by circles with a dot) and other chromatic cells (not shown
here) have low optimal spatial and temporal frequencies.

MULTI UNIT RESPONSES AND FIELD POTENTIALS

Multi-unit responses and FPs both coincide with threshold responses (when extrapolated to zero amplitude), but they are different functions of suprathreshold contrast (Kulikowski and Vidyasagar 1983). Multi-unit spike responses are close to power functions with an exponent close to one, similar to human suprathreshold contrast sensation (Kulikowski 1976), but do not reflect contrast constancy (Hess and Lillywhite 1980). Conversely, FPs, like VEPs, are best fitted by the function: log(C/threshold). Hence it is possible to compare either FPs or VEPs for various spatial frequencies at equal suprathrathreshold contrasts since the amplitudes are similar, as Fig. 3D,E indicates for a parafoveal representation of V1, around lamina 4. It is interesting that, in spite of similar amplitudes, the FP latency increases with spatial frequency, thereby indicating activity of different cell types. A similar effect in human VEPs to fine gratings (15-30 c/deg see Russell et al. 1988) was interpreted as the evidence for increasing involvement of slow units, probably of parvo-type. This notion is substantiated by the recordings from lamina 4 of V1 and from the responses of both (A) phasic (4Ca) and (B) parvo (4Cb) neurons (transient and sustained respectively in Fig.3). It is likely that cells with parvo inputs dominate the detection of finest gratings at a site, in Fig.3 above 8 c/deg, in human central vision above 45 c/deg.

VISUAL EVOKED POTENTIALS IN MAN AND MACAQUE: SIMILARITIES AND SPECIFICITIES

Macaque and man have two slightly different topographies of V1, although their visual performance and basic visual cortical architecture may be similar. The macaque V1 has the foveal representation placed laterally and the lower field to 6 deg is represented on the upper surface (operculum), whereas in man only the central foveal representation emerges at the occipital tip. This partly accounts for relative weak human VEPs elicited by coarse sinusoidal gratings. Thus 2 c/deg may be used as a compromise and chromatic VEPs can be recorded for gratings up to 7 c/deg (Murray et al. 1987). In macaques, coarser gratings may conveniently be used, e.g. 0.5 c/deg (Fig. 4A,B), which minimises effects of inaccurate fixation.

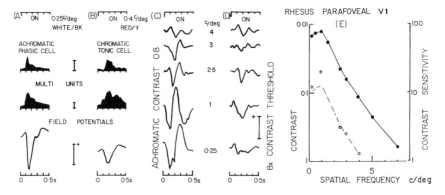

Fig. 3. Single-unit, multi-unit and field potential (FP) responses evoked by (A) achromatic and (B) chromatic (red-yellow) gratings (centered on the receptive field). FP responses at either a fixed contrast (C) or (D) 8 times above thrseshold (see also circles in F). Thresholds verified with drifting gratings and reducing contrast until the modulated dischage of many units just ceases to be noticed (F). Calibration: 10 spikes/s and 10 uV, respectively (Kulikowski and Vidyasagar 1987b).

Two features characterise VEPs elicited by low contrast coarse gratings:
1) At low temporal frequencies luminance-modulated and chromatic grating (A)
elicit VEPs of different properties, transient and sustained respectively.
2) At high temporal frequencies contrast-reversal VEPs show a distinct
minimum at isoluminance (Fig. 4B). Fig. 4C shows the corresponding recording
in man (Kulikowski and Russell 1989) combined with a psychophysical task of
setting minimum flicker. Minimal sensation coincides with a minimal VEP.
Conversely, the pattern electroretinogram (Fig. 4D) elicited by the same
stimulus has no minimum, in agreement with retinal recordings.

In summary the VEP data suggests that few or no chromatic units respond to
fast flickering gratings, although such units exist at precortical stages.
Human VEP studies can also illustrate that there are no units responding to
low contrast chromatic gratings above 7 c/deg in a manner characteristic of
linear processing (i.e. responding differently to onset than to offset).

CONCLUDING REMARKS
A. Occipital VEPs can be used to record the activity of isolated mechanisms.
It is possible to examine the presence or absence of neural responses in man
and animals without anaesthesia under normal and pathological conditions.
B. If the VEPs are to serve as an index of responses from separate parallel
pathways, investigators should choose visual stimuli which selectively
activate a single channel with minimal interference, but at the expense of
the VEP magnitude. A paradigm case of stimulus specificity is the use of low
contrast and low spatial frequency, chromatic or achromatic gratings, which
dichotomise the VEP responses into two classes:
1: Parvo-cellular specific, linear, non-transient and chromatically sensitive
2: Magno-cellular specific, nonlinear, transient and luminosity sensitive.

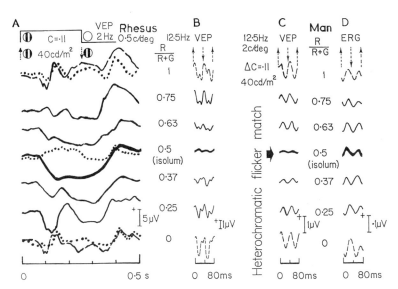

Fig. 4. Occipital Visual Evoked Potentials (VEPs in macaque and man) and
human pattern Electroretinogram (ERG) elicited by gratings with different
chromatic contents, determined by luminance ratio of red (R) and green (G)
phosphors, R/(R+G). Pure red (R=1) and green (G=0) gratings are luminance-
modulated, in fact achromatic: the corresponding records are broken lines.
For R=0.5 gratings are isoluminant (equal luminance of red/green stripes).
(A) VEPs elicited by on-off and reversal presentations (2 Hz) at the same
change in contrast (0.11); note the similarity of onset and reversal
achromatic VEPs and the difference at isoluminance. (B) VEPs to 12.5 Hz
reversal. (C) and (D) A comparison of human VEPs and ERGs for different
chromatic contents: unlike VEPs, the ERG is not reduced at isoluminance.

C. When VEPs are closely correlated with thresholds and unit responses, they can be considered functionally and physiologically meaningful. Moreover, VEPs provide information about suprathreshold responses and separate visual mechanisms can be isolated by combining the VEP and behavioural responses.

ACKNOWLEDGEMENT. These studies were supported by The Wellcome Trust and SERC.

REFERENCES

Bain, R., 1977, The evoked potential: Its response characteristics and uses in visual analysis. M.Sc. thesis. University of Manchester.

Campbell, F.W. and Maffei, L. 1970, Electrophysiological evidence for orientation and size detectors in man. J. Physiol., 207: 635-652.

Carden, D., Kulikowski, J.J., Murray, I.J. and Parry N.R.A., 1985, Human occipital potentials evoked by the onset of equiluminant chromatic gratings, J. Physiol., 369: 44P.

Dow, B. and Gouras, P., 1973, Colour and spatial specificity of single units in rhesus monkey foveal striate cortex, J. Neurophysiol. 36:79-99.

Dreher, B., Fukuda, Y. and Rodieck, R.W., 1976, Identification, clasification and and anatomical segregation of cells with X-like and Y-like properties in the LGN of old-world primates. J. Physiol., 258: 433-452.

Enroth-Cugell, C. and Robson, JG., 1966, The contrast sensitivity of retinal ganglion cells of the cat, J. Physiol. 187: 517-552.

Gouras, P and Zrenner, E, 1979, Enhancement of luminance flicker by color-opponent mechanisms, Science, 205: 587-589.

Granger, E. M. and Heurtley, J.C., 1973, Visual chromaticity modulation transferfunction, J. Opt. Soc. Am. 63: 73-74.

Hess, R.F. and Lillywhite, P.G., 1984, Effect of luminance on contrast coding in cat visual cortex, J. Physiol., 300: 56P.

Hicks T.P., Lee B.B. and Vidyasagar T.R. (1983), The reponses of cells in macaque lateral geniculate nucleus to sinusoidal gratings, J. Physiol., 337, 183-200.

Hubel D.H. and Wiesel T.N., 1968, Receptive fields and functional architecture of monkey striate cortex, J. Physiol., 195: 215-243.

Kaplan E. and Shapley R., 1982, X and Y cells in the lateral geniculate nucleus of macaque monkeys. J.Physiol., 330,125-43.

King-Smith, P.E., 1978, Analysis of the detection of a moving line, Perception, 7: 449-454.

King-Smith, P.E. and Kulikowski, J.J., 1981, The detection and recognition of two lines. Vision Res., 21: 235-250.

Kulikowski, J.J., 1971, Effect of eye movements on the contrast sensitivity of spatio-temporal patterns, Vision Res., 11: 83-93.

Kulikowski, J.J., 1972, Relation of psychophysics and electrophysiology, Trace, 6: 64-69.

Kulikowski, J.J., 1974, Human averaged occipital potentials evoked by pattern and movement, J. Physiol., 242: 70-71P.

Kulikowski, J. J., 1976, Effective contrast constancy and linearity of contrast sensation, Vision Res., 16: 1419-31.

Kulikowski, J.J., 1978, Pattern and movement detection in man and rabbit: Separation and comparison of occipital potentials, Vision Res., 18: 183-189.

Kulikowski, J.J. and Bishop, P.O., 1981, Linear analysis of the responses of simple cells in the cat. Exp. Brain Res., 44: 386-400.

Kulikowski, J.J. and Gorea, A., 1978, Complete adaptation to patterned stimuli: Weber law for contrast, Vision Res., 18: 1223-27.

Kulikowski, J.J. and Kozak, W., ERG, visual evoked responses and pattern detection in man, in: "Electrophysiology and Pathology of the Visual System", E. Schmoger, ed., VEP Thieme, Leipzig (1967).

Kulikowski, J.J. and King-Smith, P.E., 1973, Spatial arrangement of line, edge and grating detectors revealed by subthreshold summation. Vision Res., 13:1455-1478.

Kulikowski J.J., Murray I.J. and Parry N.R.A., 1989a, Electrophysiological correlates of chromatic-opponent and achromatic stimulation in man, in: "Colour Vision Deficiencies IX", G. Verriest and B. Drum, eds., Kluwer Acad. Publ., Dordrecht.

Kulikowski J.J. and Russell M.H.A., 1989, Electroretinograms and visual evoked potentials elicited by chromatic and achromatic gratings, in: "Seeing Contour and Colour", J.J. Kulikowski, I.J. Murray and C.M. Dickinson, eds., Pergamon Press.

Kulikowski, J.J. and Tolhurst, D.J., 1973, Psychophysical evidence for sustained and transient detectors in human vision. J. Physiol., 232: 149-162.

Kulikowski, J.J. and Vidyasagar, T.R., 1983, Single-unit, multi-unit and field potential responses in the striate cortex of macaque and cat as a function of contrast, J. Physiol., 334: 19-20P.

Kulikowski J.J. and Vidyasagar T.R., 1984, Macaque striate cortex: Pattern, movement and colour processing, Ophth.Physiol.Opt.4: 77-81.

Kulikowski, J.J. and Vidyasagar, T.R., 1986, Space and spatial frequency: Analysis and representation in the macaque striate cortex, Exp. Brain Res., 64: 5-18.

Kulikowski, J.J. and Vidyasagar T.R., 1987a, Linear antagonism and suppression in the visual cortex of rhesus monkey and cat. J. Physiol., 390: 200P.

Kulikowski, J.J. and Vidyasagar, T.R., 1987b, Neuronal responses and field potentials evoked by gratings in the macaque striate cortex, J. Physiol., 392: 56P.

Kulikowski, J.J., Vidyasagar, T.R. and Carden, D., 1989b, Linear/nonlinear analysis of chromatic information, in: Seeing Contour and Colour", J.J. Kulikowski, I.J. Murray and C.M. Dickinson.,ed, Pergamon Press.

Lee, B.B., Martin, P.R. and Valberg, A., 1988, The physiological basis of heterochromatic flicker photometry demonstrated in the ganglion cells of the macaque retina, J. Physiol., 404: 323-347.

Lee, B.B., Martin, P.R. and Valberg, A., 1989, Sensitivity of macaque retinal ganglion cells to chromatic and luminance flicker. J. Physiol., 414: 223-243.

Livingstone M.S. and Hubel D.H., 1984, Anatomy and physiology of a color system in the primate visual cortex, J. Neuroscience, 4: 309-356.

Livingstone M.S. and Hubel D.H., 1987, Psychophysical evidence for separate channels for the perception of form color movement and depth. J. Neuroscience 7:3416-3468.

Logothetis, N.K., Schiller, P.H., Charles, E.R. and Hurlbert, A.C., 1990, Perceptual deficits and the activity of the color-opponent and broad-band pathways at isoluminance, Science 247: 214-217.

Michael C., 1989, The origin of double opponency in the monkey striate cortex, in: "Seeing Contour and Colour", J.J. Kulikowski, I.J. Murray and C.M. Dickinson, eds., Pergamon Press.

Mullen, KT., 1985, The contrast sensitivity of human colour vision to red-green and blue-yellow chromatic gratings, J. Physiol., 359:381-400.

Mullen, K.T. and Kulikowski, J.J., 1990, Wavelength discrimination at detection threshold. J. Opt. Soc. Am., 7: 733-742.

Murray, I.J. and Kulikowski, J.J., 1984, Movement detection and spatial phase. Ophthal. Physiol. Opt., 4: 73-76.

Murray, I.J., Parry, N.R.A., Carden, D. and Kulikowski, J.J., 1987, Human visual evoked potentials to chromatic and achromatic gratings. Clin. Vision Sci., 1: 231-244.

Regan, D., 1970, Objective method of measuring the relative spectral luminosity curve in man, J. Opt. Soc. Am., 60: 856-859.

Regan D. and Spekreijse H., 1974, Evoked potential indications of colour blindness, Vision Res., 14: 89-96.

Russell, M.H.A., Kulikowski, J.J. and Murray, I.J., 1988, Spatial
frequency dependence of the human visual evoked potential, in:
"Evoked Potentials 3", C. Barber and T. Blum, ed., Butterworth.

Schiller, P.H., 1982, Central connections of the retinal ON and OFF
pathways. Nature, 297: 280-283.

Sharp, C.R., 1972, The visibility and fading of thin lines visualized by
their controlled movement across the retina. J.Physiol., 222:113-34.

Thorell, L.G., De Valois, R.L and Albrecht, D.G., 1984, Spatial
mapping of monkey V1 cells with pure color and luminance stimuli,
Vision Res., 24: 751-769.

Tolhurst, D. J., 1975, Sustained and transient channels in human vision.
Vision Res., 15: 1151-1155.

Tootell, R.B.H., Hamilton, S.L. and Switkes, E., 1988, Functional
anatomy of macaque striate cortex. IV: Contrast and Magno/Parvo
streams. J. Neurosci., 8: 1594-1609.

Ts'o D.Y. and Gilbert C.D., 1988, The organization of chromatic and
spatial interactions in the primate striate cortex,
J. Neurosci., 8: 1712-1727.

Wiesel, T.N. and Hubel, D.H., 1966, Spatial and chromatic interactions in
the lateral geniculate body of the rhesus monkey,
J. Neurophysiol., 29: 115-1156.

Zeki, S., 1980, The representation of colours in the cerebral
cortex, Nature, 284: 412-18.

Zrenner E. (1983). "Neurophysiological aspects of Colour Vision in
Primates". Springer-Verlag, Berlin.

APPENDIX: TERMINOLOGY

Psychophysical experiments make it possible to formulate certain concepts
of mechanisms whose physiological basis are discussed. To avoid confusion
these concepts will be explicitly defined, as used in this study.

* Chromatic opponent mechanisms are those which lead to identification of
unitary hues in a stimulus. Thus, for example, at least two extreme colours
have to be identified in a hue-modulated isoluminant grating. Isoluminance is
determined by heterochromatic flicker photometry (minimum flicker).

* Transient and sustained mechanisms are those which lead the contrast
thresholds being defined, respectively, by a transient change in contrast or
a standing contrast (revealed by a comparison of thresholds obtained with
ONSET and CONTRAST REVERSAL presentations of gratings, see Kulikowski and
Tolhurst 1973). These mechanisms are also called according to their band-pass
and low-pass temporal frequency characteristics (by analogy to amplifiers).

* Essentially linear mechanisms are those which show (substantial) summation
at threshold, and preserve polarity, or phase sensitivity (e.g. light versus
dark, green versus red) and whose threshold sensitivities to bars, edges and
gratings were interrelated according to standard mathematical equations
(Kulikowski and King-Smith 1973). Linear units are associated here with the
X-units of Enroth-Cugell and Robson (1966), giving modulated responses to
drifting gratings, predominantly first-harmonic response to ON-OFF grating
presentations and a null response at a specific spatial phase.

* Psychophysical detectors are defined here as processing units whose
sensitivity is highest for a given stimulus. In some cases, certain stimulus
features can be identified at the detection threshold (King-Smith and
Kulikowski 1981; Mullen and Kulikowski 1990). Note that the above units are
not trigger-feature detectors (except in colour vision). For example, visual
detection of 25 Hz flicker is normally determined by a specific type of unit
(M) with luminosity spectral sensitivity; however, when in a disease, these
units are damaged, completely different types of units must fulfill this
role, since the spectral sensitivity curve has a second peak at 450 nm.

* <u>Correlation between global responses (multi-unit discharges, FPs and VEPs)</u>
<u>and psychophysics</u> is claimed only if a specific threshold can be predicted
(by extrapolation) from an electrophysiological threshold of a distinct
response component; both thresholds must be elicited by the same stimulus.
The VEP response component must be functionally related to the stimulus being
processed, e.g. only onset VEPs (but no offset or reversal VEPs) can reflect
sustained responses involved in processing of colour and fine pattern
information (Kulikowski 1972, 1978; Russell et al. 1988; Kulikowski et al.
1989a). Occasionally, temporal filtering of the VEPs is needed to reveal
correlation with psychophysics (Regan 1970). So far, no electrophysiological
measure of suprathreshold sensation of contrast has been found; the closest
fit was provided by multi-unit responses (Kulikowski and Vidyasagar 1983).

LOCALIZATION OF THE ELECTROMAGNETIC SOURCES OF THE PATTERN ONSET

RESPONSE IN MAN

Henk Spekreijse

The Netherlands Ophthalmic Research Institute and The
Laboratory of Medical Physics and Informatics
P.O. Box 12141, 1100 AC Amsterdam-ZO, The Netherlands

INTRODUCTION

The recent technological developments in PET, SPECT and NMR allow
for detailed images of the anatomical structures or metabolical
activity of the human brain. Given these non-invasive imaging
facilities, one may wonder whether there remains a future for mapping
the electromagnetic fields produced by the brain. Their spatial (3-D)
resolution for localizing activity is comparable to e.g. NMR; in
temporal resolution they will, however, remain superior to the above
mentioned imaging techniques. Only recording of the electromagnetic
fields enables to study the function of the brain with a time
resolution comparable to the processing speed of the brain, i.e. of
the order of 10 to 100 ms. I will illustrate in this review by means
of the visual pattern onset response that such a time resolution is
needed to discriminate between the various sources of activity within
the human brain. In man these fields can be recorded by either
measuring the gradient of the magnetic flux near the scalp surface or
by metal macro-electrodes attached to the scalp. Examples of such
recordings to left half field checkerboard onset stimulation are
given in Fig. 1, which is taken from the Ph.D.-study of Stok (1986).
The visual evoked potentials (VEPs) and the magnetic fields (MEFs)
were recorded from the same subject to answer the question whether
MEFs can be used to validate or extend results based on VEPs. The
responses are displayed in accordance to the recording sites on the
back of the head. Since the head is magnetically transparent a finer
spatial sampling is needed for magnetic than for electric recording.
In this example the spacings are 3 cm for the VEPs and 2 cm for the
MEFs. To have a comparable sampling of the field shape, a grid of 1
cm for the latter would have been preferable (Stok, 1987).
    If the same current sources underlie both sets of responses and if

From Pigments to Perception, Edited by A. Valberg and
B.B. Lee, Plenum Press, New York, 1991

the part of the head from which the recordings are made is rotationally symmetric, then on theoretical grounds 3 gross differences can be expected:

a) the magnetic and electric response sets are rotated 90° to each other for sources oriented tangentially with regard to the surface of the head;

b) the gradient of the magnetic response set is about 3 times steeper than that of the electric one;

c) the magnetic response set contains contributions of tangentially oriented current sources only.

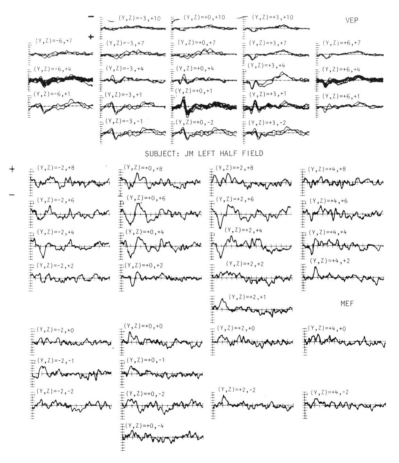

Fig. 1. Visual evoked potentials (VEPs) and fields (MEFs) of subject JM to the onset of a checkerboard pattern of 3° with 19' checks of 40% contrast and a mean luminance of 60 cd.m^{-2}. Horizontal scale: 50 ms per division. Vertical scale VEPs: 1 μV per division, negativity upwards; scale MEFs: 75 fT per division. The y and z coordinates are in cm; with the inion at y = 0 and z = -2. The MEFs were recorded with a symmetric second order gradiometer oriented vertically at a distance of 12 mm from the back of the head. Positivity at a location near the inion indicates that the field lines enter the head. For further details see Stok et al., (1990).

212

Let's now inspect Fig. 1 keeping these 3 rules of thumb in mind:
the VEPs show a positive deflection at 100 ms after checkerboard
onset and a negative one at 130 ms. Within this time window also the
MEFs show a clear peak which inverses in polarity going from the back
of the head to the front. Since the polarity inversion occurs in the
electric response set going from left to right, the two fields are
rotated 90°. Secondly, the magnetic response amplitude reduces by
about a factor of 2 between neighbouring recording sites, whereas for
this a step of 6 cm is needed in the electric response set. Thirdly,
the polarity inversion in the electric map is more outspoken for the
130 ms than for the 100 ms peak, indicating that the 130 ms peak has
to be attributed to a more tangentially oriented current source than
the 100 ms peak. Since the 3 rules of thumb seem to hold
qualitatively, one has to conclude that the same generators underlie
both fields, and that in this example electric and magnetic maps are
equivalent for localizing bio-electromagnetic sources. Therefore
preference for a particular type of map is mainly based on practical
grounds (e.g. no electrodes needed for MEFs) or costs (low for VEPs).
For both maps holds that an infinite number of sources can account
for each field measured, since always source configurations are
possible that yield zero fields. These silent sources are different
for the magnetic and electric fields. Remember, for example, that a
radial dipole in a rotation symmetrical volume conductor does not
produce a MEF.

SOURCE LOCALIZATION

To localize brain activity from a set of electric or magnetic
potential values, the medium that contains the sources of the
activity has to be described, and the number and the character of the
sources to be postulated. For a recent overview, see Van Dijk and
Spekreijse, 1990.

Since it is difficult to characterize the volume conductance
properties of the head realistically, one has to rely on model
descriptions that cover as adequately as possible the most important
features of the volume conductor. However, for visual evoked activity
no realistic shape of the volume conductor model is needed since it
can be proven that only the shape and the conductivity of the medium
between the sources and the recording sites plays a role in
localization (De Munck, 1989). Since both the operculum and the
occiput are spherical (Fig. 2), a concentric sphere model or at worst
a confocal spheroid model is sufficient for visual evoked source
localization. Usually the head is described by three concentric
spheres, each homogeneous and isotropic but with a different
conductivity. The innermost representing the brain, the second the
skull and the outermost compartment the skin.

For localization of electromagnetic activity also assumptions have
to be made about the character of the sources. The sources of brain

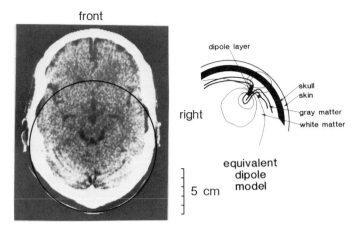

front

right

dipole layer

skull
skin
gray matter
white matter

equivalent
dipole
model

5 cm

Fig. 2. Left: CT scan showing an approximate transversal section 2 cm
above the inion of subject HS. The drawn line depicts the
outer contour of the best fitting sphere used for the ED
locations, as presented in Fig. 3.
Right: the equivalent dipole (ED) model. Activation of an
area in the cortex gives current loops that are radial to the
cortical surface.

activity are locally impressed current loops which result from the
release and uptake of potassium by neurons and glial cells. Since the
main cells of the cortex, the pyramidal neurons, have elongated forms
constituted by the apical dendritic tree, the current sinks and
sources may be distributed at different sites along the soma-
dendritic tree. These current sinks and sources are organized in
discrete layers since a) the pyramidal cells have the same
orientation within the cortex with the somata in deep layers and the
dendritic arborizations near the cortical surface, and b) the
synapses that are activated by sensory inputs are also in well
defined layers along the soma-dendritic membrane. This is the reason
why a dipole layer is an adequate description for electromagnetic
fields of the brain. In most localization procedures this dipole
layer (right hand Fig. 2) is replaced by an equivalent dipole (ED) in
the centroid of the patch of cortex that is assumed to be activated
homogeneously (Nunez, 1981; De Munck et al., 1988). More generally,
it can be shown that at each time instant an ED can represent the
activity in a layer with a continuous current source density, and
that the position of the ED is determined by the contour. An electric
field caused by such a dipole has 6 degrees of freedom: 3 for
position, 2 for orientation and one for strength. Since in a
spherically symmetric conductor a radial dipole does not produce a
magnetic field, such a field has only 5 degrees of freedom. So, if
the signal-to-noise ratio of the magnetic and electric recordings are
identical, and if the same number of derivations is used, then a
localization procedure based upon the magnetic field may give a more
accurate answer for the tangential ED.

There is another difference between the electric and magnetic fields due to a dipole source in a rotationally symmetric conductor: the magnetic fields do not depend on the conductivity profile in the conductor i.e. the head is magnetically "transparent". In that situation magnetic fields have been said (Hari & Ilmoniemi, 1986) to be superior for source localization since these fields are 1) invariant with volume conductor properties, 2) less dependent on derivations between the real volume conductor and the model, and 3) insensitive to radial currents.

This implies that the depth of a dipole can be estimated directly from the locations of the maximum and minimum of the magnetic field recorded. Depth is equal to the distance between these extremes/$\sqrt{2}$, and the dipole lies in between. Application of this rule on the magnetic response set of Fig. 1 reveals immediately the existence of an ED lying near the midline of the head, pointing from the right into the left hemisphere and situated about 2 cm above the inion. Since the left visual field projects on the medial wall of the right hemisphere, this activity seems to originate in the primary visual cortex, Brodman area 17. However, this is not the only patch of cortex being activated by the pattern onset stimulus, since visual inspection of the electric response set of Fig. 1 shows that the peaks at 100 and 130 ms have a different distribution across the back of the head. Note that this is the first moment in our analysis so far that time is taken into consideration. It is important to stress that cortical activity is not only described by *spatial* but also by *temporal* parameters.

MULTIPLE SOURCES

If a single ED is responsible for a set of VEP or MEF recordings, then at all recording sites the same time function (except for sign) should be found. When this is not the case, as in Fig. 1, then more sources have to be assumed to be active, either simultaneously or not. The question to be answered is how many sources are minimally responsible for a response set? Two main strategies are commonly used. One is based on the assumption that the source can be modeled by a single dipole that may change position, orientation and strength with time. Therefore for each time sample an ED is estimated. If the successive EDs form a cluster within a circumscribed brain area, the successive EDs may be represented by a single average ED. However, a discontinuity in the series of EDs indicates that two or more EDs with distinct localizations may be needed to account for a given response set. Another strategy is to estimate the dimensionality of the response space on the basis of the signal-to-noise ratio of the map. In that case one determines the minimal number of time functions needed to account for the significant power of the data set (Kavanagh et al., 1976; Maier et al., 1987). Principal component analysis is one way to find this minimum number. By making linear combinations of the factor loading distributions of these principal components such that they yield on a least square basis spatial distributions that are in accordance with dipole sources, the significant power in the response set is explained in terms of EDs. In this analysis the EDs are assumed to have fixed positions and orientations with variable strengths in time. The present state of the art allows localization of no more than two simultaneously active sources with unknown a priori positions, since the inverse solutions become unstable with increasing number of sources (Scherg and Von Cramon, 1986; Achim et al., 1988; De Munck, 1990).

Both the above procedures have been applied to the pattern onset response sets of Fig. 1. The clustering of successive single ED estimates yielded two main EDs for the onset pattern VEP: one was radially oriented and mainly active within the interval 80-115 ms, while the other was mainly tangentially oriented and appeared at 140-160 ms. The principal component analysis showed also a radial and a tangential ED but, in this case, both were simultaneously active, although the radial one more in the first interval and the tangential one more in the second interval of the time window chosen. In both cases two dipoles were necessary to account for the electric and magnetic fields generated by the pattern onset stimulus. A detailed analysis can be found in Lopes da Silva and Spekreijse (1991).

TOPOGRAPHIC REPRESENTATION OF THE PATTERN ONSET EDS

The analysis discussed so far dealt with the first two peaks in the pattern onset response. The VEP to the appearance of a checkerboard consists, however, of 3 successive peaks -positive, negative, positive- which are commonly labelled C_I, C_{II} and C_{III} (Jeffreys and Axford, 1972a,b). The strength of each of these peaks varies in a characteristic way with stimulus condition. For example, blurring of the pattern reduces mainly C_{II}, and binocular instead of monocular presentation enhances the C_{III} peak (Spekreijse et al., 1973). The presence of the 3 peaks varies also with checksize: C_I dominates the response to coarse checks, whereas C_{II} is most outspoken upon stimulation with fine checks of low contrast. So, by manipulation of the stimulus always situations can be created in which only two peaks dominate the VEP response set. This is important since our dipole localization algorithms can handle no more than 2 simultaneously active sources. Furthermore, stimulation should be restricted to one hemifield so that the evoked activity is restricted to a single hemisphere, since 2 sources at different locations with identical time functions cannot be distinguished, leading to erroneous localization results. The results of such a step-by-step analysis (Ossenblok and Spekreijse, 1991) of the complete pattern onset EP are summarized in Fig. 3. The left side gives the positions of the EDs that dominate the positive peaks in the response sets recorded in 5 subjects. The right side shows for subject HS the

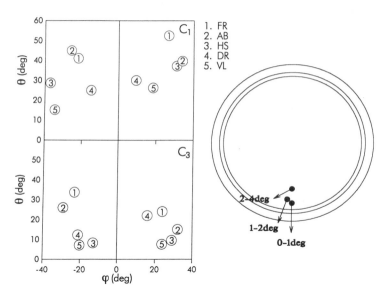

Fig. 3. Left: the positions of the EDs dominating respectively the
early (top) or the late positivity (bottom) of the EP to the
onset of a hemifield checkerboard of 8° with 48' checks of
80% contrast and a mean luminance of 65 cd.m^{-2}. Phi and theta
are the azimuth and the elevation coordinates of the best
fitting sphere with the inion at (0,0). The theta values of
the late positivity are for all five subject smaller than
those of the early one. As an indication for the phi scale,
the EDs in the two hemispheres (of each subject) are about 8
cm apart.
Right: projections of the EDs to stimulus fields presented to
subject HS at the eccentricities indicated. A transversal
plane through the three sphere model volume conductor is
shown.

location of the ED that dominates the negative peak in the response
set generated by left half stimulus fields presented at the
eccentricities indicated. This part of the figure is based on data
described in Van Dijk and Spekreijse (1989).

Since the EDs accounting for the two positive peaks to half field
stimulation are situated far apart and clearly contralateral with
respect to the midline of the head, it can be concluded that these
peaks have a mainly extrastriate origin. Their intersubject
variability is an indication of the variability in the geometry of
the extrastriate cortices amongst subjects, and the intrasubject one
reflects the inter-hemispheric differences. The finding that the EDs
accounting for the early and late positive peaks are displaced
systematically, and that the first one has a retinotopic
representation and the later one not, is indicative that these EDs
represent activity from different extrastriate regions.

As stated before, ED localizations are strongly dependent on the assumed properties of the volume conductor model used (Stok et al., 1987). The most important parameters are: conductivity of brain tissue, including its anisotropia, and the conductivity of the skull. Fortunately, the relative positions of the sources are much less influenced by these errors, except in case of anisotropia of brain tissue (Van Dijk et al., 1991). So, if position and orientation of one source is known, then the other sources can be localized with an accuracy of a few millimetres. Since relative positions can be determined accurately, retinotopic maps of the brain surface can be constructed, as illustrated in the right hand of Fig. 3. For the most foveal left half stimulus field (0-1°) the ED is radial and lies on the occipital lobe. For more peripheral stimuli the orientation becomes tangential and the ED moves over the medial wall into the interhemispheric cleft. These data are in agreement with the retinotopic map of area 17, so that the negative peak in the pattern onset response has a mainly striate origin. It is surprising that in man the activity originating in the extrastriate cortex has a shorter peak latency than the striate activity, since in rhesus monkey the subdural pattern onset responses from area 18 have a 15 ms longer peak latency than from area 17 (Dagnelie, 1986). However, visual processing occurs parallel with projections from retina to visual cortices that are strongly species-dependent and virtually unstudied in man.

MATURATION OF THE STRIATE GENERATOR

The pattern onset EP can be used to monitor maturation of the visual pathways at both distal and proximal level (Spekreijse, 1983). This becomes evident when the amplitude of the pattern onset response is plotted as a function of checksize. The results are summarized in Fig. 4 which is taken from Spekreijse and Apkarian (1986). Two main changes can be seen. 1) From birth on till about 8 months post-term there is a gradual shape invariant shift of this curve towards finer checks and a gradual reduction in amplitude. 2) From about 8 months on the maximum of the curve becomes age invariant but the tail of the curve continues to shift to smaller checks till about puberty. The initial process parallels the morphological maturation of the retina, particularly the foveal region with increasingly finer spacing of the receptors and the consequent transmission of finer details. The gradual amplitude reduction is likely to reflect changes in the volume conductor properties of the head, and therefore not relevant for the maturation of the visual system. The second process reflects maturation of the striate cortex, since it is parallelled by a gradual change in the waveform of the response. The finding that the optimal checksize does not change during this period implies that mean receptive fieldsize remains constant. The change in waveform is illustrated in the inset of Fig. 4. During the first year of life the

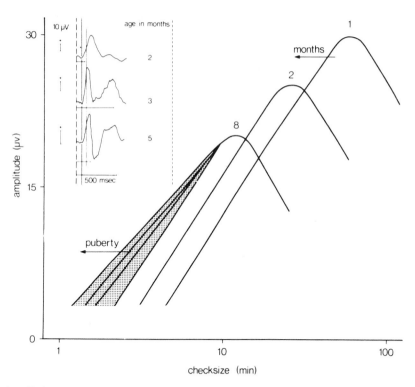

Fig. 4. Pattern EPs to the onset of a checkerboard as a function of
age (left insert). Responses depicted were recorded with the
checksize which yielded the maximum amplitude and a
presentation duration as indicated by the bar underlying each
record. At 2 months, the wave shape of the response is
simple, consisting of a single positive peak with a latency,
as seen here, of about 190 ms. By 5 months of age there is an
increase in wave shape complexity. However, the incidence and
adult-like latency of particularly the C_{II} component (second
solid vertical line) are not fully developed before puberty.
Onset latency of the response (left most solid vertical line)
seems independent of age. The amplitude vs checksize
functions illustrate the optimum and minimum pattern sizes as
a function of age.

pattern onset response consists mainly of a single positive peak with
a relatively long gradually decreasing peak latency. From about 2
years on the negative peak becomes manifest, reaching an incidence of
nearly 100% around 6 years of age, and the adult peak latency near
puberty (Spekreijse, 1983). This development of a positive striate
response into one with a negative polarity coincides with the slow
improvement phase of acuity as estimated both subjectively and
objectively. Although it remains an intriguing question which neural
and/or synaptic changes account for this reversal in polarity, it
demonstrates clearly that for studying the function of the human
brain maps with a time resolution of 10 to 100 ms are essential.

REFERENCES

Achim, A., Richer, F., and Saint-Hilaire, J.M., 1988, Methods for separating sources of neuroelectric data, Brain Topogr., 1: 22-28.

Dagnelie, G., 1986, Pattern and motion processing in primate visual cortex, PhD thesis University of Amsterdam.

De Munck, J.C., Van Dijk, B.W., and Spekreijse, H., 1988, Mathematical dipoles are adequate to describe realistic generators of human brain activity, IEEE Trans. Biomed Eng., BME-35: 960-966.

De Munck, J.C., 1989, A mathemetical and physical interpretation of the electromagnetic field of the brain, PhD thesis University of Amsterdam.

De Munck, J.C., 1990, The estimation of time varying dipoles on the basis of evoked potentials, Electroenceph. and clin. Neurophysiol. (Evoked Potentials), 77: 156-160.

Hari, R. and Ilmoniemi, R.J., 1986, Cerebral magnetic fields, CRC Crit.Rev. in Biomed. Eng., 14: 93-126.

Jeffreys, D.A. and Axford, J.G., 1972a, Source location of pattern specific components of human visual evoked potentials. I. Component of striate cortical origin, Exp. Brain. Res., 16: 1-21.

Jeffreys, D.A. and Axford, J.G., 1972b, Source location of pattern specific components of human visual evoked potentials. II. Components of extrastriate cortical origin, Exp. Brain. Res., 16: 22-40.

Kavanagh, R.N., Darcey, T.M., and Fender P.H., 1976, The dimensionality of the human visual evoked potential, Electroenceph. clin. Neurophysiol., 40: 633-644.

Lopes da Silva, F.H. and Spekreijse, H., 1991, Localization of brain sources of visually evoked responses using single and multiple dipoles. An overview of different approaches, in "Event-related brain research", Electroencephalogr. and clin. Neurophysiol., suppl. 42.

Maier, J., Dagnelie, G., Spekreijse, H., and Van Dijk, B.W., 1987, Principal Component Analysis for source localization of VEPs in man, Vision Res., 27: 165-177.

Nunez, P.L., 1981, "Electric fields of the brain," Oxford University Press, New York.

Ossenblok, P. and Spekreijse, H., 1991, The extrastriate generators of the patterns onset EP, Electroenceph. and clin. Neurophysiol. (Evoked Potentials), in press.

Scherg, M. and Von Cramon, D., 1986, Psychoacoustic and electrophysiologic correlates of central hearing disorders in man, Psychiatry and Neurol. Sci., 236: 56-60.

Spekreijse, H., 1983, Comparison of acuity tests and pattern evoked potential criteria: two mechanisms underly acuity maturation in man, Behav. Brain Res., 10: 107-117.

Spekreijse, H., Van der Tweel, L.H., and Zuidema, Th., 1973, Contrast evoked responses in man, Vision Res., 13: 1577-1601.

Spekreijse, H. and Apkarian, P.A., 1986, The use of a system analysis approach to electrodiagnostic (ERG and VEP) assessment, <u>Vision Res.</u>, 26:195-219.

Stok, C.J., 1986, The inverse problem in EEG and MEG with application to visual evoked responses. PhD thesis University Twente.

Stok, C.J., 1987, The influence of model parameters on EEG/MEG single dipole source estimation, <u>IEEE Trans. Biomed Eng.</u>, BME-34: 289-296.

Stok, C.J., Meijs, J.W.H., and Peters, M.J., 1987, Inverse solutions based on MEG and EEG applied to volume conductor analysis. <u>Physics in Med. and Biol.</u>, 32: 99-104.

Stok, C.J., Lopes da Silva, F.H., Spekreijse, H., Peters, M.J., and Boom, H.B.K., 1990, A comparative EEG/MEG equivalent dipole study of the pattern onset visual response, in "New Trends and Advanced Techniques in Clinical Neurophysiology", <u>Electroencephalogr. and clin. Neurophysiol.</u>, suppl. 41.

Van Dijk, B.W. and Spekreijse, H., 1989, Localization of the visually evoked response: The pattern appearance response, <u>in</u>: "Topographic brain mapping of EEG and Evoked Potentials," K. Maurer, ed., Springer-Verlag, Berlin, 360-365.

Van Dijk, B.W. and Spekreijse, H., 1990, Localization of electric and magnetic sources of brain activity, <u>in</u> "Visual Evoked Potentials", J.E. Desmedt, ed., Elsevier Science Publishers B.V. Amsterdam, 57-74.

Van Dijk, B.W., De Munck, J.C., and Spekreijse, H., The effects of conductivity parameters on estimates of the dipole sources of electric brain activity, <u>Electroencephalogr. and clin. Neurophysiol.</u>, submitted.

DISCUSSION: EVOKED POTENTIALS

Neville Drasdo

Department of Vision Sciences
Aston University
Birmingham, U.K.

Drasdo: My colleagues have shown formidable techniques which are perhaps difficult for those without direct experience of evoked potentials. Nevertheless, we have a unique opportunity with the assembled expertise in neurophysiology at this meeting to consider evoked potentials in detail. Spekreijse has expounded new techniques, which can examine virtually every generator in the brain, because the activity in fissures is largely concealed during conventional evoked potential recording. His contribution with principal component analysis has revealed a number of interesting phenomena which I dare say will attract some questions. It is unusual to have in an audience where we are dealing with evoked potentials, authorities on such matters as the cortical projections in the macaque, which have greatly modified our thoughts on where the evoked potentials might come from. It is a remarkable thing that when somebody places an electrode on the scalp, knowing how many processing areas there are in the brain, and being unable at this moment to very precisely select those areas with our stimuli, that we expect to obtain an intelligible result. Clearly it requires considerable expertise and insight to do this. Coming to Kulikowski's contribution, his experiments have revealed to us the value of evoked potentials. One often says when looking at a single unit recording, how does this relate to perception. We have no computer models, for example, to add together effective receptive fields. The very concept of difference of Gaussian receptive field has not really advanced very far, and perhaps people in psychophysics and also in microelectrode recording could relate their results to evoked potentials a little more frequently. Kulikowski has shown that the summated effect of neurones can be sensibly related to psychophysics, although the experiment must be carefully designed, and by this I mean designed with insight.

I would like to point out that whereas Spekreijse has been able to investigate generators in the most inconvenient or convenient positions by use of magnetometry, it is a fairly basic situation when looking at evoked potentials related to complex phenomena in foveal vision. The projection on the human cortex, though variable in individuals, is not very variable in the sense that all the exposed association areas are involved in processing a pattern. We get a different distribution of activity with different types of stimulation. This can only arise from the nature of the stimulus and therefore must relate to the different cortical areas. When we are dealing with stimuli projected on a localised area of exposed cortical surface, we don't need to venture very far into principle component analyses. For example, with local ($1°×1°$) flash and pattern onset checkerboard stimuli on the same area of retina, responses may be quite differently lateralized. We can simulate these responses with equivalent electrical fields based on single dipoles. So it seems to me that the use of evoked potentials, even to investigate subtle aspects of perception, may be pursued without too much difficulty. Future development will depend very much on the nature of the stimuli that are designed to look at these higher-level areas. We can relate these results in many ways to primate physiology, and we ought not to ignore the possibility that in the temporal region there must be more advanced processes, because it has been claimed that face recognition can be recorded as an evoked potential in humans relatively easily, and although I was sceptical initially, I think I have been converted to that point of view. We need now to consider the scope of evoked potential recording, and the origin of these signals.

van Essen: I have a couple of comments relating to this curious set of observations about an apparently extrastriate onset of the patterned evoked potential. First of all, it is really quite well established in monkeys that the striate cortex is the first cortical area to respond to a significant degree, with latencies of the order of 30 or 40 msec. In the human, the expected latency may be 60 msec so the proposal that the human striate cortex doesn't respond substantially until 130 msec or so is rather incongruous. Drasdo showed slides that showed simpler stimuli to give a response latency of 90 msec, more consistent with striate origin. Given that and given the effects of human striate lesions, it would seem to me that the most parsimonious explanation might be that a large field complex stimulus may simply lead to vigorous striate cortical responses, but a cancellation of the measurable evoked potentials.

Spekreijse: The latency I mentioned was the peak latency. If you look at the onset latency that fits in with what you expect from monkey data, starting at about 60 msec for all responses including the striate. The problem is that when I look at the responses, you have two simultaneous activities starting at 60 msec after stimulus onset. One reaches its maximum rather fast, at 100 msec, then it slows down, while the other reaches its maximum at 130 msec and then dies out again, and this late one is the striate component. So the puzzling thing is that the largest response is obtained preceding the largest response from the striate area. I do not know what response amplitude means, of course that is always a problem.

Drasdo: Could you comment on actual input to the area of V2 in humans? The classical idea is that we only have direct geniculate input to V1.

van Essen: It has been established in monkeys that there is a very weak projection from a tiny population of geniculate cells to extrastriate cortex, and also that the area MT can give very weak responses on the absence of striate projections. It is rather difficult to believe that such weak projections could give rise to the magnitude of evoked potentials. So I am heartened to hear that the latencies are adequate for striate cortex to be necessary for transmission of extrastriate responses.

Spekreijse: The problem I have is if you take into account the amplitude of the response, if the extrastriate cortex is being activated through the striate cortex, then you would assume that striate activity would resemble the size of the extrastriate component. That would be the more simple model.

van Essen: It might be the most simple model, but there may be, over the course of the first 130 msec, literally dozens of different areas being activated, so then even a 1, 2, 3 or 4 dipole model is going to be a great oversimplification. The alternative that the dipole associated with striate cortex could be much smaller would seem to me to be plausible.

Drasdo: Could you comment on the relative magnification of area V4 and the striate area. In humans the classical idea is that we have greatly expanded association areas compared with the macaque. Yet I believe in the macaque that foveal magnification is greater in V4 than in V1.

van Essen: I don't think that is well established. In different cortical areas there are not dramatic differences in the central versus peripheral magnification for V1, V2, V4 or MT. If you take the evidence from the Gross lab, there doesn't look to be a dramatic difference.

Drasdo: If the generator is suitable placed to give a big signal, it might dwarf this striate signal.

Dow: We have unpublished data on foveal magnification in V4, and it is substantially smaller then in V1. With regard to latencies, perhaps the magnocellular fibers and systems deep in V1 are not seen very much on the surface, but they activate something like MT or the human equivalent and maybe you see that activity.

Kulikowski: There are many parallel pathways, but they are all of different strength. Somehow I cannot accept that the small pathway from LGN to V2 could sustain these responses. Nevertheless there are stronger projections from LGN to V3 and especially V5. Unfortunately we don't know the location of V5 in man.

The various types of potentials contribute differently from different depths. If there is a very good heterochromatic flicker photometry correlation from the surface of the cortex, or even the surface of the skull, why don't the synaptic potentials of LGN axons mess it up. Obviously this is a much weaker component. So deep potentials do not generate very much signal on the surface. I am rather worried that the same happens to magnetic potentials, though magnetic potentials supposedly penetrate deeper.

Spekreijse: Magnetic potentials also decrease with the square of the distance.

Van Essen: With regard to the activation of MT or V5 in the absence of striate input, again Charly Gross's lab has reported that MT cells can respond after striate ablations, but they have also reported that this disappears after additional superior colliculus ablations, making a geniculate to V5 projection unlikely as a driving source for MT. I would also point out that there is some evidence from John Maunsell that in the macaque the increments in latency of response from one area to the next, e.g. from V1 to MT, or V1 to V4 is only 5 or 10 msec per stage, and so over the say 30 msec between 100 and 130 msec peaks, there is room for quite a few additional areas, particularly if you allow for parallel processing. I just re-emphasize that one needs a many dipole model to really think about the details of localization.

Shapley: To Henk Spekreijse, in view of your striking findings of the absence of pattern onset extrastriate potentials up to 10 years old, you think there should be psychophysical investigations of 10 year olds to find out what the extrastriate cortex might be doing?

Spekreijse: I am mystified just as much as you are, but with children up to 4, none shows the extrastriate component, and then you see the incidence growing. There are visual changes in a simple measure like visual acuity up to puberty. What it means I don't know, but it is only in the pattern onset response. The pattern reversal response, reflecting motion mechanisms, matures rather early and is normal.

Fiorentini: There may be a contribution form callosal fibres to the evoked potential, because in children callosal fibres have a very long maturation period.

Drasdo: There are examples, for example from the clinical normative data, of changing VEP without apparently changing perception. Even the flash evoked potential changes with age and yet we don't observe any correlation in perception.

Kulikowski: A very important point is to engage some kind of a cognitive task to maintain attention. From our experiments on V4 we know that this is a very important centre which modifies attention. You can get zero cognitive potentials from a subject who is completely passive and bored, and you can immediately make his potentials jump when you reward him for every good potential with a few dollars. I wonder whether young children are just not suitable subjects and do not overcome boredom. Therefore the activity of secondary visual areas which are so much modified by attention and motivation is just too weak to be recorded.

Creutzfeldt: I am quite taken by these methods of developing equivalent dipoles. You have to make certain assumptions. The evoked potentials are of course related to cellular activity. The phase relationship is not always as one would expect from these dipole models. It is true for certain cases where you have a straightforward dipole, this is of course afferent fibre activity. As soon as you come into the cortex, as we don't know even how the synapses are distributed across the pyramidal cell, we can't say very much. We were glad when we saw excitation going together with surface negativity. This is against the simplified dipole or the equivalent dipole, but you can use large numbers of dipoles through principle component analysis. However, the degrees of freedom are reduced if you have the magnetoencephalogram.

Drasdo: Any further comments from our two contributors, anything left unsaid which you regret?

Spekreijse: The absolute localization of our dipoles only makes sense if the number of underlying dipoles is rather restricted, and there can be substantial error, though error in relative position is much smaller. By combining human and monkey studies we hope to improve the physiological parameters for isotropia and anisotropia. We are interested in this kind of approach for clinical use to localise epileptic foci. We can update our localization procedures on those few patients that have intracortical electrodes, which can be used as a reference point. Then you can improve your procedures for the other areas.

PROBING THE PRIMATE VISUAL CORTEX:

PATHWAYS AND PERSPECTIVES

David C. van Essen[1], Daniel J. Felleman[2], Edgar A. DeYoe[3]
and James J. Knierim[1]

[1] Biology Division
Caltech
Pasadena CA 91125

[2] Dept. of Neurobiology and Anatomy
University of Texas Medical School
Houston TX 77030

[3] Dept. of Anatomy and Cellular Biology
Medical College of Wisconsin
Milwaukee, Wisconsin

INTRODUCTION

Most primates are highly visual creatures, capable of a wide variety of difficult tasks that must be carried out in a complex visual environment. It is therefore hardly surprising that a very large expanse of cerebral cortex is devoted to analyzing and interpreting the relatively raw messages transmitted from the retina. Over the past three decades, much progress has been made in elucidating the organization and function of visual cortex. In this chapter we briefly review recent progress in understanding several aspects of visual processing in the macaque monkey, as revealed by both anatomical and physiological studies. The first topic concerns the nature of information flow through the visual cortex as revealed by analysis of the numerous pathways that interconnect different visual areas. The guiding hypothesis is that the cortex is arranged as a distributed hierarchical system, in which there are many distinct levels of analysis. The second topic concerns parallel processing streams and their relationship to the M and P pathways established within the retina. The third topic deals with the functional significance of the feedback pathways that form a prominent characteristic of the cortical hierarchy. The function of feedback pathways in sensory processing has been an intriguing but elusive issue. There are probably many such functions, and here we emphasize their possible role in modulating neural responses according to the broader context of the visual environment.

HIERARCHICAL ORGANIZATION OF VISUAL CORTEX

Visual areas. Visual cortex in the macaque can be subdivided into a remarkably large number of distinct visual areas, based on a combination of anatomical, physiolog-

From Pigments to Perception, Edited by A. Valberg and
B.B. Lee, Plenum Press, New York, 1991

ical, and/or behavioral evidence obtained from studies in many different laboratories. The distribution of these areas can best be appreciated by displaying them on an unfolded map of the cortex, as shown in Figure 1. The shaded regions on the map and on the lateral and medial brain views include a total of 32 visually related areas - 25 areas that are predominantly or exclusively involved in vision plus 7 visual association areas that are implicated in other tasks as well, such as polysensory, visuomotor, or limbic functions. Based on surface area measurements on the map, we estimate that these 32 visually related areas occupy 54% of the cerebral neocortex, including all of the occipital lobe, the posterior half of the parietal lobe, the ventral half of the temporal lobe, and even part of the frontal lobe (Felleman and Van Essen, 1991).

The specific criteria used in their identification vary for the different areas, but in descending order of practical importance they include: i) a specific pattern of connections as revealed by various pathway-tracing techniques; ii) a distinct topographic map of the contralateral visual hemifield (albeit an irregular and/or incomplete map for some areas); iii) a distinctive architecture as revealed by classical cell or myelin stains or by more modern histochemical or immunocytochemical reagents; and iv) a distinctive behavioral deficit after localized lesions. Ideally, each area would be independently identifiable using any of the above approaches, but at present the identification of many areas is based on only one or two criteria. Moreover, some of the subdivisions are tentative or controversial, and plausible alternative partitioning schemes exist for some regions, such as the inferotemporal complex. Hence, the overall scheme should be regarded as a progress report that will surely be subject to refinement in the future. Nonetheless, the general conclusion that visual cortex in the macaque comprises an impressively large number of distinct visual areas is likely to be secure.

This principle also applies to other species that have been appropriately examined, such as the owl monkey (Kaas,1988; Sereno and Allman, 1990), cat (cf. Rosenquist, 1985), and even much smaller animals such as the rat and mouse (Olavarria and Montero, 1984, 1989). Much less is known about the layout of visual areas in the human brain. However, it is surely relevant that areas V1 and V2 in humans show striking similarities with other primates in many specialized characteristics such as ocular dominance stripes and cytochrome oxidase patterns (Horton and Hedley-Whyte, 1984). There are also basic similarities in connectivity patterns that have been revealed by the recent application of tracers that work on fixed, post-mortem tissue (Burkhalter and Bernando, 1989). Thus, it seems highly likely that homologies will be found for many other visual areas besides V1 and V2 (cf. Sereno and Allman, 1990).

Connectivity. Even more striking than the increase in number of identified visual areas has been the explosion in the number of identified pathways that interconnect these areas. In the early 1970's it was widely presumed that each visual area received at most a few cortical inputs and had only a few outputs to other areas. During the intervening period, a powerful arsenal of highly sensitive pathway-tracing techniques has arisen, and dozens of studies have applied these techniques to the macaque visual system. As of mid-1990, a total of 305 connections among the 32 identified visual areas had been reported (Felleman and Van Essen, 1991). This represents nearly a third of the number of connections that would be found in a fully interconnected network. Given that only about two thirds of the possible pathways have been tested systematically, it is likely that the actual degree of connectivity is closer to 40 - 50%. However, different pathways vary enormously in strength. Although quantitative data are scarce, it is likely that the great majority of physical connections (i.e., numbers of axons and of synapses) arise from a small minority (perhaps 20 or 30%) of the identified pathways.

The sheer number of known cortical connections raises the specter of a network so

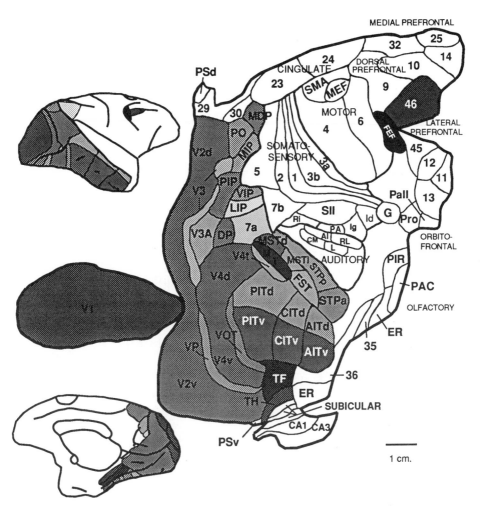

Figure 1. A two-dimensional map of cerebral cortex in the right hemisphere of the macaque monkey. Stippling indicates cortical areas implicated in visual processing. Copied, with permission, from Van Essen et al. (1990).

complex and chaotic as to be virtually impossible to decipher. Hence, it is critical to look for organizing principles that might help to understand the nature of information flow in the visual system. One such principle is that connections between areas tend to be reciprocal. Nearly all of the pathways that have been appropriately examined in the macaque visual system are in fact arranged as reciprocal pairs. Moreover, the great majority of these reciprocal connections show a striking asymmetry in laminar organization. This asymmetry, first reported by Tigges et al. (1973) and Rockland and Pandya (1979), has subsequently been used as a basis for generating an orderly hierarchy of visual areas (Maunsell and Van Essen, 1983). By our current criteria, pathways of one type (the forward or ascending direction), terminate predominantly in layer 4 of the target area, and they originate from cell bodies that either are concentrated in the superficial layers or are roughly evenly mixed in superficial and deep layers. In the

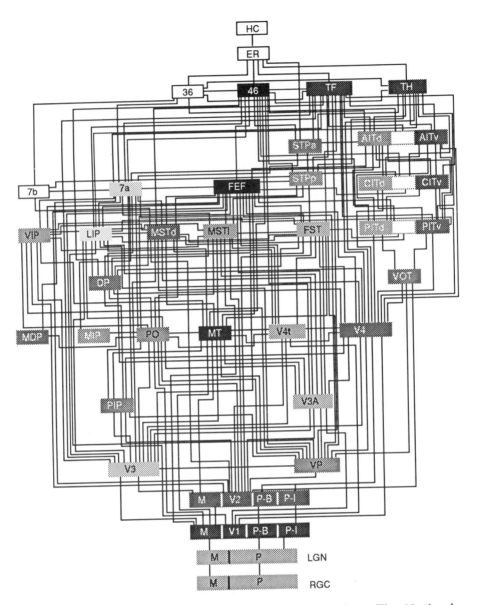

Figure 2. A hierarchy of visual areas in the macaque monkey. The 32 visual areas shown in Fig. 1, plus several limbic and other areas to which they are connected, have been placed in a hierarchical arrangement on the basis of the laminar patterns of connections between areas (see text). Each line represents an identified linkage between areas; in most cases, the linkage is known to involve reciprocal connections between areas. Modified, with permission, from Felleman and Van Essen (1991).

reverse direction (the feedback or descending direction), the terminations preferentially avoid layer 4, and the cells of origin either are concentrated mainly in the deep layers or are distributed in the same bilaminar fashion that characterizes some forward pathways. In a third pattern (lateral, or intermediate linkages), which is found in a significant minority of pathways, the terminations occur in a columnar fashion involving most or all layers, while the cells of origin are distributed in a bilaminar fashion. Altogether, the termination patterns are the most reliable index of hierarchical relationships; the patterns of cell origin are ambiguous for those pathways showing a bilaminar pattern but can be a valuable indicator for the remaining pathways.

Using these criteria, we have recently updated the visual hierarchy to include all 32 cortical areas and all 305 connections in the macaque. The resultant scheme, illustrated in Figure 2, retains the essential characteristics of its rather simpler predecessors (Maunsell and Van Essen, 1983; Van Essen, 1985), even though its complexity by now suggests a superficial resemblance to a Manhattan subway map or to a modern VLSI chip design. The key characteristic is that each visual area occupies a well-defined position in the hierarchy. To reiterate, the positioning of each area is derived strictly from anatomical information, based not simply on which other areas it is connected to, but rather on how it is connected, in terms of the laminar asymmetries described above.

Approximately 90% of the known pathways fit unequivocally into this scheme. However, there are significant irregularities associated with the remaining minority of pathways. Many of these irregularities are likely to reflect inaccurate assignments, because the anatomical data on which our analysis is based often have uncertainties in areal assignments or other sources of possible error. These assignments will presumably change when additional experimental data become available. On the other hand, some of the irregularities are likely to be biologically genuine. Hence, the principle of hierarchical organization may reflect a strong statistical bias in the laminar patterns of cortical connectivity rather than a rigid rule to which exceptions never occur.

PROCESSING STREAMS IN VISUAL CORTEX

The organizational complexity revealed by the plethora of cortico-cortical connections has been further compounded by the discovery of compartmental organization within individual visual areas, particularly V1 and V2. These compartments were initially identified with the aid of cytochrome oxidase histochemistry (Livingstone and Hubel; Tootell et al.; Carroll and Wong-Riley). Their functional importance has been established by showing they have highly specific patterns of interconnections and distinctive constellations of receptive field characteristics. Figure 3 summarizes these properties in a hierarchical scheme that includes only the subset of visual areas that have been characterized most thoroughly, but shows the compartmental organization as currently understood for each of these areas (DeYoe and Van Essen, 1988). The icons within each compartment or area represent a high incidence of cells selective for wavelength (prism icon), orientation (angle icon), direction of motion (pointing finger), and/or binocular disparity (eyeglasses).

In brief, a dichotomy between magnocellular (M) and parvocellular (P) streams is established within the retina and preserved within the lateral geniculate nucleus (LGN). Within V1, there is a rearrangement into three distinct streams: the parvo-blob (P-B) stream, the parvo-interblob (P-I) stream, and the magno (M) stream. Within V2, the P-B, P-I, and M streams are represented by the thin stripe, interstripe, and thick stripe compartments, respectively. At higher levels, areas associated with the M stream include V3 and MT, while V4 appears to be segregated into separate subregions respectively representing the P-B and P-I streams (DeYoe et al., 1988; Zeki and Shipp, 1989; Van Essen et al., 1990). The top level of the figure includes the inferotemporal complex, a collection of areas that receives massive inputs from area V4, and a separate collection of areas in the posterior parietal lobe that receives major inputs by way of area MT.

In describing the skeletal framework for this scheme, it is natural to focus first on the anatomical pathways that are most robust and specific. However, we believe

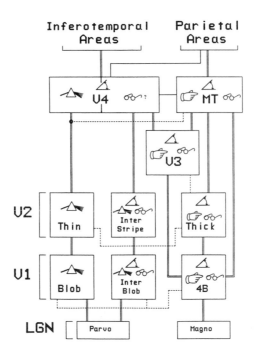

Figure 3. Concurrent processing streams in visual cortex. The preceding version of the cortical hierarchy has been simplified in one respect, to show only the best-studied areas and connections, but includes an added level of complexity by showing the compartmental organization of areas V1, V2, and V4. Icons indicate selectivities for basic stimulus characteristics that are particularly prominent in each compartment or area (prism = wavelength; eyeglasses = binocular disparity; angle = orientation; pointing finger = direction). Modified, with permission, from DeYoe and Van Essen (1988).

it is crucial not to lose sight of the fact that there is extensive cross-talk between streams occurring at all levels of the hierarchy. Some of the cross-talk pathways are illustrated explicitly in the figure (e.g., from area V3 to both MT and V4, and from V4 to both inferotemporal and parietal areas). Other such examples are discussed in more detail elsewhere (DeYoe and Van Essen, 1988; Felleman and Van Essen, 1991). Overall, it appears that inferotemporal cortex, which is implicated in pattern recognition (Ungerleider and Mishkin, 1982), receives preferential inputs from the P-I and P-B streams, but also has substantial inputs from the M-stream. Conversely, the posterior parietal complex, which is implicated in the analysis of spatial relationships, receives strong inputs from the M-stream, but also has significant inputs from the P-B and P-I streams.

The anatomical combination of highly specific connections coupled with a substantial degree of cross-talk has a significant parallel in terms of the distribution of receptive field characteristics in the different processing streams (Fig. 3). Subdivisions associated with the M-stream have a high incidence of direction selectivity, but they also include many cells that are selective for orientation and disparity. Overt wavelength selectivity is rare, but there are nonetheless many cells that respond to the presence of isoluminant borders (Lee et al., 1989; Logothetis et al., 1989; Saito et al., 1989). The P-I stream is characterized by a high incidence of orientation selectivity, but there are also many cells selective for wavelength and probably also binocular disparity. Finally, in the P-B stream, wavelength selectivity is common, whereas selectivity for orientation and direction is much less common. Tuning for spatial frequency occurs in all streams, but there are nonetheless systematic differences in preferred spatial frequencies, most notably in the lower preferred frequencies for cells in blobs vs. interblobs of V1 (Tootell et al., 1988).

In short, the available physiological and anatomical evidence suggest the presence of concurrent processing streams that are distinct in many respects, yet closely intertwined in their operation. We suggest that the cross-talk between streams is advantageous given the computational requirements of vision in the real world (DeYoe and Van Essen, 1988; Van Essen and Anderson. 1990). Many visual tasks can be carried out using several different low-level cues operating independently or in concert. For example, the perception of depth in the 3- dimensional world can be efficiently mediated by cues of binocular disparity, motion parallax, and/or geometrical perspective. Information about these different cues are represented to different degrees and in different ways in the M, P-B, and P-I streams, presumably because of the different computational requirements involved in generating highly selective tuning curves. Consequently, higher-level centers involved in depth perception are likely to need inputs from more than a single stream.

CONTEXT DEPENDENCE OF NEURAL RESPONSES

During the first two decades of the electrophysiological revolution launched by Hubel and Wiesel, the great majority of studies concentrated on the properties of what is now known as the classical receptive field, namely, the region within which the firing of a neuron could be directly increased or decreased. It was known relatively early on that stimuli outside the classical receptive field could affect neuronal firing, even in retinal ganglion cells (McIlwain, 1964). However, relatively little attention was paid to such modulatory influences until the past decade. More recently, though, an increasing number of investigators have explored the modulatory effects provided by stimuli located outside the receptive field (cf. Allman et al., 1985). We have investigated this issue using visual texture patterns.

Objects in the natural environment that differ from their neighbors in characteristics such as shape, size, color, or relative movement tend to be particularly salient - that is, they 'pop out' and easily capture our attention (Beck, 1966; Treisman and Gelade, 1980; Julesz, 1984). In our experiments we recorded from area V1 of alert monkeys and used texture patterns of the type shown in Figure 4A. Each stimulus included a central region positioned within the classical receptive field of the cell under investigation and a surround region lying completely outside the classical receptive field. The central target was a short line segment in either of two orthogonal orientations, and the surround pattern was a texture field in which the elements were either parallel to or orthogonal to the central target. Figure 4B shows results from one particular cell whose responses were especially interesting. The bar graph and the histograms below it show responses to each of the eight stimulus conditions illustrated in a simplified form at the bottom.

In brief, the cell responded well to an isolated target in either orientation (stimuli 1 and 5). The response was markedly suppressed when the central target was part of a uniform texture field (stimuli 2 and 6), whereas responses remained vigorous when the target was surrounded by a texture field in the orthogonal orientation (stimuli 3 and 7). Hence, the responses of the cell correlate with the perceptual salience of the target: weak when there is no contrast in orientation or luminance, but strong in the presence of either luminance contrast or orientation contrast. Note that the suppressive effect of the surround depends on its orientation relative to the center, rather than on its absolute orientation in space

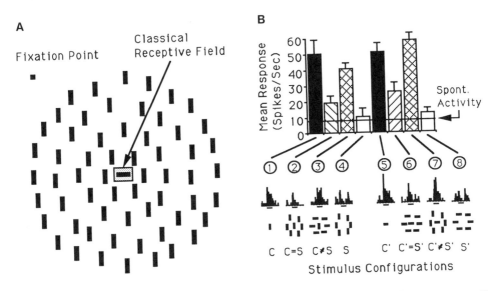

Figure 4. Orientation-dependent surround effects in area V1 of the alert macaque. Responses to a small bar presented within the classical receptive field (conditions 1 and 5) were suppressed by a surrounding texture field of the same orientation (conditions 2 and 6) but not by a texture field of orthogonal orientation (conditions 3 and 7).

In the population of 122 cells studied in two alert monkeys, we found considerable diversity in response properties. Nearly one third of the sample showed a differential surround effect that correlated with perceptual salience. An additional 27% showed a non-specific suppression by the surround. Similar results were obtained in recordings from areas V1 and V2 of anesthetized monkeys (Van Essen et al., 1989). Altogether, we suggest that information useful for figure-ground discrimination and visual popout effects is represented in a significant minority of cells in V1, but it obviously will take a different type of approach to determine whether and how these cells actually mediate in any direct sense such perceptual processes.

Whatever the significance for perception, it is now apparent that pronounced surround effects are common in all areas of visual cortex that have been examined systematically. It is natural to suspect that the massive feedback pathways prevalent throughout the cortical hierarchy may play an important role in this phenomenon. This seems very plausible, given that the surround effects arise from a large portion of the visual field (Allman et al., 1990). Alternatively, surround effects might be partially or completely mediated by the intrinsic (intra-areal) connections that are known to extend over distances of many millimeters (Gilbert). More direct tests are needed to resolve the relative importance of these two sources of influence from beyond the classical receptive field.

GENERAL CONCLUSIONS

Neurobiologists, like other scientists, generally strive to find simple and clear explanations for the phenomena and processes that they study. However, the search for simplicity must be pitted against the harsh reality that the brain is a fantastically complicated device. We believe that this complexity arises not from the sheer perverseness of nature, but rather from the extraordinary difficulty of the real-world computational tasks confronting living animals at every moment. Here, we have discussed several levels of complexity in the visual system that were hardly suspected as recently as two decades ago. There are many more areas, more connections, more complex internal heterogeneity, and greater complexity to receptive fields even in primary visual cortex.

Will these descriptions ever reach a level of sophistication adequate for a genuine, detailed understanding of the neural basis of visual perception? Obviously, we cannot know the answer with certainty, but if progress continues to accelerate, then there is good reason to be optimistic.

ACKNOWLEDGEMENTS

Supported by NIH Grant EY-02091 and ONR Grant N00014-89- 1192 to DVE.

REFERENCES

Allman, J.M., Miezin, F., and McGuinness, E., 1985, Stimulus specific responses from beyond the receptive field: Neurophysiological mechanisms for local-global comparisons of visual motion, *Ann. Rev. Neurosci.*, 3:532-548.

Allman, J.M., Miezin, F., and McGuinness, E., 1990, The effects of background motion on the responses of neurons in the first and second cortical visual areas, in: "Signal and Sense," ed., Neuroscience Research Program, New York.

Beck, J., 1966, Effect of orientation and of shape similarity on perceptual grouping, *Percep. and Psychophy.*, 1:300-302.

Burkhalter, A., and Bernardo, K.L., 1989, Organization of corticocortical connections in human visual cortex, *Proc. Natl. Acad. Sci. USA*, 86:1071-1075.

Carroll, E., and Wong-Riley, M., 1984, Quantitative light and electron microscopic analysis of cytochrome oxidase-rich zones in the striate cortex of squirrel monkeys, *J. Comp. Neurol.*, 222:1.

DeYoe, E.A., Felleman, D.J., Knierim, J.J., Olavarria, J., and Van Essen, D.C., 1988, Heterogeneous subregions of macaque visual area V4 receive selective projections from V2 thin-stripe and interstripe subregions., *Invest. Ophthal. Vis. Sci.*, 29:115 (Suppl.).

DeYoe, E.A., Van Essen, D.C., 1988, Concurrent processing streams in monkey visual cortex, *Trends Neurosci.*, 11:219.

Felleman, D.J., and Van Essen, D.C., 1991, Distributed hierarchical processing in primate cerebral cortex, Cerebral Cortex, in press.

Gilbert, C.D., and Wiesel, T.N., 1989, Columnar specificity of intrinsic horizontal and corticocortical connections in cat visual cortex, *J. Neurosci.*, **9**:2432-2442.

Horton, J.C., and Hedley-Whyte, T.E., 1984, Mapping of cytochrome oxidase patches and ocular dominance columns in human visual cortex, *Phil. Trans. R. Soc. London Ser. B* **304**:255-272.

Julesz, B., 1984, Toward an axiomatic theory of preattentive vision, in: "Dynamic aspects of neocortical function," G.M. Edelman, W. E. Gall, W. M. Cowan, eds., Neurosciences Research Foundation, Inc.

Kaas, J.H., 1988, Why does the brain have so many visual areas? *J. Cogn. Neurosci.*, **1**:121-135.

Lee, B.B., Martin, P.R., and Valberg, A., 1989, Nonlinear summation of M- and L-cone inputs to phasic retinal ganglion cells of the macaque, *J. Neurosci.*, **9**:1433-1442.

Logothetis, N.K., Schiller, P.H., Charles, and E.R., Hurlbert, 1989, Perceptual deficits and the activity of the color-opponent and broad- band pathways at isoluminance, *Science*, **247**:214-217.

Livingstone, M.S., and Hubel, D.H., 1984, Anatomy of physiology of a color system in the primate visual cortex, *J. Neurosci.*, **4**:309-356.

McIlwain, J.T., 1964, Receptive fields of optic tract axons and lateral geniculate cells: peripheral extent and barbituate sensitivity. *J. Neurophysiol.*, **27**:1154-1173.

Maunsell, J.H.R., and Van Essen, D.C., 1983, The connections of the middle temporal visual area (MT) and their relationship to a cortical hierarchy in the macaque monkey, *J. Neurosci.*, **3**:2563-2586.

Olavarria, J., and Montero, V.M., 1989, Organization of visual cortex in the mouse revealed by correlating callosal and striate-extrastriate connections, *Vis. Neurosci.*, **3**:56-69.

Olavarria, J., and Montero, V.M., 1984, Relation of callosal and striate- extrastriate cortical connections in the rat: morphological definition of extrastriate visual areas, *Exp. Brain Res.*, **54**:240-252.

Rockland, K.S., and Pandya, D.N., 1979, Laminar origins and terminations of cortical connections of the occipital lobe in the Rhesus monkey, *Brain Res.*, **179**:3-20.

Rosenquist, A.C., 1985, Connections of visual cortical areas in the cat, in: "Cerebral cortex," A. Peters, and E.G. Jones, eds., Plenum, New York.

Saito, H., Tanaka, K., Isono, H., Yasuda, M., and Mikami, A., 1989, Direction selective response of cells in the middle temporal area (MT) of the macaque monkey with the movement of equiluminous opponent color stimuli, *Exp. Brain Res.* **75**:1-14.

Sereno, M.I., and Allman, J.M., 1990, Cortical visual areas in mammals, in: "Neural basis of visual function," A. Leventhal, ed., McMillan, London.

Tigges, J., Spatz, W.B., and Tigges, M., 1973, Reciprocal point-to-point connections between parastriate and striate cortex in the squirrel monkey (Saimiri), *J. Comp. Neurol.*, **148**:481.

Tootell, R.B.H., Silverman, M.S., Switkes, E., and DeValois, R.L., 1983, Functional organization of the second cortical visual area of primates, *Science*, **220**:737.

Tootell, R.B.H., Silverman, M.S., Hamilton, S.L., Switkes, E., and DeValois, R.L., 1988, Functional anatomy of macaque striate cortex. V. Spatial frequency, *J. Neurosci.*, **8**: 1610.

Treisman, A.M. and Gelade, G., 1980, A feature-integration theory of attention, *Cognit. Psychol.*, **12**:97-136.

Van Essen, D.C., and Anderson, C.H., 1990, Information processing strategies and pathways in the primate retina and visual cortex, in: "Introduction to neural and electronic networks," Zornetzer, S.F., Davis, J.L, Lau, C., eds., Academic Press, Orlando.

Van Essen, D.C., DeYoe, E.A., Olavarria, J.F., Knierim, J.J., and Fox, J.M., 1989, Neural responses to static and moving texture patterns in visual cortex of the macaque monkey, in: "Neural Mechanisms of Visual Perception," D.M-K. Lam and C. Gilbert, eds., Portfolio, Woodlands, TX.

Van Essen, D.C., Felleman, D.F., DeYoe, E.A., Olavarria, J., and Knierim, J.J., 1990, Modular and hierarchical organization of extrastriate visual cortex in e macaque monkey, Cold Spring Harbor Symp. Quant. Biol., in press.

Zeki, S.M., and Shipp, S., 1989, Modular connections between areas V2 and V4 of macaque monkey visual cortex, *Eur. J. Neurosci.*, **1**:494.

LATERAL INTERACTIONS IN VISUAL CORTEX

Charles D. Gilbert, Daniel Y. Ts'o and Torsten N. Wiesel

The Rockefeller University
1230 York Avenue
New York, NY 10021-6399 U.S.A.

INTRODUCTION

In drawing relationships between visual psychophysics and physiology, one point of view has held that there is a closer correspondence between the function of higher order visual areas and perception, with primary visual cortex representing precognitive events. Because of the small size of receptive fields and their response to simple stimuli such an oriented line or edge, it appeared that the primary cortex served to decompose an object into a series of short oriented line segments, which presumably had to be reintegrated at a later stage in order to form a unified percept. In this paper we present evidence that such an integrative process may begin at an early stage, involving the primary visual cortex. A related problem applies to the different aspects of the visual stimulus, the submodalities of form, color, movement and depth. One line of experimental evidence, in both physiology and psychophysics, argues that these modalities are treated separately, by different populations of cells. Once again, the obvious question is how these properties come to be assigned to a single object, the so-called "binding" problem. Within the submodality of form, the binding of different component contours of an object is also referred to as segmentation. The neural mechanisms underlying binding or segmentation are still unknown, but both anatomical and physiological evidence, some of which will be described in this paper, points towards ways in which information from different visual submodalities and from different parts of the visual field interacts. Our understanding of the interactive mechanisms of cortex is based on examining the relationship between cortical microcircuitry, functional architecture and receptive field structure.

Functional architecture in V1 and V2

The classic studies of Hubel and Wiesel (1962, 1963, 1968, 1977) showed that cells having a certain property in common, such as orientation preference or ocular dominance, are grouped together in columns running from the pia to the white matter. These "columns," when seen in surface view, appear to be a system of branching and anastomosing bands, though the precise spatial relationships of different columnar systems is still under investigation. The distribution within the cortex of cells having different receptive field properties is known as functional architecture. An important development furthering the understanding of the functional architecture of visual cortex came from the histochemistry of the enzyme cytochrome oxidase. This enzyme, which participates in the energy metabolism of cells, is distributed in a non-uniform fashion in cortex. As a result, it fortuitously provides a framework against which one can compare the distribution of cells having different receptive field properties.

From Pigments to Perception, Edited by A. Valberg and
B.B. Lee, Plenum Press, New York, 1991

239

In tangential sections through the superficial layers of area V1, regions staining heavily for cytochrome oxidase are arranged in a system of rows of "blobs" (Wong-Riley, 1979; Horton and Hubel, 1981). At a first order of approximation, the cells within the blobs tend to be more specific for color, and lack orientation specificity. Cells in the interblob regions, those with the lightest cytochrome oxidase staining in V1, are orientation specific, and fewer of these are specific for color (Livingstone and Hubel, 1984; Ts'o and Gilbert, 1988). The blobs are located in the center of ocular dominance columns, and contain cells that are mostly monocular (Horton and Hubel, 1981). However, the staining pattern is quite irregular, with blobs of different size and spacing, and with bridges of dense staining connecting blobs along and between ocular dominance stripes. The interconnecting bridges of cytochrome oxidase staining have cells with properties that are similar to those in the blobs themselves. Cells in bridges running along ocular dominance columns are also unoriented and color specific. Bridges between blobs in different ocular dominance columns have unoriented cells which are binocular, and frequently color specific.

The receptive field properties of cells found in the blobs come in several varieties. Many reflect the properties of cells in the parvocellular layers of the LGN. These include type I, color and spatial opponent cells, type II cells with color but no spatial opponency, and type III cells, broadband in the color domain and having spatial opponency (Wiesel and Hubel, 1966). An additional type found in the blobs are a variant of the type II cells. These cells, which we call modified type II cells, have color opponency in the center and a broad band inhibitory surround. In effect, these cells combine information about color with information concerning luminance boundaries (Ts'o and Gilbert 1988). It is likely that the modified type II cells were earlier mistaken for double opponent cells; cells with opponent color mechanisms in both center and surround, with the opponency of the center having opposite sign of that in the surround. Upon closer examination, however, it appears that true double opponent cells are rarely found in V1. Whereas double opponent cells would respond most vigorously to an isoluminant color boundary, the modified type II cells are suppressed by light of a broad range of wavelengths in the surround.

We have found, in agreement with earlier studies, that a majority of the cells in the blobs are color specific. Beyond that specificity, however, we observed that individual blobs specialize in a particular color opponency: red/green or blue/yellow. The connections between the blobs are specific for opponency as well, with red/green blobs connected to other red/green blobs, and blue/yellow blobs connected to blobs of like color opponency. Thus there is considerable specialization in the functional properties of the blobs, and in the connections between them (Ts'o and Gilbert, 1988).

Beyond the blob/interblob distinction of color versus broadband and circularly symmetric fields versus oriented fields, there are cells having both color and orientation specificity in the superficial layers of V1 (Livingstone and Hubel, 194). These we believe to lie in a "periblob" distribution, at the boundary between the blob and interblob compartments (Ts'o and Gilbert, 1988). The interaction between the color and orientation domains is seen not only in the presence of color specific oriented fields, but also from evidence that these cells receive input from cells with color specific non-oriented fields. We have found, using cross-correlation analysis, that a cell preferring oriented bars of a particular color, receives input from a modified type II cell selective for the same color, and that modified type II cells receive input from standard type II cells of the same center type (Ts'o and Gilbert, 1988).

The interactions seen in V1 between the domains of form and color may fulfill the requirements of visual perception: in the well known "binding" problem, it may be necessary to associate the color of an object with its form at the level of single cells. The McCullough effect demonstrates that one can differentially adapt colored contours of a particular orientation and orthogonally oriented contours of the opponent color (McCullough, 1965). In the experiments of Karen DeValois described in this volume, the tendency to perceive two moving gratings of different orientation as a checkerboard pattern moving in a single direction versus two sliding surfaces moving in different directions is influenced by coloring the gratings differently.

Area in V2 and the cytochrome oxidase bands

The pattern of cytochrome oxidase staining in V2 also provides a framework for understanding the functional architecture of V2. In tangential sections of V2 this pattern appears as stripes of different width (Livingstone and Hubel, 1982, 1984; Tootell et al, 1983; Horton, 1984). It is popular to refer to these as thin and thick stripes, though the histological distinction between different stripes is more clearly seen in squirrel monkeys than in the macaque. Several studies showed that one set of alternating stripes has a concentration of color specific cells and the intervening dark stripes have a concentration of cells which are binocular and disparity selective. Cells in the interstripes are orientation specific. (DeYoe and VanEssen, 1985; Shipp and Zeki, 1985; Hubel and Livingstone, 1987; Tootell and Hamilton, 1989; Ts'o et al, 1989). Upon closer inspection, however, there is an overlap between the populations, as seen in the existence of cells coded for both color and depth, or color and orientation, or even all three (Ts'o et al, 1989). By applying a combination of optical and electrophysiological recording, we have revealed the distinction between, and functional specificity of, different stripes in V2.

Optical recording, as employed in our studies, is a technique that makes use of the observation that the reflectance of the cortex decreases in regions that are electrically active. This can be observed with an array of photodiodes (Grinvald et al, 1986), or, at higher resolution with a CCD camera (Ts'o et al, 1990). Prior to the advent of optical recording, we established functional architecture by making numerous closely spaced electrode penetrations. This was rather time consuming, and could make it difficult to include more detailed examination of receptive field properties and the fine structure of functional architecture. Optical recording affords the ability to image the architecture over a much larger area, and with high resolution. It has the advantages over deoxyglucose autoradiography that one can observe the spatial distribution of activity to different visual stimuli, and that one obtains the maps in the living cortex, which is then available, after optical recording, for detailed electrophysiology.

In V2, optical recording reveals a pattern of stripes that coincides with the stripes seen with cytochrome oxidase histochemistry (Ts'o et al, 1990a). The optical recording reveals a further detail in the striping pattern. With monocular stimuli, there are regions that are unresponsive, and consequently appear white with optical recording. When examined electrophysiologically, these areas have cells that are disparity selective. The cells do not respond to monocular stimuli, but when presented with binocular stimuli at the appropriate disparity, respond quite vigorously. These cells are termed "obligatory binocular." The white patches are aligned with one set of cytochrome oxidase stripes. When doing a finer scale mapping of these areas with microelectrodes, one finds clusters of cells responsive to stimuli in front of the plane of fixation (near cells), behind the plane of fixation (far cells) or on the plane of fixation (tuned excitatory cells). Within a single penetration, the cells often show a regular shift in optimum response at different points relative to the plane of fixation. Thus there is a fine grained substructure in the V2 stripes (Ts'o et al, 1989; 1990b).

Other stripes of V2 have numerous cells specific for color opponency. In optical recording the color selective areas are seen as dark patches. These stripes are not always spatially segregated from the stripes with disparity selective cells, separated from each other by pale stripes. Instead, optical recording shows occasional places where the dark patches merge with the light patches. The histochemistry alone would not allow one to distinguish between a single broad stripe representing a particular functional domain and two "merged" stripes, each with a distinct functionality. Electrode penetrations at the points of contact reveal the presence of cells coded for both color and disparity. Whether the disparity selectivity seen here, or even in the more classic broadband disparity and orientation selective regions, underlies the perception of depth *per se* is not known. But it opens the possibility for interaction between visual submodalities as seen in V1, and in particular may represent the mechanism for obtaining 3-dimensional structure from color cues. Here too, the interaction can be seen at the boundaries between compartments specializing in one submodality. One value of organizing the cortex in this way is to allow binding between different kinds of information when it pertains to the same object. In addition, it is perhaps more efficient to do it this way rather than permuting every visual parameter by every other, which could be prohibitive in the number of cells that would be required. Having one cell population specializing in a particular parameter allows the visual system to have the required resolution of that parameter, and having a smaller number of cells that are multiply coded allows the association of one parameter with another.

Because of the compartmentalization of cells according to their functional type, as represented by cortical functional architecture, the interaction between visual submodalities may be mediated by cross-talk between the compartments, which is one manifestation of lateral interactions in cortex.

Horizontal connections and spatial interactions

The lateral interactions between visual submodalities is not the only form of lateral interaction observed in cortex. Even within a submodality, such as orientation, there is another variety of lateral interaction--one between different positions in visual space. The first evidence for this was obtained from intracellular recording and dye injection (Gilbert and Wiesel, 1979, 1983). In distinction to the classical view of cortical connectivity, based on Golgi impregnations, which pointed towards a pattern of vertical connectivity within a column, with relatively little lateral spread, the intracellular injections showed that pyramidal cell axons extended for long distances parallel to the cortical surface, reaching up to 6 mm. This is a long distance in cortical terms, and has interesting functional implications. In visual cortex, in particular V1, the retinotopic map points towards a relationship between visual field position and distance traveled in the cortex. Vertical electrode penetrations in the cortex encounter cells with overlapping receptive fields, with a certain amount of scatter. With electrode penetrations parallel to the cortical surface, if one moves what corresponds to two hypercolumns, a distance of about 1.5 mm, one reaches a point where none of the receptive fields overlap with those at the beginning of the penetration (Hubel and Wiesel, 1974). Therefore, with 6 mm connections, the target cells integrate information over an area of cortex representing a visual field area which is much larger than their receptive fields!

The classical receptive field is supposed to represent the area of visual space within which the appropriate stimulus can activate a cell. One might expect that receptive field area would match the visuotopic area that provides input to the cell, but there is a considerable discrepancy between these two measures. This seeming contradiction can be resolved because of the stimulus dependency of receptive field definition. A simple visual stimulus, such as a light bar or edge, will activate a cell over a fairly limited area. Stimuli outside this area will not be effective, but they can influence the response of the cell to stimuli placed within the receptive field. Thus the functional properties of cells are dependent on the context within which a feature is presented. Contextual dependency is a familiar concept in psychophysics, where it has long been known that perception of a visual attribute can be influenced by the presence of additional features in the visual environment (Butler and Westheimer, 1978; Badcock and Westheimer, 1985; Westheimer, 1986; Westheimer et al, 1976; Westheimer and McKee, 1977). In the domain of orientation this is seen in the tilt illusion (Gibson and Radner, 1937; see Howard, 1986 for review; Westheimer, 1990). One may have thought that this seemingly highly integrative characteristic of visual perception was produced by higher order visual areas. The nature of the horizontal connections, however, and their relationship to the functional architecture of cortex raises the possibility that the substrate for such an illusion may be present even at as early a stage as area V1. Before discussing the context dependency of receptive fields, therefore, it is worth first describing the relationship between the horizontal connections and the cortical columns.

Though the horizontal connections are quite broad in their lateral extent, they do not uniformly innervate the cortical area that they cover. Rather, the axon collaterals are distributed in discrete clusters, suggesting a specificity in the functional relationships between cells connected across such distances (Gilbert and Wiesel, 1979, 1983; Martin and Whitteridge, 1984). We explored these relationships using two approaches: cross-correlation analysis and anatomical labeling of the horizontal connections and orientation columns.

Cross-correlation analysis is a statistical technique relating the pattern of firing between a pair of neurons. A simplified description of the technique is that any pair of neurons that are synaptically connected or that share a common input will have a peak in their correlogram, with upward going peaks indicative of excitation and dips in the correlogram representing inhibition. By examining a population of cells in the superficial layers of cortex, at varying horizontal separations, one can correlate firing for cells with varying separation in receptive field position and with different relationships in functional specificity, such as orientation preference. We have found that correlated firing exists between cells having similar orientation preference, and could be seen between cells with receptive fields separated by up to several receptive field diameters. The waxing and waning in the strength of correlation as one moves across orientation columns with varying specificity is reminiscent of the clustering of the axon collaterals seen in the horizontal connections, and the ability to see peaks in the correlograms at a couple of millimeters horizontal separation between the cells is consistent with the extent of these connections (Ts'o et al, 1986; Ts'o and Gilbert, 1988).

The anatomical approach to identifying the specificity of the horizontal connections involved labeling the horizontal connections by retrograde transport of extracellularly applied tracers and labeling the columns by 2-deoxyglucose autoradiography. The tracer used was rhodamine filled latex microspheres, which are taken up by axon terminals at the injection site and transported retrogradely to the cells projecting to the injection site. After allowing sufficient time for transport, a tiny injection, about 150 μm in diameter (which would therefore include orientation columns covering a narrow range of orientations), results in labeling of a large area of cortex, about 8 mm in extent. In the area of cortex where the injections were made, this would represent approximately 10 degrees of visual angle, in contrast to the receptive field diameter of cells at this site, which is only about 1 to 2 degrees. Within the area of labeling the cells have a clustered distribution. At the end of the experiment the orientation columns were labeled by 2-deoxyglucose autoradiography: cells responding to the stimulus, vertical stripes, have higher metabolic requirements, take up more glucose (including the radioactively labeled glucose), and active regions will appear as dark patches and stripes after the cortex is sectioned and exposed to autoradiographic film. When we injected a vertical orientation column with the dye, the resultant clusters of retrogradely labeled cells were superimposed with the labeled vertical orientation columns (Gilbert and Wiesel, 1989). This then confirmed the idea that the horizontal connections run between columns of similar orientation specificity.

This characterization of the horizontal connections would suggest that the functional role of the connections might be related to the contextual dependence of the perception of the attribute of orientation: The connections are widespread, and therefore could mediate interactions between disparate visuotopic loci, and the connections are orientation specific, so could play a role in perceptual phenomena that are dependent on orientation.

The contextual dependence of orientation specificity

The property of orientation selectivity, originally described by Hubel and Wiesel (1959, 1962), is seen for most cells in visual cortex, and is a property generated in visual cortex. Even though the description of the phenomenon is some 30 years old, the neural mechanism underlying it is not known, though the original suggestion by Hubel and Wiesel that it originates from convergence of excitatory input from the LGN is still a very viable model. Inhibitory mechanisms might also play a role. According to classical views, one would expect the orientation tuning of cells to be fixed. The question we addressed was whether the presence of more complex contextual stimuli, surrounding the receptive field, might have an influence on the orientation tuning of a cell.

For a subset of cells in visual cortex we found a range of effects of placing an array of oriented lines surrounding the receptive field. Effects included suppression or facilitation of the firing of cells having the same orientation as the surround, suppression of cells oriented differently from the surround, shifts in peak orientation tuning towards or away from the orientation of the surround, and changes in the bandwidth of tuning. To determine which of these effects may bear a relationship to the perceived shifts in orientations quantified by Westheimer (1990), we created a model relating the characteristics of an ensemble of oriented cells to perceived orientation. This model showed that isorientation suppression, attractive shifts in peak orientation, and changes in tuning bandwidth could all result in the "repulsive" effect seen in the tilt illusion. Further experiments are required to determine which effects are most common and which relate to other properties of the tilt illusion. At this stage, however, one can say that rather than having receptive fields that are restricted in their extent, there is a sensitivity of cells to stimuli placed over much larger regions of visual space. Furthermore, rather than having cells with fixed filter characteristics, orientation tuning may be a dynamic property, changing according to context (Gilbert and Wiesel, 1990). This context dependency may be seen for other visual submodalities: color constancy in the domain of color, which involves comparing surface reflectances at different visuotopic locations and figure-ground separation generated by differential movement between a moving figure and its background. Color constancy at the single cell level has been observed at the single cell level by Zeki (1983) and interactions in the domain of movement have been observed in a number of areas (Allman et al, 1985; Tanaka et al, 1986; Gulyas et al, 1987; Orban et al, 1987).

The relationship between contextual sensitivity and the horizontal connections is suggestive, but much needs to be done to establish the connection. For cells in V1, convergent connections are provided by feedback connections from other visual areas, such as V2, in addition to the horizontal connections within V1. It is by no means clear how the relationship between the clustered connections and orientation columns would lead to changes in orientation tuning. The synaptic mechanisms of the changes suggest that the horizontal connections, if involved, might gate other inputs to the cell. Ultimately one would like to establish the relationship between patterns of connectivity and the response characteristics of cells to complex visual stimuli, and between these higher-order receptive field properties and perceptual phenomena.

The purpose of the contextual dependence of orientation selectivity and of tilt illusions in normal visual perception is not clearly established. In one sense, it may represent a "contrast" mechanism, analogous to that seen in the spatial domain, to highlight orientation differences or changes in the visual environment. Additionally, it may be a biproduct of the process of segmentation, where different contours are linked to form a unified percept, though as yet the algorithm for such linking has not been worked out.

Changes in cortical topography following retinal lesions

The above context dependency must be a fast process, with the properties of cells adjusting to each visual scene as the eyes move around the visual environment. We have observed longer term changes by making permanent perturbations in sensory input. These experiments were modeled on work done in the somatosensory system, where finger removal leads to a reorganization of the somatotopic maps of cortex (Merzenich et al, 1984, 1988) and spinal cord (Wall and Werman, 1976; Devor and Wall, 1978).

We made retinal lesions using an argon laser, which destroyed cells in the photoreceptor layer over a predetermined area. Recording in the cortex on the day of the lesion revealed a scotoma, an area of cortex that could not be driven by visual stimuli. A clear boundary between visually activated and silent cortex was mapped by multiple electrode penetrations. Over the next several weeks the boundary shifted into the originally silent cortex, until at 8 weeks we found that virtually the entire cortical scotoma was once again driven by visual stimuli. The receptive fields of the cells in the "scotoma", however, did not respond to stimuli covering the retinal lesion but instead areas of retina immediately outside the lesion. These receptive fields were somewhat larger than normal, but were clearly oriented. The shift in the cortical map was on the order of 3 mm, which is consistent with the extent of the horizontal connections in cortex. We believe that the horizontal connections are a good candidate for the source of visual input into the scotoma, but further experiments are required to differentiate contributions of LGN (Eysel et al, 1981) or other cortical areas to the topographical reorganization (Gilbert et al, 1990). Similar experiments on the effect of retinal lesions combined with eye removal have recently been reported by Kaas et al (1990). Changes in cortical topography resulting from retinal lesions may have consequences in spatial perception, as reported by Craik (1966).

Thus cortical topography is capable of alteration, even in adult animals. The retinal lesion experiments might represent a model for the recovery of function that occurs after other lesions of central nervous system. Furthermore, though removal of retinal input is clearly an extreme and abnormal form of sensory intervention, it might be a useful model for changes occurring in cortical areas responsible for storing novel visual images.

The dynamic nature of cortical organization and receptive field properties raises new perspectives for the processing capabilities of visual cortex.

References

Allman, J.M., F. Miezin, E. McGuinnes. 1985. Direction and velocity specific surround in three cortical visual areas of the owl monkey. *Perception* **14**: 105-126.

Badcock, D.R. and G. Westheimer. 1985. Spatial location and hyperacuity: the centre-surround localization function has two substrates. *Vision Research* **25**: 1259-1269. *J. Neurophysiol.* **60**: 1053-1065.

Butler, T. and G. Westheimer. 1978. Interference with stereoscopic acuity: spatial, temporal and disparity tuning. *Vision Res.* **18**: 1387-1392.

Craik, K.J.W. 1966. Localized aniseikonia following eclipse blindness Press, pp.102-103.

Devor, M. and P.D. Wall. 1978. Reorganisation of spinal cord sensory map after peripheral nerve injury. *Nature* **276**: 75-76.

DeYoe, E.A. and D.C. Van Essen. 1985. Segregation of efferent connections and receptive field properties in visual area V2 of the macaque. *Nature* **317**: 58-61.

Eysel, U.T., F. Gonzalez-Aguilar, and U. Mayer. 1981. Time-dependent decrease in the extent of visual deafferentation in the lateral geniculate nucleus of adult cats with small retinal lesions. *Exp. Brain Res.* **41**: 256-263.

Gibson, J.J. and M. Radner. 1937. Adaptation, after-effect and contrast in the perception of tilted lines. *J. Exp. Psychol.* **20**: 453-467.

Gilbert, C.D. and T.N. Wiesel. 1979. Morphology and intracortical projections of functionally identified neurons in cat visual cortex. *Nature* **280**: 120-125.

Gilbert, C.D. and T.N. Wiesel. 1983. Clustered intrinsic connections in cat visual cortex. *J. Neurosci.* **3**: 1116-1133.

Gilbert, C.D. and T.N. Wiesel. 1989. Columnar specificity of intrinsic connections in cat visual cortex. *J. Neurosci.* **9**: 2432-2442.

Gilbert, C.D. and T.N. Wiesel. 1990. The influence of contextual stimuli on the orientation selectivity of cells in primary visual cortex of the cat. *Vision Res.* in press.

Gilbert, C.D., J.A. Hirsch and T.N. Wiesel. 1990. Lateral interactions in visual cortex. *Cold Spring Harbor Symp. Quant. Biol.* 55: in press.

Grinvald, A., E. Lieke, R.D. Frostig, C.D. Gilbert and T.N. Wiesel. 1986. Functional architecture of cortex revealed by optical imaging of intrinsic signals. *Nature* **324**: 361-364.

Gulyas, B., G.A. Orban, J. Duysens, and H. Maes. 1987. The suppressive influence of moving texture background on responses of cat striate neurons to moving bars. *J. Physiol.* **57**: 1767-1791.

Horton, J.C. 1984. Cytochrome oxidase patches: A new cytocarchitectonic feature of monkey cortex. *Phil. Trans. R. Soc. Lond. B* **304**: 199-253.

Horton, J.C. and D.H. Hubel. 1981. A regular patchy distribution of cytochrome-oxidase staining in primary visual cortex of the macaque monkey. *Nature* **292**: 762-764.

Howard, I.P. 1986. The perception of posture, self-motion and the visual vertical. in *Handbook of Perception and Human Performance*, **v. 1** ed. K.R. Boff, New York: John Wiley.

Hubel, D.H. and M.S. Livingstone. 1987. Segregation of form, color, and stereopsis in primate area 18. *J. Neurosci.* **7**: 3378-3415.

Hubel, D.H. and T.N. Wiesel. 1959. Receptive fields of single neurones in the cat's striate cortex. *J. Physiol.* **148**: 574-591.

Hubel, D.H. and T.N. Wiesel. 1962. Receptive fields, binocular interaction and functional architecture in the cat's visual cortex. *J. Physiol.* **160**: 106-54.

Hubel, D.H. and T.N. Wiesel. 1963. Shape and arrangement of columns in cat's striate cortex. *J. Physiol.* **165**: 559-568.

Hubel, D.H. and T.N. Wiesel. 1968. Receptive fields and functional architecture in monkey striate cortex. *J. Physiol.* **195**: 215-243.

Hubel, D.H. and T.N. Wiesel. 1974. Uniformity of monkey striate cortex: a parallel relationship between field size, scatter and magnification factor. *J. Comp. Neurol.* **158**: 295-306.

Hubel, D.H. and Wiesel, T.N. 1977. Functional architecture of macaque monkey visual cortex. *Proc. R. Soc. Lond. B.* **198**: 1-59.

Kaas, J.H., L.A. Krubitzer, Y.M. Chino, A.L. Langston, E.H. Polley, and N. Blair. 1990. Reorganization of retinotopic cortical maps in adult mammals after lesions of the retina. *Science* **248**: 229-231.

Livingstone, M.S. and D.H. Hubel. 1982. Thalamic inputs to cytochrome oxidase rich regions in monkey visual cortex. *Proc. Natl. Acad. Sci. USA* **79**: 6098-6101.

Livingstone, M.S. and D.H. Hubel. 1984. Anatomy and physiology of a color system in the primate visual cortex. *J. Neurosci.* **7**: 3371-3377.

Martin, K.A.C. and D. Whitteridge. 1984. Form, function and intracortical projections of spiny neurones in the striate visual cortex of the cat. *J. Physiol.* **353**: 463-504.

McCollough, C. 1965. Color adaptation of edge-detectors in the human visual system. *Science* **149**: 1115-1116.

Merzenich, M.M., R.J. Nelson, M.P. Stryker, M.S. Cynader, A. Schoppmann, and J.M. Zook. 1984. Somatosensory cortical map changes following digital amputation in adult monkeys. *J. Comp. Neurol.* **224**: 591-605.

Merzenich, M.M., G. Recanzone, W.M. Jenkins, T.T. Allard, and R.J. Nudo. 1988. Cortical representational plasticity in *Neurobiology of Neocortex* ed. Rakic and Singer Chichester: John Wiley and Sons, 41-68.

Orban, G.A., B. Gulyas, and R. Vogels. 1987. Influence of a moving textured background on direction selectivity of cat striate neurons. *J. Neurophys.* **57**: 1792-1812.

Shipp, S. and S.M. Zeki. 1985. Segregation of pathways leading from area V2 to areas V4 and V5 of macaque monkey visual cortex. *Nature* **315**: 322-325.

Tanaka, K., K. Hikosaka, H. Saito, M. Yukie, Y. Fukada, and E. Iwai. 1986. Analysis of local and wide-field movements in the superior temporal visual areas of the macaque monkey. *J. Neurosci.* **6**: 134-144.

Tootell, R.B.H. and S.L. Hamilton. 1989. Functional anatomy of the second visual area (V2) in the Macaque. *J. Neurosci.* **9**: 2620-2644.

Tootell, R.B.H., M.S. Silverman, R.L. De VaLois, and G.H. Jacobs. 1983. Functional organization of the second cortical visual area in primates. *Science* **220**: 737-739.

Ts'o, D.Y., C.D. Gilbert, and T.N. Wiesel. 1986. Relationships between horizontal connections and functional architecture in cat striate cortex as revealed by cross-correlation analysis. *J. Neurosci.* **6**: 1160-1170.

Ts'o, D.Y. and C.D. Gilbert. 1988. The organization of chromatic and spatial interactions in the primate striate cortex. *J. Neurosci.* **8**: 1712-1727.

Ts'o, D.Y., C.D. Gilbert, R.D. Frostig, A. Grinvald and T.N. Wiesel. 1989. Functional architecture of visual area 18 of Macaque monkey. *Abs. Soc. Neurosci.* **15**: 161.

Ts'o, D.Y., C.D. Gilbert and T.N. Wiesel. 1990b. Functional architecture of color and disparity in visual area 2 of Macaque monkey. *Abs. Soc. Neurosci.* **16**: in press.

Ts'o, D.Y., R.D. Frostig, E. Lieke, and A. Grinvald. 1990. Functional organization of primate visual cortex revealed by high resolution optical imaging. *Science* **249**: 417-420.

Wall, P.D. and R. Werman. 1976. The physiology and anatomy of long ranging afferent fibres within the spinal cord. *J. Physiol.* **255**: 321-334.

Westheimer, G. 1986. Spatial interaction in the domain of disparity signals in human stereoscopic vision. *J. Physiol.* **370**: 619-629.

Westheimer, G. 1990. Simultaneous orientation contrast for lines in the human fovea. *Vision Res.*, in press.

Westheimer, G., and S.P. McKee. 1977. Spatial configurations for visual hyperacuity. *Vision Res.* **17**: 941-949.

Westheimer, G., K. Shimamura, and S. McKee. 1976. Interference with line orientation sensitivity. *J. Opt. Soc. Am.* **66**: 332-338.

Wiesel, T.N. and D.H. Hubel. 1966. Spatial and chromatic interactions in the lateral geniculate body of the rhesus monkey. *J. Neurosci.* **29**: 1115-1156.

Wong-Riley, M. 1979. Changes in the visual system of monocularly sutured or enucleated cats demonstrable with cytochrome oxidase histochemistry. *Brain Res.* **171**: 11-28.

Zeki, S.M. 1983. Colour coding in the cerebral cortex: The reaction of cells in monkey visual cortex to wavelengths and colours. *Neuroscience* **9**: 741-765.

THE PERCEPTUAL SIGNIFICANCE OF CORTICAL ORGANIZATION

Peter Lennie

Center for Visual Science
University of Rochester
Rochester, NY 14627

BACKGROUND

Neurophysiologists are well aware of the limitations of single-unit recording in elucidating the functional organization of visual cortex: the technique provides us with information about the behavior of only one component in a complex network, and is subject to sampling biases. Moreover, there are good reasons to fear that even when we have learned a lot about the behavior of an individual cell, we will not understand much about its role in vision. A well-worn analogy would be trying to understand the role of a particular transistor in a computer by observing the voltage changes at its emitter. The recognition of these limitations has not diminished our reliance on single-unit recording as a source of ideas about the organization of cortex; indeed, the many impressive similarities between the visual sensitivities of individual neurons and those of psychophysical observers reinforce the notion that perception is simply and directly related to the activity of single units (Barlow, 1972). Two recent developments have prompted me to worry that we may have become a little too willing to overlook the limitations of single-unit recording.

First, a study John Krauskopf, Gary Sclar and I made of mechanisms of color vision in striate cortex (Lennie, Krauskopf & Sclar, 1990) resulted in an account of chromatic mechanisms that differs in important respects from other recent ones: for example, we saw no concentrations of chromatically interesting cells in the cytochrome-oxidase blobs, and we saw no clear signs of distinct chromatic groups. A dull reason for the discrepant results is that different workers use different terminology to describe the same thing (and I think that this contributes to the disagreement). A much more interesting reason for the disagreement might be that by using different visual stimuli, different workers actually see different behaviors that cannot be related through the concept of a linear receptive field. In our work we excited cells with spatially extended stimuli and weak signals (generally low-contrast chromatic modulations about a neutral background). Most other workers have used discrete and usually more intense, more saturated stimuli.

The second development has to do with observations Derryl DePriest, John Krauskopf and I recently made on the properties of cells in area V2. One of the major tenets of recent work on visual cortex has been that the system is hierarchically organized (albeit loosely). It therefore came as a considerable surprise to us to discover that the visual characteristics of cells in V2 were much like those of cells in V1. Receptive fields in V2 are indeed larger, and almost all are complex, and neurons may be a little easier to drive binocularly, but the overall impression one obtains is that cells in V2 are remarkably like those in V1. This suggests either that our notions of hierarchy are wrong, or (more likely) that our methods—those well-suited to the analysis of linear systems—were ill-suited to the study of nonlinear cells, and that by using them we ensured that we observed only certain kinds of behavior. Were this the source of the problem, one might expect to solve it by using different stimuli and methods of analysis. The difficulty lies in finding the right method.

From Pigments to Perception, Edited by A. Valberg and
B.B. Lee, Plenum Press, New York, 1991

Two features of ganglion cells and neurons in the mammalian LGN make it easy to characterize receptive fields. First, most cells behave linearly, so different methods of exploring receptive fields (e.g., with spots, lines or sinusoidal gratings) yield similar accounts of receptive field organization. Even when such cells behave nonlinearly, the non-linearities are often simple enough that they can be understood through the application of linear methods (Shapley & Lennie, 1985). Second, there seem to be only a few distinct types of cells in the retina and LGN, and different populations of them are quite homogeneous. The upshot is that single-unit physiology has been able to provide a solid account of the visual characteristics of these neurons. Because in several domains (e.g., measurement of contrast sensitivity, the identification of opponent processes in color vision) the behavior of the individual neuron parallels that of the psychophysical observer, the behavior of the single-unit has come to be associated very directly with perception (Barlow, 1972).

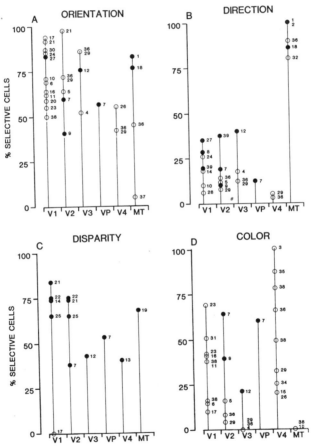

Figure 1. Relative frequency with which neurons selective for orientation (A), direction of movement (B), binocular disparity (C) and color (D) are encountered in different cortical areas (from Felleman & Van Essen, 1987). Each panel shows, for six different visual areas, the estimates obtained from various studies. The number beside each circle identifies the study from which the estimate was obtained. Felleman and Van Essen provide a key. Open circles represent informal estimates, filled circles represent measured frequencies of occurrence.

The population of cortical cells is more heterogeneous, and neurons have complicated receptive fields whose properties are often hard to characterize. This has left us able to establish fewer (and perhaps less secure) connections between the behavior of individual units and that of the observer, although parallels have been drawn in the domains of stereopsis (Barlow, Blakemore & Pettigrew, 1967), contrast adaptation (Movshon & Lennie, 1979), and detection of movement (Barlow, Hill & Levick, 1964). Why is it hard to characterize receptive fields of cortical neurons, and how well do we succeed in doing it? Some of the complications (at least in anesthetized preparations) may stem from the susceptibility of the cortex to anesthesia, and in the awake animal fluctuations in attention are clearly important. However, I think the most important difficulty lies in the way we go about characterizing receptive fields.

Even if we know little in advance about the organization of receptive fields, as long as neurons behave linearly we can use the standard techniques of linear systems analysis to characterize them. Most cells in cortex are manifestly nonlinear, and for these we have available no general methods analogous to those used to study linear systems. Unless we know in advance what properties to look for, we are likely to waste our time exploring the wrong universe. The history of discovery in visual cortex is littered with examples of this: there were many failed attempts to describe the receptive fields of neurons in striate cortex before Hubel and Wiesel used stimuli that matched the requirements of neurons. It is often hard to find stimuli that excite cortical neurons; indeed the fact that we can drive them at all with visual stimuli tends to encourage the (often mistaken) belief that we have found the "right" stimulus. I think that this has a lot to do with the fact that physiological studies show the properties of neurons varying rather little from one visual area to another. Fig. 1, from Felleman & Van Essen (1987) indicates that, with the possible exceptions of V4 and MT, the different areas are not distinguished by the responses of cells to the visual stimuli commonly used to explore receptive fields. This apparent "sameness" of cells in different areas is a remarkable feature of cortex. It might mean that cells in different areas really are similar, although I think it is more likely to reflect the limitations of the techniques we use to characterize neurons—in most cases we have not found the stimuli that drive the cells best, and which would distinguish different areas.

In understanding the properties of non-linear cortical cells we are greatly helped by having in advance a clear idea about how they transform visual signals. Spitzer & Hochstein's (1985) analysis of the behavior of complex cells in cat shows the power of this approach. However, even when cells are well characterized, it is often very difficult to discern their contributions to perception. The general problem is that we cannot easily infer the role of a small component in a complicated machine just by observing the signals in that component. Thus, although we know a good deal about the transformation of signals undertaken by a complex cell, and have in some sense found its 'trigger feature' (Barlow, 1972), we know little about how it contributes to perception. (In the case of complex cells, rectified responses make the neurons potentially useful as motion detectors [Emerson, et al., 1987], but there are so many such neurons that this is unlikely to be their only role.) We always need some design blueprint—some theory about the functional organization of the system—before we can understand the roles of components.

Visual physiologists depend on theory to guide their explorations. In the past, psychophysics provided much of this, in the shape of often sharply articulated models that made explicit statements about the properties of underlying mechanism. The physiological analyses of trichromacy, of opponent processes, of light- and dark-adaptation and of lateral inhibition have benefitted immeasurably from the psychophysical work. Psychophysics has provided similar inspiration to physiological studies of phenomena that involve visual cortex—disparity sensitivity, motion sensitivity, and spatially- and orientationally-selective aftereffects. However, physiologists have received rather little guidance about how to study processes that underlie the perception of images more complicated than those typically used in psychophysical experiments. The upshot is that the burgeoning physiological and anatomical work on higher visual mechanisms has developed with very little of the context formerly provided by psychophysics. We now know an enormous amount about aspects of organization (for example, the magnocellular and parvocellular pathways through LGN, the multiple representations of the visual field on cortex) about which psychophysics had fostered no prior expectations, and we find ourselves sometimes seeking psychophysical indicators of mechanisms that have been distinguished anatomically and/or physiologically.

With psychophysics providing a weak context for their observations, physiologists have turned inwards, and have drawn heavily on physiological measurements and anatomical observations for ideas about the perceptual significance of cortical organization. I want to argue that this emphasis on physiology and anatomy has yielded few theoretical insights.

PROCESSING STREAMS IN CORTEX

A dominant idea during the last decade (originating with Ungerleider & Mishkin, 1982, and developed by others) has been that, within visual cortex, there are two major anatomical divisions, each responsible for a different kind of visual analysis. The anatomical work shows that one system—called the occipitotemporal system—originates in V1 and runs through visual area 2 (V2), then principally to area V4, which projects to inferotemporal cortex (IT); the other—called the occipitoparietal system—originates in V1, and runs through (and around) V2, through V3, the middle temporal area (MT) and the medial superior temporal area (MST). Fig. 2 is a simplified diagram of the principal projections of V1 and its neighboring extrastriate areas. Although they may receive substantially segregated inputs from the magnocellular and parvocellular pathways through the LGN (the magnocellular pathway making a substantial contribution to the occipitoparietal stream), the systems are plainly not independent.

Information about the functions of these pathways comes principally from studies that have characterized the receptive fields of individual neurons, and from studies of the perceptual effects of clinical and experimental lesions. Two components in the pathways have drawn particular attention: in the occipitoparietal pathway, area MT, and in the occipitotemporal pathway, area V4.

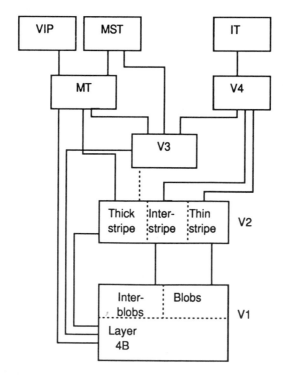

Figure 2. Simplified diagram of the major connections between striate cortex (V1) and neighboring extra-striate visual areas. The hierarchical sequence is indicated by the bottom-to-top ordering of areas. See text for further details.

Zeki's single unit studies first pointed to distinctive characteristics of cells in V4 and MT. He and his colleagues showed that the bulk of neurons in MT were strongly directionally selective (Dubner & Zeki, 1971), and that many neurons in V4 were chromatically selective (Zeki, 1973). More recent work has examined the physiology of what are thought to be the earlier stages of the pathways that feed these areas. Hubel and Livingstone (1987) observed that the properties of cells in V2 vary with their location: directionally selective units are found mostly in 'thick' stripes—wider bands of cortex rich in cytochrome oxidase—that send a projection to MT (DeYoe & Van Essen, 1985; Shipp & Zeki, 1985); chromatically-selective units are found mostly in thin stripes, which project to V4; orientation-selective units, many with 'end-stopped' receptive fields, are found in the 'inter-stripe' regions, and project to V4, among other destinations. These and related observations led Livingstone & Hubel (1987) to argue that there were three fundamental streams, one specialized to handle information about motion and binocular disparity, one concerned with the chromatic attributes of images, and the third specialized for high-resolution form perception. Parts of the analysis are contentious. For example, DeYoe & Van Essen (1988) argue that there is no strict segregation of function in the anatomically distinct pathways, and Levitt & Movshon (1990) find that although motion-selective cells are most common in the thick stripes, and color-selective cells are most common in the thin stripes, both types of cell form a minority even in the places where they are most often encountered. Nonetheless, the idea of distinct functional streams, with broadly the properties adduced by Livingstone & Hubel, is widely accepted as a principle of cortical organization.

Why would one want this modular organization of the visual system? Parallel and hierarchical organizations are known in general to be well-suited to the analysis of complex problems (Ballard, Hinton & Sejnowski, 1983), but the benefits of any particular organization are less easy to understand. Why should we analyze the dynamic aspects of a scene independently of the chromatic aspects, and why should information about contour orientation be dealt with by one mechanism and information about surface shading by another? I believe that the major force in shaping our notions about specialized analyses that are undertaken in the different regions has been the physiological evidence on the visual preferences of individual cells. I have already argued that the properties of most cells are not well-understood, and in any case physiologists disagree about them so I think we cannot draw strong inferences about the modular organization of cortex from work on single-units. The evidence from clinical and experimental lesions is not a great deal more encouraging. Experimental lesions of MT and V4 in macaques ought perhaps to provide the clearest evidence, but that is equivocal: lesions in MT impair capture and pursuit eye-movements (Newsome, Wurtz, Dursteler & Murikami, 1985), and bring about a rise in thresholds for detecting motion (Newsome & Paré, 1988), but both effects are transitory; lesions of V4 seem to disrupt color vision (Wild, Butler, Carden & Kulikowski, 1985), but they also markedly disrupt many other aspects of vision (Heywood & Cowey, 1987). The clinical literature is also not very encouraging: although there are well-documented cases of modality-specific deficits, these are mostly not the cases where lesions have been identified precisely. Where lesions have been identified, I know of no case where the deficit has been confined to a single modality. This kind of anatomical evidence has so far provided only weak guidance about function.

Imaging techniques such as Positron Emission Tomography (PET) hold great promise for informing us about the functions of different regions, but have not so far contributed much to our understanding of cortical organization: the nature of the PET scan means that human observers can be studied for a few minutes every several months, and the small amount of information we obtain as a result leads us to worry about the scope and reliability of measurements. We need to use PET on monkeys (where the frequency and duration of measurements need not be an issue) to establish that visual stimuli specially designed to drive color or motion mechanisms can differentially excite the regions (V4 and MT) suspected of being heavily involved in their analysis.

By any measure, we have a vastly richer knowledge of the architecture and function of visual of visual cortex than we had ten years ago. There is clear support for the notion of hierarchical modular organization, but we have not got far in answering the larger questions about perceptual significance that encouraged the abandonment of psychophysics as the major impetus to theory. The physiological work has not been able to tell us why we need modules of particular types, nor has it been able firmly to identify the kinds of analyses undertaken at different levels of the hierarchy. Moreover, we still know little about the properties of many cells, and even when cells are well-characterized, we know little about how they contribute to the function of a cortical area. These limitations of physiology compel us to look elsewhere for more substantial intuitions about perceptual significance.

ANALYZING IMAGES

In trying to understand how visual cortex works it is helpful to ask: what is the system designed to do, what are the constraints within which it has to operate, and, given these constraints, what is the most efficient design for achieving the goals?

The goal (loosely stated) is to identify objects and their positions in the world. All recent attacks on the problem of recognition, whether psychological or computational, rest upon the idea that in identifying objects some parts of the image are more important than others—for example, one of the more widely held positions is that recognition proceeds through the extraction of features.

There are good reasons to pay more attention to some parts of the image than to others: most images are redundant. It saves much storage capacity and perhaps computational effort in analyzing them if they can be described economically. It therefore seems useful to draw a distinction between the *description* of the image and the *analysis* of the image. The description of an image is an account of it from which the original image could be more or less fully reconstructed (e.g., a pixel map). The analysis of the image results in descriptions of objects and their positions, and almost certainly discards information that would be needed for full recovery of the image.

A lot of evidence from different disciplines suggests that early mechanisms in cortex (in conjunction with pre-cortical mechanisms) work to provide an economical description of the image. I want to examine that briefly to glean its implications for physiological organization. I then want to examine what kind of analyzers might lie beyond mechanisms that describe the image. There is little evidence that the description and analysis are dissociated, though it seems to me sensible to design the system that way because an economical (full) description can be used in any number of different kinds of analyses.

Economical Description

Barlow (1972) argued that "the sensory system is organized to achieve as complete a representation of the sensory stimulus as is possible with the minimum number of active neurons." A great deal of analysis undertaken before cortex can be characterized as providing economical description. For example, the trichromatic organization of color vision seems to be what we need to represent the reflectances of natural surfaces; mechanisms of light-adaptation preserve information about contrast, without having to represent absolute levels of illumination. Further principles may be brought to bear in cortex.

One recognized principle for describing an image economically is to create a set of differently filtered copies that each represent it on a different spatial scale (Dyer, 1987). The series of filtered images is often known as a pyramid. In computational work the most widely used decompositions are Gaussian pyramids, made up of a series of Gaussian (low-pass) filtered copies of the original image, and Laplacian pyramids, comprised of a series of band-pass filtered copies of the original image. In many applications (e.g., Burt & Adelson, 1983) the Laplacian operator is approximated by a difference-of-Gaussians operator.

The scheme devised by Burt & Adelson (1983), and later developments of it, result in remarkably economical descriptions of the image. A superficial attraction of such schemes as biological models is that the spatial properties of their Laplacian operators resemble the receptive fields of retinal ganglion cells and neurons in LGN. However, at any one eccentricity, LGN probably samples the image on only two scales (those of P- and M-cells). In visual cortex, though, neurons seem to have the wide range of spatial preferences that would be required for representation on multiple scales. Moreover, the orientation-selectivity of cortical cells may enable them to support an even more economical representation than could be provided by cells with concentrically organized receptive fields (Adelson, Simoncelli & Hingorami, 1987; Watson, 1987).

Psychophysical studies of spatially-selective channels (see Graham, 1989, for a full account) strongly endorse the idea that the human visual system forms multi-scale representations, and work in computer vision has shown such representations to be helpful in tasks such as edge-finding (Marr & Hildreth, 1980) and extracting depth from disparate images (Marr & Poggio, 1979).

The multi-scale sampling schemes mentioned above make no assumptions about the properties of the scenes being analyzed. If the coding mechanism knows something of the properties of the scenes to be described, it can construct more efficient representations. Field (1987) analyzed the (spatial) spectral content of some natural scenes and showed that a coding mechanism with broadly the properties found in V1 neurons could represent the scenes efficiently. Derrico & Buchsbaum (1990) used principal components analysis to identify the fundamental dimensions of variation in a small sample of color photographs of natural scenes. Although one might worry about the sample of photographs used and how well it represented real scenes, the analysis yielded the interesting finding that the first 10 or so components contain nearly all the signal energy. These can be described (in descending order of importance) as spatially low-pass filters for achromatic, red-green, yellow-blue dimensions, plus spatially band-pass, oriented, achromatic filters at 3 different orientations and 3 different spatial scales. The principal components analysis thus bolsters an already strong case that the oriented, spatially-selective filters are used to describe scenes economically.

The foregoing arguments encourage us to look in cortex for analyses at several orientations, at several spatial scales, and along selected chromatic directions. Of course, we already know that cortex undertakes analyses of this general kind, but more formal examination of the issues motivates us to ask: does cortex produce a complete description of the image as a prelude to analysis; exactly what kind of representation is created by linear mechanisms, and is it specialized for the representation of natural scenes? Watson (1987) and Field (1987) have made very explicit suggestions about the numbers and distributions of spatial filters needed to account for performance on different tasks. These suggestions need to be examined physiologically, through more detailed analysis of the filter characteristics of individual neurons and the sampling properties of populations of them. We also need to know more about the spatio-chromatic content of natural scenes. Until we do, we will find it hard to assess the significance of physiological findings.

Having accepted the case that the visual system uses some kind of multi-scale sampling to describe images economically, we need to tackle the less tractable issue of how to analyze the description of the 2-D image to yield a 3-D description of objects.

Analysis of three-dimensional structure

The move from a 2-D to a 3-D description is inherently ambiguous, and one might have thought that it would require high-level perceptual mechanisms that know a lot about objects in the world. However, as J.J. Gibson first pointed out, the image contains a tremendous amount of information that, together with assumptions about regularities ("natural constraints") in the world, but no knowledge of particular objects, might be used to obtain a relatively rich description of the shapes and positions of objects. The regularities that might be exploited include shading, specular reflections, contour shapes, texture gradients and motion parallax. A good deal of work, almost entirely in computer science, shows how these regularities can be used in forming a description of the surfaces in a scene and their orientations and positions. Most of this has dealt with particular problems (e.g., Pentland's [1989] theory of shape from shading), and has generally not tackled the broader question of how the visual system goes about assembling a full 3-D representation of objects. The two fullest treatments of this problem, by Marr (1982) and Barrow and Tenenbaum (1978, 1981) suggest that in going from a description of the image to a full representation of 3-D objects we need an intermediate representation of surfaces and their spatial relationships in depth. I am not persuaded that any analysis undertaken in visual cortex takes us as far as the recognition of objects, but it does seem worthwhile to take up the idea that visual cortex constructs an intermediate 3-D representation of surfaces in a scene. We can try to deduce properties of the mechanisms that support such a representation, with a view to guiding physiological experiments.

There are two problems to pursue: we need to formulate an account of *what* should be done (i.e., stipulate computational operations), and we need to formulate an account of *how* it should be done (i.e., put these processes in mechanism).

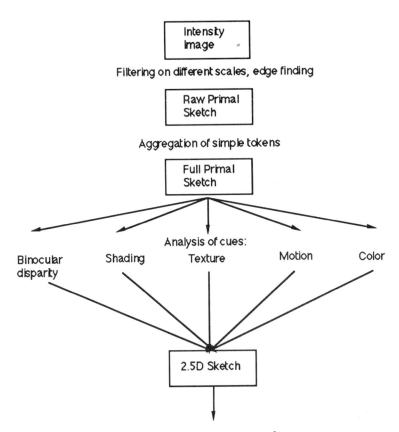

Figure 3. Organization of systems giving rise to the $2\frac{1}{2}$D sketch. The intensity image is filtered on different spatial scales, and signals that appear at multiple scales are taken to represent visually significant features. These significant signals identify discontinuities, which are then aggregated into more elaborate tokens such as oriented bars and blobs, to form the "raw primal sketch". At the next level, the spatial relations among tokens in the raw primal sketch are analyzed to identify units that might characterize larger entities in the image (e.g., clusters of tokens that have the same orientation). The outcome of this process is the "full primal sketch," whose primitives provide information to modular systems that analyze different sorts of information about the organization of surfaces in the scene. The outputs of these modules are used to construct the $2\frac{1}{2}$D sketch, which describes the 3-D arrangement of surfaces in the scene.

Computational operations. Marr's (1982) $2\frac{1}{2}D$ *sketch* and Barrow and Tenenbaum's (1978, 1981) *intrinsic images* are perhaps the best known examples of intermediate representations that reveal the arrangement of surfaces in a scene but stop short of describing objects. Although neither scheme is formulated in a way that provides simple guidance to physiologists, both provide hints about what physiologists might look for.

Marr first develops a 2-D representation (the "primal sketch") that describes the image in terms of tokens that represent edges, corners, and higher primitives that are aggregates of these. Once formed, this 2-D representation is subjected to secondary analyses that extract information about such things as disparity, motion parallax, shading and texture gradients. This information is used to construct a 3-D description of surfaces and their arrangement in depth (the $2\frac{1}{2}D$ sketch). Fig. 3 shows roughly how the $2\frac{1}{2}D$ sketch is thought to rest upon separate underlying analyses of cues to the structure of the scene.

Barrow & Tenenbaum's intrinsic images are representations that "directly support tasks such as navigation, manipulation, and material identification." The four principal intrinsic images represent the surface reflectances, distances, surface orientations and incident illumination in a scene. (The distance and orientation images contain the information that would be embodied in Marr's $2\frac{1}{2}D$ sketch). Barrow & Tenenbaum emphasize the analysis of information about illumination and reflectance, and show (for a model world) that constraints such as: surfaces are continuous and have uniform (lambertian) reflectance, incident illumination varies smoothly, and step changes in intensity occur at surface boundaries or shadow boundaries, provide much of the information needed to recover 3-D surface structure from an intensity image. Binocular disparity and motion parallax provide additional information about the structure of a scene, but are given less weight than other cues. Barrow & Tenenbaum consider the early identification of edges to be an important step in the formation of intrinsic images. The different images are formed initially independently, but are constrained through co-operative interactions to provide a mutually consistent account of a scene. Fig. 4 shows a loose representation of the analyses undertaken to yield the intrinsic images.

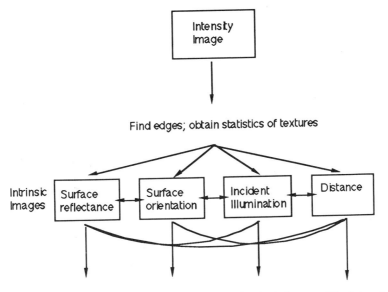

Figure 4. Organization of systems giving rise to intrinsic images. The intensity image is analyzed to discover edges, which are tagged with descriptors of such things as orientation and contrast. By analyzing the distribution of these primitives, elementary textures can be distinguished. Information about local occlusion, texture and shading gradients and local contour shape in the scene is discovered through analysis of the local patterns of image features. Arrays representing surface orientation, distance, reflectance and illumination (the "intrinsic images") are estimated from an analysis of these local cues.

In some respects the intermediate representations suggested by Marr and by Barrow & Tenenbaum are similarly organized: both postulate early edge-finding mechanisms, both identify distinct parallel operations that divide the task into domains that each yield a coherent description of the scene, and both organize their stages hierarchically. However, Barrow & Tenenbaum require nothing as elaborate as the full primal sketch (a feature-rich description of the 2-D structure of a scene) in going from the distribution of intensities in the image to a 3-D representation of the scene; the analyses of cues, such as disparity and motion parallax, that in Marr's scheme are organized as distinct stages, are absorbed into the computation of the intrinsic image. For Barrow and Tenenbaum the 3-D representation is a only a step beyond edge-finding.

Relevance for physiology. Although the computational schemes provide general suggestions about visual organization, it is hard to derive specific predictions from them. Neither says much about how the outputs of its various modules are bound together. In Marr's model, in particular, little is said about the relative importance of the analyses undertaken by the different modules (e.g., disparity vs. color) in creating the $2\frac{1}{2}$D sketch—in fact, there appears to have been no full-scale computational attempt to create a $2\frac{1}{2}$D sketch. Barrow and Tenenbaum have created a working set of intrinsic images, but for a very restricted world.

We would also like guidance about mechanism. We would like to identify modules and to deduce from the function of each module something about the properties of its constituent mechanisms; in the context of visual cortex, we would like to be able to say something about the behavior of single neurons. Here the computational work provides few constraints, because it deals with process and requires no particular machinery. Despite these substantial limitations, the work does have something to offer the physiologist.

First, by emphasizing the distinct analyses that can be brought to bear in parsing a scene, it suggests what kinds of operations might be undertaken in relatively self-contained modules. In Marr's scheme, an analysis that can yield a representation of the 2-D structure of a scene (for example, segregation by texture or color) is a candidate for incorporation in a module. For Barrow & Tenenbaum an analysis that can give rise to a 3-D description of a scene might be incorporated in a module (e.g., extracting shape from shading, but not analyzing color). The hierarchical, modular organization of both schemes resembles that favored by physiologists and anatomists, but the functions of the modules differ, and those identified in the computational work (especially that of Barrow & Tenenbaum) are not obviously related to any machinery postulated by biologists.

Second, the computational work emphasizes the early extraction of information about edges, and relies heavily on edge detection guiding subsequent analysis. Although some higher-level operations (for example extracting depth from disparity, and shape from shading) could be undertaken directly on the image representation, it does seem worthwhile to look for what might be generalized edge-finding mechanisms (for example, mechanisms sensitive to texture edges or color edges or occlusion edges such as illusory contours). Some work of this kind is already under way: von der Heydt and Peterhans (1989) have shown that some neurons in macaque V2 respond to patterns that (for human observers) define illusory contours in ways that suggest they may be encoding the position and orientation of the virtual edges; Van Essen et al. (1990) have examined the responses of cells in V1, V2 and MT to some static and moving texture patterns and have found (particularly for moving patterns) that cells are sensitive to texture boundaries. We need more work of this kind, especially studies that examine how cells respond to statistically described textures of the kinds represented in normal scenes.

Third, the computational work prompts us to study the relative importance of different kinds of information. To direct our efforts sensibly we need to know whether texture gradients contain more information than shading changes or color changes. Neither perceptual nor computational work gives us clear guidance about the relative importance of different kinds of information about scene structure, and I suspect that the effort spent studying different kinds of information does not very adequately reflect their true importance. Much attention has been devoted to the information provided by binocular disparity and motion parallax, yet there must be few circumstances in normal life where either sort of information makes much difference to our capacity to recognize objects. Shading and color are probably very important, but we need to develop ways of measuring just how important they are.

CONCLUSIONS

I have drawn attention to two major propositions about the organization of visual cortex: first, that the earliest operations in cortex work to yield a full (but economical) representation of the image; second, that the principal function of extrastriate cortex is to deliver a 3-D representation of a scene. Neither proposition is readily examined experimentally, but both provide context for the physiological exploration of cortex.

ACKNOWLEDGEMENTS

This work was supported by NIH grants EY04440 and EY01319.

REFERENCES

Adelson, E. H., Simoncelli, E. and Hingorami, R., 1987, Orthogonal pyramid transforms for image coding, Proc. SPIE 845:50.

Ballard, D. H., Hinton, G. E. and Sejnowski, T. J., 1983, Parallel visual computation, Nature, 306:21.

Barlow, H.B., 1972, Single units and sensation: a neuron doctrine for perceptual psychology? Perception, 1:371.

Barlow, H.B., Blakemore, C. and Pettigrew, J.D., 1967, The neural mechanism of binocular depth discrimination, J. Physiol., 193:327.

Barlow, H.B., Hill, R.M. and Levick, W.R., 1964, Retinal ganglion cells responding selectively to direction and speed of image motion in the rabbit, J. Physiol., 173:377.

Barrow, H. G. and Tenenbaum, J. M., 1978, Recovering intrinsic scene characteristics from images, in: "Computer vision systems," A. Hanson and E. Riseman, eds., Academic Press, New York.

Barrow, H. G. and Tenenbaum, J. M., 1981, Computational vision, Proc. IEEE, 69:572.

Burt, P. J. and Adelson, E. H., 1983, The laplacian pyramid as a compact image code, IEEE Trans Commun., COM-31:532.

Derrico, J. B. and Buchsbaum, G., 1990, Spatial and chromatic properties of cortical mechanisms in V1 correspond to the principal components of natural color images, Invest. Ophthal. Vis. Sci., 31:264(suppl.)

DeYoe, E. A. and Van Essen, D. C., 1985, Segregation of efferent connections and receptive field properties in visual area V2 of the macaque, Nature, 317:36.

DeYoe, E. A. and Van Essen, D. C., 1988, Concurrent processing streams in monkey visual cortex, Trends Neurosci., 11:219.

Dubner, R. and Zeki, S. M., 1971, Response properties and receptive fields of cells in an anatomically defined region of the superior temporal sulcus in the monkey, Brain Res., 35:528.

Dyer, C. R., 1987, Multiscale image understanding, in: "Parallel computer vision," L. Uhr, ed., Academic Press, Orlando.

Emerson, R.C., Citron, M.C., Vaughn, W.J. and Klein, S.A., 1987, Nonlinear directionally selective subunits in complex cells of cat striate cortex, J. Neurophysiol., 58:33.

Felleman, D. J. and Van Essen, D. C., 1987, Receptive field properties of neurons in area V3 of macaque monkey extrastriate cortex, J. Neurophysiol., 57:889.

Field, D.J., Relations between the statistics of natural images and the response properties of cortical cells, J. Opt. Soc. Am. A, 4:2379.

Graham, N. V., 1989, "Visual Pattern Analyzers," Oxford University Press, Oxford.

von der Heydt, R. and Peterhans, E., 1989, Mechanisms of contour perception in monkey visual cortex. I. Lines of pattern discontinuity, J. Neuroscience, 9:1731.

Heywood, C. A. and Cowey, A., 1987, On the role of cortical area V4 in the discrimination of hue and pattern in macaque monkeys, J. Neurosci., 7:2601.

Lennie, P., Krauskopf, J., and Sclar, G., 1990, Chromatic mechanisms in striate cortex of macaque, J. Neurosci., 10:649.

Levitt, J. B. and Movshon, J. A., 1990, Homogeneity of response properties of neurons of macaque V2, Invest. Ophthal. Vis. Sci., 31:89(suppl.)

Livingstone, M. S. and Hubel., D. H., 1987, Psychophysical evidence for separate channels for the perception of form, color, movement and depth, J. Neurosci., 7:3416.

Marr, D. and Hildreth, E., 1980, Theory of edge detection, Phil. Trans. R. Soc. Lond. B., 290:199.

Marr, D. and Poggio, T., 1979, A computational theory of human stereo vision, Proc. R. Soc. Lond. B., 204:301.

Marr, D., 1982, "Vision," Freeman, San Francisco.

Maunsell, J. H. R. and Van Essen, D. C., 1983, The connections of the middle temporal area (MT) and their relationship to a cortical hierarchy in the macaque monkey, J. Neurosci., 3:2563.

Movshon, J.A. and Lennie, P., 1979, Pattern-selective adaptation in visual cortical neurones, Nature, 278:850.

Newsome, W. T and Paré, E.B., 1988, A selective impairment of motion perception following lesions of the middle temporal area (MT), J. Neurosci., 8:2201.

Newsome, W. T., Wurtz, R. H., Dursteler, M. R., and Mikami, A., 1985, Deficits in visual motion processing following ibotenic acid lesions of the middle temporal area of the macaque monkey, J. Neurosci.,5:825.

Pentland, A., 1989, Shape information from shading: a theory about human perception, Spatial Vision, 4:1655.

Shapley, R. M. and Lennie, P., 1985, Spatial frequency analysis in the visual system, Ann. Rev. Neurosci., 8:547.

Shipp, S. and Zeki, S. M., 1985, Segregation of pathways leading from area V2 to areas V4 and V5 of macaque monkey visual cortex, Nature, 315:322.

Spitzer, H. and Hochstein, S., 1985, A complex-cell receptive field model, J. Neurophysiol., 53:1266.

Ungerleider, L. G. and Mishkin, M., 1982, Two cortical visual systems, in "Analysis of visual behavior," D. J. Ingle, M.A. Goodale and R.W.J. Mansfield eds., MIT Press, Cambridge.

Van Essen, D.C., DeYoe, E.A., Olavarria, J.F., Knierim, J.J., Fox, J.M., Sagi, D. & Julesz, B., 1990, Neural responses to static and moving texture patterns in visual cortex of the macaque monkey, in "Neural mechanisms of visual perception," D. M-K. Lam and C.D. Gilbert, eds., Portfolio, Woodlands, Texas.

Watson, A.B., 1987, Efficiency of a model human image code, J. Opt. Soc. Am. A, 4:2401.

Wild, H. M., Butler, S. R., Carden, D., and Kulikowski, J. J., 1985, Primate cortical area V4 important for colour constancy but not wavelength discrimination, Nature, 313:133.

Zeki, S. M., 1973, Color coding in rhesus monkey prestriate cortex, Brain Res., 53:422.

ORIENTATION AND SPATIAL FREQUENCY SELECTIVITY:

PROPERTIES AND MODULAR ORGANIZATION

Russell L. De Valois

University of California at Berkeley
Berkeley, CA 94720 USA

Striate Cell Properties

There is by now overwhelming evidence that two of the primary characteristics of visual cells first seen at the striate level are those of orientation and spatial frequency selectivity (see De Valois & De Valois, 1988, for summary and discussion). Although lateral geniculate neurons show some minor degree of selectivity for both orientation and spatial frequency, they are rather broadly tuned along both parameters. Striate cortex neurons vary widely in their bandwidths along these dimensions, but on average they are much more narrowly tuned than geniculate cells. The average orientation bandwidth (full bandwidth at half amplitude response), in both cat and monkey striate cells, is about 45°, 1/4 of the total 180° range of possible orientations. The average spatial frequency bandwidths of cells are quite comparable: about 1.3 octaves, which is about 1/4 of the total ca. 5 octave range of spatial frequencies to which we are sensitive at a given eccentricity (De Valois, Hepler & Yund, 1982; De Valois, Albrecht & Thorell, 1982).

An important aspect of this striate organization is that the same cells show both orientation and spatial frequency selectivity; that is, there are not some cells narrowly tuned for orientation and different ones narrowly tuned for spatial frequency. Rather, selectivity along these parameters is positively correlated. Those cells that have narrow orientation tuning also tend to have narrow spatial frequency tuning, and those broadly tuned along one dimension tend to be broadly tuned along the other as well (De Valois, Albrecht & Thorell, 1982). There is thus a sizable subpopulation of cells narrowly tuned for both spatial frequency and orientation.

Looked at from the perspective of filtering of images, (one-dimensional) spatial frequency and orientation are not separate parameters, but different aspects of two-dimensional (2-D) spatial frequency. That is, a one-dimensional pattern can be analyzed into the sum of sine waves of different spatial frequency, amplitude, and phase; a two-dimensional pattern can be analyzed into sine waves of different spatial frequency, amplitude, phase, and orientation. Thus the above description of striate cells can be rephrased to state that striate cells are relatively narrowly tuned for 2-D spatial frequency, the average cell having a bandwidth of about 1/16 (that is, 1/4 times 1/4) of the total range of 2-D spatial frequencies to which one is sensitive.

The spatial RFs of simple cells narrowly tuned for spatial frequency consist of several alternating excitatory and inhibitory regions, not just the 2 or 3 regions classically described (Mullikin, Jones & Palmer, 1984; De Valois, Albrecht & Thorell, 1985). Because of the low or even negative maintained rates of most simple cells, RF mapping as usually carried out often fails to see these sidebands, although they in fact make major contributions to the cells' responses to extended stimuli. The relative

From Pigments to Perception, Edited by A. Valberg and
B.B. Lee, Plenum Press, New York, 1991

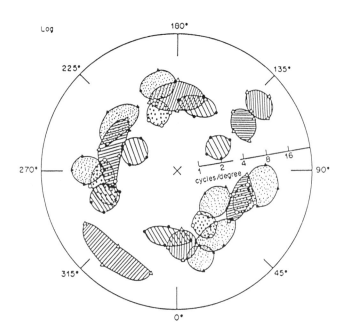

Fig. 1.   The receptive fields in the spatial frequency domain of all the
          narrowly cells recorded from in one location in macaque striate
          cortex. These cells all had the same RF locus.   In this 2-D
          frequency plot, spatial frequency increases out from the center,
          and orientation is represented by the angle.   Each shaded area
          represents the range of orientations and spatial frequencies to
          which one cell responded with at least half-amplitude responses.
          It can be seen that the cells vary in their orientation tuning and
          also in their spatial frequency tuning: note the three cells all
          tuned to about 130°, but with non-overlapping spatial frequency
          tuning.   Presumably if not just these few cells but all of the
          thousands of cells in the region had been recorded from, the whole
          2-D spatial frequency area would have been paved with the RFs of
          the cells.

strengths of these sidebands drop off with distance from the RF center.
Thus the receptive field of a striate simple cell closely approximates a 2-D
Gabor filter -- a sine wave tapered in X and Y by a Gaussian -- (De Valois &
Webster, 1985; Jones & Palmer, 1987), or perhaps as a similar-shaped
difference-of-Gaussians filter (Young, 1987).

       Correlated with the selectivity of units is the presence within each
striate region of cells with different orientation and spatial frequency
preferences and thus different locations in 2-D spatial frequency space.
The above comments about the average bandwidths of striate cells might
suggest that only 16 cells -- ones tuned to each of four spatial frequencies
at each of four orientations -- would be required at each location to
completely pave the local 2-D spatial frequency space, to ensure that there
be cells responsive to every frequency component in any visual pattern.
Some psychophysical models (e.g. Wilson & Bergen, 1979) adequately fit much
visual data with just 4 spatial frequency channels. However, the evidence
from recording the responses of numerous cells within a limited striate
region indicates that the cells show a great deal of overlap in their 2-D
spatial frequency RFs, with cells tuned not just to four but to dozens of
orientations, and not just to four or six but to dozens of spatial
frequencies. In Fig. 1 are shown the 2-dimensional receptive fields, in the
spatial-frequency domain, of a sample of narrowly-tuned cells recorded in a
single small region of macaque cortex.

       It is also likely, at least in the central representation, that there
are pairs of simple cells in approximate quadrature phase relation to each
other. Simple cell RFs have been classically described as either symmetric,
e.g., excitatory center with inhibitory flanks on both sides (plus perhaps
additional sidebands on either side) and thus in cosine phase, or

262

asymmetric, e.g., main excitatory area on one side and main inhibitory one on the other, in sine phase. Quantitative studies of populations of striate cells do not support this notion of there being just two RF profiles (Field & Tolhurst, 1986; Hamilton, Albrecht, & Geisler, 1989), but they do not preclude the possibility of pairs of cells in approximate quadrature phase. That is, there could be one pair in sine-cosine (0° and 90°) phase and another in, say, 30° and 120° phase. The absolute phases could be scattered all over, but there could nonetheless be local pairs in at least an approximate quadrature relation. The presence of quadrature pairs would be very attractive for linear models of striate function, and there is some direct physiological evidence for neighboring simple cells being in a quadrature arrangement (Pollen & Ronner, 1981).

Anatomical Arrangement of Functional Cell Types

Early studies of the anatomical arrangement of cells in the striate cortex revealed a systematic grouping by ocular dominance and by orientation (Hubel & Wiesel, 1974). We have examined the anatomical arrangement of cells with different spatial frequency tuning, and have found evidence for a systematic organization along this dimension as well. In addition, there is of course also a retinotopic organization to the striate cortex (Tootell, Silverman, Switkes, & De Valois, 1982).

Staining the macaque striate cortex for cytochrome oxidase (cytox) reveals some 5000 blobs: columnar regions of high cytox concentration. We and others have shown that these blobs are located in the center of each ocular dominance region of each cortical module (Switkes, Tootell, Silverman & De Valois, 1986; Horton & Hubel, 1981). Using the blobs as landmarks, one can ask how various functional cell types are arranged within each cortical module. We have done this using both the 2-deoxyglucose (2DG) and unit-recording techniques.

In 2DG experiments, we found that stimulation with a low spatial frequency grating (of varying orientation over time, to control for that variable) produces 2DG uptake on the blobs. This is true whether it is a luminance-varying or a color-varying grating (Tootell, Silverman, Hamilton, De Valois & Switkes, 1988; Tootell, Silverman, Hamilton, Switkes & De Valois, 1988). On the other hand, a high-spatial-frequency pattern, either luminance- or color-varying, produces 2DG uptake in annuli around the cytox blobs. Note that what constitutes a "low" or "high" spatial frequency, in terms of whether cells tuned to it occur in or outside the blobs, depends on eccentricity: a mid-spatial-frequency pattern was found to produce large 2DG uptake on the blobs in the foveal projection (thus being treated as a "low" frequency in the fovea) but outside the blobs -- and thus a "high" spatial frequency -- in cortical regions related to the peripheral visual field.

Unit recordings from long tangential probes through the supragranular layers support this basic organization of spatial frequency within cortical modules: cells tuned to low spatial frequencies were found in the blobs, and those tuned to high frequencies were found to lie outside the blobs (Silverman, Grosof, De Valois & Elfar, 1989). Furthermore, the variation in spatial frequency with distance from the blob centers was continuous, with units tuned to higher and higher spatial frequencies increasingly far from the blob centers, see Fig. 2.

The arrangement within cortical modules of cells tuned to different orientations is not so clear, despite the fact that this was the first functional organization found (Hubel & Wiesel, 1974). The fact that they found a regular sequence of orientation steps with tangential probes through the cortex suggested to Hubel & Wiesel, 1974, that the orientation columns were arranged in rectangular slabs within each module. However, such a rectangular organization, combined with a radial spatial frequency organization would not account for the fact that one encounters cells with all the possible orientation and spatial frequency combinations, since the spatial frequency and orientation organizations would not be orthogonal. A spoke-like arrangement of orientation columns, on the other hand, would provide for an intersection of all the spatial frequency and orientation columns. With single tangential probes through the cortex, it would be very hard to distinguish between these alternate orientation models, because each would produce a regular sequence of orientation tuning.

Our model of the retino-striate processing is that visual information is

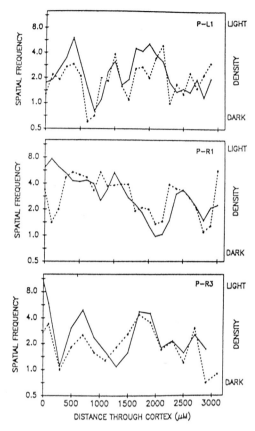

Fig. 2.   The relation between the spatial frequency tuning of cells recorded
           from at regular intervals during three long tangential probes
           through macaque striate cortex and the density of cytochrome
           oxidase (cytox) at the same locations. It can be seen that there
           is a high correlation between these: cells tuned to low spatial
           frequencies tend to be a cytox-dense region (blobs) while those
           tuned to high spatial frequencies are in cytox-light regions.
           Note, however, that both the peak spatial frequencies and the cytox
           densities vary continuously, not in an all-or-none manner,
           suggesting a continuum rather than a dichotomous blob-interblob
           organization.

transformed from a pure space representation at the level of the receptors
to a combined space and 2-D spatial frequency representation at the striate
level. Within each cortical module, dealing with information from a small
region of visual space, are cells narrowly tuned to numerous 2-D spatial
frequency ranges. Our evidence suggests that cells with these various 2-D
tuning properties are systematically arranged anatomically, by peak tuning,
as shown in Fig. 3.

     There is a distinctive cytox organization to V2 as well V1, with cytox-
dense stripes running orthogonal to the V1-V2 border (Tootell, Silverman, De
Valois & Jacobs, 1983). The different parts of the V1 modules project
systematically onto the V2 stripes, with the blobs going to the thin stripes
and the interblobs to the interstripes (Livingstone & Hubel, 1983).
Reflecting this anatomical interrelation, the thin strips contain cells
tuned to low spatial frequencies, and the interstrips those tuned to high
(Tootell & Hamilton, 1989). Both of these regions then send their primary
output to area V4 (Shipp & Zeki, 1985). We interpret this to indicate that
the systematic organization within the V1 modules of cells tuned to
different 2-D spatial frequency ranges is maintained through V2, and that
all of these cells then project to V4.

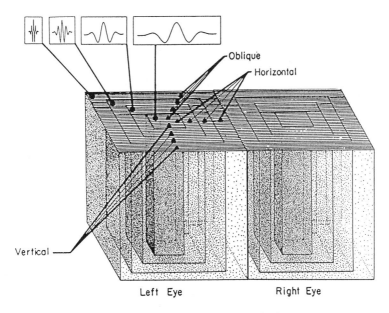

Model of Striate Module in Monkeys

Fig. 3. Our model of the organization of primate striate cortex modules.
Cells tuned to low spatial frequencies occupy the module centers
(region of highest cytox density) and those tuned to increasingly
higher frequencies regions increasingly further out. Cells tuned
to a particular orientation lie along a radius out from the center,
intersecting each of the spatial frequency rings, so that
altogether all combinations of spatial frequency and orientation
tuning occur within each half module.

A very different model of both the anatomical and the functional
organization of striate cortex has been put forth by Livingstone & Hubel
(1984). They propose, first, that the blob-interblob differences constitute
a dichotomy, rather than lying along a continuum. Secondly, they propose
that what distinguishes the cells in the blobs from those in the interblobs
is not spatial frequency tuning but color and orientation tuning -- that is,
that the cytox blobs, but not the interblobs, contain color-selective cells,
and the interblobs but not the blobs contain cells with orientation tuning.
Thirdly, they conclude that the blob-interblob anatomical dichotomy reflects
a dichotomous functional subdivision of the visual system. Thus, they
propose that there are separate functional systems, presumably subdivisions
of the parvo system, processing color and form information, respectively.

Our experimental evidence does not support any of these three positions.
First, the actual densitometric measurements on cytox stained brain sections
show that the cytox density varies continuously across the module, with no
evidence for an all-or-none, blob-or-interblob dichotomy. Fig. 2 shows a
few such measures, made along our electrode tracks; we have made numerous
others along random tracks through the cortex which reinforce the evidence
that variation in cytox staining is continuous. We have suggested,
incidentally, that the metabolic enzyme cytochrome oxidase is probably
concentrated in the most highly activated cells, and that cells tuned to low
spatial frequencies might well have a chronically higher activity level
because natural stimuli contain most of their energy at low spatial
frequencies (Silverman, Grosof, De Valois & Elfar, 1989).

Secondly, our evidence indicates that what differentiates cells in the
blobs vs those outside is not color selectivity but peak spatial frequency
tuning. As pointed out above, low spatial frequency patterns based on
luminance variations alone produce 2DG uptake just in the blobs, hardly what
one would expect if the blobs were just involved in color processing. One

also gets exactly the same uptake pattern with low frequency isoluminant color gratings. High spatial frequency patterns, whether based on color or luminance, produce 2DG uptake in the interblobs. Both of these facts stand in contradiction to a color vs form (orientation) distinction. How, then, do we account for their finding of more color cells in the blobs? If cells were in fact lined up, as we believe them to be, by their spatial frequency tuning, not by color selectivity, one would in fact expect more color cells in the blobs, because our chromatic sensitivity is shifted to lower frequency for color relative to luminance. (Note that Livingstone & Hubel actually find many color cells in the interblobs too, just not as many as in the blobs). There are other problems for their theory. For instance, cytox blobs are if anything even more prominent in the nocturnal owl monkey, which is almost completely lacking in color vision, than they are in macaque or humans.

Thirdly, we find no evidence for a dichotomy in the functional properties of cells in cytox-rich and cytox-poor regions of the module, nor psychophysical evidence for a lack of sensitivity to orientation or other form characteristics in the color system. We have shown (De Valois, Hepler & Yund, 1982) that orientation tuning varies continuously, and that cells tuned to low spatial frequencies tend to have broader orientation tuning. Although we have not directly tested this point, we would thus expect, based on our other evidence, that orientation selectivity should increase gradually with distance from the blob centers, but with no dichotomy. We predict that this would be equally true whether the cells are responding to luminance-varying patterns or to color-varying patterns. We have shown (Thorell, De Valois & Albrecht, 1984) that most striate cells respond to both color and luminance patterns, and have spatial frequency tuning to both. Finally, the psychophysical evidence indicates that one has very similar spatial frequency channel bandwidths for color and luminance (Bradley, Switkes & De Valois, 1988), and our sensitivity for spatial frequency and orientation differences is extremely good with isoluminant patterns (Webster, De Valois & Switkes, 1990).

In summary, we conclude that the basic organization within striate cortex modules is based on a systematic arrangement of cells by the 2-dimensional spatial frequency tuning, and that it is unlikely that there are separate systems involved in the processing of color and form information.

ACKNOWLEDGEMENTS

The research reported in this paper was supported by grants from the National Eye Institute (EY00014) and the National Science Foundation (BNS 8819867).

REFERENCES

Bradley, A., Switkes, E. & De Valois, K.K., 1988, Orientation and spatial frequency selectivity of adaptation to colour and luminance gratings. Vision Res. 28: 841-856.

De Valois, R.L. & De Valois, K.K., 1988, Spatial Vision. New York: Oxford University Press.

De Valois, R.L. & Webster, M.A., 1985, Relationship between spatial-frequency and orientation tuning of striate-cortex cells. J. Opt. Soc. Amer. A 2: 1124-1132.

De Valois, R.L., Albrecht, D.G. & Thorell, L.G., 1982, Spatial frequency selectivity of cells in macaque visual cortex. Vision Res. 22: 545-559.

De Valois, R.L., Albrecht, D.G. & Thorell, L.G., 1985, Periodicity of striate-cortex-cell receptive fields. J. Opt. Soc. Amer. A 2: 1115-1123.

De Valois, R.L., Yund, E.W. & Hepler, N., 1982, The orientation and direction selectivity of cells in macaque visual cortex. Vision Res. 22: 531-544.

Field, D.J. & Tolhurst, D.J., 1986, The structure and symmetry of simple-cell receptive-field profiles in the cat's visual cortex. Proc. Roy. Soc. Lond. B 228: 379-400.

Hamilton, D.B., Albrecht, D.G. & Geisler, W.S., 1989, Visual cortical receptive fields in monkey and cat: Spatial and temporal phase transfer function. Vision Res. 29: 1285-1308.

Horton, J.C. & Hubel, D.H., 1981, Regular patchy distribution of cytochrome oxidase staining in primary visual cortex of the macaque monkey. <u>Nature</u> <u>292</u>: 762-764.

Hubel, D.H. & Wiesel, T.N., 1974, Sequence regularity and geometry of orientation columns in the monkey striate cortex. <u>J. Comp. Neurol</u>. <u>158</u>: 295-306.

Jones, J.P. & Palmer, L.A., 1987, An evaluation of the two-dimensional Gabor filter model of simple receptive fields in cat striate cortex. <u>J. Neurophysiol</u>. <u>58</u>: 1233-1258.

Livingstone, M.S. & Hubel, D.H., 1983, Specificity of cortico-cortical connections in monkey visual system. <u>Nature</u> <u>304</u>: 531-534.

Livingstone, M.S. & Hubel, D.H., 1984, Anatomy and physiology of a color system in the primate visual cortex. <u>J. Neurosci</u>. <u>4</u>: 309-356.

Mullikin, W.H., Jones, J.P. & Palmer, L.A., 1984, Periodic simple cells in cat area 17. <u>J. Neurophysiol</u>. <u>52</u>: 372-387.

Pollen, D.A. & Ronner, S.F., 1981, Phase relationship between adjacent simple cells in the visual cortex. <u>Science</u> <u>212</u>: 1409-1411.

Shipp, S. & Zeki, S., 1985, Segregation of pathways leading from area V2 to areas V4 and V5 of macaque monkey visual cortex. <u>Nature</u> <u>315</u>: 322-325.

Silverman, M.S., Grosof, D.H., De Valois, R.L. & Elfar, S.D., 1989, Spatial frequency organization in primate striate cortex. <u>Proc. Natl. Acad. Sci</u>. <u>USA 86</u>: 711-715.

Switkes, E., Tootell, R.B.H., Silverman, M.S. & De Valois, R.L., 1986, Picture processing techniques applied to autoradiographic studies of visual cortex. <u>J. Neurosci</u>. <u>Methods 15</u>: 269-280.

Thorell, L.G., De Valois, R.L. & Albrecht, D.G., 1984, Spatial mapping of monkey V1 cells with pure color and luminance stimuli. <u>Vision Res</u>. <u>24</u>: 751-769.

Tootell, R.B.H. & Hamilton, S.L., 1990, Functional anatomy of the second visual area (V2) in the macaque. <u>J. Neurosci</u>. <u>9</u>: 2620-2644.

Tootell, R.B.H., Silverman, M.S., De Valois, R.L. & Jacobs, G.H., 1983, Functional organization of the second visual cortical area in primates. <u>Science</u> <u>220</u>: 737-739.

Tootell, R.B.H., Silverman, M.S., Switkes, E. & De Valois, R.L., 1982, Deoxyglucose analysis of retinotopic organization in primate striate cortex. <u>Science</u> <u>218</u>: 902-904.

Tootell, R.B.H., Silverman, M.S., Hamilton, S.L., De Valois, R.L. & Switkes, E., 1988, Functional anatomy of macaque striate cortex. III. Color. <u>J. Neurosci</u>. <u>8</u>: 1569-1593.

Tootell, R.B.H., Silverman, M.S., Hamilton, S.L., Switkes, E. & De Valois, R.L., 1988, Functional anatomy of macaque striate cortex. V. Spatial frequency. <u>J. Neurosci</u>. <u>8</u>: 1610-1624.

Webster, M.A., De Valois, K.K. & Switkes, E., 1990, Orientation and spatial-frequency discrimination for luminance and chromatic gratings. <u>J. Opt. Soc. Amer</u>. <u>A 7</u>: 1034-1049.

Wilson, H.R. & Bergen, J.R., 1979, A four mechanism model for threshold spatial vision. <u>Vision Res</u>. <u>19</u>: 19-32.

Young, R.A., 1987, Simulation of human retinal function with the Gaussian derivative model. <u>Spatial Vision</u> <u>2</u>: 273-293.

ORIENTATION AND COLOR COLUMNS IN MONKEY STRIATE CORTEX

Bruce M. Dow

Departments of Psychiatry and Anatomy/Neurobiology
University of California
Irvine, CA 92717

The parallel stripe model of orientation columns in monkey striate cortex was first proposed by Hubel and Wiesel (1972) as part of an anatomical lesion study of projections from parvo and magnocellular layers of the lateral geniculate nucleus (LGN) to the cortex. This paper demonstrated ocular dominance columns (in the form of parallel stripes) in striate cortex, a given stripe receiving input from either layer 5 or layer 6 of the LGN, whose inputs in turn derive from the ipsilateral and contralateral eyes, respectively. There was no evidence in the paper for any particular arrangement of orientation columns, but it was proposed that orientation columns might take the form of parallel stripes running orthogonal to the ocular dominance stripes. There has been no evidence since then to support the existence of the parallel stripe orientation column model in monkeys, but it has somehow persisted, in the absence of any better model to replace it. Hubel still chose to include it, with a new name (the 'ice cube' model), in his recent book (Hubel, 1988). In fact, the model has not changed in 16 years.

When I began working on the problem of color and orientation processing in monkey striate cortex in 1970, there was no orientation column model. Hubel and Wiesel (1968) had proposed that color was processed according to the same simple-complex-hypercomplex hierarchy as they had proposed for noncolor information. They had suggested that color columns (each column presumably containing all colors) might simply be interspersed among the orientation columns.

My early work on color processing in monkey cortex (Dow, 1974) did not support Hubel and Wiesel's contention that color was processed in orientation columns. Rather, there seemed to be patches of nonoriented color-rich zones in the uppermost layers of striate cortex. One test of the idea of separate color and orientation processing systems in striate cortex was to look at prestriate cortex, in particular, area 18, or V2 as it is now called. Baizer et al. (1977) did so in awake monkeys, and found separate groups of color, orientation, and direction selective cells, which seemed to be arranged in columns.

Another perplexing early finding was that direction- selective cells in striate cortex were clustered in a single layer, the layer now known as 4B (Dow, 1974). It made no sense that directional cells should be clustered in one layer, if striate cortex seemed to be organized in columns. A few years later when layer 4B was found to be the main source of the projection from striate cortex to area MT (Lund et al., 1975; Tigges et al., 1981), the laminar clustering in striate cortex became meaningful, since it provided a mechanism for the apparent segregation of directional information seen in area MT

From Pigments to Perception, Edited by A. Valberg and
B.B. Lee, Plenum Press, New York, 1991

of the monkey (Dubner and Zeki, 1971), and noted earlier in the suprasylvian gyrus of the cat (Dow and Dubner, 1971).

The color/orientation dichotomy in striate and prestriate cortex was 'rediscovered' in the early 1980's, with the discovery of cytochrome oxidase blobs in V1 and stripes in V2 (Wong-Riley, 1979; Horton and Hubel, 1981; Hendrickson et al., 1981; Tootell et al., 1983; Carroll and Wong- Riley, 1984), and the finding that cells in CO blobs and thin dark CO stripes tended to be nonoriented and color sensitive (Livingstone and Hubel, 1984; Shipp and Zeki, 1985; De Yoe and Van Essen, 1985; Hubel and Livingstone, 1987). Clearly the original orientation column model proposed by Hubel and Wiesel (1972) was no longer viable. Modifications were required for both color and direction of movement.

Studies conducted in Buffalo beginning in the late 1970's (Bauer et al., 1980; Bauer et al., 1983; Dow et al., 1984) provided evidence that a new model was required for the processing of orientation itself. We noted (Bauer et al., 1980) and later confirmed more quantitatively (Bauer et al., 1983) that lower-layer striate cells in macaque monkeys participate in a rather different columnar arrangement from upper layer cells. The major evidence for this was our finding of orientation 'shifts' between supragranular and infragranular cells as a recording electrode advanced vertically or nearly vertically through the tissue. The latter paper includes data on orientation shift sizes and locations for 57 such microelectrode penetrations. A substantial number of the shifts were 70 degrees or more, with 90 degrees being the maximum possible. The shift occurs typically across the granule cell layer, i.e., between supra and infragranular layers. The essential features of the orientation shift phenomenon have been confirmed in monkeys by Kruger and Bach (1982).

A possible explanation for the orientation shift is that upper layer cells follow a radial bias for preferred orientation, while lower layer cells follow a tangential bias (Bauer and Dow, 1989). That is, there is a tendency for the preferred orientation of upper layer cells to point toward the center of the fovea (center of the visual field), while the preferred orientation of lower layer cells tends to point tangentially in relation to center fovea (i.e., tangent to a circle around the fovea).

Up to this point we have identified 3 problems with the parallel stripe model: 1) the existence of color columns which lack orientation selectivity, 2) direction processing in layers rather than columns, and 3) separate upper and lower layer orientation processing systems.

A fourth problem concerns 'reversals', that is, changes in the direction of orientation 'drift' as an electrode makes its way horizontally or obliquely through the striate cortical tissue (Hubel and Wiesel, 1974; Livingstone and Hubel, 1984). The parallel stripe model does not predict reversals, and cannot account for them.

A few years ago my colleagues and I became interested in alternate types of orientation column models. Braitenberg and Braitenberg (1979) were the first to propose 'radial' and 'concentric' orientation column models, even before the discovery of cytochrome oxidase blobs. It occurred to us that cytochrome oxidase blobs might serve as nonoriented 'singularities' within an orientation column network. Our first radial model (Dow and Bauer, 1984) had a checkerboard appearance, with the 2 diagonal orientations represented in alternating squares across the cortical surface, at a periodicity corresponding to CO blob separation. Other radial models have been proposed by Horton (1984) and Gotz (1987, 1988).

Baxter and Dow (1989) took a closer look at the issue of radial models, and found 4 types, which are distinguishable by the kinds of singularities found at the CO blobs.

Singularities can be either clockwise (+) or counterclockwise (-) depending on whether orientation changes in the same or in the opposite direction as one circumnavigates the singularity. Singularities can have index 1 or 1/2 depending on whether 360 degrees or 180 degrees of orientation are mapped radially around the center. In E type models (E1 and E 1/2) every (E) blob is a positive singularity (index 1 and 1/2, respectively). In A type models (A1 and A 1/2) blobs are alternately (A) positive and negative singularities (index +1 and -1 or +1/2 and -1/2, respectively).

Computer simulations of horizontal microelectrode penetrations through the different models reveal both reversals (not achievable in the parallel stripe model) and long, essentially linear, orientation drift curves. The slopes of the orientation drift curves vary from model to model. Average slopes for several thousand simulated penetrations through each model were obtained. There is a pretty good fit of the E1 model to empirical data obtained from other laboratories, both electrophysiological recordings (Hubel and Wiesel, 1974, Livingstone and Hubel, 1984; Michael, 1985), and voltage sensitive dye studies (Blasdel and Salama, 1986)), but the results are not conclusive. The A 1/2 model gives the worst fit to the electrophysiological recording data.

An interesting issue raised by the 4 radial models described by Baxter and Dow (1989) is whether all CO blobs are the same. There is at the present time no anatomical evidence to suggest more than one type of blob. However, physiological data from our own laboratory (Dow and Vautin, 1987; Yoshioka, Dow and Vautin unpublished) have suggested that there may be 2 different kinds of blobs in the domain of color processing.

Vautin and Dow (1985) determined the color selectivity of 218 striate cells on the basis of complete color tuning curves, each cell undergoing systematic color testing with its optimal spatial stimulus (either an oriented bar or a square of the appropriate size). Other orientations were also presented to each cell, to establish its orientation selectivity. Computer analysis of color tuning curves revealed 4 distinct profiles, corresponding to the 4 Hering (1878) primary colors, red, yellow, green, and blue.

Subsequently we noted (Dow and Vautin, 1987) that red cells and blue cells showed a strong tendency to occur in penetrations with nonoriented upper layer cells (N penetrations), while yellow cells and green cells (and achromatic cells) showed a strong tendency to occur in penetrations with oriented upper layer cells (O penetrations). There were 2 types of N penetrations, NR, containing mostly red cells, and NB, containing mostly blue cells. Lesion studies in our laboratory have more recently established NR and NB penetrations as being in blob zones, with O penetrations in interblob zones (Yoshioka, Dow and Vautin, unpublished). Furthermore, there is a close correlation between the preferred color and the orientation selectivity of 246 quantitatively studied foveal striate cells (Yoshioka et al., 1988), which has the shape of a smooth curve with a midspectral peak (high orientation selectivity) and symmetrical endspectral troughs (low orientation selectivity).

More recently, we have been looking at luminance-contrast sensitivity, and have found that cells with preferences for midspectral colors like yellow and green (as well as achromatic cells) have higher luminance contrast thresholds than cells with preferences for endspectral colors like red and blue (Yoshioka et al., 1988, and unpublished). The relationship between luminance- contrast threshold and color, like that between orientation and color, has the appearance of a smooth curve with a midspectral peak and symmetrical endspectral troughs. Why should interblob cells have higher luminance contrast thresholds than blob cells? Presumably they are subjected to higher levels of spatial inhibition. This is suggested not only by their orientation selectivity, but also by their preferences for higher spatial frequencies (Silverman et al., 1989).

The color data are consistent with a color column model (Dow and Vautin, 1987; Dow, 1990a,b) in which CO blobs are alternately red and blue, with the spectrum represented as a series of concentric rings, beginning either in the short wavelengths (e.g., 400nm) in a 'blue' blob, or in the long wavelengths (e.g., 700nm) in a 'red' blob. The middle of the spectrum (e.g., 550nm) is represented in the interblob centers. The model shows the luminance-contrast threshold function superimposed on the spectral map. An important perceptual implication of the color data is that visual acuity should be higher for luminance contrast gratings composed of black and a midspectral color as compared with black and an endspectral color. There is some psychophysical evidence for this (Van der Horst et al., 1967; Williams et al., 1981; Mullen, 1987). A second perceptual implication is that midspectral colors should require more lightness to achieve maximal saliency than endspectral colors. There is also psychophysical evidence for this (Purdy, 1931; Priest and Brickwedde, 1938; Graham and Hsia, 1969; Boynton and Olson, 1987; Uchikawa et al., 1989). If color columns form concentric rings, and if orientation is organized in radial columns, as proposed here, the resulting orthogonality of the 2 systems maximizes the possibilities for combinations of particular orientations and particular colors.

The proposed models are readily testable with existing anatomical and physiological methodologies. Establishment of a definitive model will probably require the combined use of single unit recording, one of the global anatomical techniques (2- deoxyglucose, voltage sensitive dye, optical reflectance), and computer simulation.

REFERENCES

Baizer, J.S., Robinson, D.L. and Dow, B.M. (1977) Visual responses of area 18 neurons in awake, behaving monkey. J. Neurophysiol. 40: 1024-1037.

Bauer, R. and Dow, B.M. (1989) Complementary global maps for orientation coding in upper and lower layers of the monkey's foveal striate cortex. Exp. Brain. Res. 76: 503-509.

Bauer, R., Dow, B.M., Snyder, A.Z. and Vautin, R. (1983) Orientation shift between upper and lower layers in monkey visual cortex. Exp. Brain. Res. 50: 133-145.

Bauer, R., Dow, B.M. and Vautin, R.G. (1980) Laminar distribution of preferred orientations in foveal striate cortex of the monkey. Exp. Brain. Res. 41: 54-60.

Baxter, W.T. and Dow, B.M. (1989) Horizontal organization of orientation-sensitive cells in primate visual cortex. Biol. Cybern. 61: 171-182.

Blasdel, G.G. and Salama, G. (1986) Voltage-sensitive dyes reveal a modular organization in monkey striate cortex. Nature 321: 579-585.

Boynton, R.M. and Olson, C.X. (1987) Locating basic colors in the OSA space. Color Res. Appl. 12: 94-105.

Braitenberg, V. and Braitenberg C, (1979) Geometry of orientation columns in visual cortex. Biol. Cybern. 33: 179-186.

Carroll, E.W. and Wong-Riley, M.T.T. (1984) Quantitative light and electron microscopic analysis of cytochrome oxidase-rich zones in the striate cortex of the squirrel monkey. J. Comp. Neurol. 222: 1-17.

DeYoe, E.A. and van Essen, D.C. (1985) Segregation of efferent connections and receptive field properties in visual area V2 of the macaque. Nature 317: 58-61.

Dow, B.M. (1974) Functional classes of cells and their laminar distribution in monkey visual cortex. J. Neurophysiol. 37: 927-946.

Dow, B.M. (1990a) Color vision. In: The Neural Basis of Visual Function ed. A. Leventhal) MacMillan: London (in press).

Dow, B.M. (1990b) Nested maps in macaque monkey visual cortex. In: The Science of

Vision: A Convergence of Disciplines (ed. K.N. Leibovic) Springer-Verlag: New York (in press).

Dow, B.M. and Bauer, R. (1984) Retinotopy and orientation columns in the monkey: A new model. Biol. Cybern. 49: 189-200.

Dow, B.M. and Dubner, R. (1971) Single-unit responses to moving visual stimuli in the middle suprasylvian gyrus of the cat. J. Neurophysiol. 34: 47-55.

Dow, B.M., Bauer, R., Snyder, A.Z. and Vautin, R.G. (1984) Receptive fields and orientation shifts in foveal striate cortex of the awake monkey. In: Dynamic Aspects of Neocortical Function. (Eds., GM Edelman, WE Gall, WM Cowan) New York: Wiley and Sons, pp. 41-65.

Dow, B.M. and Vautin, R.G. (1987) Horizontal segregation of color information in the middle layers of foveal striate cortex. J. Neurophysiol. 57: 712-739.

Dubner, R. and Zeki, S.M. (1971) Response properties and receptive fields of cells in an anatomically defined region of the superior temporal sulcus in the monkey. Brain. Res. 35: 528-532.

Gotz, K.G. (1987) Do 'd-Blob' and '1-Blob' hypercolumns tessellate the monkey visual cortex? Biol. Cybern. 56: 107-109.

Gotz, K.G. (1988) Cortical templates for the self-orientation of orientation-specific d- and l-hypercolumns in monkeys and cats. Biol. Cybern. 58: 213-223.

Graham, C.H. and Hsia, Y. (1969) Saturation and the foveal achromatic interval. J. Opt. Soc. Am. 59: 993-997.

Hendrickson, A.E., Hunt, S.P. and Wu, J.Y. (1981) Immunocytochemical localization of glutamic acid decarboxylase in monkey striate cortex. Nature Lond 292: 605-607.

Hering, E. (1878) Outlines of a Theory of the Light Sense. (trans LM Hurvich and D Jameson, 1964), Cambridge, MA: Harvard Univ Press.

Horton, J.C. (1984) Cytochrome oxidase patches: a new cytoarchitectonic feature of monkey visual cortex. Phil. Trans. Roy. Soc. Lond. 304: 199-253.

Horton, J.C. and Hubel, D.H. (1981) Regular patchy distribution of cytochrome oxidase staining in primary visual cortex of macaque monkey. Nature 292: 762-764.

Hubel, D.H. (1987) Eye, Brain and Vision, New York, WH Freeman (Scientific American Library).

Hubel, D.H. and Livingstone, M.S. (1987) Segregation of form, color and stereopsis in primate area 18. J. Neurosci. 7:3378-3415.

Hubel, D.H. and Wiesel, T.N. (1968) Receptive fields and functional architecture of monkey striate cortex. J. Physiol. 195: 215-243.

Hubel, D.H. and Wiesel, T.N. (1972) Laminar and columnar distribution of geniculo-cortical fibers in the macaque monkey. J. Comp. Neurol. 146: 421-450.

Hubel, D.H. and Wiesel, T.N. (1974) Sequence regularity and geometry of orientation columns in the monkey striate cortex. J. Comp. Neurol. 158: 267-294.

Kruger, J. and Bach, M. (1982) Independent systems of orientation columns in upper and lower layers of monkey visual cortex. Neurosci. Lett. 31: 255-230.

Livingstone, M.S. and Hubel, D.H. (1984) Anatomy and physiology of a color system in the primate visual cortex. J. Neurosci. 4: 309-356.

Lund, J.S., Lund, R.D., Hendrickson, A.E., Bunt, A.H. and Fuchs, A.F. (1975) The origin of efferent pathways from the primary visual cortex, area 17, of the macaque monkey as shown by retrograde transport of horseradish peroxidase. J. Comp. Neurol. 164: 287-304.

Michael, C.R. (1985) Laminar segregation of color cells in the monkey's striate cortex. Vision Res. 25:415-423.

Mullen, K.T. (1987) Spatial influences on colour opponent contributions to pattern detection. Vision Res. 27: 829-839.

Priest, I.G. and Brickwedde, F.G. (1938) The minimum perceptible colorimetric purity as a function of dominant wave-length. J. Opt. Soc. Am. 28: 133-139.

Purdy, D.M. (1931) On the saturations and chromatic thresholds of the spectral colors. Br. J. Psychol. 21: 283-313.

Shipp, S. and Zeki, S. (1985) Segregation of pathways leading from area V2 to areas V4 and V5 of macaque monkey visual cortex. Nature 315: 322-325.

Silverman, M.S., Grosof, D.H., DeValois, R.L. and Elfar, S.D. (1989) Spatial frequency organization in primate striate cortex. Proc. Natl. Acad. Sci. 86: 711-715.

Tigges, J., Tigges, M., Anschel, S., Cross, N.A., Letbetter, W.D. and McBride, R.L. (1981) Areal and laminar distribution of neurons interconnecting the central visual cortical areas 17, 18, 19 and MT in squirrel monkey (Saimiri). J. Comp. Neurol. 202: 539-560.

Tootell, R.B.H., Silverman, M.S., DeValois, R.L. and Jacobs, G.H. (1983) Functional organization of the second cortical visual area in primates. Science 220: 737-739.

Uchikawa, H., Uchikawa, K. and Boynton, R.M. (1989) Influence of achromatic surrounds on categorical perception of surface colors. Vision Res. 29: 881-890.

Van der Horst, G.M.C., de Weert, C.M.M. and Bouman, M.A. (1967) Transfer of spatial chromaticity-contrast at threshold in the human eye. J. Opt. Soc. Am. 57: 1260-1266.

Vautin, R.G. and Dow, B.M. (1985) Color cell groups in foveal striate cortex of the behaving macaque. J. Neurophysiol. 54: 273-292.

Williams, D.R., MacLeod, D.I.A. and Hayhoe, M.M. (1981) Punctate sensitivity of the blue-sensitive mechanism. Vision Res. 21: 1357-1375.

Wong-Riley, M.T.T. (1979) Changes in the visual system of monocularly sutured or enucleated cats demonstrable with cytochrome oxidase histochemistry. Brain Res. 171: 212-214.

Yoshioka, T., Dow, B.M. and Vautin, R.G. (1988) Close correlation of color, orientation and luminance processing in V1, V2, and V4 of the behaving macaque monkey. Soc. Neurosci. Abs. 14: 457.

DISCUSSION: CORTICAL PROCESSING

Otto Creutzfeldt

Max–Planck–Institute for Biophysical Chemistry
Göttingen
F.R.G.

Creutzfeldt: Charles Sherrington once wished that the cortex were equipped with light bulbs, and each bulb should light up if a certain region were activated. He hoped for a clear picture as to what was really being represented in cortex. Optical recordings of electrical activity almost correspond to such light bulbs, but somehow cortical function still escapes us. The data presented to us this morning show that, if anything, we find a system of transformed maps of the retina and thus transformations of the visual environment, represented in the various visual areas of the cortex. David van Essen described the complexity and multiplicity of these areas, but it seems that the original view, as expressed in his pioneering studies by Semi Zeki, that each area represents a specific visual feature has to be modified to some extent. As Peter Lennie pointed out, the distribution of feature sensitive or cue-representing cells is nearly the same between these areas, only the proportions may change. This concurs with my own experience from recordings in V1 and V4. One major difference is of course the increase in receptive field size the farther you go from area 17.

Donald MacKay once asked what was represented in the various visual areas. Are they like maps of Europe, one according to geography, another to population density, elevation, minerals and so forth. We may go on asking whether the organization of maps is hierarchical in nature, though perhaps not in the conventional sense of hierarchy? The view that the cortex is organized in such a way that inputs from sensory organs are broken up into different features or cues, which are then distributed in different areas whose output finally converges into an executive, motor response system, has a strongly philosophical bias as it corresponds to traditional cognitive models based on introspection. We have learned in recent years, however, that there is much forward and backward interplay between the different visual areas. These and other data argue at least for a network of representations, in which no single feature is represented in a single area.

In considering the inputs into the various visual areas of the cortex, I wonder whether we should not pay more attention to the extra-geniculate, subcortical afferent input into extrastriate visual areas. There exists a strong and topographically organized input into visual association areas from the pulvinar. How important is this input functionally? Another aspect is that feature specific neuronal populations are topographically represented in V1, V2 and V4 with colour, spatial frequency and orientation tuned neurons arranged in regular matrices, such as shown here by DeValois and Dow for V1, and Charles Gilbert for V2. The actual sorting of afferent inputs for example from the LGN into such a matrix, is not at all clear, and there is also the problem of how this functional topography is matched by the multiple interconnections between these areas.

Finally, besides sending efferents directly into V2, MT and other areas, each point in area

17 sends axons directly into an 'action' system, that is the tectum. This is also the case with other areas, which each have a direct output into motor control systems such as the basal ganglia, the tectum/pretectum, midbrain nuclei etc. This output is organized in a topographical manner, like the inputs to the various visual areas. This might open another view for understanding multiple representations in the brain.

Any general model of the brain must take into account that the brain is a control system which allows the organism to behave appropriately in a given environment. This could be analyzed in model terms, such as described by Peter Lennie. Instead of saying the visual system constructs perceptual features from physical features, we may say that it constructs the appropriate response or behaviour in a given visual environment. In fact, every visual stimulus implies an intention to respond to it. If we restrict our concepts just to maps we tend to forget that vision is not a passive recipient system like a set of screens, but an active process in which the subject rapidly responds to physical features and thus changes the internal representation of the visual environment. Each of these actions, or even the intention to act, which can be easily related to attention and shifting of attention, alters the subject's relation to his environment and thus creates a new object-subject relation. The percept corresponds to or is the consequence of the responses it induces.

Considering this we may conceive the cortical visual system as a parallel, distributed but interconnected sensory–motor control system in which each visual subarea represents a different functional relationship between the subject and his environment. In such a model, it is not necessary to consider where different perceptual features are represented as such, nor is one forced to assume a hierarchical organization from the specific to the general where the various features might be recombined into a unitary percept, since the various functional loops are unified in a unitary action or behavior. In such a model, the mechanisms of conscious perception i.e. how "we read" or extract the perceptual features which are inherently contained in the activity patterns of such a distributed control system, may escape us. But unitary perception is not answered either by the classical model of feature dissection and hierarchical recombination. Instead it is contained in the responses, intentional or overt, the visual environment induces.

In the following I suggest that we should discuss in the following hour, in the context of such general considerations, first what is in fact represented in different areas, secondly what sorting mechanisms may underly the feature or cue matrices in area 17 and 18, and thirdly cooperative network models in connection with the parallel input and output organization of the visual system.

Van Essen: With regard to Gilbert's mapping of disparity tuning of V2 cells, can you relate that to the cytochrome oxidase staining, in particular was disparity tuning restricted to the thick stripes?

Gilbert: It seems that they tend to be concentrated in these stripes, but we also see it in the other stripes, even in colour-selective cells, but more toward the boundary.

Dow: With respect to the connectivity part of your study, you show pyramidal cells connect like regions. Have you identified GABA cells to see what connections they make?

Gilbert: We have seen the longest connections mediated by pyramidal cells but others have shown the longest connections of basket cells do project for a certain distance, maximally about 1 mm. Pyramidal cells can go up to about 6mm, so the excitatory connections can go a longer distance. However, a significant proportion of their postsynaptic cells are inhibitory stellate cells, so the actual effect of the projection can be inhibitory by going through an interneurone.

Creutzfeldt: What is the source of the signal in your optical recordings?

Gilbert: One prominent source is local vasodilation of capillaries, so a limit to its resolution would then be given by inter-capillary distance, about 50 microns or so. There are other sources, the light scattering signal which is dependent on extracellular space, and changes in the oxidative state of haemoglobin and other chromophores. So it is a complex set of sources, the ultimate resolution of which has yet to be explored. Substantial averaging is also required which can take some time, but the technique is still being refined.

Kulikowski: In Russ DeValois' model with Tootel, in the middle of the blob you had a projection from the magno–cellular geniculate layers. Is this part of your current model?

Russ DeValois: I think this is obscure. There is anatomical evidence for magno– and parvo–cellular input to the upper layers, but their distribution is less clear than we wrote then.

Rodieck: I think of the cortex as a feature map, but one should distinguish between van Essen's series of mapping regions and Gilbert's sub-divisions of these regions with respect to features. One problem with the notion of features is that when certain stimuli are used on the cortex, it is easy to say, 'All the cortex participates in this or that'. There may well be a feature there, but we have failed to use the right parameters to identify it. A given region might be quite pure, functionally, but bringing with us the notion it is all done by Fourier analysis we may fail to discover what these features are; the discrete regions shown anatomically should perhaps lead us to look for more specific functions or features.

Creutzfeldt: Would you extend that to extra-striate cortex as well as to area 17?

Rodieck: Gilbert has shown us there is sub-division into different regions within area 18 as well as 17; once we understand them we may go further. With reference to van Essen's talk, he says that 40% of possible connections are known to exist between cortical areas, and then a certain number of connections are known not to exist, but there must be a lot of possibilities that have not been tested.

van Essen: One can massage the data in different ways to take this into account, and get between 30% and 45%.

Kulikowski: Probably different cells in different areas overlap in their functional properties, but if as a psychophysicist I can demonstrate, for example, a certain transient system in a threshold task, it means that some fraction of the cells in the pathway must be capable of following these kind of features. Obviously, there are other cells with more complex properties. For instance, instead of saying that fast information from P–cells in LGN is discarded, we should look for texture cells in the striate and extrastriate visual areas, responding to small spots with high temporal resolution.

Lennie: I object to Rodieck's notion that anatomy can lead physiology. Anatomy tells us that the system is complicated but nothing about function. The problem for the physiologist in my view is that psychophysics has historically provided physiologists with much inspiration, but psychophysics stops at the description of what Graham calls analyzers; psychophysics tells us a lot of what to look for at early stages in cortex, but almost nothing about what to look for beyond that. The only people to consider these questions seem to be computer scientists, and the problem there is to translate algorithm into hardware, and that might be difficult.

van Essen: I argue strongly against the notion of a feature map as a description of any cortical areas or compartments we know about. To me, a feature map means a population of cells sharing tuning or selectivity for a particular feature, usually a low level dimension or cue. Up to now, there has been within each area or compartment such a degree of heterogeneity of response properties and tuning along more than one dimension in cortical areas, that some kind of multiplexing of information in individual cells with the population as a whole representing more than one kind of information seems to me more likely.

Williams: I'd like to address the question of what V1 might be doing. Peter Lennie described computer science models. In the pyramid model you save processing capacity by representing low spatial frequencies with fewer spatial samples than high frequencies. Does physiology offer any evidence that pyramid structures might be used in V1?

Russ DeValois: There are more cells tuned to middle spatial frequencies than low, not so many at high; but there may be sufficient. Incidentally, as far as the blobs are concerned, I think they are just an epiphenomenon, a sort of slow deoxyglucose mapping.

Lennie: Two points; there is no evidence below cortex of a representation at multiple spatial scales, for example, in retina or LGN. The mere existence in cortex of cells tuned to different spatial frequencies is a start. Secondly, you get much more economic representation of visual scenes with oriented filters than isotropic ones. So two substantial changes have occurred in area 17. Whether these changes meet the requirements for the representation of real scenes is a question for which we have insufficient evidence.

Rodieck: Back to features. I think there is a confusion between the idea that a cell responds to a stimulus and that it codes for that stimulus. Let us take the example of the region which Charles Gilbert was pointing out was coding for disparity. If you wade into a cell coding for disparity with bars and spots and gratings and evaluate that area in terms of what you thought it was doing and missed its disparity selectivity, then you would characterize it in the wrong terms; on finding the right parameter, other characteristics are revealed as epiphenomena, and the region as a unit may be doing what the individual cells are doing.

Mollon: Neurological patients with specific deficits, often attributed to damage to areas of cortex devoted to particular attributes, seem almost to contradict what the neurophysiologists perceive. It is almost impossible to conceive of a sufficiently specific lesion to cause these deficits. There may be another origin for these effects. We have a patient with acquired tritanopia which developed over a few weeks, binocularly spreading across the visual field. It seems a lesion is here an unlikely explanation, but a specific immunological reaction of a tiny sub-set of cells widely spread through cortex might cause such an effect.

Creutzfeldt: We have heard several suggestions as to ordering in area 17. Perhaps different speakers might compare their approaches.

Gilbert: With reference to the way in which Lennie's computational models may inform us about the way to look at cortical cells, there have been efforts from information theory as to what cells may be good for, but they may be a restatement of the biology. For example, the use of centre-surround structure was demonstrated neurophysiologically and interpreted as a way of signalling contrast. Others looked at this from an information-theory standpoint using the zero-crossing theorem to indicate it was an efficient way of treating patterns in visual space. Also, you said no one knows what a complex cell is for, but the two models you discussed use oriented filters for the original decomposition of the stimulus.

Lennie: That formal analyses only show what biologists knew already is partially true, but these formal analyses have given us deeper understanding of why things are organized as they are. Biology did not tell us why it would be desirable to have multiple-scale representation. Informal conversation does not give us solid justification of this sort. Secondly, you are wrong to suppose that any of these accounts have given us a role for complex cells; the analyses you mention give us a role for linear operators, which complex cells are not. I feel complex cells are a substantial mystery. One possibility is that they are used for detecting texture discontinuities. Another is that they detect motion energy, but good evidence is lacking. What I wanted to emphasize was that single unit physiology is a bad way of discerning what the system is about, but a good way of testing some idea of how the system might be designed. We need independent information as to how the system may be designed to steer the physiology properly.

Cavanagh: I would argue against efficiency being the reason for the development of orientation filters or multiple scales; there are more cells in area 17 than in the geniculate nucleus. I think their emergence must be useful for orientation, size and pattern analysis.

Creutzfeldt: It should be realized that orientation specific cells are also important for extracting texture information. Our environment is not so much made up of lines but structures of different texture, which have certain oriented characteristics.

Shapley: Problems with ideas of hierarchical streams of visual processing and the difficulties in interpreting single unit recording from extrastriate cortex, as pointed out by Peter Lennie, partly stem from the fact we treat the visual system as a passive device, only driven by the environment, and we use techniques suitable for linear systems or passive non-linear systems, having no memory. For areas representing memory, involving predictions, which are precisely what extra-striate cortex is all about require different kinds of models. These might be testable on single units, if the right paradigm were found to test them.

Hayhoe: Is a use for complex cells that they are the beginnings of a generalisation of the response across different retinal locations? This would seem a desirable property.

Lennie: Firstly, you never do lose the information about where things are. On the other hand, if studied in the right way, maybe one could show that complex cells are very good at localising certain things, for example texture discontinuities, rather than location information being lost in their responses.

van Essen: Many elements of perception are relativistic rather than absolute, so you do lose absolute track of where things are. The way in which we make transformations from absolute retinal coordinates to a variety of other reference frames is a very important question, which may need new paradigms for analyzing extrastriate and possibly striate cortex, perhaps by studying directed visual attention. Psychophysics does guide us in an analysis where we scrutinize restricted portions of visual space in temporal sequence. Understanding these dynamics is also extremely important.

Williams: David van Essen made the point of a benefit of cortical magnification in giving scale invariance for looming objects, but in practice it is more common to move ones eyes around with the same object distance. Then you change the spatial sampling rate enormously during the observation period but you can still do object recognition. So I would suggest an alternative view, that is that the linear relationship with eccentricity is just a nice way of incorporating a specialized region into the visual field without discontinuities.

van Essen: I don't think these views are mutually inconsistent. Both phenomena occur, we zoom in on objects while maintaining our attention, but the eyes also jump around a great deal, and we need to have a system that can handle both smoothly. But how we maintain object centered representations in the face of changes in spatial and temporal sampling remains a problem.

Russ DeValois: With Pete Lennie's point about complex cells, there have been models for motion and texture in which they have turned out to be very useful. Since these are among the few serious models that have been developed of cortical function based upon physiological data, perhaps we should develop more models for other functions. Secondly, even the largest receptive fields in area 17 are small compared with the distances over which object integration must occur during form perception of nearby objects, and perhaps later cortical regions allow such long-range connections. This is consistent with the larger fields in later areas.

Drasdo: With respect to spatial scaling of cortical magnification, this has been related to sampling uniformity and scale invariance but it is also related to flow-field sampling, which is evolutionarily important for primates moving in a varied environment. Secondly, the brain is a learning instrument requiring activity in neighbouring neurones to secure suitable development of the circuitry. Therefore, if we perform such a process on a moving or coloured object, some local concentration of active neurones may be required, thus forming the different cortical areas.

Dow: I want to comment about the notion of nested maps. Within the retinotopic map, the idea that other features are also maps is convenient. Let's consider a stimulus with the colour orange, a horizontal orientation and a particular direction of movement. Shifting the colour from orange to yellow is a slight change in the colour map but there is still coherence within the other maps of the object. If each parameter is mapped, with a slight change in one, coherence of cell activity is maintained because activity in only one map might change. Away from a retinotopic area, things might be different.

Gilbert: What should guide the experiment? In studying the brain we cannot all expect to be as lucky as Hubel and Wiesel, who, with their broken slide, discovered orientation selectivity. On the other hand, as Lennie said, we don't have suggestions from a particular systematic model as to what hardware to look for, and even if we had, it might not be unique. Perhaps an intermediate level may lead to progress. Oriented line segment filters may lead to suggestions of how things might work. On the other hand, there may be some need for generalisation of position. In V4, where one has large-field, complex-like cells, Desimone has found that by expectation one can actually focus the area over which the cell is responsive, so there are contextual and expectation-driven properties in cells. Some of these approaches might be driven by psychophysics, especially as to suggestions of suitable stimuli of intermediate complexity between spots and real objects.

Schiller: To get back to feature extraction, as you move up from the retina to cortical areas, the brain breaks up and then recombines various features. Rod and cone signals break up and are then recombined, increments and decrements are broken up and then recombined in complex cells, so these processes are very dynamic in nature. This fits poorly into the idea of specific feature regions in the cortex. For example, in extracting shape many cues, luminance, chrominance, motion, may be used, or even stereo. So there are at least four ways of seeing objects, which would probably not be processed in separate regions. In fact, with a V4 lesion, feature extraction deteriorates whatever cue is used to test it, whether the cue is luminance, colour, motion or whatever. What we see is initially a break up of features and then their recombination in higher areas in ways that are difficult to predict unless you go in and study them. I don't think there is any kind of modelling which can predict these kinds of things. They can define constraints, but models have never predicted on and off systems, colour opponent and broad-band channels, or most of the things we have found in the brain; physiologists, anatomists and psychophysicists have made in comparison quite remarkable contributions.

Creutzfeldt: That was a good last word. In conclusion, perhaps we should remember that the task of the visual system is not just to detect something but rather to transform the visual environment into an appropriate act, taking into account also the inner state and demands of the organism. These tell us that we should go and have lunch now.

VISUAL PHOTOMETRY: RELATING PSYCHOPHYSICS TO SOME ASPECTS OF

NEUROPHYSIOLOGY

Peter K. Kaiser

Department of Psychology, York University
North York, Ontario, Canada

INTRODUCTION

Light measurement requirements are like other physical measures because when measurements are made on the same physical quantities it is necessary that they be identical regardless of who makes them and where they are made. When measuring light in radiance units this type of measurement consistency is expected when using a well calibrated radiometer. When measuring light using physical photometry such consistency is expected when using a well calibrated photometer. But visual photometry usually does not follow suit. In this paper I would like to explore some possible physiological substrates underlying visual photometry, especially as it relates to flicker. Particular attention will be paid to the double duty hypothesis which says that the parvocellular pathway (tonic cells) processes both chromatic and luminance information. Specifically, I will look at the change from spectral opponent to spectral non-opponent sensitivity as a function of temporal frequency.

Photometry

The CIE (1987) defines photometry as "Measurement of quantities referring to radiation as evaluated according to a given spectral luminous efficiency function, eg. $V(\lambda)$ or $V'(\lambda)$." In plain language this means that photometry is the physical measurement of light modified by the spectral sensitivity of an internationally accepted standard observer commonly referred to as the V lambda function. There are two classes of photometry: 1. physical photometry which depends on physical detectors and 2. visual photometry which depends on an observer's eye as the detector. It is with visual photometry that we will be concerned.

The visual system is not a spectrally flat detector and herein lies one problem. An important question concerns the appropriate way to evaluate its spectral sensitivity. There are a number of methods for determining spectral sensitivity depending on the attributes of the observer and the use to which the information will be put. For the sake of simplicity I will focus primarily on human psychophysics and monkey

From Pigments to Perception, Edited by A. Valberg and
B.B. Lee, Plenum Press, New York, 1991

electrophysiology, using photopic vision with stimulus sizes of at least 0.5 degrees visual angle and the criteria of minimum flicker and minimally distinct border.

In every day language we refer to the brightness of a source. More properly we should distinguish between brightness and, the equally well known term, luminance. Brightness refers to the subjective experience associated with the amount of light. Luminance is defined by the CIE according to Equation (1).

$$L = K_M \int L_{e,\lambda} V(\lambda)\, d\lambda \qquad\qquad (1)$$

where L = luminance, K_m = maximum luminous efficacy, $L_{e,\lambda}$ = spectral radiance, $V(\lambda)$ = CIE photopic luminous efficiency function. The key to the relation between photometry and luminance lies with the $V(\lambda)$ function. In physical photometry the standard photopic CIE $V(\lambda)$ function is used. In visual photometry a more liberal view of Equation (1) is taken; specifically, the spectral sensitivity of the observer is used for $V(\lambda)$ and then reference to 'sensation luminance' is made.(Kaiser, 1988)

It has been shown that the spectral sensitivity measured by heterochromatic flicker photometry and minimally distinct border are similar and that both are narrower than that obtained by brightness matching(Wagner & Boynton, 1972). Arguments have been made that brightness matching is mediated by a combination of the spectrally opponent channels and the spectrally non-opponent channels, whereas flicker photometry and minimally distinct border tasks are mediated only by the non-opponent channel (Lee,Martin & Valberg, 1988; Kaiser, Lee, Martin & Valberg, 1990). Considering the recent literature, some might prefer the concepts parvocellular and magnocellular pathways respectively for the opponent and non-opponent channels. Psychophysical evidence for this pathway dichotomy includes multipeaked spectral sensitivity functions and non-additivity when brightness matching and increment threshold sensitivity are the criteria. On the other hand when minimum border or minimum flicker are the criteria to measure spectral sensitivity, the functions are unimodal and additivity of luminances is obtained. Physiological evidence includes finding retinal ganglion cells and LGN cells that are spectrally opponent or spectrally non-opponent.

PARVO PATHWAY DOUBLE DUTY(?)

Although there is considerable evidence for a spectrally opponent (parvo, tonic cell) channel which processes hue information and a spectrally non-opponent (magno, phasic cell) channel which processes luminance information, this is not the only view. Gouras and Zrenner (1979) reported responses recorded from tonic R-G ganglion cells which at low temporal frequencies gave sensitivity functions like spectrally opponent neurons and at 33 Hz gave sensitivity functions like spectrally non-opponent neurons. They concluded that it is not necessary to postulate a separate luminance channel. Zrenner(1983) elaborates this double duty hypothesis.

282

Ingling & Grigsby (1990) recently restated a model which exemplifies a double duty activity in the parvocellular pathway. They contend that simple R-G cells carry both a chromatic and an achromatic signal and specifically assert that, "The channel signals hue (R-G) at low spatial frequencies and luminance (R+G) at high." (p. 824) Ingling & Martinez-Uriegas (1985) attribute similar characteristics to the temporal domain.

Klingaman et al. (1980) reported psychophysical, increment threshold spectral sensitivity functions which seemed to support this double duty hypothesis. Their subjects performed a flicker detection experiment on stimuli whose frequencies were varied from 3 to 40 Hz. At the slowest frequencies their flicker and hue detection data looked like typical Sperling & Herwerth (1971) trimodal chromatic functions. At 40 Hz the flicker detection function appeared more like an achromatic spectral sensitivity curve. These results are similar to those reported by King-Smith (1975) who obtained trimodal increment threshold spectral sensitivity functions with 200 ms flashed stimuli and a unimodal function when observers detected the flicker of stimuli temporally modulated at 25 Hz.

I am puzzled by Klingaman et al. observers' ability to perform the hue detection task. If at 40 Hz all the R-G neurons are now R+G like how can hue detection be possible in the mid to longwaves? Merigan (this volume) and Schiller (Schiller & Logothetis, 1990 and this volume) report that when the parvo pathway is lesioned, animals no longer discern color differences. From this I assume that when the visual system's opponent activity is eliminated color discrimination is gone. So the when the parvo pathway is lesioned, leaving only the magno pathway with its R+G cone input color discrimination is lacking as Merigan and Schiller tells us. But Klingaman et al's observers presumably could do their task up to 40 Hz without an apparent major deficit in the three peaked spectral sensitivity functions. It is possible to explain this result and still hang on to the assumption of R-G neurons becoming R+G? One possibility is that the conversion of R-G to R+G found by Gouras & Zrenner (1979) does not occur in a uniform manner to all tonic ganglion cells. Perhaps the transformation occurs only in some neurons but not others. If this were the case, then perhaps a sensitivity function would move towards being more like an achromatic function but sufficient numbers of R-G neurons would remain to permit hue detection. This could work in the following way. Suppose that the synergistic shift between center and surround that Gouras and Zrenner report occurs at different frequencies for different cells. Then some neurons may not have completed the shift and still operate more like R-G. Yet others may have shifted at much lower frequencies and are now unable to resolve the higher flicker frequency and therefore response similarly as if looking at a steady light. While yet others are synergistically shifted and operate as R+G. If this were the case, then observers could yield results like those reported by Klingaman et al. I must admit, however, that I am unaware any physiological evidence to support this view other than the reported range of center-surround latencies. (Lee, this volume) Also, making decisions based on the shapes of spectral sensitivity functions is a risky business.

There is a problem with analyzing spectral sensitivity data to decide whether the chromatic or achromatic channel is being activated. Kaiser & Ayama (1986) have shown that under stimulus conditions which yield luminance like spectral sensitivity functions, it is still possible to obtain additivity failures. It would appear that looking at the shape of a spectral sensitivity function is not as precise a method of analysis as quantitatively evaluating luminance additivity.

ADDITIVITY

The additivity experiment provides a good test for differentiating between the involvement of the opponent (parvo) pathway and the non-opponent (magno) pathway. Since the opponent pathways are formed by subtraction of the longwave from middlewave receptors and also by a subtraction of the longwave+middlewave from shortwave receptors one would not expect the visual system to respond additively where the opponent pathways are involved. On the other hand it is also commonly accepted that the sum of the longwave and middlewave receptors form the nonopponent channel.[1] Thus an experiment where only the non-opponent, luminance system is involved could be expected to produce additivity.[2] Furthermore, the integral in Equation (1) requires additivity if we are going to call a particular pathway a *luminance* channel. When the spectrally opponent, parvo pathway is active non-additivity is expected. Although I must admit that Kremers has recently tried to persuade me otherwise. (personal communication).

As noted above, the results of a spectral sensitivity experiment (King-Smith, 1975; Gouras & Zrenner, 1979; Klingaman et al., 1980) do not always allow the interpretation that a particular channel has been silenced. We can come somewhat closer however with an additivity experiment.

Let's take a closer look at a typical additivity experiment by assuming we are interested in the luminance additivity associated with adding red and green light and using the minimum flicker criterion. Three steps are involved: 1. The observer adjusts the amount of red light to obtain minimum flicker with respect to a reference light. 2. Repeat this with a green light. 3. The experimenter puts some fraction of the amount of red light from step one in the test field and the observer adds a sufficient amount of green light to the test field so that the criterion of minimum flicker is re-established. Equation (2) shows how the additivity index is computed. If the additivity index equals unity additivity is obtained. With the criteria of minimum flicker and minimally distinct border, the typical result is additivity. If heterochromatic brightness matching is the criterion additivity failure is obtained.[3]

---

[1] Recent evidence by Stockman et al. (1987) and Vos et al. (1990) suggests that there is a very small shortwave, negative component in the luminance signal.

[2] But see Ingling et al. (1978).

$$Additivity\ Index\ =\ \frac{r}{R}\ +\ \frac{g}{G} \qquad \textbf{(2)}$$

where R and G are the amount of light required to match the
reference, r and g are the amounts in the mixture required to
match the reference.

Since additivity failures are almost always obtained when
using a brightness matching paradigm, Kaiser, Vimal, & Hibino
(1988) attempted to psychophysically test the Gouras & Zrenner
physiological result and Ingling & Martinez' model. We tested
it by performing a brightness matching additivity experiment
using a bipartite, field that was temporally modulated from 1
to 30 Hz. According to the double duty hypothesis, one would
expect additivity failure at low frequencies which would progressively become more additive as frequency was increased.
Then at the highest frequencies, when flicker is no longer
apparent additivity failures should again be obtained. As can

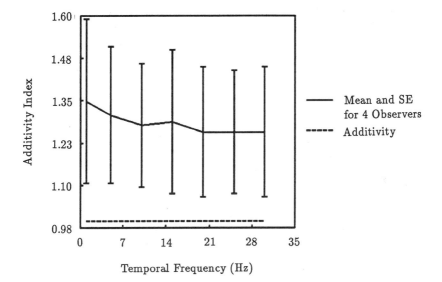

Fig. 1. Brightness matching additivity as a function of
temporal modulation of the bipartite field. Means and S.E.
of four observers.

---

[3]Ingling et al (1978) argue that because of cone response
compression, additivity failure should be expected. Further they
show a way of performing an additivity experiment which yields
additivity failures with minimum flicker and only very small
deviations from additivity with minimally distinct border.

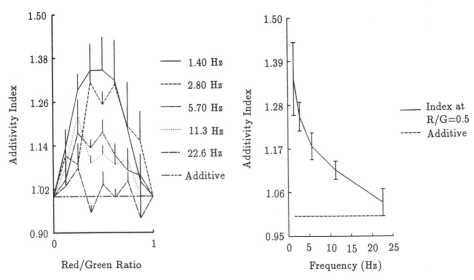

Fig. 2A. Additivity as a function of red/green ratio and temporal frequency. Means and S.E. for 4 O's (Nakano, personal commuincation)

Fig. 2B. Additivity as a function of temporal frequency for a 0.5 red/green ratio (Nakano, personal communication)

be seen in Figure 1, additivity was not found at any of the frequencies tested. Pandey & Vimal (1989) performed a modified version of this experiment. Instead of a brightness matching criterion, they used an absolute threshold criterion and found additivity failures below 5 Hz, additivity between 5 and 15 Hz and additivity failure above 15 Hz.

Nakano (personal communication) wondered why Pandey & Vimal found additivity at these frequencies and not higher ones. He thought that additivity should be obtained at frequencies closer to 30 Hz because that was the frequency near which Gouras and Zrenner reported R-G neurons behaving like R+G neurons. Further, at 10 Hz, the sensitivity functions which Klingaman et al (1980) reported still showed clear evidence of chromatic (R-G) activity. Nakano wondered if the use of an absolute threshold criterion could be the reason why Pandey and Vimal found additivity at the lower frequencies. Since temporal frequency seemed to be a critical variable in Gouras and Zrenner's electrophysiological investigation, Nakano hypothesized that perhaps flicker detection, not absolute threshold would yield data more in line with their results. So he repeated the Pandey and Vimal experiment using a detection of flicker criterion. Sure enough, he found additivity failures up to 11 Hz and additivity at 22 Hz as can be seen in Figure 2A and B. Figure 2A shows data collected at a variety of red/green ratios. Figure 2B shows the data just for a red green ratio of 0.5 and plotted as a function of temporal frequency. From these two graphs one can see that, when approximately equal amounts of red and green light are used, additivity is obtained above 20 Hz.

Let us now consider the evidence for the possible physiological substrates underlying visual photometry. Recently Lee and his colleagues (Lee et al. 1988; Kremers et al., 1990; Kaiser et al. 1990) reported evidence that visual responses to flicker photometry and minimally distinct border have their physiological substrate in the phasic retinal ganglion cells. They showed by single unit recordings that activity in these phasic ganglion cells is minimized at equal luminance. Lee, Martin & Valberg (1988) reported that under minimum flicker conditions phasic retinal ganglion cell activity reaches a minimum when the alternating fields are equal in luminance. Likewise, Kaiser et al. (1990) showed that when a border generated by precisely juxtaposed hemifields was moved back and forth across the receptive field of phasic ganglion cells, a minimum response was obtained when the luminances of the hemi fields were equal. At the 1989 Optical Society meeting we reported that the requisite photometric properties of additivity, transitivity and proportionality were demonstrable with these physiological MDB data. (Kaiser, Lee & Valberg, 1989) Further when spectral sensitivity functions were calculated for these neurons under the minimum flicker and minimally distinct border conditions they closely approximated human psychophysically measured photopic spectral sensitivity functions. In Figure 3 (Kaiser, et al., 1990) we see the spectral sensitivity function obtained under the minimum border condition. These data were calculated from phasic retinal ganglion cell responses from the macaque monkey. Lee et al. (1988) showed essentially the same function for temporally modulated stimuli

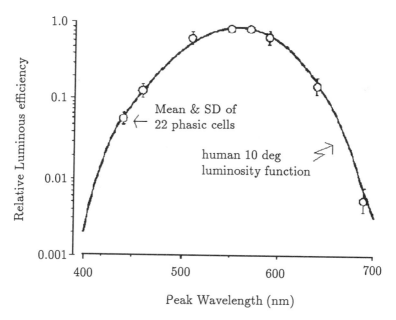

Fig. 3. Open circles: Spectral sensitivity derived from 22 phasic retinal ganglion cells, with MDB. Solid curve: human 10 deg. luminosity function (Kaiser, Lee, Martin & Valberg, 1990)

when recording from phasic retinal ganglion cells. Thus it seems reasonable to conclude, that the photometric criteria of MDB and minimum flicker are a function of activity in the phasic or magnocellular pathway. However, for a different view regarding MDB see Ingling (this volume).

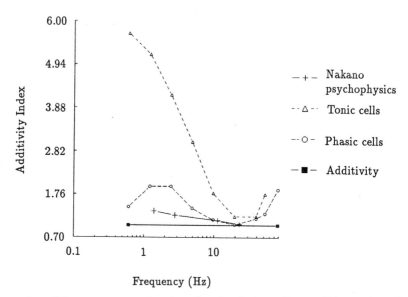

Fig. 4. Pluses: psychophysical data from Fig. 2. Open triangles: mean 6 red on-center, 9 green on-center & 2 red off-center tonic cells; open circles: mean 8 on and 4 off phasic cells (Kremers, personal communication)

Lee (this volume) concludes that whereas parvo cellular ganglion cells respond to luminance flicker at 20 Hz and above this information must be filtered at some cortical level leaving the cortex to respond only to the information relayed by the magnocellular pathway. Although he provides circum- stantial evidence to support this view, an experiment has not yet been devised to critically test his conclusion.

Let us now consider how the additivity experiment relates to what is occurring in the ganglion cell. On this point we have data from Kremers. (Figure 4, personal communication) He performed an additivity study on single retinal ganglion cells. In Figure 4 we can see that Kremers' tonic cells exhibit larger additivity failures at all frequencies than do the phasic cells. If the Gouras and Zrenner hypothesis of R-G neurons becoming R+G at specific temporal frequencies is to receive full support from these data, we should find that the tonic cells become additive at near 30 Hz. As seen in Figure 4, tonic cells come near additivity at about 20 Hz, but do not reach it. Phasic cells, on the other hand reach additivity at about 20 Hz which is close to Nakano's result which I have added to Kremers' graph. The fact that additivity failures are obtained at the Kremers' highest frequencies is not surprising. At these higher frequencies the cells probably have difficulty resolving the flicker frequency.

We should now consider why additivity failures are found in phasic cells if the input to them is the sum of the middle and long wavelength receptors? A subsidiary consideration is why, since phasic cells are non-additive at low temporal frequencies and if luminance is dependent on phasic cells can we find psychophysical additivity with the minimum flicker and minimally distinct border criteria? Some observations related to these questions are possible. Ingling et al. (1978) report that because of receptor compression super-additivity by any psychophysical criterion ought to be expected. They provide an analysis which shows that additivity failure should be the rule and also show by increasing the amount of red and green in the mixture over a two log unit range super-additivity is obtained above about 50 td. They further report that, "For particular test wavelengths and a particular spectral distribution of the standard it is possible to get perfect additivity regardless of the compression."(p. 380) What is surprising, however, is that additivity tested physiologically (Kaiser, et al., 1989) and psychophysically by the method described earlier in this paper yields additivity without apparent selective attention to the precise wavelength and spectral distribution conditions. I can think of one possible reason why psychophysically additivity is found by the method described above and evaluated by Equation (2). Perhaps it is because any system appears linear if evaluated over a short enough range of values. Ingling et al.'s approach of progressivly increasing the amount of light overcomes this short range limitation. However, the physiological additivity failure of phasic cells found by Kremers goes in the opposite direction. He found subadditivity, not super additivity. But, now we must ask the question why do phasic cells become additive near 20 Hz? It is possible that response compression is dependent on a temporal factor. Boynton (1979) reminds us that the response compression calculations he performed (with Whitten) apply strictly to the steady state.

SUMMARY AND CONCLUSIONS

Let's summarize what we seemed to have learned. Super threshold brightness matching additivity experiments conducted between 1 and 30 Hz show no evidence of additivity. Clearly under these conditions the physiological results of Gouras and Zrenner or of Kremers are not supported. However, when the observer's task is to detect the presence of light, additivity is obtained between 5 and 15 Hz which is clearly below the 33 Hz, the frequency at which Gouras and Zrenner report tonic retinal ganglion cells to change from R-G to R+G. When the psycho-physical task is detection of flicker, additivity failures are observed at 11 Hz and below, with additivity being reached at 22 Hz. This result is closer to the physiological data obtained by Kremers and approaches the temporal conditions of Gouras and Zrenner. It should be noted that Gouras and Zrenner did not do an additivity experiment as Kremers did.

Their criterion was the change in spectral sensitivity from that describing red-green opponency to red-green non-opponency.

Clearly, the parvo pathways respond to flicker. However, whereas we have seen research which quantitatively points to magno pathway involvement in visual photometry, I know of no quantitative data which shows how the parvo pathway psycho-physically contributes to visual photometry information.

Finally, although the shift of center surround antagonism to synergism (Gouras & Zrenner, 1979) may occur for some R-G tonic ganglion cells, we can not from these data infer the absence separate luminance channel. The psychophysical and physiological additivity experiments to which I've referred support the contention that the physiological substrate underlying visual photometry is the magnocellular pathway. However, Ingling (this volume) argues that while flicker photometry is mediated by the magnocellular pathway minimally distinct border is mediated by the parvocellular pathway.

ACKNOWLEDGMENTS

Support by the Natural Sciences and Engineering Research Council of Canada is acknowledged. Drs. B. Lee & J. Kremers gave helpful criticisms and I thank Drs. Kremers and Nakano for permitting the use of their unpublished additivity results.

REFERENCES

Boynton, R.M., 1979 Human Color Vision, Holt, Rinehart, Winston, N.Y.

CIE, 1987, International Lighting Vocabulary, CIE Publ. No. 17.4

Gouras, P. & Zrenner, E., 1979, Enhancement of luminance flicker by color opponent mechanisms, Science, 204, 587-589.

Ingling, C.R., Tsou, B.H., Gast, T.J. Burns, S.A., Emerick, J.O. & Riesenberg, L., 1978, The achromatic channel. 1. The non-linearity of minimum border and flicker matches, Vision Res., 18, 379-390.

Ingling, C.R. & Martinez-Uriegas, E., 1985, The spatiotemporal properties of the r-g X-cell channel. Vision Res., 25, 33-38.

Ingling, C.R. & Grigsby, S.S., 1990, Perceptual correlates of magnocellular and parvocellular channels: seeing form and depth in afterimages, Vision Res., 30, 823-828.

Kaiser, P.K., 1988, Sensation luminance: a new name to distinguish CIE luminance from luminance dependent on an individual's spectral sensitivity, Vision Res., 28, 455-456.

Kaiser, P.K. & Ayama, M., 1986, Small, brief foveal stimuli: an additivity experiment. J. Opt. Soc. Am. A, 3, 930-934.

Kaiser, P.K., Vimal, R.L.P. & Hibino, H., 1988 Psychophysical test of R-G channel becoming R+G at high temporal frequency. Inves. Ophthal. & Visual Sci., Supplement, 29, 328.

Kaiser, P.K., Lee, B.B., Martin, P.R. & Valberg, A., 1990, The physiological basis of the minimally distinct border demonstrated in the ganglion cells of the macaque retina, J. Physiol. 422, 153-183.

King-Smith, P.E., 1975, Visual detection analyzed in terms of luminance and chromatic signals, <u>Nature</u>, 255, 69-70.

Klingaman, R.L., Zrenner, E. & Baier, M., 1980, Increment flicker and hue spectral sensitivity functions in normals and dichromats: the effect of flicker rate, <u>Colour Vision Deficiencies</u> V G. Verriest, Ed. Adam Hilger Ltd. 240-243.

Kremers, J.J.M , Lee, B.B. & Kaiser, P.K., 1990, Sensitivity of macaque ganglion cells and human subjects to mixed luminance and chromatic modulation. Paper presented at ECVP, Paris, 1990.

Lee, B.B., Martin, P.R. & Valberg, A., 1989, Nonlinear summation of M- and L-cone inputs to phasic retinal ganglion cells of the Macaque, <u>J. Neurosci</u>, 9, 1433-1442.

Lee, B.B., Martin, P.R. & Valberg, A., 1988, The physiological basis of heterochromatic flicker photometry demonstrated in the ganglion cells of the macaque retina. <u>J. Physiol.</u>, 404, 323-347.

Pandey, R. & Vimal, R.L.P., 1989, Transition from luminance additivity at threshold to subadditivity above threshold at intermediate temporal frequencies. <u>Inves. Ophthal..</u> & <u>Visual Sci.,</u> <u>Supplement</u>, 30, 127.

Schiller, P.H., Logothetis, N.K. & Charles, E.R., 1990, Functions of the colour-opponent and broad-band channels of the visual system. <u>Nature</u>, 343, 68-70.

Sperling, H.G. & Herwerth, R.S., 1971 Red-green cone interactions in the increment-threshold spectral sensitivity, <u>Science</u>, 172, 180-184

Stockman, A., MacLoed, D.I.A. & Priest, D.D., 1987 An inverted S-cone input to the luminance channel: evidence for two processes in S-cone flicker detection. <u>Invest. Ophthal.</u> & <u>Visual Sci.(Suppl.)</u> 28, 92.

Vos, J.J., Estevez, O. & Walraven, P.L., 1990 Improved color fundamentals offer a new view on photometric additivity. <u>Vision Research</u>, 30, 937-943.

Wagner, G. & Boynton, R.M., 1972 Comparison of four methods of heterochromatic photometry, <u>J. Opt. Soc. Am.</u>, 62, 1508-1515.

Zrenner, E., 1983, <u>Neurophysiological Aspects of Color Vision in Primates</u>, Springer-Verlag, N.Y.

SENSORY AND PERCEPTUAL PROCESSES IN SEEING BRIGHTNESS AND
LIGHTNESS

Paul Whittle

Experimental Psychology
Cambridge University, UK

INTRODUCTION AND SUMMARY

This chapter is concerned with the distinction between brightness
and lightness and with their dependence on luminance and
contrast. The perception of luminance is often thought, with good
reason, a prime candidate for a satisfactory meeting of
psychophysics and neurophysiology, but there are certain
persistent problems that stand in the way of a lasting solution.
I try to clarify some of these.

I discuss uniform sharp-edged patches that are increments or
decrements on large steady backgrounds, rather than other contrast
effects such as Mach Bands. 'Brightness' means the apparent
amount of light reaching the eye from some region. 'Lightness'
means apparent reflectance, ranging from black to white. When it
would be pedantic to distinguish the two, I use 'brightness' as
the inclusive term.

The chapter has three parts.
(1) The simple story. If measured under certain conditions (an
important qualification), brightness fits neatly into current
knowledge of retinal function. It relates simply to what we know
about the psychophysics and physiology of increment threshold,
adaptation and contrast-coding. 'Simultaneous brightness
contrast' and 'light adaptation' are just two names for the same
set of phenomena.
(2) Complications. Not all measurements of brightness fit this
simple story. This is shown by quantitative comparisons of data
from different studies. I link the pattern of results with the
need for different types of lightness constancy in different
situations.
(3) Lightness and brightness. What is the difference? Are the
'complications' linked to it?

From Pigments to Perception, Edited by A. Valberg and
B.B. Lee, Plenum Press, New York, 1991

THE SIMPLE STORY: BRIGHTNESS, ADAPTATION AND CONTRAST

Brightness matching experiments under appropriate conditions show that brightness behaves like increment or decrement thresholds. Fig 1 shows data from Whittle and Challands, 1969. The lowest curve is the increment threshold, the higher ones are constant-brightness contours on which each point represents a match to a standard, constant for each curve. Brightness is constant if an expression roughly of the form $\Delta L/(L_b + L_o)$ is constant. (L and L_b are patch and surround luminances, $\Delta L = |L-L_b|$, and L_o is a constant.)

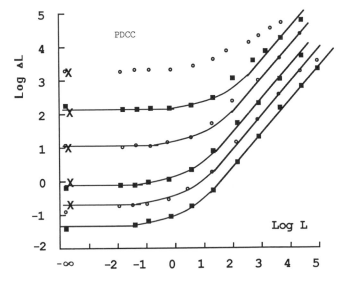

Fig. 1   The lowest set of points is the Increment threshold. The others are constant-brightness loci: brightness matches to the corresponding standard, constant for each set, shown by the X's. The curves, drawn to fit the threshold points, are all the same shape. Test stimuli were seen with the right eye, standards with the left. Symbols alternate only for clarity. Units are trolands.

In later experiments I found that for short flashes (40ms) seen by cones, brightness was determined by the most active Stiles π mechanism (Wyszecki and Stiles, 1982) and that for a given mechanism the constant-brightness curves were exactly the same shape as the increment threshold curve and vertically above it, for increments up to 100 times threshold intensity (Whittle, 1973). The same was true for rods. Data for decrements were different in some respects, but the overall pattern was similar.

These results were obtained with a haploscopic display (Fig 2) in which the patches to be matched were presented in surrounds that were binocularly superimposed and fused. Subjectively, matching is particularly easy in this set-up. A striking and theoretically important feature of brightness in this display, implicit in the results just described, is that increments are always brighter than decrements, whatever the four luminances involved. This result can be produced in other ways too, as for example in the work of Glad and Magnussen (1974), where the surround of the test patch was alternated, in square-wave fashion, between dark and bright.

I argue that these results reflect retinal processes because
(1) the increment threshold curve almost certainly does so and the constant brightness curves are similar.
(2) They depend on the (monocular) <u>luminance</u> of the surround not its <u>appearance</u>, which is obscured in the haploscopic display. This is in contrast to the type of assimilation studied by Shapley and Reid (1985), which does depend on the appearance of the surround (Shevell, Holliday and Whittle, 1988).

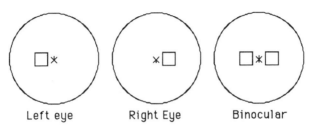

Left eye   Right Eye   Binocular

Fig. 2   Display for the brightness-matching experiments of Whittle et al 1969. Backgrounds are 11° in diameter, squares of side 1°.

These results fit well into a whole body of current evidence about retinal processes. They show that <u>in this situation</u> brightness depends on the extraction of luminance differences (ΔL), and the attenuation of the difference signal by a function of the absolute luminance. These are the subtractive and multiplicative components into which light-adaptation can be analysed (see Walraven et al, 1990). They are also the two components of the coding of physical contrast. Another way of putting it is to say that in this situation, 'simultaneous brightness contrast' and 'light adaptation' are just different names for the same set of phenomena. Such results give strong pointers towards a model of brightness perception.

However, this cannot be the whole story. First, even if these results were generally true, they provide only 'pointers towards a model'. There are large gaps, such as the notorious 'filling-in' problem. If brightness depends on contrast signals, and those are generated at edges, as the fading of stabilized images shows, how is it that we see uniform areas as uniform? Second, although this pattern of results is very appealing to physiologists and psychophysicists with certain preconceptions, it cannot be generally true.

COMPLICATIONS

A little reflection on how things look shows that brightness is not usually so extremely dependent on contrast as these results imply. If, for instance, I find a light stain (an increment) on the floor in a dark corner, and compare its brightness with that of a light grey smudge (a decrement) on a brightly lit white wall, I cannot convince myself that the former looks the brighter. Or, pick up an object and move it above a variegated background. Does it appear to fluctuate strikingly in brightness as it passes over objects which are sometimes brighter than it and sometimes darker? It may fluctuate a little, but nothing like as much as it should if all increments are brighter than all decrements.

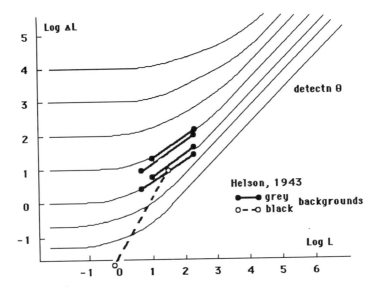

Fig. 3   A comparison of lightness matches reported by Helson (1943) with constant-brightness curves (continuous lines) from the family of Fig.1. There is good agreement when Helson's backgrounds were the same reflectance, and vision was photopic.

Further, we can take published quantitative results rather than anecdotes, and carefully compare the results I've described with those from other experiments. Figs 3 and 4 show two examples. There are plenty of comparable studies, all using brightness or lightness matching in similar centre-surround displays. Some were conceived as studies of 'simultaneous contrast' and some, of 'lightness constancy'. What we find is that my results are at one extreme: they show the biggest simultaneous contrast effect, the greatest influence of physical contrast on brightness. They are not alone. Heinemann (1955) got very similar results, particularly with himself as subject (Fig 7 of Whittle et al 1969); and Fig. 3 shows some data of Helson (1943) that agree well. At the other extreme are results like some from Jacobsen & Gilchrist (1988), in Fig.4, which show virtually no simultaneous contrast effect. There are many studies in between. From the studies that I have examined so far, an interesting pattern emerges.

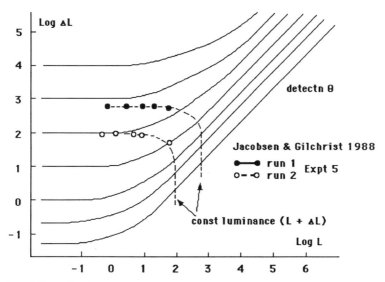

Fig. 4   Like Fig.3, but this time showing data that do not agree with the constant-brightness curves of Fig.1.

(a) <u>Brightness</u> matches sometimes agree with mine (e.g. Heinemann, 1955, subject EGH), but others deviate towards luminance-matching (Heinemann's other subject; Arend et al, 1987).

(b) <u>Lightness</u> matches between patches on <u>backgrounds of the same reflectance</u> but differently illuminated agree well with my data. (Helson, 1943; Jacobsen et al, 1988, for decrements; Arend & Goldstein, 1987). I shall call this 'contrast-matching'.

(c) <u>Lightness</u> matches when the <u>backgrounds have different reflectance</u> deviate towards (and sometimes reach) luminance-matching. (This is so whether the illumination is apparently constant (Gilchrist, Delman & Jacobsen, 1983), or not (Helson, 1943)).

I interpret this pattern of results as follows.

(a)   The instructions in studies of brightness contrast, and their usually impoverished stimulus displays, allow subjects to take different attitudes; for example, those corresponding to the two types of constancy described below. Results vary accordingly.

(b) is the traditional form of lightness constancy with respect to changes in overall illumination. The contrast at the edge of the object is invariant under illumination changes that affect surround as well as object, so if lightness is a function of contrast, it too would be invariant. This is a familiar argument (E.g. Shapley and Enroth-Cugell, 1984). Results in this type of situation agree with mine because my data exemplify this dependence of brightness on contrast.

(c) shows another form of lightness constancy, now with respect to changes in the background against which an object is seen, rather than with respect to changes in illumination. I gave anecdotal evidence for the existence of such constancy above (that an object moved over a variegated background does not change much in brightness), and the studies cited under (c) provide more formal evidence. If this kind of constancy is to be mediated by a system that codes contrast at its input stage, then the later stages must perform something like the edge integration that has been suggested in various contexts (e.g. Gilchrist et al, 1983; Land's retinex) to restore perception of luminance. This process may also be connected with 'assimilation' as described classically by von Bezold and Helson, or more recently by Shapley and Reid (1985). Assimilation is a phenomenon showing the reversal of 'simultaneous contrast': just what is needed here. Such an integrative process might also be involved in 'filling-in'.

It is likely that such integration can counteract contrast-coding only over a quite limited range. The contrast-coding Weber behaviour describes retinal function over 9 log units. But the counter-processes have only been shown to work within the range of reflectances in one scene, about 1.5 log units.

It is the high-level integrative processes that must be kept out of play if brightness is to follow the physiologically 'simple story'. I infer that they are 'high-level' because some depend on perceiving the global structure of the scene, and because in (a) they are quasi-voluntary. The failure to appreciate the importance of these different levels in brightness/lightness perception seems to me responsible for much continued confusion in the subject. Since lightness perception behaves in an ecologically sensible way, including the two constancies just described and also other abilities such as seeing transparency or illumination, or through a veil, such flexible processes must exist, and indeed the differences between the results of the studies I have cited provide experimental evidence for them.

My experiments escape the fate described in (a) because the display is even more impoverished. So much so that the edge integration necessary to (c) is prevented, because the edges of the backgrounds, which have to be involved in such integration, cannot be separately seen because of the binocular superimposition (Fig 2). A similar result can be produced by using stabilized backgrounds, so that their edges disappear altogether. The brightnesses of superimposed unstabilized increments and decrements then seem to behave as in my experiments (Gilchrist, pers. comm.; example in Yarbus, 1965, Fig 51).

In the last section I want to say more about this distinction. It is a long-standing problem in visual science. I find it difficult to stay clear about it. I think this is partly because it straddles two cognitive frameworks: a classical psychophysical one of measuring sensations and a contemporary one of ecological optics and machine vision.

The current consensus definitions of the terms in visual science are, as I said, to use 'brightness' to mean the subjective intensity of light and 'lightness' to mean the apparent reflectance of surfaces. The CIE definitions are along these lines, though more cumbersome.[1]

The paradigmatic example showing the need for the distinction is this. Imagine two pieces of white paper in the same room. One is in a shadowed corner, and one in sunlight coming through the window. Suppose you can see clearly that both are white paper; that is, that you would mistake neither for grey paper. Therefore, they have the same <u>lightness</u>, the maximum possible. But the one in sunlight is enormously <u>brighter</u> than the one in shadow. You can argue about details, but that observation in some form or other is indubitable. It shows without any doubt that we must distinguish lightness and brightness, and that the difference is large, not a matter of splitting hairs.

If we think of vision as seeing the world, there seems at first nothing puzzling about this. It just reminds us that we can see both objects and their lighting: reflectance and illumination. In any natural scene we can ask subjects about both: 'Are they cut from the same grey paper?', 'Are they equally brightly lit?'. (Arend and Goldstein did something like this. Previous work has done it surprisingly rarely, though with hindsight we can see that some experimenters were asking the first question and some inclined towards the second.)

In fact, it raises two kinds of serious problem: first, scientific, second, conceptual. The scientific problem is well-

---

[1]Wyszecki and Stiles (1982, after the CIE): 'Brightness is the attribute of a visual sensation according to which a given visual stimulus appears to be more or less intense; or, according to which the area in which the visual stimulus is presented appears to emit more or less light. Variations in brightness range from "bright" to "dim".'
     'Lightness is the attribute of a visual sensation according to which the area in which the visual stimulus is presented appears to emit more or less light in proportion to that emitted by a similarly illuminated area perceived as a "white" stimulus. In a sense, lightness may be referred to as relative brightness. Variations in lightness range from "light" to "dark".'
     Note that these definitions are in terms of 'sensation' and do not refer to objects, surfaces or illumination at all.

known. How is it done? Reflectance and illumination are
confounded in the retinal image. How do we manage to perceive
them separately? What I've been saying makes some suggestions
towards this ('we must distinguish different types of lightness
constancy, one may be mediated by retinal contrast-coding, etc'),
but we all know it's a difficult problem into which much effort is
currently being put See, for example, the shape-from-shading work
collected in Horn and Brookes, 1989.

The conceptual problems are less well-known, but they are what
make this distinction a perpetual irritant. First, brightness is
not just seeing illumination. It is closer to seeing luminance:
illumination times reflectance. White paper in sunlight looks
lighter than black paper in the same light, but also brighter.
Further, 'brightness' is also the term we apply to light sources.
Indeed, the distinction is sometimes taken to refer primarily to
different types of object (reflecting versus self-luminous) rather
than to co-existing aspects of the same object (reflectance versus
illumination). A simple example occurs in my experiments. There,
'brightness' seems the right word for the increments, and
'lightness' (or darkness) for the decrements, which look like
pieces of grey paper. Thus in my set-up brightness and lightness
are two non-overlapping parts of a single signed continuum;
brightness the positive part and lightness the negative (though
lightness has to increase towards the origin, which is zero
contrast, or invisibility). The object quality is produced simply
by the luminance ratio.

Now of course seeing luminance and seeing light sources are both
seeing 'light intensity', and they are thus brought together in
the accepted definition of brightness. We do seem to have that
ability, to a certain degree. It is shown also in what Katz
called the 'subjective attitude', in which we can see the world as
a mosaic of colours and brightnesses rather than objects and
lighting. These facts, together with metaphysical assumptions,
which may have been more influential, have led to the idea that
each area of our visual field has a value that we call
'brightness' associated with it. This is implicit in the CIE
definition, and is taken for granted in much current work.[2] It is
a familiar type of metaphor in thought about perception, in which
an 'inner world' is modelled on an outer; in this case on the
luminance distribution over the retina.

However, this seemingly innocent idea leads us perilously away
from conceptually manageable engineering problems back towards
'sensations' (cf the definitions in footnote 1). This is an
unsatisfactory hybrid concept, compounding Aristotle's
straightforward notion of sensations as bodily feelings, such as
pains and tickles, with metaphysical entities like 'redness' which
are so problematic that even cognitive scientists' favourite
philosophers (such as Daniel Dennett, 1988) can still argue that
they don't exist at all. Others have made powerful critiques of

---

[2]Eg attempts to model brightness perception by convolving the
retinal luminance distribution with suitable receptive field
profiles. These are a necessary component in explanations of
brightness, but will not, if my arguments are right, yield the
whole story.

the metaphor of a 'mental world'.[3] I want to join these critics, but with a specifically visual-science argument.

I argue that if such a quantity exists, it is so badly behaved mathematically as to be virtually useless. Its bad behaviour consists in the relation 'brighter than' being radically intransitive. You can set up light patches a, b and c as in Fig.5 such that a is brighter than b, b is much brighter than c, but c is brighter than a. You have to choose them as in the figure. Heinemann's data, like mine, show that brightness follows contrast. It follows that you can choose luminances a and c such that the low-contrast increment a looks less bright than the higher contrast increment c.[4]

One way to understand this is to reflect that in judging brightness the indubitable psychological reality is the act of judging, rather than brightness. Further, that these three judgements ('a>b' etc) impose different psychological sets, different perceptual (figure-ground) organisations on the stimuli, and brightness judgements depend on such organisations. Of course they do. It's the main lesson of Gestalt psychology: the act of perceiving determines a perceptual organisation, and the perception is a function of that.

If you doubt the importance of this in the context of brightness and lightness, I recommend setting up a demonstration that I owe to Alan Gilchrist. Find a black-and-white slide with some white print on a completely black background. Gilchrist uses one bearing the question 'Is the background black or white?'. Project it onto a white screen in a room with sufficient ambient light to see clearly that the screen is white, and ask yourself Gilchrist's

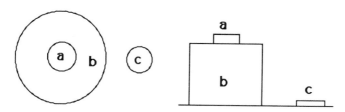

Fig.5 Display to demonstrate the intransitivity of brightness. The figure on the right shows the luminance 'cross-section'.

---

[3]Such as Morgan, 1979, arguing against the idea of a subjective space with an associated geometry. He specifically opposed this to the case of colour, but I think that many of his arguments apply to that too.

[4]Less bright to Heinemann or to me. It is not so easy to get into the right 'frame of mind', though the results of Wallach (1948) and Heinemann show that it can be achieved. If you set it up you may have to be content with the observation that a is only a little brighter than c, and somewhat ambiguously so, even though the other differences are large and clear.

question. I hope you will agree that it shows that perceptual organisation affects lightness judgements, and here can shift you right from one end of the scale to the other. If you also ask yourself whether you are seeing brightnesses or lightnesses, for the various areas concerned, you will see what I mean by saying that the distinction is tricky.[5]

I'm arguing that the notion of a 'brightness value' which is task-independent, and which judgements are 'of', is mainly theoretical prejudice. If you add the philosophical arguments to those I've produced, the notion is clearly in trouble. It's not totally useless. I've pointed out that we can see 'light intensity' to some degree. And you can take a robust engineer's attitude, and think of it as approximate read-out from low levels of the system. In perception we _can_ always attend to different levels. In speech, we can attend to meaning or to voice quality. In vision, to objects or to the quality of light. Maybe one can partially rehabilitate something like 'sensation'. But it's going to be very context-dependent, that is, task-dependent. If you use it, you had better be careful. It can get you into trouble.

Another way of putting it, is to say that talk about 'sensations' in the generalised sense, introduces a type of description that is unnecessary and a positive hindrance. We need to talk about the world, experience, behaviour and neural processes. But experience is best described in terms of the objective world. If I want to explain to a subject what I mean by judging brightness, I say something like 'Make the two patches look like windows into the same lit room'. It is in such terms that we make descriptions of our experience more precise. Further, such language links in a natural way the domains of experience, behaviour, ecological optics, computer vision and physiology. Whereas talk about 'sensations' impedes such links, and doesn't even have the vocabulary to yield good descriptions of experience.

The brightness-lightness distinction and the results of matching experiments

In the previous section I classified the results of brightness- and lightness-matching experiments into contrast-matching and luminance-matching. That was a distinction between different _mathematical descriptions_ of the data. The brightness-lightness distinction is orthogonal to it. It has nothing to say to it. You can have 'contrast-matches' of lightness (classical constancy under changing illumination) or of brightness (judgement of the darkness of shadows). You can have 'luminance-matches' of lightness (varying background) or of brightness (a window onto a lit scene, in a wall on which the illumination is changing). So all four combinations can occur. This is true in the experimental data, though I illustrate it here in terms of the ecologically relevant situations.

---

[5]Projected slides, television screens etc, can also introduce a whole extra dimension of complexity, because we so compellingly see them as representing objects of a different physical type. The brightness-lightness distinction is between _psychological_ object types, not physical ones.

CONCLUSIONS

1 The simple story

If measured under certain conditions, brightness behaves like increment threshold. It fits into current knowledge of adaptation and contrast-coding, that is, of retinal function. 'Simultaneous brightness contrast' and 'light adaptation' become two names for the same set of phenomena.

2 Complications

Not all measurements of brightness fit this simple story. The results of brightness- and lightness-matching experiments lie between two extremes: 'contrast-matching' (the simple story), and 'luminance-matching'. These behaviours can be understood in terms of different tasks facing the visual system: lightness constancy under changing illumination or with changing background (and many other tasks). If the retina codes contrast, luminance-matching will require something like integration.

3 Brightness and Lightness

We must distinguish between these two. Lightness is seeing reflectance. Brightness can be described in a shorthand way as seeing light intensity, and is our route to judging sources and illumination. This distinction is orthogonal to the distinction in Section 2 between the results of matching experiments.

Beware of the notion of brightness as a 'sensation' or of the idea that we have a visual field of brightnesses independent of how we judge them. These are only dubious metaphors, and they break down. We would be better advised to restrict ourselves to thinking of vision as seeing objects and light.

REFERENCES

Arend, L., and Goldstein, R., 1987, Simultaneous constancy, lightness and brightness, J. Opt. Soc. Am., A4:2281.

Dennett, D., 1988, Quining qualia, in "Consciousness in Contemporary Science", A.J. Marcel and E. Bisiach, eds, Oxford.

Gilchrist, A., Delman, S., and Jacobsen, A., 1983, The classification and integration of edges as critical to the perception of reflectance and illumination, Percept. Psychophys., 33:425.

Glad, A., and Magnussen, S., 1974, Simultaneous Contrast: Effect of Temporal Modulation of the Inducing Field, Physica Norvegica, 7: 211.

Heinemann, E.G., 1955, Simultaneous brightness induction as a function of inducing and test-field luminances, J. Exp. Psychol. 50:89.

Helson, H., 1943, Some factors and implications of color constancy, J. Opt. Soc. Am., 33:555.

Horn, B.K.P., and Brooks, M.J., 1989, "Shape from Shading", MIT, Cambridge, Mass., 1989.

Jacobsen, A., and Gilchrist, A., 1988, Hess and Pretori revisited: resolution of some old contradictions, *Percept. Psychophys.*, 43:7.

Morgan, M.J., 1979, The two spaces, *in* "Philosophical Problems in Psychology", N. Bolton, ed., Methuen, London.

Shapley, R., and Enroth-Cugell, C., 1984, Visual adaptation and retinal gain controls, *Prog. Retinal Res.*, 3:263.

Shapley, R., and Reid, R.C., 1985, Contrast and assimilation in the perception of brightness, *Proc. Natnl. Acad. Sci. USA*, 82:5983.

Shevell, S.K., Holliday, I., and Whittle, P., 1988, paper to November *Opt. Soc. Am. Meeting*.

Walraven, J., Enroth-Cugell, C., Hood, D.C., MacLeod, D.I.A., and Schnapf, J.L., 1990, The control of visual sensitivity, *in* "Visual Perception: Neurophysiological Foundations," L. Spillman and J.S. Werner, eds, Academic Press, San Diego.

Wallach, H., 1948, Brightness constancy and the nature of achromatic colors. *J. Exp. Psychol.* 38:310.

Whittle, P., and Challands, P.D.C., 1969, The effect of background luminance on the brightness of flashes, *Vision Res.*, 9:1095.

Whittle, P., 1973, The brightness of coloured flashes on backgrounds of various colours and luminances. *Vision Res.*, 13:621.

Wyszecki, G., and Stiles, W.S., "Color Science," Wiley, New York, 2nd Edtn 1982.

Yarbus, A.L., 1965, "Eye Movements and Vision," Eng. transltn, Plenum, New York, 1967.

ASSIMILATION versus CONTRAST

Charles M.M. de Weert

NICI
University of Nijmegen
The Netherlands

By stating the subject in this way, I have more or less taken up the position of the devil's advocate. Amidst these physiologically (exclusively bottom-up) oriented papers, I will defend the viewpoint that many perceptual phenomena cannot satisfactorily be explained by such bottom-up processes. The first statement I would make is directed towards the possibly misleading role of physiology, when simple and successfully applicable mechanisms have been found. An example is lateral inhibition. There is hardly a mechanism which has more strongly influenced developments in perceptual theories: contrast, Mach-bands, spatial contrast transfer functions, Marr's theory of zero-crossings and so forth. There is no doubt that contrast is a very important phenomenon, but is it so important that other phenomena, which seem of a contradictory nature, must be ignored or neglected?

If contemporary textbooks on visual perception are reviewed as to their treatment of assimilation (or similitude) effects, contrast effects, or both, it turns out that much less than half mention assimilation, whereas all deal with contrast phenomena rather extensively. Is it because the phenomenon does not occur? That cannot be the reason, because presumably assimilation occurs in daily life at least as frequently as contrast. In their book on Spatial Vision, R. and K. DeValois (1988) point out that assimilation could quite well be a more common phenomenon than contrast. A few examples* are presented here which are very well known. One is the difference in perceived colour of bricks in a wall with or without mortar in the joints, another is (in the colour example shown at the meeting) the strongly different appearance of a (simulated) coloured carpet, as a function of the colour of the small frame surrounding the individual dots. A black/white version is shown in Fig. 1. This makes clear how influential assimilation can be, and that to simulate a design (for example, the one for a carpet) on a screen in a veridical way, one has to take into account that shadows, and/or the colour of the matrix has its influence on the total percept.

It is very easy to give examples which can be made oneself with paper and pencil. Any piece of text gives a good specimen of assimilation. The space in-between the text lines looks darker than the space outside the printed area; see also Fig. 1 for other

---

* Note that the examples presented at the workshop were nearly all in colour. Here, only achromatic pictures are shown, and the filling in of the colours will be left to the reader where necessary.

From Pigments to Perception, Edited by A. Valberg and
B.B. Lee, Plenum Press, New York, 1991

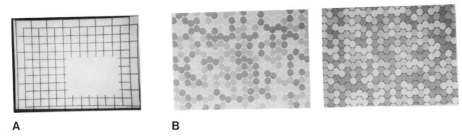

A   B

Fig. 1. Examples of assimilation effects. In A, the darkness of the lines spreads (homogeneously) over the background. In B, the patterns in the left and right part are identical in form and the luminances of the circles are equal in the two parts. Only the small areas between the circles differ in luminance.

examples. In the illustrations in Fig. 2, a number of particularly strong artifactual examples are presented in which assimilation effects occur. One should attend to the robustness of the effect. A change in spatial frequency, either by changing the viewing distance or by changing the width of the bands in the picture, does not really disturb the assimilation impression. The number of bands is of importance, as is the width of the bands, but even with large variation in these parameters the effect of assimilation is very conspicuously present. In a further study the effect of a number of spatial variations is investigated (De Weert, Troost and Van Kruysbergen, in preparation). If the black and the white lines are replaced by red and green lines respectively, and the gray background by a yellow, a similar effect occurs; the yellow surrounded by the red lines appears reddish, and the yellow within the green lines appears greenish, also when red, green and yellow are about equiluminous. Much more could be said about this particular assimilation picture, but it is more important to emphasize here that assimilation is not just a negligible side effect, but instead a very common effect in normal images. Later in this paper simpler versions of the assimilation figures will be presented.

The preceding demonstrations lead to the following conclusion. The mechanism of lateral inhibition is such a convenient mechanism and fits so well with contrast phenomena that acceptance of a second, perhaps simpler, mechanism (integration) only complicates our view of the system.

One objection against this point of view is that that the phenomenon of assimilation can perhaps be explained starting from only the lateral inhibition mechanism. Until now, I have not encountered such a model which explains the strong assimilation effects in Fig. 2 to any satisfactory extent.

Although the mechanism of lateral inhibition seems to lead to a direct relationship with simultaneous contrast, it is possible to find examples which point to the contrary. Kanizsa (1979), and Burnham, Hanes and Bartleson (1963) give examples of Benary's cross and of the Maltezer cross (a combination of a yellow and a blue cross with a gray central area) respectively. The colour impression of the central gray area is different for the perception of a blue cross on a yellow background than for the perception of a yellow cross on a blue background). In the Benary Cross physically identical areas in different positions in a stimulus have different appearances, depending upon their function as figure or as ground. Helmholtz already described the observation that the occurrence of colour contrast effects was dependent upon the 'meaning' of a scene, and the related habit of some painters to turn their head or turn the scene in order to enhance

306

Fig. 2. Assimilation effects for varying spatial arrangements in a single type of pattern organization.

perceived colour contrasts in it. I would like to emphasize the idea that figure-ground distinctions may play a role in the occurrence of contrast effects, which is not easily reconciled with a pure bottom-up process. If the occurrence of contrast and of assimilation is dependent on segmentation and on figure-ground distinctions, there is no *a priori* reason to put assimilation in another domain than contrast. This might be important, because assimilation has often been ascribed to more special, possibly cognitively influenced conditions. If we say that contrast is related to the existence of receptive fields with some center-surround structure which can mediate spatial differentiation, and assimilation is in some way related to a mechanism of integration (perhaps receptive fields without center-surround interaction), a general problem becomes apparent. This leads to a second statement: one of the problems of finding feature detectors in physiological systems is that one also has to find out how responses to all other features, which might be evoked by similar types of stimulus, are silenced or not 'read'. To be more explicit, a small circular stimulus not only activates a cell with a receptive field with a center just matching the size of the spot (perhaps signalling contrast), but also cells with a receptive field without a center-surround structure and with (much) larger receptive areas (perhaps involved with integration). How are these cells silenced? Are they silenced? Is this necessary at all, or are these data simply not 'read' at a more central level?

ASSIMILATION AND CONTRAST IN SUBJECTIVE CONTOURS, THE NEON-EFFECT AND TRANSPARENCY

There is clearly a strong relation between contrast and assimilation effects on the one hand and the formation of subjective contours on the other. I would like to include these effects in this discussion, together with the combined effects of assimilation, contrast and subjective contours in transparency effects, because they may help us to arrive at a less vague concept of what can be meant by the 'central' processes involved.

Are subjective contours a consequence of higher order processes or are they simple and necessary consequences of the existence of certain feature detectors (endpoint or end-stopping detectors) as some physiologists maintain? I would propose that this is not a fair question. There are no cognitive effects without the necessary sensory data, and the fact that these effects occur so predictably for all observers suggests a rather simple

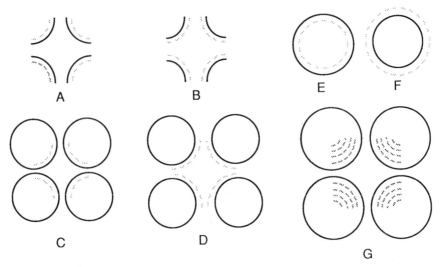

Fig. 3. Simplified versions of the complex patterns depicted in Fig. 2. Use a colour-felt tip pen to make a continuous, coloured line of the dotted lines in the figures to see the colour spreading effects. See text for explanation of 3A to 3F. 3G is a stronger form of 3C.

and reproducible basis. The discussion in terms of simple physiology versus higher order psychology seems to be a form of escapism. The approach of Von der Heydt (1987) looks very promising. He sought for neurophysiological localization of the detection of subjective contours and combined useful concepts from physiology (the existence of endpoint detectors) with the possible significance of such detectors, i.e. detectors of changes in illumination or changes in depth. The important point he makes is that some form of 'knowledge' influences the process. No pure bottom-up process is adequate, or at least decision rules are built in. This 'knowledge' is not at all of a high level. In fact, the 'knowledge' corresponds to the type of rules presently used in computational vision. The following citation is from Von der Heydt (1987): "Thus a neuronal mechanism derived from single unit experiments with anomalous contours can be interpreted as the computational solution to the problem of finding contours in images in three dimensions without *a priori* knowledge about the objects in the scene, a problem that can be solved only by exploiting statistical properties of occluding contours".

As with all perception theories it is very unlikely that all phenomena can be explained at only one level. In the simple series of drawings in Fig. 3, essentially the same effects are obtained as in the stimuli of Fig. 2. Some personal participation using a coloured felt-tip pen will provide you with coloured stimuli (perhaps best made on photocopies of Fig. 3). If you draw with, for example, a red pen, a line over the dotted lines you will probably notice rather strong differences between the different spatial configurations. Fig. 3A and 3B show that the colour of the innermost lines is spread over the central area. The colour of the outer lines, however, is not spread over this area or at least to a much smaller extent. It seems as if the inner lines act as barriers for the inward spreading of the colour of the outer lines; the reverse probably also occurs. In Fig. 3C the barrier function of the inner lines, which are now full circles, is much less apparent. In fact, the difference in the colour impressions at the centers of Figs. 3C and 3D is less than that between Figs. 3A and 3B. Some observers even mention that in 3A a colour contrast effect occurs in the center, rather than a reddish assimilation colour. The interesting aspect of these differences for the previous discussion of possible figure-ground influences is the speculation that it is as if the closed lines in Figs. 3C and 3D are analysed in another image plane than the 'endpoint' lines, as if they belong to different objects, whereas all the endpoint lines in Fig. 3A and 3B are dealt with as belonging to the same object. In the further examples E and F it is shown that barrier functions as described for A and B are also visible in a closed figure.

I would not like to leave this discussion without stating that the examples make clear that both assimilation and contrast effects occur simultaneously, and that the weights of the two effects are not exclusively determined by the relative sizes of the center-surround and nonopponent receptive fields at that particular place in the visual field.

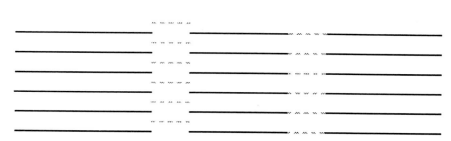

Fig. 4. The Neon-Colour spreading effect. After colouring the dotted lines with a coloured pen, a difference can be seen in the amount of colour spreading in the left (connected) and right (unconnected) lines. Furthermore, the perceived sharpness of the lines will differ. See text.

THE NEON-EFFECT: A COLOUR SPREADING EFFECT RELATED TO THE BEZOLD EFFECT

The neon colour spreading effect (Van Tuijl, 1975) is related to the Bezold spreading effect, which in turn is an example of assimilation. It is interesting to see in the illustration of Fig. 4, in which one may again colour the dotted lines with a red-felt tip pen, how strongly with the different appearances of the same lines in this picture, you see all the effects mentioned earlier: assimilation, subjective contours, the occurrence of the so-called neon effect and the occurrence of transparency. Note how the perceived sharpness of the lines differs in the two patterns. It seems as if the amount of spreading, expressing the involvement of large receptive-field cells, seems to have grown at the cost of the activity of the cells with more sharply tuned receptive fields. How is this relative weight determined, if this is a reasonable type of explanation? Physiologically based accounts of this effect are lacking. Explanations based on endpoint detectors do not take into account that what constitutes an endpoint is context dependent. It could be that the strong difference in effect between abruptly ending lines and lines which are interrupted only by a change in light (a shadow, for example) indicates a different significance for the visual (perceptual) system and that as a consequence a different selection is made of the available data. This indicates a difficulty in defining what is meant by an endpoint. Changes in line thickness are endpoints, as are changes in noise density or changes in regularity. In other words, the character of these endpoints is context dependent.

Fig. 5 A Simple example of transparency. A change in the luminance of the centralmost part of the picture leads to a big difference in perceptual organization. In the left picture the central rectangle is part of a larger rectangle, in the right picture it acquires an object character of its own.

ASSIMILATION, CONTRAST, SUBJECTIVE CONTOURS AND TRANSPARENCY

The study of transparency effects will be dealt with very shortly here (De Weert and Van Kruysbergen, 1987), De Weert, 1984). We have essentially defended the view that physical rules underlie perceptual effects; what we mean by 'knowledge 'is always implicit knowledge of physical rules. An excellent example is the occurrence of transparency. We know from Metelli's elegant studies the conditions under which transparency occurs. The pictures in Fig. 5 show how slight differences in the luminance relations can cause the small central square to be seen as a separate opaque object, or as part of a larger, transparent object. The rules are, in our view, physical rules. When stimulus conditions fulfil certain physical requirements, transparency is perceived. The reverse is not true; not every perceived transparency corresponds to physical reality, for the simple reason that physical conditions can be generated by experimentation in displays or drawings, which do not occur in nature. In Fig. 6 a number of illustrations

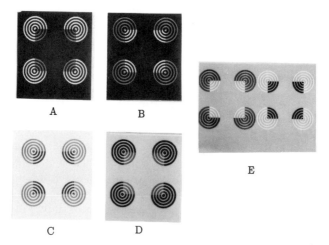

Fig. 6. Example of the combined effects of subjective contours and transparency in the pictures A-D and an example of the absence of both transparency and subjective contours in E. In E, assimilation still occurs (see De Weert and Van Kruijsbergen (1987)).

demonstrate how the effect of transparency is only apparent in those pictures where the composing parts have similar contrasts with respect to the background. In fact the conditions for the occurrence of transparency are similar to those for the occurrence of neon-colour spreading (Van Tuijl and de Weert, 1979). The interesting feature of Fig. 6 is that it also shows that the same conditions seem to be necessary for the occurrence of subjective contours. Especially in those figures where reversed contrasts are present between the lines, the impression of subjective contours is absent, and so is the impression of transparency. This also throws some doubt on the validity of the explanation of subjective contours on the basis of the existence of endpoint detectors alone. Sensory data are available in these cases (activity of center surround receptive fields, contrast, and of nonopponent types, assimilation, but that it depends on local computational rules how these 'data' are to be used.

REFERENCES

Burnham, R. W., Hanes, R. M. and Bartleson, C., 1963, "Color:a guide to basic facts and concepts," pp. 84-110, Wiley, New York.

DeValois, R. and DeValois, K., 1988, "Spatial Vision," Oxford University Press, Oxford.

Heydt, R. van der, 1990, Processing of contour and form in monkey visual cortex. Preprint.

Kanizsa, G., 1979, "Organization in Vision" p. 171, Praeger, New York.

Tuijl, H. F. van, 1975, A new visual illusion: neon-like color spreading and complementary color induction between subjective contours. *Acta Psychologica,* **39**:441-445.

Tuijl, H. F. van and Weert, Ch. M. de, 1979), Sensory conditions for the occurrence of the neon spreading illusion. *Perception,* **8**:211-215.

Weert, Ch. M. de, 1984, Veridical perception; a key to the choice of colours and brightnesses in multicolour displays, in "Monochrome vs Colour in Electronic Displays," P. Gibson, ed., RAE, Farnborough.

Weert, Ch. M. de and Kruysbergen, N. van, 1987, Subjective contour strength and perceptual superimposition: transparency as a special case, in "The perception of illusory contours," S. Petry and G. Meyer, eds., Springer, New York.

ON ACHROMATIC COLORS

Paul Heggelund

Institute of Neurophysiology
University of Oslo
Oslo, Norway

INTRODUCTION

Achromatic colors are qualities like white, grey, black, and the luminous qualities seen in stars and in lamps emitting "white" light. Although they are the least spectacular among the color impressions, the achromatic colors are probably the most interesting ones from a physiological point of view. They occur in scotopic as well as in photopic vision, and an achromatic aspect is involved in all kinds of chromatic colors of both normal and defective color vision. In fact they are involved in all kinds of visual perceptions.

The most common view concerning the structure of the achromatic colors is that they are unidimensional, being specifiable by a single variable usually called brightness. Brightness is considered to be the third perceptive color variable beside hue and saturation. The meaning of the term brightness has, however, often been ambiguous, and this ambiguity demonstrates that more than one perceptive variable are needed to specify the achromatic colors. Sometimes brightness is used for the variation along the grey scale between black and white, and sometimes it is used to specify the variation of color intensity between dim, just noticeable colors, and strong and dazzling ones. To avoid this ambiguity the term lightness is usually used to specify the variation along the grey scale, and the term brightness is reserved for the intensity specification of the colors (Judd 1951). This shows that at least two different variables are needed to specify the achromatic colors.

In addition to these two kinds of achromatic color variation two types of colors called light and object colors are usually distinguished. The light colors are normally seen on light sources, and they are perceived as a property of the emitted light. The object colors are normally seen in reflecting surfaces, and they are perceived as a constant property of the object surfaces. The black-white variation occurs only in the object colors (Evans 1964). The dim-dazzling variation is usually associated with the light colors. Thus, the achromatic colors of a single type (light or object color) might be specifiable with one variable (Judd and Wyszecki 1963).

In contrast to this view, it was shown already by Hering (1875) that the object colors themselves are bidimensional. He showed that the different grey colors can occur in different strength or weight, as he called it. To illustrate this, consider for instance a given medium grey quality seen in low illumination, and the same grey seen in high illumination. Despite the matching grey quality the two colors are different. The one in lower illumination looks dimmer or weaker than the one in higher illumination, exemplifying the difference of color weight.

A few more multivariable systems for achromatic colors have been developed (Katz 1930, Evans 1964), but in none of them have the color variations of both the light and the object colors been incorporated into a single multidimensional structure. Heggelund (1974a) analyzed various types of achromatic color variations and showed that both the light and the object colors can be treated within a simple bidimensional system. The main results from that study are summarized below.

THE STRUCTURE OF ACHROMATIC COLORS

Many achromatic colors are contrast colors. The simplest situation for studying them is therefore a two-field configuration like a disc (test field) surrounded by a contiguous ring (inducing field). Heggelund (1974a) used this configuration to study the types of color variation that occur at different combinations of test and inducing luminance. To simplify the analysis, two pairs of fields were used as illustrated in the inset of Fig. 1A. In this way different colors in the two discs could be directly compared. The two pairs were presented to different eyes to minimize mutual influences. The disc presented to the left eye was termed the comparison field. The fields were seen in otherwise completely dark surroundings to simplify conditions as much as possible.

Four different kinds of color change were observed. They could all be seen when the luminance of the test field was fixed while the luminance of the inducing field was varied. Firstly, a change in the mode of appearance was observed. When the inducing luminance was much lower than the test luminance, the test field appeared in the light mode as a luminous-whitish object, like the color of a lamp bulb. This character was maintained as the inducing luminance was increased up to a value where the test and inducing luminances were about equal. When the inducing luminance was further increased the color changed to the object mode, and appeared as a light grey surface. Secondly, a change in luminous-whitish qualities between predominantly luminous to predominantly white colors was observed. This occurred when the inducing luminance was varied between a value far below the test luminance and the value of the fixed test luminance. At the lowest inducing luminance the color quality of the test field was predominantly luminous. As the inducing luminance increased, the test field became less and less luminous and more and more whitish. Close to the point where the test and inducing luminances were equal, the test field lost its luminous quality and acquired instead a very light grey color quality. This shift from luminous to grey occurred simultaneously with the shift from the light to the object mode. Near the point where the test and inducing luminances were equal, the test field was closest to a pure white, similar to the color of a white paper. Thirdly, the color could change through the grey series between white and black. This occurred when the inducing luminance was increased from about the same value as the test luminance to a value far above the test luminance. Fourthly, the color of the test field could change in intensity or strength (Hering's weight) between weak and just noticeable colors and very strong and dazzling colors.

All these changes can be accounted for within a simple 2-dimensional vector space (Heggelund 1974a) as illustrated in Fig. 1A. In the upper quadrant the light colors are represented. They are assumed to be completely specifiable by the strength of a luminous and a white component. In the lower quadrant the object colors are represented. They are assumed to be completely specifiable by the strength of a white and a black component, as originally demonstrated by Hering. The radial lines from the origin represent colors of the same achromatic quality, i.e.

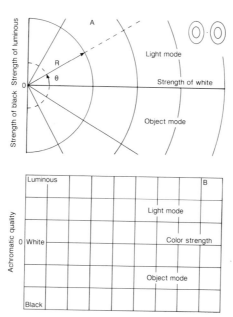

Fig. 1. Bidimensional system for representation of achromatic colors. **A:** Representation with reference to strength of white and strength of luminous/black. The radial lines represent constant achromatic quality (θ). The circles represent constant color strength (R). **B:** Representation with reference to θ and R. Horizontal lines represent constant achromatic quality, the vertical lines represent constant color strength.

the same ratio of strength of white and black. The circles around the origin represent colors of constant strength.

Fig. 1B shows an alternative representation of the system. Here the colors are specified by the two variables achromatic quality (vector angle, θ) and color strength (vector length, R). The relationship between these variables and the strength of luminous/black (B) and the strength of white (W) is defined by the equations:

$$W = R\cos\theta$$
(1)
$$B = R\sin\theta$$

The specification of the light colors by the two orthogonal variables luminous and white differs radically from the treatment in other color systems. The basic idea is that these colors can be characterized by a simultaneous similarity to a luminous and a white quality in the same sense that an orange color can be characterized by a simultanoeus similarity to red and yellow. When, for instance, the color quality of a lit lamp bulb is compared to the color of a white paper seen in good illumination it is evident that the two colors are similar in that both have a white aspect. Furthermore, both colors may have a luminous aspect. A main difference between the colors is that the white component is almost completely dominant in the paper whereas the luminous component is dominant in the light bulb.

It might be argued that the luminous quality is basically an intensity characteristic occurring at high luminances. This can, however, not be correct. The color qualities seen on stars against the dark night sky, e.g. are predominantly luminous with only a weak whitish aspect. Nevertheless such colors can be very weak, only just above the absolute thereshold.

White and black are treated as orthogonal rather than as opponent variables. Orthogonal variables are involved simultaneously in some colors as with red and blue in the violet colors. Opponent variables like red and green are mutual exclusive and are not simultaneously involved in a given color. Accordingly, white and black must be orthogonal since these variables are simultaneously involved in the grey qualities. Nevertheless, white and black are often treated as opponent variables (Hurvich and Jameson 1966). Hering was inconsistent in his treatment of these two variables. In his color system he treated them as orthogonal, whereas in his color theory they were treated as opponent (Hering 1875, 1920).

The three variables luminous, white and black can be treated in a two-dimensional system because luminous and black are regarded as opponent. A blackish or greyish aspect occurs only in the object colors and not in the light colors (Evans 1964). On the transition between light and object colors the luminous aspect disappears when the greyish aspect becomes apparent. Furthermore, colors which are simultaneously luminous and blackish may be just as rare or nonexistent as colors which are simultaneously reddish and greenish. Luminous and black seem therefore to be opponent variables, just as red and green or blue and yellow.

One important advantage of this color system is that both the light and the object colors are incorporated into a single system, and the perceptive relationship between the two types of color is specified. Since the work of Helmholtz (1867) and Hering (1874) it has generally been assumed that the types of color are determined by experiential factors. However, even if the differences between the perception of light and object colors are largely dependent on experience there must be perceptual cues that an individual can use to discriminate between them. Thus, the consistent presence of the luminous quality and lack of the black quality in the light colors, and the presence of the black and absence of the luminous quality in the object colors, could serve as a basic cue for learning when the color is a property of illumination, and when it is a constant property of reflecting surfaces.

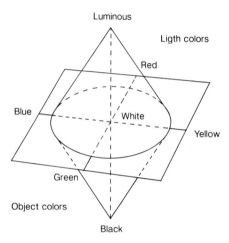

Fig. 2. Representation of the color qualities in a 4-dimensional color space under the condition of constant color strength. The shape of the color quality body is unknown, and the double cone was arbitrarily selected for the illustration.

The system might be extended to incorporate also the chromatic colors by adding the two opponent variables red/green and yellow/blue. This 4-dimensional system would comprise both the light and object colors of the chromatic colors. The system would also account for the different strengths of each chromatic color quality. Considering only the color qualities at a fixed color strength, the system can be represented in 3-dimensional space (Fig. 2). Here achromatic quality (vertical axis in Fig. 1B) is used as z-axis. The structure of such a color quality body is unknown. The double cone was arbitrarily chosen. The actual structure probably depends to a large extent on the selected value for the constant color strength.

The quantitative relationship of the achromatic color of a test field to the test and inducing luminance (configuration as in inset of Fig. 1A) was measured by Heggelund (1974b). The measurements were made by two sets of psychophysical matching experiments. In both experiments the observer adjusted the luminance of a comparison field until it matched the color of the test field with respect to a specific aspect. In the first set of experiments the color of the two fields were matched with respect to achromatic quality. In the second they were matched with respect to color strength. The results from these experiments were then transformed to values on equidistant perceptive scales for achromatic quality and color strength, respectively. This was done by psychophysical scaling experiments where the relationship between the comparison luminance and the two perceptive variables were determined.

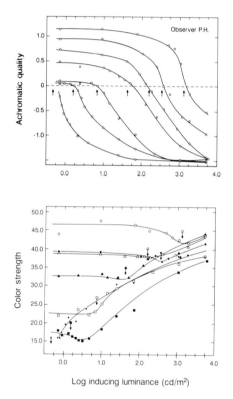

Fig. 3. Data from two sets of simultaneous contrast experiments that show how the achromatic quality and color strength varied with test and inducing luminance. For each curve the test luminance was fixed. The value of the inducing luminance that was equal to the test luminance is indicated by an arrow. The data are replotted from Heggelund (1974a).

The results for one observer are shown in Fig. 3. Each curve in the upper display shows the changes in achromatic quality for a fixed value of test luminance when the inducing luminance was varied. The inducing luminance that was equal to the fixed test luminance is indicated by an arrow. Notice that all the curves pass through the white point (achromatic quality equal to zero) close to the value where the test and inducing luminance were equal (zero contrast). This implies that the value of the luminous/black variable was small compared to the white variable at minimal contrasts.

The curves in the lower panel show the changes of color strength for the same set of test luminances. Notice that the color strength changed relatively little in the first part of the curve where the inducing luminance was below the test luminance, compared to the second part where inducing luminance increased above the test luminance. The color strength seemed to depend most heavily on the one of the two variables test and inducing luminance that was highest.

NEUROPHYSIOLOGICAL CORRELATES

Following the logic of opponent color theory, one might expect that the grey colors are produced by simultaneous ac-

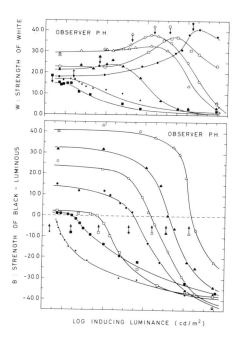

Fig. 4. The curves show how the strength of white (upper panel) and the strength of luminous/black (lower panel) varied with inducing luminance for a fixed value of test luminance. The curves were obtained by transformation of the data in Fig. 3 according to the Eq. (1).

tivity in a white- and a black-mediating process, and that the luminous colors are produced by simultaneous activity in a luminous- and a white-mediating process. Furthermore, the opponent relationship between luminous and black indicates that the processes mediating these two qualities are operating in a mutually antagonistic manner.

To determine basic properties of these processes, the results from the measurements of color changes in the test/inducing field configuration were transformed (Eq. 1) from achromatic quality and color strength coordinates to luminous-black and white coordinates, i.e. from the coordinate system seen in Fig. 1B to the one seen in Fig. 1A. The results for one observer are illustrated in Figs. 4. and 5. Each of the curves in the upper panel of Fig. 4 shows how the strength of white varied in a test field of constant luminance as the luminance of the contiguous inducing field was varied. The curves in the lower panel show the variation of luminous/black for the same set of test and inducing luminance values. Each curve in Fig. 5 shows how the strength of white (upper panel) and strength of luminous/black (lower panel) varied with the luminance of the test field for a fixed value of inducing luminance. The same set of data was used to generate the curves in Figs. 4 and 5.

The data show that the luminous/black and the white process have quite different properties. The luminous/black process was primarily related to contrast as indicated by several properties apparent from the curves in the lower panel of Figs. 4 and 5. Firstly, under all the conditions of fixed inducing lumi-

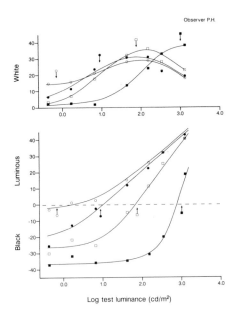

Fig. 5. The curves show how the strength of white (upper panel) and the strength of luminous/black (lower panel) varied with the test luminance for a fixed value of the inducing luminance.

nance (fig. 4) or fixed test luminance (Fig. 5) the zero value of the luminous/black variable occurred close to zero contrast, i.e. around the points where the test and inducing luminance were equal (marked by arrow). Secondly, the strength of luminous and the strength of black increased rapidly with increasing positive and negative contrast, respectively. At moderate to high test luminances (Fig. 4) or inducing luminances (Fig. 5) most of the change occurred within a range of one log unit around zero contrast. Thirdly, the change of luminous/black was most rapid around zero contrast, showing that this process had very high sensitivity for low contrasts.

The white process was not dependent on contrast in a sim-ilar manner. The strength of white was quite high at minimal contrast. Comparing the strength of white at zero contrast on the various curves, it appears that the value increased with increasing test luminance. Thus, the strength of white depended to a large extent on the local luminance value. Fig. 5 shows that the strength of white increased with test luminance over 2-3 log units. For the two curves with the highest inducing luminances the maximum strength of white occurred when the test luminance was slightly higher than the inducing luminance. For the two curves with the lowest inducing luminances the maximum strength occurred at test luminances 1-2 log units above the inducing luminance.

These differences indicate that the two processes are me-diated by two different kinds of neurophysiological mechanisms. The strong dependence of luminous/black on contrast indicates that this property is mediated by cells with strong center/sur-round antagonism of their receptive fields. The luminous qu-ality could be mediated by on-center cells, and the black quality by the off-center cells. This would extend upon the suggestion that the two cell types mediate brightness and darkness, re-spectively (Barlow 1957, Baumgartner 1961, Jung 1961). The fact that the white process depended to a large extent on local lumi-nance could indicate that this process is mediated by cells without a strong center/surround antagonism, e.g. by cells like the Type II cells of the lateral geniculate nucleus (Wiesel and Hubel 1966) which lack an antagonistic surround of the recep-tive field.

Fig. 4 shows, however, that the strength of white is de-pendent also on the inducing luminance. It is interesting to notice that there seemed to be a consistent relationship be-tween strength of white and the strength of luminous/black during changes of inducing luminance (cf. Fig. 4). Under con-ditions where the strength of luminous/black increased, the strength of white decreased. Furthermore, the magnitude of the changes of luminous/black were reflected in the magnitude of the changes in the strength of white. For example, at low test luminances where there was only little increase in strength of luminous as the inducing luminance decreased below zero con-trast, there was also little decrease in the strength of white. On the other hand, at high test luminances when there was a marked increase in the strength of luminous with decreasing in-ducing luminance, there was also a more marked decrease in the strength of white. This could indicate that the physiological mechanisms involved in the two processes are not independent. One possible explanation is that the mechanisms involved in the luminous/black process are inhibiting the mechanisms involved in the white process. Accordingly, it might be suggested that the luminous/black process is mediated by cells with center/surround organization of the receptive fields, while the white process is mediated by cells lacking center/surround organization. The contrast influence on the white process is assumed to depend on suppression from the mechanisms involved in the luminous/black process.

REFERENCES

Barlow H.B., Fitzhugh R. and Kuffler S.W.,1957, Change of organization in the receptive fields of the cat's retina during dark adaptation. J. Physiol. (Lond.), **137**: 338-354.

Baumgartner G.,1961, Kontrastlichteffekte an retinalen Gangliezellen: Ableitungen vom Tractus Opticus der Katze. In: Neurophysiologie und Psychophysik des visuellen Systems. (Eds. Jung R. and Kornhuber H.), pp. 45-53. Springer, Berlin.

Evans R.M.,1964, Variable of perceived color. J. opt Soc. Am., **54**: 1467-1474.

Heggelund P.,1974a, Achromatic color vision. I: Perceptive variables of achromatic colors. Vision Res. **14**: 1071-1079.

Heggelund P.,1974b, Achromatic color vision. II: Measurement of simultaneous achromatic contrast within a bidimensional system. Vision Res. **14**: 1081-1088.

Helmholtz H. v.,1867, Handbuch der physiologischen Optik. Voss, Leipzig.

Hering E.,1874, Zur Lehre vom Lichtsinne. IV. Über die sogenannte Intensität der Lichtempfindung und über die empfindung des Schwarzen. Sber. Akad. Wiss. Wien, Math. nat. Kl. Abth. III, **69**: 85-104.

Hering E.,1875, Zur Lehre vom Lichtsinne. VI. Grundzüge einer Theorie des Farbensinnes. Sber. Akad. Wiss. Wien, Math. nat. Kl. Abth. III **70**: 169-204.

Hering E.,1920, Outlines of a Theory of the Light Sense. (Translated by Hurvich L.M. and Jameson D.). Harvard University Press, Cambridge, Mass. (1964).

Hurvick L.M. and Jameson D.,1966, Theorie der Farbwahrnehmung. In: Handbuch der Psychologie. (Eds. Metzger W. and Erke H.), **Vol I/1**: pp. 131-160. Verlag fur Psychologie, Göttingen.

Judd D.B.,1951, Basic correlates to the visual stimulus. In: Handbook of Experimental Psychology. Ed. S.S. Stevens, John Wiley, New York, pp. 811-867.

Judd D.B. and Wyszecki G.W.,1963, Color in Buisness, Science, and Industry. 2nd. Ed. John Wiley and Sons, New York.

Jung R.,1961, Neuronal integration in the visual cortex and its significance for visual information. In: Sensory Communication. (Ed. Rosenblith W.A.), pp. 627-674. M.I.T. Press, Cambridge, Mass.

Katz D.,1930, Der Aufbau der Farbwelt. 2nd. Ed. J.A. Barth, Leipzig.

Wiesel T.N. and Hubel D.H.,1966, Spatial and chromatic interactions in the lateral geniculate body of the rhesus monkey. J. Neurophysiol., **29:** 1115-1156.

COLOR OPPONENCY FROM EYE TO BRAIN

Steven K. Shevell and Richard A. Humanski

Departments of Psychology and Ophthalmology and
Visual Science, University of Chicago
5848 S. University Avenue, Chicago, Illinois 60637, U.S.A.

INTRODUCTION

The opponent colors theory of Ewald Hering (1878; 1920) has proved to be a remarkable insight into the neural signals of many visual pathways. Physiological and psychophysical evidence for opponency has accumulated rapidly since Svaetichen and MacNichol (1958) reported spectrally selective S-potentials recorded from the retina of fishes. The concept of opponency is so well accepted now that one easily forgets the controversy and skepticism surrounding it in the first part of the century (Jacobs, 1986; Sperling, 1986).

The psychophysical evidence of the last 40 years goes far beyond Hering's observation that "redness and greenness, or yellowness and blueness are never simultaneously evident in any color, but rather appear to be mutually exclusive (1920, p.50)". Modern studies of opponency reveal a true bipolar neural response by showing that added light can either increase or decrease the response, depending on the emission spectrum of the light. Yet, most perceptual evidence for opponency in <u>central</u> visual pathways (at or beyond the locus of binocular fusion) is even weaker than Hering's original observation. While a long-wavelength 'red' light in one eye when fused with middle-wavelength 'green' light in the other eye can result in a percept that is neither reddish nor greenish, this phenomenon does not require central color opponency (De Weert and Levelt, 1976). In fact, binocular fusion of yellow was intended as a test of retinal receptor systems, and taken as evidence for the Young-Helmholtz three-receptor theory (Hecht, 1928; Hurvich and Jameson, 1951).

This paper describes psychophysical evidence for opponency in a central visual pathway that mediates color perception. The approach is modeled on modern studies of opponency in monocular pathways, which directly demonstrate a bipolar neural response. In this approach, one first measures the adapting effect of a particular stimulus presented alone. Then additional light is added to the adapting field, and the adapting effect of the admixture is measured. Consider, for example, the experiments of Thornton and Pugh (1983), who measured increment thresholds of a 430 nm test light. They demonstrated opponency by showing (1) that a short-wavelength (470 nm) adapting background increases the amount of 430 nm test light

From Pigments to Perception, Edited by A. Valberg and
B.B. Lee, Plenum Press, New York, 1991

required for threshold detection, and (2) that adding 578 nm light to the 470 nm adapting field <u>reduces</u> the amount of test light required for threshold (Fig. 1). With the short-wavelength background present, adding 578 nm light reduces polarization at a chromatically opponent neural site and thus reduces the loss of sensitivity caused by the opponent response.

The approach taken here to examine opponency in a central visual pathway is analogous to monocular studies except that neural responses rather than physical lights are combined. The effect of a <u>fixed</u> stimulus (and thus fixed neural response) in one eye is measured as a function of a binocularly fused stimulus presented to the other eye. The measurements show that the effect of adapting the contralateral eye can be either increased or decreased, depending on the spectral composition of the light with which the contralateral stimulus is fused.

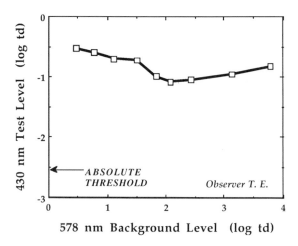

Figure 1.   Increment threshold measurements for a 430 nm test upon a background that is an admixture of 470 and 578 nm light. The 470 nm component is fixed at 1.8 log td; the level of 578 nm light is varied (horizontal axis). Note that adding 578 nm background light reduces threshold. Arrow indicates the absolute threshold for the 430 nm test (replotted from Fig. 1 of Thornton and Pugh, 1983).

EXPERIMENTAL DESIGN AND METHODS

The experiments examine the chromatic adapting effect of a background presented to one eye (the contralateral eye) on the color of a test field presented to only the other eye (the test eye). First, the action of a central mechanism is established by demonstrating the adapting effect of a contralateral chromatic background. These measurements, with the test eye dark adapted, define the baseline level of the central effect due to the contralateral background. Next, the effect of the same contralateral background is remeasured while various wavelengths of adapting background are presented to the test eye. The background in each eye is identical in size and retinal location, resulting in a fused percept of a single uniform field. A chromatically opponent visual pathway at or beyond the locus of binocular combination should have the following properties: (1) a particular test-eye adapting wavelength (say, 540 nm) can increase the contralateral adapting effect above baseline, (2) a different test-eye adapting wavelength (say, 660 nm) can reduce the contralateral effect below baseline, and (3) results (1) and (2) depend on test-eye adapting wavelength but not luminance. The conclusion of chromatic opponency would be further strengthened if (4) an intermediate test-eye adapting wavelength (say, 577 nm) causes no change from baseline (that is, an intermediate adapting wavelength in the test eye is equivalent to darkness with respect to contralateral chromatic adaptation). All four of these features are observed in psychophysical measurements presented below.

The test field is a thin annulus (1-1.5°) composed of an admixture of 540 nm and 660 nm light (denoted Δ_{540} and Δ_{660}, respectively). With the radiance of the 660 nm test light held fixed, the observer adjusts the radiance of the 540 nm test component so that the test appears a 'perfect' yellow (that is, neither reddish nor greenish). This procedure provides precise and reliable measurement of color perception (typical standard errors of 0.1 log unit, based on replication of measurements over days). A concentric 4.8° background may be presented to the test eye, contralateral eye, or both eyes (Fig. 2). In the present experiments, the contralateral background is always 660 nm; the wavelength of the test-eye background is varied. Stimuli are generated with a binocular Maxwellian-view six-channel optical apparatus (3 channels for each eye) that incorporates three-cavity interference filters to control spectral composition of the stimuli. Descriptions of the apparatus and typical procedures are in Shevell and Humanski (1984).

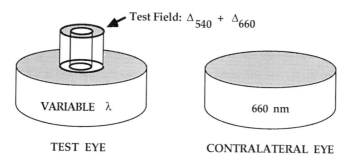

Figure 2. Schematic diagram of the stimuli. The binocularly fused percept is an annular test superimposed upon a single fused background.

CONTRALATERAL LONG-WAVELENGTH ADAPTATION

Monocular (test-eye) and contralateral long-wavelength (660 nm) adaptation are compared in Fig. 3. Each panel shows results for a different observer. The change in color appearance of the test caused by adaptation is measured by the amount of Δ_{540} required to maintain a neither reddish nor greenish percept. Results are shown at three adapting levels that range from dim to moderate (the highest is 100 td). Photopigment bleaching is negligible. At the lowest adapting level (3 td), there is virtually no difference between monocular and contralateral adaptation (compare open and filled triangles). As expected, these dim backgrounds are ineffective adapting stimuli. With more intense backgrounds, the monocular results (open symbols) show (i) a typical shift toward greenness (less 'greenish' 540 nm light in the test), which is accounted for by selective receptoral desensitization, and (ii) curvature expected from the redness of the long-wavelength background on which the test is superimposed (Shevell, 1978; 1982). Contralateral adaptation (filled symbols) is qualitatively different. Raising the contralateral adapting level shifts the test toward redness (more Δ_{540} required to maintain a neither-reddish-nor-greenish percept). The shift toward redness does not decline with test level, which implies the contralateral field does not simply add redness to the test. In summary, contralateral adaptation has a centrally mediated influence on color perception that is quite unlike the effect of adapting the test eye.

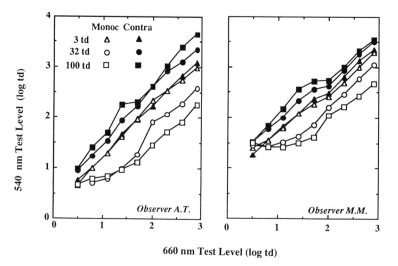

Figure 3. Measurements of the amount of 540 nm test light (Δ_{540}, vertical axis) required to maintain a neither-reddish-nor-greenish percept, at various levels of 660 nm test light (Δ_{660}, horizontal axis). Open symbols are results with a 660 nm monocular (test-eye) background; filled symbols are results with a 660 nm contralateral background. Measurements are shown for backgrounds of 3, 32 and 100 td (see legend). The standard error of most points is less than 0.1 log td (over 90% of standard errors less than 0.15 log td).

The experimental design requires a baseline level of centrally mediated chromatic adaptation. Conceptually, the central effect is established by introducing a fixed contralateral background, but this has the undesirable property of confounding the effect of the contralateral light itself with possible effects of changing from monocular to binocular viewing. Therefore we defined baseline as the effect of raising a 660 nm contralateral field from 3 to 100 trolands. The 3 troland field is clearly visible but has a negligible contralateral effect (filled triangles, Fig. 3).

BINOCULARLY FUSED CHROMATIC ADAPTATION

The change in color appearance due to contralateral adaptation next was determined with the test-eye adapted as well. The test-eye and contralateral-eye fields fused binocularly. Measurements with a 660 nm background presented to both eyes are shown in the upper panels of Fig. 4. With a 3 td test-eye background, the effect of raising the contralateral 660 nm background from 3 to 100 td is indicated by the separation between open squares and triangles. With a 100 td test-eye background, the contralateral effect is given by the separation between filled squares and triangles. While raising the contralateral adapting level still tends to shift the appearance of the test toward redness for observer A.T. (more 540 nm test light required with the more intense contralateral field), the 660 nm adapting light in the test eye substantially reduces the contralateral effect. For observer M.M., the contralateral effect is small or negligible (squares and triangles overlie each other). Clearly, raising the level of the contralateral background is much less effective when the test-eye also is adapted to 660 nm light. Note that this is found even with a dim 3 td test-eye background (open symbols, Fig. 4), which has a feeble adapting effect as a monocular stimulus (open triangles, Fig. 3).

The specific effect of contralateral adaptation is plotted in the lower panels of Fig. 4, which show the change in the amount of 540 nm test light due to raising the contralateral 660 nm background from 3 to 100 td. The vertical axis is labeled 'Change in $\log(\Delta_{540}/\Delta_{660})$', indicating the change in the ratio $\Delta_{540}/\Delta_{660}$ when the contralateral background is increased to 100 td. In Fig. 4 this is simply the change in Δ_{540} due to raising the contralateral background, because Δ_{660} is identical at both contralateral background levels. The more general notation of the ratio $\Delta_{540}/\Delta_{660}$ allows other measurements (presented below), for which Δ_{660} varies, to be plotted in the same units. The contralateral effect with a 3 td (100 td) test-eye background is indicated by the open (filled) symbols. The dashed line shows the contralateral effect when the test eye is dark adapted (calculated from the filled triangles and squares in Fig. 3). The comparison with dark-adapted values emphasizes the reduced effect of contralateral adaptation when a 660 nm test-eye background is present.

The dampening influence of test-eye adaptation might suggest that contralateral adaptation is significant only when the test eye is dark adapted. The visual system might be uncharacteristically susceptible to chromatic signals from the opposite eye when the contralateral eye alone is adapted. This was evaluated by replacing the 660 nm test-eye background with 540 nm light. The observer now fused the 540 nm background with the 660 nm contralateral field. Because the 540 nm test-eye background can add greenness to the superimposed test (Humanski and Shevell, 1985), the observer controlled the level of the 660 nm test light (Δ_{660}) while the 540 nm test component (Δ_{540}) was held fixed. Measurements in the upper panels of Fig. 5 show the settings of Δ_{660} required for a neither-reddish-nor-greenish percept (vertical axis) at various levels of Δ_{540} (horizontal axis). Results at two test-eye background levels are plotted (5 and 100 td). They show that the 540 nm test-eye background accentuates contralateral adaptation rather than reduces it. There is a clear

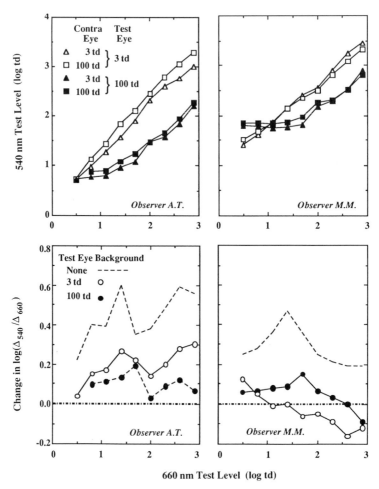

Figure 4. <u>Upper panels</u>: Measurements with a 660 nm background presented to both eyes (axes as in Fig. 3). Each background was either 3 or 100 td, and all combinations were tested (see legend). Each panel shows results for a different observer. Most standard errors are less than 0.1 log td (over 95% are less than 0.15 td). <u>Lower panels</u>: The change in the observer's setting of Δ_{540} caused by raising the level of the contralateral background from 3 to 100 td (see text for description of unit on vertical axis). Results are shown with the test-eye background at 3 td and at 100 td (circles, calculated from values in upper panels), and with no test eye background (dashed line, from values in Fig. 3).

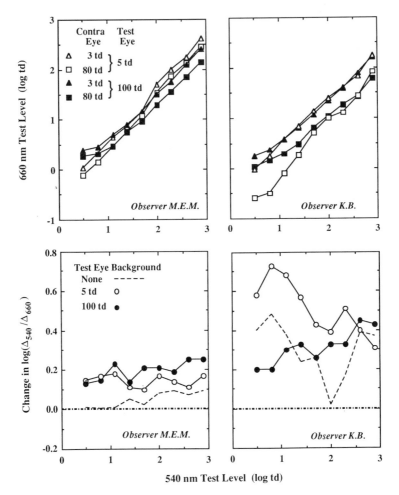

Figure 5. Upper panels: Measurements with a 540 nm test-eye background (5
or 100 td) and a 660 nm contralateral-eye background (3 or 80 td).
All combinations were tested (see legend; contralateral-eye
backgrounds for observer M.E.M. were 0.1 log td lower). The axes
are reversed from previous figures because the observer now
controlled Δ_{660} rather than Δ_{540}. Each panel shows results for a
different observer. Most standard errors are less than 0.1 log td
(90% are less than 0.15 td). Lower panels: The change in the
observer's setting of Δ_{660} caused by raising the level of the
contralateral 660 nm background from 3 to 80 td (unit of vertical
axis as in Fig. 4). Results are shown with a 5 td and 100 td
test-eye background (circles, calculated from values in upper
panels), and with no test eye background (dashed lines, from
measurements on these observers comparable to those in Fig. 3).

separation between open squares and triangles (5 td test-eye background) and
between filled squares and triangles (100 td test-eye background). As
before, raising the level of the contralateral 660 nm background (from 3 to
80 td in this experiment) shifts the appearance of the test toward redness
(less 660 nm test light required with the 80 td contralateral field). The
lower panels of Fig. 5 explicitly show the change in the test field due to
raising the contralateral background from 3 to 80 td. The vertical axis is
the same as in Fig. 4 so that effects of 540 nm and 660 nm test-eye
adaptation can be compared. The dashed line is the effect of contralateral
adaptation with the test eye dark adapted. The dashed line is below most of
the other values, indicating that a 540 nm background in the test eye
increases the contralateral effect. Therefore changes in color perception
due to contralateral adaptation are not a consequence of adapting only the
opposite eye. Rather, the contralateral effect depends on properties of the
adapting stimulus in the test eye.

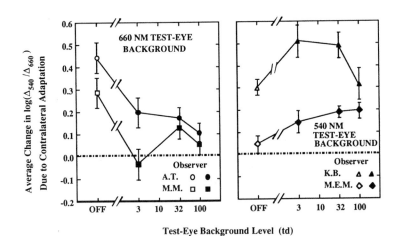

Figure 6.   Average change in the test-field ratio $\Delta_{540}/\Delta_{660}$ (vertical axis)
caused by raising the level of the 660 nm contralateral
background. Results are shown at three test-eye background
levels (filled symbols) and with no test-eye background (open
symbols). Left panel: 660 nm test-eye background. Right panel:
540 nm test-eye background. Error bars indicate ±1 standard
error of the mean.

     A convenient way to summarize the measurements is to take the average,
over all test-field retinal illuminances, of the change in the test-field
ratio, $\Delta_{540}/\Delta_{660}$, due to raising the level of the contralateral 660 nm
field. This value is simply the mean of a connected set of points in the
lower panels of Figs. 4 and 5. The averages are plotted as a function of
test-eye background level (horizontal axis) in Fig. 6, which includes
results for an intermediate level of test-eye background not shown in Figs.
4 and 5. Connected points are values for one observer. The leftmost point
of each set is the purely contralateral (baseline) condition with the test
eye dark adapted (open symbols). Note that the effect of contralateral 660

nm adaptation is reduced at every level of the 660 nm test-eye background (filled symbols, left panel), relative to a dark adapted test eye. In contrast, the effect of contralateral adaptation is increased, relative to test-eye dark adaptation, at nearly every level of the 540 nm test-eye background (filled symbols, right panel). The only exception, for one observer (K.B.), is a smaller effect of contralateral adaptation at the highest level of the 540 nm test-eye adapting field. This can be explained by monocular effects of test-eye adaptation, which are stronger at higher levels and thus tend to dominate the fixed contralateral influence. The dimmer test-eye backgrounds, which minimize retinal adaptation, are the most informative. Overall, wavelength is the critical property of the test-eye background that regulates the effect of contralateral adaptation: the centrally mediated adapting effect can be increased or decreased, relative to a dark adapted test eye, depending on the wavelength of light with which the contralateral background is fused. This is evidence for opponency analogous to monocular increment-detection studies.

BINOCULARLY FUSED CHROMATIC ADAPTATION AS A FUNCTION OF TEST-EYE ADAPTING WAVELENGTH

Chromatic opponency in a central pathway was examined further by measuring the contralateral adapting effect at 10 wavelengths of test-eye background (540, 550, 560, 570, 577, 590, 600, 620, 640 and 660 nm). The test-eye background was always dim (3 td) to minimize retinal adaptation; a single level of background in the test eye is sufficient because level is not a critical factor (Fig. 6). Measurements were also taken with no test-eye background (baseline). The contralateral background was 660 nm at either 3 or 100 td. The test field was an admixture of 540 and 656 nm light (Δ_{540} plus Δ_{656}). In this experiment the observer always controlled the level of the 540 nm test component, as dim middle-wavelength backgrounds contribute minimal greenness to a superimposed test. Measurements of Δ_{540} were taken with Δ_{656} set at 0.5, 1.0, 1.5, 2.0 and 2.5 log td.

The raw data are 22 sets of measurements (all combinations of 11 test-eye background conditions and 2 contralateral-eye background levels). Each set is comprised of the results for the 5 retinal illuminances of Δ_{656}. As before, the centrally mediated effect is determined by the change in Δ_{540} caused by raising the contralateral field from 3 to 100 td. This change, averaged over the 5 levels of Δ_{656}, is plotted in Fig. 7 as a function of test-eye adapting wavelength (horizontal axis). The label on the vertical axis maintains the ratio notation ($\Delta_{540}/\Delta_{656}$) for consistency with earlier plots but is equivalent here to Δ_{540}. Each panel shows results for a different observer.

The magnitude of the contralateral effect with no test-eye background is indicated by the dashed line. The results show that the contralateral effect is larger at test-eye wavelengths below 560 nm (points above dashed line) and smaller at wavelengths beyond 620 nm (points below dashed line). At intermediate wavelengths, which cover a range that includes equilibrium yellow, the contralateral effect is comparable to that found with no test-eye background. Overall, the measurements show that introducing a test-eye background can increase, decrease, or leave unchanged the magnitude of the central effect, depending on the wavelength of test-eye background with which the contralateral field is fused. Wavelength of test-eye background has a centrally mediated influence characteristic of red-green chromatic opponency.

DISCUSSION

These experiments exploit the change in color appearance of a light, presented to one eye, caused by long-wavelength chromatic adaptation of the other eye. The baseline effect of contralateral adaptation was defined with no test-eye background. Then the same fixed contralateral backgrounds were fused binocularly with a background in the test eye. The properties of the test-eye background were varied to explore the central process that combines signals from the two eyes. The measurements show that greenish test-eye backgrounds (wavelengths below 560 nm) increase above baseline the change in color appearance caused by long-wavelength contralateral adaptation, while reddish test-eye backgrounds (wavelengths longer than 620 nm) reduce it. Intermediate test-eye wavelengths, near equilibrium yellow, had little effect. This is evidence for chromatic coding in a central pathway, analogous to monocular increment-detection results that show the adapting effect of a light can be either enhanced or reduced by introducing an additional adapting stimulus.

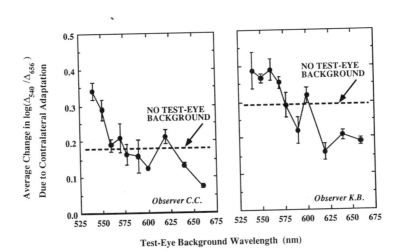

Figure 7.   Average change in the observer's setting of Δ_{540} (which is identical here to the change in test-field ratio $\Delta_{540}/\Delta_{656}$, vertical axis) caused by raising the level of the 660 nm contralateral background from 3 to 100 td. Results are shown for 10 test-eye background wavelengths at 3 td (circles) and with no test-eye background (dashed line). Each panel shows results for a different observer. Error bars indicate ±1 standard error of the mean.

Contralateral adaptation can influence color perception but does not affect all percepts. Threshold sensitivity typically is unchanged by adapting the opposite eye (Crawford, 1940; Whittle and Challands, 1969; Yellot and Wandell, 1976), an observation we confirmed in the course of these experiments. Recently, however, Verdon, Haegerstrom-Portnoy and Adams (1990) found that detection of a test by an SWS-cone pathway can be strongly inhibited by a contralateral field of the appropriate size and spectral composition. In general, the properties of a central pathway mediating one aspect of vision should not be assumed to hold for other percepts.

A possible function of the central process examined here is to improve color contancy when there are different signals from an observer's left and right eyes. In natural viewing one rarely experiences stimulation of the two eyes as different as in these haploscopic experiments but signals from the eyes may differ because of prereceptoral factors, photopigments, receptoral alignment, and neural processes. Perceptual differences between the eyes are well known. Brightness matching shows interocular differences as large as 0.3 log unit (Wright, 1946) and color (Rayleigh) matches in a person's left eye and right eye often are not the same (Humanski and Shevell, 1990). It is noteworthy that the long-wavelength ('red') contralateral field shifts the appearance of the test toward redness, and does so at all levels of the test field. It is as though the central process increases gain in the test eye to compensate the sensitivity loss in the other eye. This central mechanism would tend to compensate for a difference in the average spectral illumination of the two eyes without defeating the usual processes of chromatic adaptation, because the central effect is much reduced when both eyes are adapted to the same wavelength.

ACKNOWLEDGEMENT — Supported by grant EY-04802 from the National Eye Institute.

REFERENCES

Crawford B. H. (1940) Ocular interaction in its relation to the measurement of brightness threshold. *Proc. R. Soc. Lond.*, **128B**, 552-559.
De Weert C. M. M. and Levelt W. J. M. (1976) Comparison of normal and dichoptic colour mixing. *Vision Res.*, **16**, 59-70.
Hecht S. (1928) On the binocular fusion of colors and its relation to theories of color vision. *Proc. Natl. Acad. Sci. U.S.*, **14**, 237-241.
Hering E. (1878) Principles of a new theory of the color sense. [English translation by K. Butler, 1961]. In R. C. Teevan and R. C. Birney (Eds.), *Color Vision: An Enduring Problem in Psychology*. New York: Van Nostrand.
Hering E. (1920) *Outlines of a Theory of the Light Sense*. [English translation by L. M. Hurvich and D. Jameson, 1964]. Cambridge, Mass.: Harvard University Press.
Humanski R. A. and Shevell S. K. (1985) Color perception with binocularly fused adapting fields of different wavelengths. *Vision Res.*, **25**, 1923-1935.
Humanski R. A. and Shevell S. K. (1990) Factors contributing to differences in Rayleigh matches of normal trichromats. In B. Drum and J. Moreland (Eds.), *Colour Vision Deficiency X*. Dordrecht: Kluwer Academic Press, in press.
Hurvich L. M. and Jameson D. (1951) The binocular fusion of yellow in relation to color theories. *Science*, **114**, 199-202.
Jacobs G. H. (1986) Cones and opponency. *Vision Res.*, **26**, 1533-1541.
Shevell S. K. (1978) The dual role of chromatic backgrounds in color perception. *Vision Res.*, **18**, 1649-1661.
Shevell S. K. (1982) Color perception under chromatic adaptation: Equilibrium yellow and long-wavelength adaptation. *Vision Res.*, **22**, 279-292.

Shevell S. K. and Humanski R. A. (1984) Color perception under contralateral and binocularly fused chromatic adaptation. *Vision Res.*, **24**, 1011-1019.

Sperling H. G. (1986) Spectral sensitivity, intense spectral light studies and the color receptor mosaic of primates. *Vision Res.*, **26**, 1557-1571.

Svaetichin G. and MacNichol E. F. (1958) Retinal mechanisms for chromatic and achromatic vision. *Ann. N.Y. Acad. Sci.*, **74**, 385-404.

Thornton J. E. and Pugh E. N. (1983) Relationship of opponent-colours cancellation measures to cone-antagonistic signals deduced from increment threshold data. In J. D. Mollon and L. T. Sharpe (Eds.), *Colour Vision: Physiology and Psychophysics*. New York: Academic Press.

Verdon W., Haegerstrom-Portnoy G. and Adams A. J. (1990) Spatial sensitization in the short wavelength sensitive pathway under dichoptic viewing conditions. *Vision Res.*, **30**, 81-96.

Whittle P. and Challands P.D.C. (1969) The effect of background luminance on the brightness of flashes. *Vision Res.*, **9**, 1095-1109.

Wright W. D. (1946) *Researches on Normal and Defective Colour Vision*. London: Henry Kimpton.

Yellot J. I. and Wandell B. A. (1976) Color properties of the contrast flash effect: Monoptic versus dichoptic comparisons. *Vision Res.*, **16**, 1275-1280.

CHROMATIC MECHANISMS BEYOND LINEAR OPPONENCY

Qasim Zaidi and Daniel Halevy

Department of Psychology
Columbia University
New York, NY 10027

BACKGROUND

The current picture of the human color system consists of a series of stages, each with a number of types of mechanisms functioning in parallel. Successive stages are composed of mechanisms that combine the outputs of the previous stage. An extensive account of the empirical and theoretical foundations of the early stages of the color system is given in Zaidi (1990). In brief, the first stage consists of three independent cone mechanisms, L, M and S, that form the substrate for trichromatic color matches (Maxwell, 1860), and whose spectral sensitivities have been derived by psychophysical (Smith and Pokorny, 1975) and electrophysiological (Schnapf et al, 1987) methods. The second stage consists of two chromatic mechanisms RG and YV, and an achromatic mechanism LD. The RG mechanism signals the difference between the responses of the L and M cones, and the YV mechanism signals the response of the S cones minus a sum of the responses of the L and M cones. The independence of the chromatic signals from these cardinal mechanisms are reflected in habituation (Krauskopf et al, 1982) and motion (Krauskopf and Farrell, 1990) experiments. In addition, parvo-neurons of primate lateral geniculate nucleus cluster into two classes with chromatic properties similar to the two chromatic cardinal mechanisms (Derrington et al, 1984).

A number of recent experiments have provided partial clues to the properties of chromatic mechanisms subsequent to the second stage. Multiple mechanisms maximally sensitive to different isoluminant colors distributed about the color circle are needed to explain the results of chromatic habituation along lines intermediate to the cardinal directions (Krauskopf et al, 1986; Webster and Mollon, 1990), the transient elevation of chromatic thresholds after changes in adaptation state (Krauskopf et al, 1986), the detection and discrimination of isoluminant changes in color (Krauskopf et al, 1986), pure chromatic induced contrast (Krauskopf, Zaidi and Mandler, 1986) and contrast detection in chromatic noise (Gegenfurtner and Kiper, 1990). In addition, pairs of color mechanisms that only respond to opposite directions of change along a color line are required to explain the desensitizing effect of sawtooth temporal modulation (Krauskopf et al, 1982; Krauskopf and Zaidi, 1986). The responses of these mechanisms can be qualitatively modeled as the rectified derivative of the opponent response with respect to time (Zaidi, 1987). The chromatic responses of simple and non-oriented neurons in the striate cortex can be described as linear combinations of the responses of parvo-neurons in the LGN. However, unlike LGN parvo-neurons, these cells do not cluster into two classes; instead their chromatic preferences are distributed around the isoluminant circle. In addition, the response of some cells in the striate cortex is confined to narrow bands of color space and cannot be described as a linear combination of LGN responses (Lennie et al, 1990).

From Pigments to Perception, Edited by A. Valberg and
B.B. Lee, Plenum Press, New York, 1991

RESPONSES TO TEMPORAL COLOR CHANGES

This paper presents measurements of responses to exclusively temporal changes in color stimuli, i.e. changes to which there were no spatial cues. To measure responses in a more extended range than the small-signal linear domain studied by means of flicker thresholds, increment thresholds were measured on top of spatially invariant backgrounds that were modulated continually in time. To analyze the chromatic properties of the mechanisms underlying these responses, the color of the background was changed along the circumference of an isoluminant color circle. As will be shown below, the increment threshold results cannot be explained in terms of conventional first and second stage mechanisms, but are compatible with mechanisms that signal the direction of color changes.

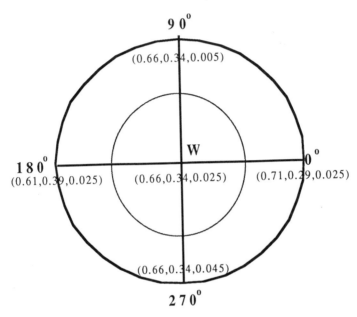

Figure 1.   *Isoluminant color disk. (L,M̄,S) cone excitations in luminance units.*

Equipment

Stimuli were displayed on the screen of a Tektronix 690SR color monitor running at 120 interlaced frames per second. The screen consisted of a 480x512 pixel display, subtending 10x11 degrees of visual angle at the observer's eye. Images were generated by an ADAGE 3000 that allowed 10 bit specification of output levels for each television gun. Each gun was calibrated by averaging the readings of a UDT photometer at each output level through 10000 up and down series. The calibration procedure also tested the independence of these guns. Linear specification of stimuli was achieved by using stored back-transform tables. The stability of the calibration was checked regularly by means of a Minolta chroma-meter. All stimulus generation and data acquisition was done automatically under computer control.

Stimuli

The range of colors used in this experiment are depicted in Figure 1 as an isoluminant color disk centered at a white point. The triplets of numbers represent Smith and Pokorny (1975) L, M and S cone excitations in luminance (L+M) units corresponding to the colors at the center and at the ends of the principal axes. Similar to the MacLeod and Boynton (1978) chromaticity diagram, along the horizontal axis, the L and M excitations are traded against each other whereas their sum and S-cone excitation are kept constant; along the vertical axis, only S-cone excitation is changed. Following Derrington et al (1984), angles are numbered in a counter-clockwise fashion around the circle. The color names that approximately correspond to different angles along the circumference are Red at 0, Yellow at 90, Green at 180, and Violet at 270°. In this experiment, the background was a uniform disk whose color was set at some angle on the circumference of the inner circle in Figure 1. The color of the disk was changed in either the clockwise or counter-clockwise direction along the circumference of the inner circle. Figure 2 shows the cone excitations at each color angle along the inner circle minus the cone excitation at the central white. Changing the color through one complete circle modulates the excitation of the three classes of cones through one complete sinusoidal cycle each. The peak of the modulation occurs at 0° for L, at 180° for M and at 270° for S-cones. The cone signals from Figure 2 are linearly combined in Figure 3 to show that, in response to color changes around the circle, signals from second stage chromatic cardinal mechanisms also vary sinusoidally. The response of higher chromatic mechanisms will reflect interaction between signals from second stage mechanisms. These interactions could be linear or non-linear, temporal or spatial. To measure higher level responses adequately requires stimuli that simultaneously modulate the responses of all second stage mechanisms in a precisely specifiable manner. As Figures 2 and 3 show, changing colors around the isoluminant circle not only achieves this aim, in addition, the sinusoidal modulation of responses provides a natural basis for tests of linearity.

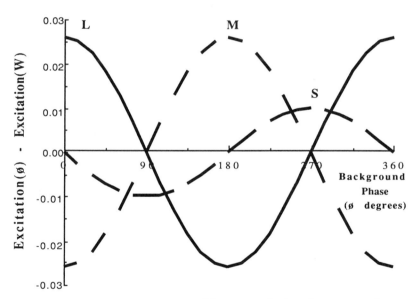

Figure 2.   *Modulation of L, M, and S cones excitations for one complete counter-clockwise excursion around the isoluminant color circle.*

The spatial configuration used in the experiment is shown in Figure 4. An 8° colored disk was displayed in the center of a 10° square white steady field of the same luminance (40 cd/m^2). A 1.5 min dark fixation spot was continuously visible on the screen. The color of the disk was modulated around the circle, and probe colors were added to the background modulation. There was thus a temporal cue when the probe was added, but no spatial cue.

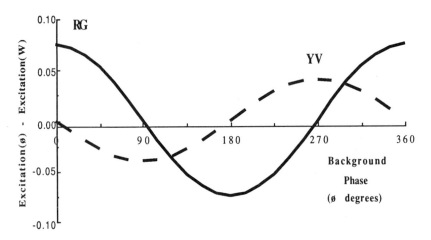

Figure 3. *Modulation of the excitation of the second stage chromatic mechanisms for a complete counter-clockwise circle.* $[RG=L-2M; \ YV=4S-0.1(L+M); \ RG(W)=YV(W)=0].$

Procedure

After adapting for a few minutes to a steady white screen at 40 cd/m^2, the observer was presented with a number of trials, each consisting of a two-interval forced choice (2IFC). On each trial, the color of the disk started modulating from a random phase angle on the circumference of a circle of half the maximum radius. In the trial depicted in Figure 5, the color of the background was changed in a clockwise direction and the observer's response to an incremental probe in the 0° direction was measured at a background phase equal to 0°. In the "No Probe" interval, the color of the disk changed around the circle uninterrupted at a constant angular velocity. In the "Probe" interval, when the color of the disk reached the chosen background phase, the color of the probe vector was added to the background color. There was thus an abrupt change in the color of the disk. As shown in Figure 5, the color of the disk continuously changed in the clockwise direction while the magnitude of the incremental vector

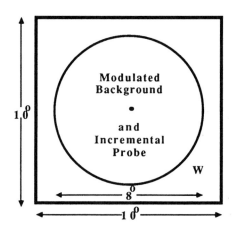

Figure 4. *Spatial configuration of stimuli.*

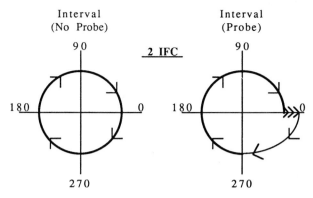

Figure 5. Chromatic sequence for a 2IFC, clockwise trial. Background phase 0°; probe direction 0°.

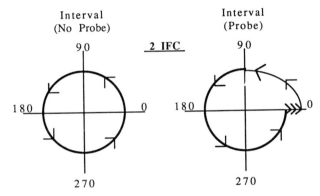

Figure 6. Chromatic sequence for a 2IFC, counter-clockwise trial. Background phase 0°; probe direction 0°.

was linearly decreased to zero in a quarter of a cycle. The observer's task was to select the interval with the abrupt change. Figure 6 shows the chromatic sequence for a trial in which the background color was modulated in the counter-clockwise direction.

The temporal sequence of a typical trial is described in Figure 7. Prior to the commencement of a trial, the color of the disk was the same as the square steady white field. At the onset of a trial the color of the disk was set to a random phase. The color was then modulated through at least one complete circle before the test interval. The modulation continued after the first interval until the second interval was reached. Subsequent to the second interval, the modulation continued until the observer responded. Three beeps signaled the onset of each interval. This particular procedure was chosen after discarding a number of other spatial configurations and temporal sequences. Pilot procedures revealed that observers responded exclusively to the temporal change only if there was no spatial cue as to which interval contained the incremental probe. In particular, if the probe and the modulated background were not spatially coincident, then the "Probe" and "No Probe" intervals could be differentiated on the basis of spatial configuration, and the results were quite different. In addition, reliable measurements were possible only if there were no abrupt temporal changes shortly before or after the test intervals.

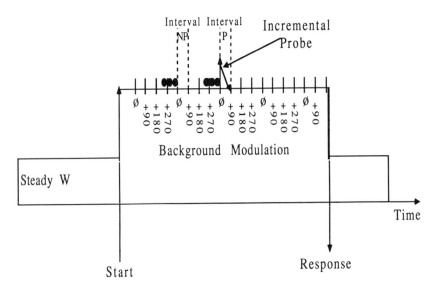

Figure 7. Temporal sequence for a 2IFC, counter-clockwise trial for a
background phase of Ø°. The filled dots refer to the three beeps
that signaled the onset of each interval.

In the present experiment, as shown in Figure 8, four background phases, 0, 90,
180 and 270°, were used, once each for clockwise and counter-clockwise modulations. At
each background phase, responses were measured to four incremental probes in the 0, 90, 180
and 270° directions, making a total of 32 conditions. A double-random staircase was run for
each condition. The magnitude of the probe was reduced after three consecutive "correct"
selections and increased after every "incorrect" selection. The 64 staircases were run
simultaneously over a period of many days in short sets of 64 randomly chosen trials. All data
points were the means of 20 reversals each.

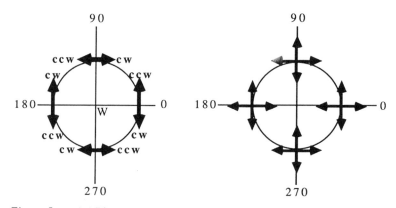

Figure 8. (a) Phases and directions of background at which probes were
added. (b) Directions of incremental probe vectors.

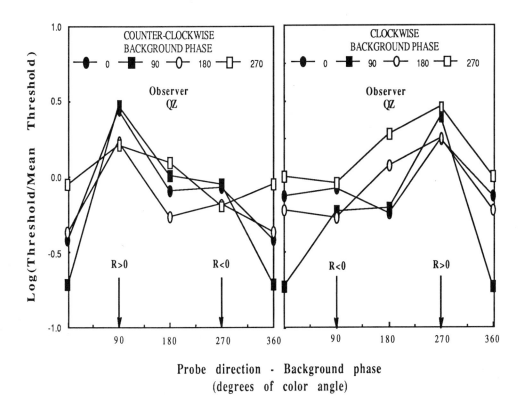

Figure 9.   Probe thresholds for (a) clockwise and (b) counter-clockwise
             background modulation (0.46 Hz). Abscissa equal to the angle of the probe
             vector minus the background phase angle, modulus 360°. Ordinate equal to
             log of the ratio of the threshold for a probe to the mean of all eight
             thresholds for that probe.

Results

Probe thresholds for one observer are shown in Figure 9. Each set of similar symbols joined by lines represent thresholds for different probes measured on the same phase of the background: Figure 9(a) for counter-clockwise and Figure 9(b) for clockwise modulation of the background color. Since the color space in Figure 1 is affine, there is no natural metric to compare threshold measurements along different probe angles to each other. In Figure 9, every threshold for a particular probe direction is expressed relative to the average threshold in that probe direction. Since the background conditions were distributed uniformly around the circle, this measure provides roughly equivalent units for different probe directions. The horizontal axes in Figure 9 represent probes in terms of the angular color difference between the probe direction and the background phase; so that 0° represents probes on backgrounds of the same color angle, 90° represents a probe direction equal to the background phase plus 90°, etc.. In the patterns formed by the data in Figure 9, three aspects are prominent. First, the counter-clockwise and clockwise figures look like mirror symmetric versions of one another. Second, both figures have one prominent peak, at background phase plus 90° for counter-clockwise and at background phase plus 270° for clockwise modulation of the color of the background. Third, on the 90° background, thresholds for probe vectors in the 90° (yellowish) direction were significantly lower than all other thresholds.

In classical increment threshold experiments (Stiles, 1959), the color of the background at the onset of a probe is the important factor. However, the results in Figure 9 show that, for purely temporal changes, thresholds on every phase of the background depend on the direction of modulation. Therefore, the important factor is not the instantaneous color of the

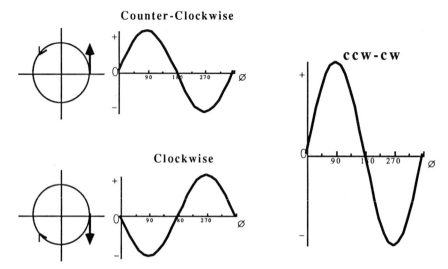

Figure 10. Instantaneous rate of change in color at an arbitrary phase represented by arrow, resolved towards directions of Ø° for counter-clockwise (Top Left) and clockwise (Bottom Left) modulation, where Ø is the Probe Direction minus Background Phase. (Right Panel) Rate of change for counter-clockwise minus rate of change for clockwise modulation at any background.

background, but the direction of change of color at that instant, i.e. the vectorial component of the color change in the direction of the probe. A physical analogy to the color modulation is a particle moving at constant angular velocity around a circle. If the particle is moving in a counter-clockwise direction as in the top left panel of Figure 10, when its position is at 0°, it is actually moving towards 90°. At that instant, the directional component of the velocity, and hence the rate of change of position, is maximum and positive towards 90°, minimum and negative towards 270°, and null towards 0 and 180°. The rate of change towards any angle can be calculated from one cycle of a sine-wave with zero crossings at 0 and 180, maximum at 90 and minimum at 270°. The same sinusoidal function describes the rate of change of color for counter-clockwise modulation at an instantaneous phase of 0°. When the background is modulated in a clockwise fashion, at any individual phase, the rate of change of color is simply the negative of that for counter-clockwise modulation (bottom left panel of Figure 10). If, similar to Figure 9, the abscissa ø represents the probe direction in terms of the angular difference from the background phase, then the two sinusoidal curves express the rate of change of color at any arbitrary background phase.

To test the hypothesis that the direction of change of color is an important factor, the results of Figure 9 have been plotted in Figure 11(a). As an index of directional influence on threshold judgments, for every probe direction on every background phase, the ordinate is expressed as the log of the ratio of the threshold on the counter-clockwise modulation to the threshold on the clockwise modulation. Similar to Figure 9, the abscissa is in terms of the angular difference between the probe direction and the background phase, and each set of similar symbols joined by lines represents probe thresholds on the same background phase. The connected lines in Figure 11(a), representing the difference between the logs of the counter-clockwise and clockwise thresholds, resemble the sinusoidal curve in the right panel of Figure 10, which represents the rate of change for counter-clockwise minus the rate of change for clockwise modulation. Four points per curve do not provide enough resolution for a formal goodness of fit test to a sinusoidal function, but it is significant that all the data curves are at a maximum at background plus 90° and at a minimum at background plus 270°. Data for a second observer in Figure 11(b) form a sinusoidal pattern similar to the data of the first observer. The sensitivity of the mechanisms that determine these thresholds seems to be controlled by the directional rate of change in the color of the background at the instant that the probe is added.

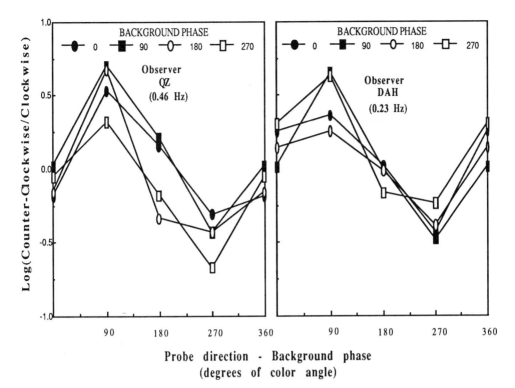

Figure 11. (a) Direction selective threshold changes for Observer QZ at a background modulation of 0.46 Hz., (b) data for Observer DAH at 0.23 Hz.

MECHANISMS THAT DETECT COLOR CHANGES

An explanation of the empirical results requires four mechanisms, ΔR, ΔY, ΔG and ΔV, each sensitive to a change in color in one direction along one of the cardinal axes. Each of these mechanisms requires three essential components: (1) a temporal filter that cuts off high frequencies, (2) a differentiator with respect to time of the response generated by one of the cardinal mechanisms RG or YV shown in Figure 3, and (3) a compressive half-wave rectifier. Symbolically, the mechanisms can be defined as in Equations 1 to 4, where "f" is some compressive positive function:

$$\Delta R = f \left[\max\left(\frac{dRG}{dt}, 0 \right) \right] \qquad - (1)$$

$$\Delta Y = f \left[\max\left(-\frac{dYV}{dt}, 0 \right) \right] \qquad - (2)$$

$$\Delta G = f \left[\max\left(-\frac{dRG}{dt}, 0 \right) \right] \qquad - (3)$$

$$\Delta V = f \left[\max\left(\frac{dYV}{dt}, 0 \right) \right] \qquad - (4)$$

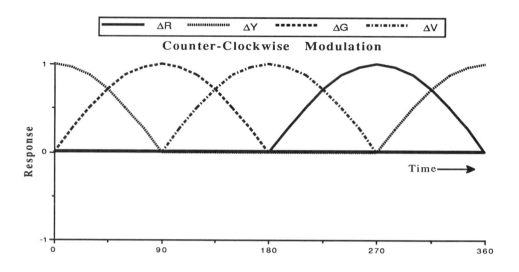

Counter-Clockwise Modulation

Clockwise Modulation

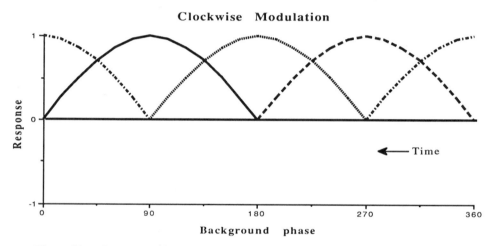

Figure 12. *Responses of mechanisms that detect color changes to counter-clockwise (Top Panel) and clockwise (Bottom Panel) modulation around the color circle at constant angular velocity. Time's arrow goes from left to right in the top panel and from right to left in the bottom panel.*

Qualitatively, these mechanisms can be said to be sensitive to color changes in the reddish, yellowish, greenish and bluish directions, respectively. The responses of these mechanisms to color modulations around the circle are illustrated in Figure 12. The response of each mechanism is a half-wave rectified and compressed sinusoid. The peaks of the responses are separated by multiples of 90°. At each of the four background phases used in the present experiment, only one mechanism has a response different from zero. Table 1 shows that the incremental changes used as probes in the experiment are each detected by one of the postulated mechanisms exclusively.

The three components that comprise these mechanisms can be justified on the following grounds: A temporal filter that rolls off high frequencies is needed to keep the output of the

Table 1.   *Detection of probes parallel to cardinal directions by mechanisms that detect color changes.*

Direction of Probe	Exclusive Detector
0°	ΔR
90°	ΔY
180°	ΔG
270°	ΔV

differentiator within bounds, and to be compatible with the greater effectiveness of the slow phase of sawtooth color modulation at 1 Hz. The temporal filter or high-frequency-integrator could possibly be located even prior to the linear opponent stage. A differentiator is one type of mechanism that responds to the rate of change of opponent signals, and that could explain the results in Figure 11. Differentiators can be easily built from resistance and capacitance elements (Horowitz and Hill, 1980), but the actual neural implementation could also involve time-lagged differencing or cross-correlation of signals from mechanisms with different temporal properties. A compressive function is one method of expressing a limited response range for a mechanism. Coupled with an assumption that all thresholds reflect an equal incremental response to the probe, a compressive non-linearity predicts that thresholds will be elevated with respect to a baseline, if a probe is detected by a mechanism whose response at the background phase is greater than its response at the baseline. For example, during counter-clockwise modulation, at 0° phase, the only mechanism whose response is greater than the baseline in Figure 12, is ΔY. From Table 1, it can be predicted that for this condition the threshold should be elevated only for the 90° probe, which is consistent with the results in Figure 9(a). Rectification is needed to create unidirectional mechanisms that respond to change in a color direction but not its complement. An alternative would be a bidirectional mechanism that responds to change in both directions along a color line with a sigmoid response function. Both types of mechanisms predict threshold elevations, compared to the mean threshold for a probe, when the rate of change of the background towards the probe direction is positive. However, when the rate of change is negative, i.e. when the background is changing in the direction complementary to the probe, bidirectional mechanisms predict threshold elevations, but unidirectional mechanisms predict no change from the mean threshold. In Figure 9, "R > 0" or "R < 0" indicates whether the rate of change of the background is positive or negative in the probe direction. The rate of change is positive at an angular difference of 90° for counter-clockwise and 270° for clockwise modulation. It is negative at 270° for counter-clockwise and at 90° for clockwise. It is obvious that thresholds are elevated when the rate of change is positive, but not when it is negative, consistent with the properties of unidirectional mechanisms.

SUMMARY

This paper shows that thresholds for temporal color changes are affected by the rate of change of the color of the background towards the direction of the probe. These changes are probably processed by mechanisms that are sensitive to the rate of change of color and respond to change in a color direction but not its complementary direction. The model presented in this paper makes strong predictions for background phases and probes in directions intermediate to the cardinal directions. Experiments currently in progress, involving a larger number of directions, will test these predictions and provide clues to the number, and the temporal and chromatic properties, of the mechanisms that detect color changes.

ACKNOWLEDGEMENTS: Helpful comments on this manuscript were made by Norma Graham, Don Hood, Ben Sachtler, Art Shapiro, Anne Sutter and Tom Wiegand. This research was partially supported by the National Eye Institute through grant EY07556.

REFERENCES

Derrington, A.M., Krauskopf, J., and Lennie, P., "Chromatic mechanisms in lateral geniculate nucleus of macaque," *J. Physiol.*, vol. 357, pp. 241-265, 1984.

Gegenfurtner, K., and Kiper, D.C., "Contrast detection in luminance and chromatic noise," *Inv. Opth. and Vis. Sc.* (ARVO supplement), vol. 31, p. 109, 1990.

Horowitz, P., and Hill, W., *The Art of Electronics*, Cambridge University Press, New York, 1980.

Krauskopf, J., Williams, D.R., and Heely, D.M., "The cardinal directions of color space," *Vis. Res.*, vol. 22, pp. 1123-1131, 1982

Krauskopf, J., Williams, D.R., Mandler, M.B., and Brown, A.M., "Higher order color mechanisms," *Vis. Res.*, vol. 26, pp. 23-32, 1986.

Krauskopf, J., and Zaidi, Q., "Induced desensitization," *Vis. Res.*, vol. 26, pp. 753-762, 1986.

Krauskopf, J., Zaidi, Q., and Mandler, M.B., "Mechanisms of simultaneous color induction," *J. Opt. Soc. Am.*, vol. A3, pp. 1752-1757, 1986.

Krauskopf, J., and Farell, B., "The influence of color on the perception of coherent motion," (In Press), 1990.

Lennie, P., Krauskopf, J, and Sclar, G., "Chromatic mechanisms in striate cortex of macaque," *J. Neuroscience*, vol. 10, pp.649-699, 1990.

MacLeod, D.I.A., and Boynton, R.M., "Chromaticity diagram showing cone excitation by stimuli of equal luminance," *J. Opt. Soc. Am.*, vol. 69, pp. 1183-1186, 1978.

Maxwell, J.C., "On the theory of compound colours and the relations of the colours of the spectrum," *Phil. Trans.*, vol. 150, pp. 57-84, 1860.

Schnapf, J.L., Kraft, T.W., and Baylor, D.A., "Spectral sensitivity of human cone photoreceptors," *Nature*, vol. 325, pp. 439-491, 1987.

Smith, V.C., and Pokorny, J., "Spectral sensitivity of the foveal cone photopigments between 400 and 700 nm," *Vis. Res.*, vol. 15, pp. 161-171, 1975.

Stiles, W.S., "Color vision: The approach through increment threshold sensitivity," *Proc. Natl. Acad. Sci.*, vol. 45, pp. 100-114, 1959.

Webster, M. A. and Mollon, J. D., "Changes in perceived color and lightness following selective adaptation of post receptoral visual channels," *Inv. Opth. and Vis. Sc.* (ARVO supplement), vol. 31, p. 109, 1990.

Zaidi, Q., "Non-linear color mechanisms: a new model and a new method," *J. Opt. Soc. Am.*, Vol. A4, p. P107, 1987.

Zaidi, Q., "Parallel and serial connections between human color mechanisms," in *Applications of Parallel Processing in Vision*, ed. J.R. Brannan, Elsevier, Amsterdam, 1990 (In Press).

DISCUSSION:

PSYCHOPHYSICS AND POST-RECEPTORAL PROCESSES

James M. Thomas

Department of Psychology
U.C.L.A.
Los Angeles, California
U.S.A.

Thomas: This session is entitled post-receptoral processes; what are these processes supposed to do? One thing they can do is to unconfound information that is confounded by the receptors themselves, for example information about intensity and wavelength, or which comes to the receptors already confounded, such as reflectance and illumination. Another is to decorrelate information shared between receptors, as is done in the colour domain by opponent processes. One can also think of these processes as extracting information about objects. One may ask what information is pulled out and how is it done. Both topics have been treated; Heggelund proposed that the supposedly simple dimension of achromatic colour is at least bidimensional. One can argue about the details, but if you have worked with these experiments it is clear that you are involved with at least two dimensions. The same theme was taken up by Paul Whittle in the case of lightness and brightness. He mentioned several cases of constancy. The most general case of lightness constancy or colour constancy is when the object is seen and recognised as having its particular surface properties, when both the illumination and the background vary. As Whittle noted, this type of constancy cannot be generated simply by responding only to edge contrast. A big question is whether human viewers actually demonstrate this type of constancy; the data on this question are varied and not conclusive. The mechanisms which have been proposed to mediate constancy vary from low-level gain control to the retinex type of mechanism mentioned by Whittle. The retinex type of model is often called a "white plate" model because it requires a white area somewhere in the scene. Another type of model is the "grey world" model in which it is assumed that all of the reflectances in a scene average to a neutral grey; constancy results from integrating the light reflected over the whole scene and comparing the light reflected by each particular object to this average. It is interesting to use these models in thinking about how we should explore the neurophysiology of cells beyond V1. For example, how would we find out if cells in V4 act like retinex-type operators, extracting contrast information from individual edges and then integrating that information over the scene in order to scale the response to a particular object. Or, how would we determine if these cells act like grey-world operators, comparing the central object to a grey soup made from the light reflected from the rest of the field.

De Weert described and discussed the phenomenon of assimilation. As DeValois commented, some part of this phenomenon may be mediated by pathways or channels tuned to low spatial frequencies. Shevell and Zaidi presented papers on the perception of chromatic properties. Shevell argues that opponency, which is well established at peripheral sites, also occurs centrally. One can ask if the central opponency differs from the more peripheral type , or whether it is just another iteration of the latter. In either case, what are the implications for the spectral properties, e.g. bandwidths and cross-over points, of higher level neurons, say cells in V4? In his

From Pigments to Perception, Edited by A. Valberg and
B.B. Lee, Plenum Press, New York, 1991

paper, Zaidi argued that there are mechanisms which sense the direction of a chromatic change, as well as the temporal change itself. One possibility is that a bipolar dimension responding to changes in both directions is rectified and transformed into two orthogonal dimensions. Klein's distinction between unipolar and bipolar mechanisms is relevant here and suggests some possible psychophysical tests.

Rodieck: To Peter Kaiser, the use of such terms as *the* luminance channel or *the* chromatic channel reflects the use of constructs, created by this rather flawed P and M notion. I think it more useful to consider individual pathways based on cone types rather than constructs derived from psychophysics. I can understand how psychophysicists have been seduced by thinking that at last the physiologists have come up with something, but physiologists came to this conclusion after treating the system according to their own predispositions. In the example you showed, on increasing flicker frequency an opponent cell would turn into a 'luminance channel' while the perception remained a yellow hue. As long as you think of *the* R-G channel that looks like a paradox.

Lee: I believe the M- and P-pathways form distinct, psychophysically relevant and separable systems, but nevertheless I agree that the neuronal populations within these systems are heterogeneous. In terms of Kaiser's argument, there is a difference between photometric measurements where you are nulling out a particular sensation, and threshold measurements. In the former, you may minimise the average activity in a pathway, but in the latter you may be using a heterogeneous population of detectors which may sum together in a detection paradigm by vector-length summation or whatever. So additivity may be present in the one case but not the other, although they both rely on a unitary mechanism. That was not so much a question, more a statement of faith.

Kaiser: Paul Whittle makes no distinction between whiteness or lightness; they seem to me different.

Whittle: I am assuming they are they same; I find it difficult enough to keep the distinction between brightness and lightness clear.

Kaiser: Black-luminous opponency would exclude luminous or shiny blacks perceived on non-matt surfaces.

Heggelund: Such reflecting surfaces are rather inhomogeneous, some parts being luminous and some black. A homogeneous area would not be like this and you would not see a shiny black.

Whittle: What about the strength or weight dimension? Does it increase monotonically with illumination?

Heggelund: With respect to illumination, there are completely different relationships for samples that are brighter than the surround adaptation field and those that are darker and look greyish, although strength increases with illumination in both cases.

Walraven: With the problem of confounding background and illuminant changes, Helmholtz noted we never do so in the real world, though it is easy to see it in the laboratory. One reason is that edges and black borders tend to protect against these contrast effects of a background since contrast is now estimated directly at the black border. Many objects in the real world have border cues, such as depth or other things.

Richter: We have heard results in which changes in appearance have been shown as the surround luminance changes. Normally, we look at different objects, for example different greys, under fixed surround luminance conditions. These different conditions do yield different sensitivities (or gain) for thresholds and for suprathreshold scaling.

Russ DeValois: When talking about the phenomenon of assimilation, one has to consider cortical systems with different spatial scales. Anything tuned to low frequencies is going to cause neural spreading, so this will always be present.

de Weert: But perhaps over not such large areas; receptive fields mentioned so far have been smaller than the spatial extent of assimilation.

Whittle: There is a lot of lability between contrast and assimilation; one can often see reversals of assimilation into contrast.

Hayhoe: Are such reversals due to attention or the lack of eye movement?

Shevell: To what extent can stray light account for assimilation effects?

de Weert: It is difficult to disprove directly that stray light has an effect. On the neural level one would need large receptive fields because of the large extent of the effects.

Shapley: The idea of neural integration being a mechanism for assimilation, the idea of different spatial filters, or different size receptive fields, has been proposed before by Jameson and Hurvich. But Arend has shown that there must be more to it, for 'artificial' contours like those in the Craik-Cornsweet effect can also be effective, and these would not be expected to activate linear spatial filters. Thus long-distance border interactions must play some role in assimilation.

Zaidi: There are at least three sorts of assimilation. One is as shown by de Weert, requiring fine lines, and it is not quite clear how much is light scatter and how much is neural integration. Second there is the sort Shapley alluded to, which is more like contrast in a distant field pervading a nearer field, and seems to have more to do with spatial integration of contrast than with assimilation. A third sort seems to be neither of these, and has to do with corner effects. There thus seems to be several sorts of assimilation, and we understand none of them.

Dow: Neurophysiologically, with single cells in V1, V2 and V4 we explored relative luminance and hue responses for both dark and bright colours in relation to a background, and some effects could be understood in relation to today's session. Some cells only responded to some kind of optimum luminance and hue combination as with a surface colour, other cells responded more along a particular axis, for example a decreasing or increasing luminance dimension, yet others responded to hue irrespective of the luminance relationship to a surround. Perhaps there are two sorts of processing at a cellular level, one very specific like feature detection and the other a more general response to several parameters.

Heggelund: The luminance relationship for colours is very similar to that for the white component in the achromatic domain and is different to that in the luminous-black domain.

Thomas: Steve Shevell, it is reasonable to compare central and monocular opponency?

Shevell: It has been done. If we have a null (a unique yellow) at a retinal level, the nature of the intensity-response function of the system doesn't matter; a zero input gives zero output. At a central level, the nature of that function becomes critically important. Some of Lee Guth's work on dichoptic presentation suggests a non-linearity is present in these functions; you don't get cancellation with the same lights that cancel at the retinal level of opponency.

Creutzfeldt: Our models suffer from one problem in quantification neurophysiologically. As well as what happens in the receptors, changing adaptation levels changes the maintained discharge level which may change the signal to noise as well as the gain of the cell. Do we see the absolute amplitude of a response or do we see the signal-to-noise ratio?

Mollon: In this respect, Zaidi's experiments are part of a large class of sensory experiments, the masking experiments. I wonder if these experiments tell you anything unless you have a model of the noise in the system. Provided you assume some noise, any detection system will do worse when the vector of the target and the noise are the same.

Zaidi: To decide this, we are now doing experiments with a number of directions of backgrounds and probes.

Kaplan: It is well known that thalamic neurons are well able to differentiate a signal. Why must it be more complicated than that?

Zaidi: I would like to see if you could show differentiation of these stimuli in chromatic thalamic cells.

Spillmann: I should like to suggest some points for discussion over lunch. Both assimilation and contrast have in common they can be induced by rather small stimuli, thin lines or whatever. Secondly, they extend over very large areas, up to 50 degrees. Thirdly, these effects work better with low or medium contrast of the inducing stimulus than with high contrast and lastly they work with fuzzy borders. I think this is a challenge to the neurophysiologist.

ADAPTATION MECHANISMS IN COLOR AND BRIGHTNESS

Mary Hayhoe and Peter Wenderoth

Center for Visual Science, University of Rochester, Rochester, NY 14627, USA and
Psychology Department, University of Sydney, Sydney, Australia 2006

In order for color to be a useful property for segmenting and identifying objects in a scene it must remain constant in spite of variations in intensity and color of the ambient light, and we know that the visual system achieves an elegant deconfounding of incident light and the reflectance of objects. The last decade has seen considerable advances in the formal analysis of color constancy. These advances have not been matched, however, by advances in understanding whether or how the nervous system carries out these computations. Models of constancy generally require some sort of 'knowledge' or computation of the illuminant (see review by Lennie & D'Zmura, 1988). This knowledge may be realized in the visual system to a large extent by mechanisms of light adaptation which can compensate for the changes in the illuminant so that objects of constant reflectance will have a constant effect following adaptational transformations (D'Zmura & Lennie, 1986). Light adaptation is generally acknowledged to play a major role in both color and brightness constancy, but the nature of the adaptational transformations and how they contribute to constancy are poorly understood. In this chapter we will report on preliminary experiments which attempt to specify how light adaptation mechanisms operate to determine the color and brightness of lights. First, we will describe what we know about light adaptation from measurements of sensitivity and then describe how this might be extended to account for color appearance and brightness.

Retinal light adaptation mechanisms transform the visual signal to one which reflects contrast, or variation around the mean, rather than the mean level itself. This is reflected in Weber's Law: a constant small contrast is required for threshold. There is fairly general agreement that the adaptational transformations involve gain control mechanisms (multiplicative adaptation) and ac coupling mechanisms (subtractive adaptation, spatio-temporal filtering), as originally proposed by Barlow and Levick (Barlow, 1965; Barlow & Levick, 1969, 1976) who used the analogy of a zero offset adjustment for the subtractive operation (Adelson, 1982; Geisler, 1981; Walraven & Valeton, 1984; Hayhoe et al, 1987). Subtractive operations remove the steady state signal without affecting added transients. Multiplicative mechanisms, on the other hand, will scale down the transients as well. Subtractive operations become important in determining overall sensitivity if there are significant non-linearities in the system, which limit the available response range. Such limits are indeed set by the limited range of responses in retinal ganglion cells, and more proximal elements such as the bipolar cells.

A schematization of these ideas is shown in Fig 1. (This reflects a combination of the work of Geisler, Hood, Adelson, Walraven, Valeton, and Hayhoe, among others.) Following linear transduction, the signal at light onset is attenuated by a multiplicative gain control mechanism (probably feedback). The signal is further reduced by subtractive lateral inhibition (center-surround antagonism), followed by another subtractive mechanism which slowly removes most of the residual signal. It is not very clear where the gain control mechanism (or mechanisms) reside. At least part of it appears to be post-receptoral. The evidence from physiology is conflicting about whether or not there are receptoral gain changes (Walraven et al, 1989). The gain changes appear to be very fast (within 50 msec, Hayhoe et al, 1987; Robson & Powers, 1988) in the low to mid-photopic range.

From Pigments to Perception, Edited by A. Valberg and
B.B. Lee, Plenum Press, New York, 1991

At higher intensities (above about 3.5 log td) a slower multiplicative process appears to come into play, probably related to pigment bleaching (Geisler, 1983). The center-surround antagonism is probably essentially instantaneous, and the third mechanism is quite slow, taking 10-15 sec for lights of moderate intensities (Hayhoe & Levin, 1990). The final non-linearity reflects the limited operating range of the stage or stages of the visual system responsible for the saturating effect of flashes. (In the absence of a representation of both photon and neural noise in the model, it also implicitly incorporates the effects of noise.) The adaptation mechanisms accomplish the transition to steady state sensitivity following light onset.

The conception of adaptational transformations outlined in Fig 1 was developed to account for sensitivity in the presence of achromatic adapting fields. However, a model of the same general form is required to account for both chromatic and brightness induction, and for sensitivity to colored flashes on colored backgrounds. For example, the Pugh & Mollon model (1979), which was proposed to account for transient tritanopia and a range of related phenomena, posited two sites of adaptation: receptor gain changes at the first site, followed by response compression at a chromatically opponent site which was moderated by a very slow subtractive process (restoring force). This is qualitatively very similar to the model outlined above. Similarly, the color of an increment on a colored background can be explained by von Kries adaptation in the receptors (corresponding to a multiplicative or gain control mechanism), followed by a subtractive process which discounts most or all of the residual background signal (Hurvich & Jameson, 1958; Jameson & Hurvich, 1972; Walraven, 1976; Shevell, 1978). Similarly the work of Whittle & Challands (1969) demonstrates that a background discounting mechanism is required in addition to changes in gain to describe the brightness of increments on white backgrounds.

If the same mechanisms were responsible for sensitivity, brightness, and color, this would result in a simple and elegant picture of retinal transformations, with immediate implications for constancy. However, this is unlikely for several reasons. One of the primary functions of adaptation is to preserve sensitivity over a wide range of illuminances. The problem for both sensitivity and brightness constancy is to cope with a wide range of absolute levels. For color constancy it is only required to compensate for changes in the distribution of excitation across the three cone types. Light adaptation requires both fast and slow mechanisms. It must be fast so that the organism is not incapacitated by sudden large changes in illumination. At the same time if it were all accomplished within a single fixation scenes would undergo rapid fading and revival with every eye movement. Color constancy, however, calls for rather slow adaptation processes. Many algorithms for constancy depend on adaptation to the average light reflected from a scene. If the space average reflectance is spectrally flat, then the average light will have the chromaticity of the illuminant, and adaptation to the mean can then allow recovery of reflectance. D'Zmura & Lennie (1986) suggest that this can be achieved with localized but sluggish adaptation processes. The rapid gain changes observed in sensitivity experiments would subvert adaptation to the space average light. That is, we do not want a system for chromatic adaptation which adapts significantly to individual objects. We might therefore expect to find different mechanisms underlying color and brightness constancy. In preliminary experiments we have investigated the adaptational mechanisms which underlie the color appearance of lights, and compared them to the adaptational mechanisms which determine sensitivity as outlined in Fig 1. The mechanisms which regulate sensitivity probably determine brightness as well, but the mechanisms for color appear to be quite different.

We will consider the color of increments on colored backgrounds in the chromatic induction paradigm. This is like a simple 'scene' with only two objects. If we are to analyze the effects of light adaptation we need a simple display like this. We will therefore avoid the higher level processes

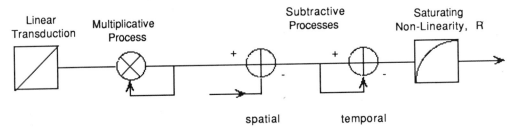

Fig 1. A model of adaptation, described in the text.

which operate in real scenes, where segregation of objects in depth, shadows, color memory, and other factors clearly affect color appearance (eg Gilchrist, 1980, 1983). The chromatic induction paradigm is a useful one to investigate since it is relatively easy to separate multiplicative and subtractive processes. Separation of the different mechanisms in chromatic induction is of interest since the two types of process affect the chromaticity of a region in different ways as the contrast varies (see Werner & Walraven, 1982). In order to reveal the mechanisms of chromatic induction and compare them with those controlling sensitivity on achromatic fields, we have measured the time course of multiplicative and subtractive mechanisms following the onset of the chromatic field.

ADAPTATIONAL MECHANISMS IN CHROMATIC INDUCTION

The question here is what processes mediate the changes in color appearance of a field following the onset of a chromatically biased background and to what extent they reflect the same processes that determine sensitivity on achromatic fields. An analysis of the time course is one way of identifying the underlying mechanisms, and is also important in predicting the consequences of eye movements on constancy. Both sensitivity regulation and chromatic induction are thought to have in common a stage of receptor gain adjustment. On the basis of the evidence from adaptation to achromatic fields one would expect this to be extremely fast at levels where there is insignificant pigment bleaching (Hayhoe et al, 1987; Hayhoe & Levin, 1989). However, experiments by Shevell (1982), who measured the changes in unique yellow settings following the offset of a red field, showed very little recovery of gain over a period of 1 sec. In this experiment we consider the changes in unique yellow following background onset, in order to facilitate comparison with the experiments on achromatic backgrounds.

The subtractive adaptation mechanisms are also of interest. The von Kries gain changes will not completely compensate for spectral bias in the illuminant (D'Zmura & Lennie, 1986). Subtractive processes will also discount some of the illuminant. In the chromatic induction situation this discounting may in fact lead to an overshoot in constancy as Walraven (1988) and Blackwell & Buchsbaum (1988a) have pointed out. Removal of all the background signal will lead to constancy to the extent that the background signal (i.e. the signal common to all fields in a scene) is due to the illuminant. However since part of the common signal may in fact be due to similarity in the reflectance of the neighboring regions, chromatic induction will result. This would correspond to a situation in which the space-average reflectance is not spectrally flat, resulting in inappropriate discounting of the illuminant, as pointed out by D'Zmura & Lennie. In a multi-object (spectrally flat) scene, such an overshoot in the discounting mechanism will not occur as long as the mechanisms are quite slow. The time course of the subtractive processes were therefore measured in addition to the gain changes. It is not clear how this will relate to the achromatic case. There are numerous instances of chromatic adaptation that take much longer than 10-15 secs (Jameson et al, 1979; Mollon & Polden, 1977; Augenstein & Pugh, 1977). Time courses of minutes are more the rule. One might therefore expect to find evidence of a much slower subtractive process that reflects the action of a chromatically opponent stage and is not activated by the achromatic field. The subtractive mechanism is indeed generally associated with a chromatically opponent stage (Larimer, 1981, Pugh & Mollon, 1979; Jameson & Hurvich, 1972).

Method

The time course of the changes in unique yellow were measured following the onset of a red background, and from this it was possible to separately infer the time course of multiplicative and subtractive components. We repeated Walraven (1976) and Shevell's (1978) classic experiments which showed that the color of a yellow increment on a red background could be explained by receptor gain changes (von Kries adaptation) plus a second mechanism which discounts (subtracts) the background. However, the experiment was extended by measuring the effect of the red background on unique yellow judgements as a function of time after onset of the red field. The yellow field was made up of a mixture of red and green lights ($\lambda = 660$ and 540). The green was adjusted to keep the test field appearing unique yellow. The background was 656nm. The background subtended 8 deg and the test 40 min arc. The test was flashed for either 500 or 150 msec. Pilot observations showed that at durations shorter that 150 msec, more green light is required for unique yellow settings when the judgements are made in the presence of a steady red background, suggesting that retinal stimulation from the red background occurring outside the time window defined by the increment influences the judgement. Unique yellow judgements were made over a range of test intensities in the dark, in a steady state of light adaptation, and at various intervals following the onset of the red background. The sequence of events is shown in Fig 2a. The red background was re-cycled at 20 sec intervals. Stimuli were presented using a 4-channel, computer-controlled Maxwellian View system, with one of the channels being used for a continuously-visible white background just below cone dark adapted threshold to minimize rod intrusion and maintain fixation.

Results

Multiplicative and subtractive components can be easily separated in this paradigm. This is shown in Fig 2b. If the amount of green in the mixture is plotted against the intensity of the red increment the ratio will be constant for settings in the dark. That is, $\Delta G / \Delta R = k$; equivalently $\log \Delta G = \log \Delta R + \log k$, a line with a slope of 1.0 on a log-log plot, as shown by the open circles in Fig 2b. In the absence of any adaptation, the settings would follow $\Delta G / (\Delta R + R) = k$ when the background, R, is present. However, because the von Kries gain changes make the red increment less effective, and because the background signal is reduced to $f(R) = R - S(R)$ by the background discounting mechanism, the settings will follow the function: $\Delta G / g(\Delta R + f(R)) = k$, where g, the relative attenuation of the red light, is less than 1.0. When the background is completely discounted, $f(R) = 0$ and $\Delta G / \Delta R = gk$ (filled diamonds in Fig 2b). This is also true for ΔR large relative to $f(R)$, so at high test intensities

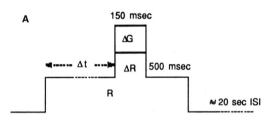

A

Fig 2a. The stimulus sequence.

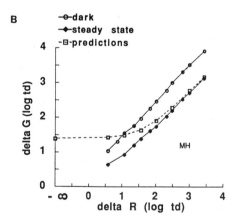

B

Fig 2b. Intensity of the green increment required for unique yellow, as a function of the intensity of the red increment. Open circles show settings in the dark adapted fovea; squares show settings in the presence of a steadily exposed 1.7 log td red background; triangles are predictions for settings shortly after background onset if the gain changes are fast and the subtractive changes are slow.

the relative gain change can be estimated from the lateral shift of the curve on a log-log plot. If there is a residual additive component (f(R) greater than zero), then the curve flattens out at low test intensities, since this signal from the background will dominate the settings, and f(R) can be estimated from the value of ΔG at $\Delta R = 0$, once the value of g is known. If the gain change is very rapid, the straight line part of the curve should shift immediately to the light adapted asymptote and only the bottom part of the curve, reflecting the subtractive process, should change during the transition to the steady state (open triangles in Fig 2b).

The measurements show that this was far from the truth. Fig 3 shows data for adaptation to a 1.5 log td background. The squares in part a of the figure show the state of adaptation 100 msec after the onset of the red background. No gain change is apparent, and the curve falls close to that

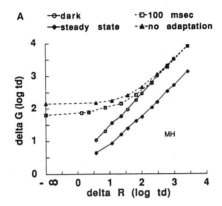

Fig 3a. Intensity of the green increment required for unique yellow, measured 100 msec after the onset of the red background (squares). The triangles show where the data would lie if there is no adaptation and the red light from the background simply adds to that in the increment. The dark adapted and steady state light adapted settings are reproduced from Fig 2.

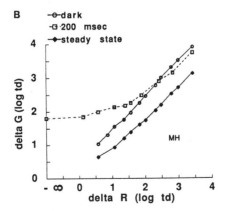

Fig 3b. Squares show measurements 200 msec after the onset of the red background.

357

expected if no adaptation at all had occurred, that is, if the light from the background, R, simply added to the light in the test, ΔR filled triangles). By 200 msec, a small gain change has occurred (Fig 3b), and it progresses slowly so that by 5secs it is only half complete (part c). The floor of the curve also sinks slowly during this period. This reflects the combined action of multiplicative and subtractive processes, since the gain change alone will effectively attenuate the signal from the background. Data for another S on a brighter, 2.5 log td background are shown in Fig 4.

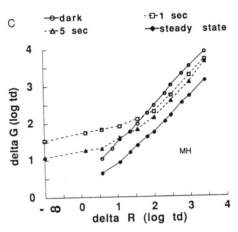

Fig 3c. Measurements made 1 sec (squares) and 5 sec (triangles) after onset.

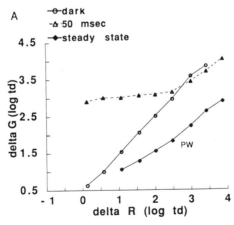

Fig 4a. As for Fig 3a, for a different subject. In this case the red background was 2.5 log td.

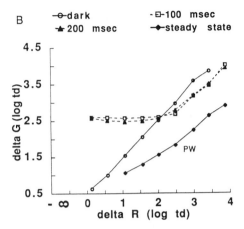

Fig 4b. As in part a, measurements made 100 and 200 msec after background onset.

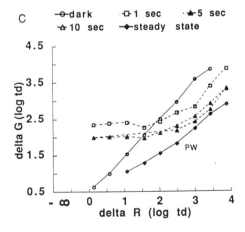

Fig 4c. Measurements made 1, 5, and 10 sec after background onset.

For this S a small gain change is apparent by 50 msec, but the time course is similarly slow, with the adaptation still incomplete by 10 sec. From these data we extracted the values for the multiplicative and subtractive components, as described above, and these are plotted as a function of time in Figs 5 and 6, from the data in Figs 3 and 4 respectively. Note that the duration of the test field was 500 msec in Fig 3 and 5, and 150 msec in Fig 4 and 6. Since some adaptation may be occurring during the 500 msec period of the test flash we would expect this data to overestimate the adaptation in the first half sec. It is clear, nonetheless, that the multiplicative changes take many secs, rather than msec. The subtractive process appears substantially slower than for adaptation to

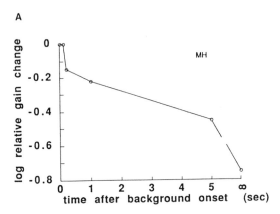

Fig 5a. Reduction in the relative gain in response to the red and green increments as a function of time after onset of the red background calculated from the data in Fig 3.

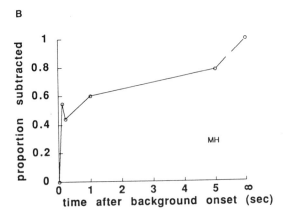

Fig 5b. Proportion of the signal subtracted from the red background (1 - f(R)/R, where f(R) is the residual background signal as discussed in the text) plotted as a function of time after background onset.

achromatic fields, although further measurements are necessary. On the brighter background (Fig 6b), there appears to be little subtractive adaptation in the first 10 sec. Since about 0.75 of the signal is discounted in the steady state, most of the subtractive adaptation must therefore be occurring after the first 10 sec. (A similar result was found on subject MH at this background). On the dimmer background the process appears to be somewhat faster. Other measurements indicated that it takes about a minute to reach the steady state after the onset of the 1.7 log td field, and about 3 min or more for the 2.5 log td field. This is true at test intensities where the effect of an additive signal from the background would be negligible, so it appears that the gain changes must take minutes to reach asymptote. At this point it is not clear whether the subtractive changes are faster than this.

A

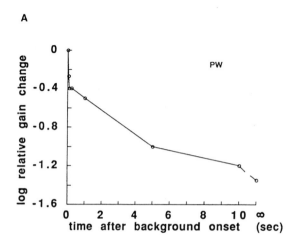

B

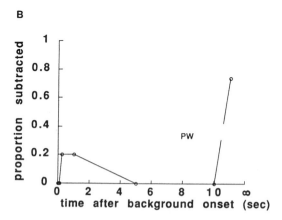

Fig 6. Time course of relative gain and proportion subtracted, calculated from the data in Fig 4.

Implications

These results are consistent with the findings of Shevell (1981) that the multiplicative adaptation in chromatic adaptation is a rather slowly acting process. The rapid local gain changes revealed in threshold measurements do not seem to affect color appearance. This is inconsistent with a common stage of receptoral (or receptor specific) gain adjustment for both sensitivity and color appearance, which is almost universally assumed in the literature on chromatic adaptation. There cannot be both fast receptor gain changes for sensitivity and slow ones for chromatic induction, although the von Kries changes are usually referred to as receptor gain changes. There must be two different gain mechanisms involved with different time course and they cannot both be in the receptors. Some support for the present claim is found in Walraven (1981) and Werner & Walraven (1982), who needed pi mechanism spectral sensitivities to predict their data on chromatic induction, although they used the absorption spectra of the receptors as the input. It may also be related to the finding by Ahn and MacLeod (1989) that chromatic adaptation had different effects on unique yellow settings and flicker photometry. If indeed luminance and chromatic adaptation have separate initial stages, this raises interesting questions about the retinal basis of the separate gain mechanisms. It is possible, for example, that the two types of cone bipolar synapse subserve different gain control mechanisms.

One alternative to this scheme of no common receptor adaptation would be if slow receptor gain changes were followed by rapid ones in the luminance pathway. The fast, proximal gain mechanism would respond immediately to a change in background intensity. A subsequent slow decrease in receptor gain might then be compensated by an increase in the proximal gain mechanism, giving a constant combined effect of the two stages. Given that the physiology is ambiguous about the existence of receptor gain changes, it is hard to evaluate this possibility. The distinction made previously in the literature between fast 'neural' and slow 'photochemical' adaptation is consistent with such a scheme (eg Baker, 1955, 1963; Boynton & Kandel, 1967; Battersby & Wagman, 1959). One way to test this hypothesis is to investigate recovery of gain at the offset of an adapting field, because in this case it is the slow mechanism which will be limiting, and recovery at offset will reflect the time course of the slow receptoral change. Indeed there are many instances in the literature where recovery of sensitivity at offset is substantially slower than at light onset (eg Sharpe & Mollon, 1982).

It is also interesting that the slow chromatic gain changes observed here are much better suited to adapting to the space-averaged chromaticity than the fast adaptive changes observed with white backgrounds. The subtractive changes, too, are probably such slower. The fast changes appear to be confined to a luminance system, and have separate implications for brightness constancy. In brief, it appears that light adaptation on white fields is very different from chromatic adaptation, despite their superficial similarity.

THE INFLUENCE OF LIGHT IN THE SURROUNDING REGIONS

Although subtractive mechanisms clearly can play an interesting role in discounting the illuminant we currently have little idea about the underlying physiological mechanisms. The lateral inhibitory mechanism in Fig 1 would subtract out some of the space-average signal instantaneously without requiring eye movements at all, so this is a potentially powerful mechanism, particularly given its importance in sensitivity control. Lateral inhibition has often been considered to underlie color contrast/induction and is referred to as a 'simultaneous mechanism' by Arend and Reeves (1986) and Blackwell & Buchsbaum (1988). In addition, many models of lightness constancy share an initial stage of spatial filtering, which would be achieved by center-surround antagonism (Hurlbert, 1986).

Retinal center-surround antagonism is a natural candidate for discounting the background in chromatic induction. With achromatic backgrounds, the influence of signals in the retinal region surrounding the test on sensitivity is quite simple. Signals in the surrounding retina are subtracted from those in the center, and this improves increment sensitivity as revealed in the Westheimer effect (Hayhoe, 1990). Moreover, there is probably very little delay in this signal. The influence of light in the surrounding retinal regions on induced color, however, seems rather complicated. First, the effects are generally rather small (Arend & Reeves, 1986; Blackwell & Buchsbaum, 1988a; Shevell, 1989; Humanski, 1988; Nick & Larimer, 1984), unlike the threshold effects (Buss et al, 1983).

In addition, multiplicative as well as subtractive effects of surrounding stimulation have been reported (Shevell, 1989; Wesner & Shevell, 1989; Shevell, 1981; Humanski, 1988). The effects are rather slow (Krauskopf et al, 1986; De Valois et al, 1986), and spatially extensive (De Valois et al, 1986; Wesner & Shevell, 1989). The lateral extent of the effects on threshold of light falling on the surrounding retina, however, are fairly small and consistent with the extent of retinal receptive field surrounds (Ransom-Hogg & Spillman, 1980). This all adds up to rather different pictures for the role of the surrounding retina in the two types of experiment. It appears that multiple non-local mechanisms may be operating in chromatic induction. At this point it is not clear whether the strong, subtractive, center-surround effects which appear to regulate psychophysical sensitivity (see Fig 1) have any role at all in chromatic induction.

Further evidence for this conclusion is given in Fig 7. This shows settings made at the onset of the red background. In the case of the open squares the background was 8 deg, as it was for the measurements above. This curve coincides with the curve expected if the light in the region of the test is simply adding to the red increment (symbols). This means that the additive effect of the large field is not attenuated relative to the 40 min test at light onset. There is no instantaneous subtractive effect induced by the large field, as there is with threshold measurements.

ADAPTATION MECHANISMS AND BRIGHTNESS

Although formal treatments of color constancy have implicitly included brightness constancy, it has received comparatively less explicit consideration. Several computational models (see Hurlbert, 1986) rely on an initial stage of spatial differentiation of the sort that would be achieved by center-surround antagonism (see Fig 1) in order to remove the common light from a scene. Other adaptation mechanisms, such as gain changes, must also be involved, however. It seems likely that the mechanisms which determine sensitivity on achromatic backgrounds will also determine brightness, even though the chromatic adaptation mechanisms are apparently unrelated. If this is so, then the adaptational mechanisms underlying brightness and color constancy must be quite different. In a classic experiment, Whittle & Challands (1969) measured the brightness of incremental flashes added to backgrounds of various intensities. The paradigm is the same as that used by Walraven an Shevell for chromatic induction, and the outcome can be described in the same terms: the brightness of an increment was determined by sensitivity set by the background, and in addition a background discounting mechanism was required, since the background did not add to the brightness of the increment. Whittle and Challands' data are reproduced in Fig 8a which shows constant brightness curves as a function of background intensity. Their data are replotted in Fig 8b in the same waya as the chromatic induction data. The brightness of the increment required to match a

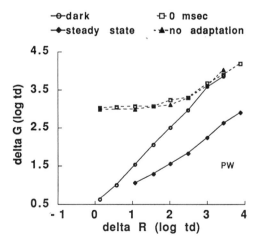

Fig 7. Unique yellow settings at the onset of the 8 deg, 2.5 log td red background compared with predictions for no adaptation.

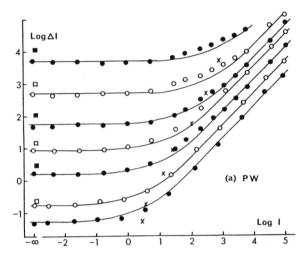

Fig 8a. Data from Whittle & Challands (1969) showing the intensity of an increment in one eye which matches a standard light in the other eye, as a function of background intensity against which the variable increment is viewed. The different curves are for different intensities of the standard.

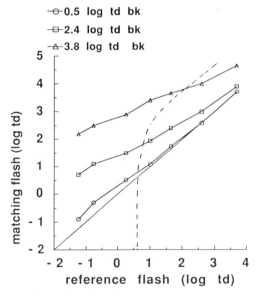

Fig 8b. The same data replotted in the manner of Figs 3 and 4. The intensity of the variable increment which matches the standard is plotted against the intensity of the standard, for three different background intensities in the variable eye. The dotted curve shows expectations if there were no background discounted (subtractive) mechanisms.

standard light in the other eye is plotted against intensity of the standard. The straight line of slope 1.0 represents the matches in the dark. The points show the matches on backgrounds of three different intensities indicated in the legend. The curves are displaced up as expected because of the gain reduction by the background. However, instead of having a slope of 1.0, they converge at high test intensities. That is, the effect of the background diminishes as the increment gets brighter. The effect of the background cannot be described as adding a neutral filter, as suggested by Walraven et al (1989). (The data of a similar study by Onley & Boynton (1962), replotted in Geisler (1978), show some convergence, although less than in Whittle & Challands data.) If the gain changes with white lights are very rapid, as the adaptation data suggest, it seems likely that the convergence in Whittle and Challands data result from adaptation to the increment itself during the 200 msec exposure period. (Presumably the gain will be determined by the sum of background plus increment.) When the increment is bright the gain will be primarily determined by the increment and the match will be closer to a physical match. In a subsequent experiment (Whittle, 1973) where the duration of the flash was only 40 msec, the slopes of the lines, plotted as in Fig 8b, are much closer to 1.0. Thus when there is insufficient time for adaptation to the increment, it appears that the background is the primary determinant of the gain. Since the duration of a single fixation in normal viewing is 200 msec or more, it seems likely that local adaptation to individual objects will influence their brightness. The final point of interest in this figure is that there is no hint of an additive signal from the background. If the background were attenuated only by a gain change, with no background discounting, the bottom part of the curve would look like the dashed curve. As Whittle & Challands point out, that would mean that an increment in the reference eye would be matched to a decrement in the test eye, which never happens. (The top part is where it would be if the gain is set only by the background).

What are the implications of this for constancy? Shapley (among others) (Shapley et al, 1989; Shapley, 1986) has argued that as the eye moves from one object to another, the effective stimulus will be the difference in amount of light coming from the two regions, that is IR_o-IR_b, where R_o and R_b are the reflectances of the first and second regions (background and object), and I is the illumination. This will be multiplied by the gain, which will be proportional to $1/IR_b$ as long as Weber's law holds, and the gain is set by the background. Thus the effective stimulus will be $(IR_o$-$IR_b)/IR_b$ = R_o-R_b/R_b, which is the stimulus contrast. Thus as long as the neural response is linear, it will be proportional to contrast and independent of illumination. On this view it will make little difference if the gain is set by IR_o instead of IR_b. That is, the constant brightness curves still follow Weber's Law (ie constant contrast means constant brightness) as Whittle & Challands observed. In Fig 8b this is reflected in the fact that the curves (eg the filled and open circles) are parallel over a wide range of flash intensities, even though the slope is less than 1.0. It can be seen from the dashed line in Fig 8b, which shows predicted matches without a subtractive mechanism, that the subtractive processes are important in preventing the background from contributing to the brightness of the increment. That is, it allows the brightness to be determined by the increment (IR_o-IR_b). If the entire background signal is discounted this will presumably lead to an overshoot in constancy to the extent that background and object are similar in reflectance, as discussed in for chromatic induction. This may be moderated by the slow subtractive mechanism discussed in Fig 1, which would tend to stabilize at a value appropriate for the scene average. Complete background discounting may only occur with large steady backgrounds.

SUMMARY

Retinal adaptation mechanisms are the first step in assigning a reflectance value to an object in a complex scene; that is, in lightness and color constancy. In this chapter we have attempted to specify the nature of the early retinal transformations. The data we have presented indicate that the mechanisms which mediate achromatic sensitivity and brightness induction, on the one hand, and color induction, on the other, are markedly different. Previous data using achromatic stimuli had shown that adapting fields produce very fast multiplicative and subtractive effects, as well as slower subtractive effects, resulting in Weber's law and constant brightness with constant contrast. We have shown that with the chromatic adaptation paradigm, there is little or no fast subtractive discounting of the background at onset; that multiplicative gain changes occur slowly even with a relatively intense adapting field; and that very little of a slower background discounting (subtractive) effect occurs prior to 10 seconds of adaptation. While these results suggest that achromatic sensitivity and brightness are mediated by mechanisms different from those underlying chromatic induction, the different mechanisms probably reflect the different functional requirements of achromatic and chromatic systems.

ACKNOWLEDGEMENTS

Supported by NIH grant EY05729 to M.H., EY01319 to the Center for Visual Science, and University of Sydney Special Study leave to P.W.

REFERENCES

Adelson, E. A., 1982, Saturation and adaptation in the rod system, Vis. Res., 22:1299.
Ahn, S. J. and MacLeod, D. I. A., 1989, Adaptation mechanisms of the chromatic and the luminance channels, Invest. Ophthal. Vis. Sci. Suppl., 30:322.
Arend, L. and Goldstein, R., 1990, Lightness and brightness over spatial illumination gradients, Opt. Soc. Am. A., in press.
Arend, L. and Reeves, A., 1986, Simultaneous color constancy, J. Opt. Soc. Am. A., 3:1743.
Augenstein, E. J. and Pugh, E. N., 1977, The dynamics of the II_1 colour mechanism: further evidence for two sites of adaptation, J. Physiol., 272:247.
Baker, H. D., 1955, Some direct comparisons between light and dark adaptation, J. Opt .Soc. Am. A., 45:839.
Baker, H. D., 1963, Initial stages of dark and light adaptation, J. Opt. Soc. Am. A., 53:98.
Barlow, H. B. and Levick, W. R., 1969, Coding of light intensity by the cat retina, Proc. Intl. Schl. Phys. - Enrico Fermi, 385-396.
Barlow, H. B., 1965, Optic nerve impulses and Weber's law, Cold Spring Harbor Symposia on Quantitative Biology, Vol. XXX:539.
Barlow, H. B. and Levick, W. R., 1976, Threshold setting by the surround of cat retinal ganglion cells, J. Physiol., 259:737.
Battersby, W. S. and Wagman, I. H., 1959, Neural limitations of visual excitability. I. The time course of monocular light adaptation, J. Opt. Soc. Am., 49:752.
Blackwell, K. T. and Buchsbaum, G., 1988a, Removing the common signal: a unifying principle for color constancy and color induction, Opt. Soc. Am., Tech. Digest, 17:104.
Blackwell, K. T. and Buchsbaum, G., 1988b, The effect of spatial and chromatic parameters on chromatic induction, Col. Res. Appl., 13:166.
Boynton, R. M. and Kandel, G., 1957, On responses in the human visual system as a function of adaptation level, J. Opt. Soc. Am., 47:275.
Derrington, A. M. and Lennie, P., 1984, Spatial and temporal contrast sensitivities of neurons in Lateral Geniculate Nucleus of Macaque, J. Physiol., Lond., 357:219.
De Valois, R. L., Webster, M. A., De Valois, K. K., 1986, Temporal properties of brightness and color induction, Vis. Res., 26:887.
D'Zmura, M. and Lennie, P., 1986, Mechanisms of color constancy, J. Opt. Soc. Am. A., 3:1662.
Fairchild, M. D., 1990, Chromatic adaptation and color appearance, Doctoral dissertation, University of Rochester.
Geisler, W. S., 1978, The effects of photopigment depletion on brightness and threshold, Vis. Res., 18:269.
Geisler, W. S., 1981, Effects of bleaching and backgrounds on the flash response of the cone system, J. Physiol., Lond., 312:413.
Geisler, W. S., 1983, Mechanisms of visual sensitivity: Backgrounds and early dark adaptation, Vis. Res., 23:1423.
Gilchrist, A. L., 1980, When does perceived lightness depend on perceived spatial arrangement, Percep & Psychophys, 28:527.
Gilchrist, A. L., Delman, S. and Jacobsen, A., 1983, The classification and integration of edges as critical to the perception of reflectance and illumination, Percep. & Psychophys., 33:425.
Hayhoe, M. M., 1990, Spatial interactions and models of adaptation, Vis. Res., 30:957.
Hayhoe, M. M., Benimoff, N. I., Hood, D. D., 1987, The time-course of multiplicative and subtractive adaptation process, Vis. Res., 27:1981.
Hayhoe, M. M. and Levin, M., 1990, Subtractive processes in light adaptation, Manuscript submitted to Vision Research.
Humanski, R. A., 1988, Adapted SWS-cone contributions to redness/greenness. Doctoral dissertation, University of Chicago.
Hurlbert, A., 1986, Formal connections between lightness algorithms, J. Opt. Soc. Am. A., 3:1684.

Hurvich, L. M. and Jameson, D., 1958, Further development of a quantified opponent-color theory, in: "Vision Problems of Color II", HMSO, London, pp. 691-723.

Jameson, D. and Hurvich, L. M., 1972, Color adaptation: Sensitivity, contrast, and afterimages, in: "Handbook of Sensory Physiology: Visual Psychophysics", Vol. VII/4, D. Jameson and L. M. Hurvich, eds., Springer, Berlin, pp. 568-581.

Jameson, D., Hurvich, L. M., Varner, F. D., 1979, Receptoral and postreceptoral visual processes in recovery from chromatic adaptation, Proc. Natl. Acad. Sci., 76:3034.

Krauskopf, J., Zaidi, Q., Mandler, M. B., 1986, Mechanisms of simultaneous color induction, J. Opt. Soc. Am. A., 3:1752.

Larimer, J., 1981, Red/green opponent colors equilibria measured on chromatic adapting fields: evidence for gain changes and restoring forces, Vis. Res., 21:501.

Lennie, P. and D'Zmura, M., 1988, Mechanisms of color vision, CRC Crit. Rev. Neurobiol., 3:333.

Mollon, J. D. and Polden, P. G., 1977, Post-receptoral adaptation, Vis. Res., 19:435.

Nick, J. and Larimer, J., 1983, Yellow-blue cancellation on yellow fields: Its relevance to the two-process theory, in: "Color Vision", J. D. Mollon and L. T. Sharpe, eds., Academic Press.

Onley, J. W. and Boynton, R. M., 1962, Visual responses to equally bright stimuli of unequal luminance, J. Opt. Soc. Am., 52:934.

Pugh, E. N. and Mollon, J. D., 1979, A theory of the II_1 and II_3 color mechanisms of stiles, Vis. Res., 19:293.

Ransom-Hogg, A. and Spillmann, L., 1980, Perceptive field size in fovea and periphery of the light- and dark-adapted retina, Vis. Res., 20:221.

Robson, J. G. and Powers, M. K., 1988, Dynamics of light adaptation, Opt. Soc. Am., Tech. Digest, 11:67.

Shapley, R., 1986, The importance of contrast for the activity of single neurons, the VEP and perception, Vis. Res., 26:45.

Shapley, R., Caelli, T., Grossberg, S., Morgan, M. and Rentschler, I., 1990, Computational theories of visual perception, in: "Visual Perception: The Neurophysiological Foundations", L. Spillmann and J. S. Werner, eds., Academic Press.

Sharpe, L. T. and Mollon, J. D. Mollon, 1982, Dynamic changes in sensitivity to long-wavelength incremental flashes, Docum. Ophthal. Proc. Series, 33:53.

Shevell, S. K., 1978, The dual role of chromatic backgrounds in color perception, Vis. Res., 18:1649.

Shevell, S. K., 1981, Two independent mechanisms of chromatic adaptation, Invest. Ophthalmol. Vis. Sci. Suppl., 20:206.

Shevell, S. K., 1982, Color perception under chromatic adaptation: equilibrium yellow and long-wavelength adaptation, Vis. Res., 22:279.

Shevell, S. K. and Wesner, M. F., 1989, Color appearance under conditions of chromatic adaptation and contrast, Col. Res. Appl., 14:309.

Walraven, J., 1976, Discounting the background: The missing link in the explanation of chromatic induction, Vis. Res., 16:289.

Walraven, J., 1981, Perceived color under conditions of chromatic adaptation: Evidence for a gain control by Π-mechanisms, Vis. Res., 21:622.

Walraven, J., 1988, Chromatic induction: a misdirected attempt at color constancy?, Opt. Soc. Am., Tech. Digest, 17:103.

Walraven, J., Enroth-Cugell, C., Hood, D. C., MacLeod, D. I. A. and Schnapf, J. L., 1990, The control of visual sensitivity: Receptoral and postreceptoral processes, in: "The Neurophysiological Foundations of Visual Perception", in press, Spillman, L. and Werner, J. S., eds., Academic Press.

Walraven, J. and Valeton, J. M., 1984, Visual adaptation and response saturation, in: "Limits in Perception", A. J. van Doorn, W. A. van de Grind and J. J. Koenderink, eds., VNU Science Press, Utrecht.

Werner, J. S. and Walraven, J., 1982, Effect of chromatic adaptation on the achromatic locus: The role of contrast, luminance and background color, Vis. Res., 22:929.

Wesner, M. F. and Shevell, S. K., 1989, Changes in color appearance caused by dark contours in a chromatic adapting background, Opt. Soc. Am., Tech. Digest, 18:215.

Whittle, P., 1973, The brightness of coloured flashes on backgrounds of various colours and luminances, Vis. Res., 13:621.

Whittle, P. and Challands, P. D. C., 1969, The effect of background luminance on the brightness of flashes, Vis. Res., 9:1095.

Zaidi, Q., Hood, D., Shapiro, A., 1989, The time course of sensitivity change in S-cone color mechanisms, Invest. Ophthalmol. Vis. Sci. Suppl., 30:221.

TESTING THE CONTRAST EXPLANATION OF COLOR CONSTANCY

J. Walraven[1], T.L. Benzschawel[2], B.E. Rogowitz[2] and M.P. Lucassen[1]

[1]TNO Institute for Perception
Soesterberg, The Netherlands

[2]T.J. Watson Research Center
Yorktown Heights, New York, USA

INTRODUCTION

The light that an object reflects towards the eye is the product of surface reflectance and incident illumination. The problem how to decompose that product, and thus separate light from matter, is one of the central issues in the study of the visual system. Research in this area has been boosted by its obvious relevance for machine vision (e.g. Horn, 1985; Hurlbert and Poggio, 1988), but as yet, with only limited success.

Computational approaches to color constancy typically focus on the problems of extracting the spectral composition of the illuminant, usually on the basis of fore-knowledge of a few basis functions that may be used to (de)compose spectral surface reflectance functions and illuminant power distributions (e.g. Buchsbaum, 1980; Maloney and Wandell, 1986). An alternative strategy, possibly that of the visual system, is to evade this problem altogether by making color processing rather insensitive to illuminant changes. This is, in essence, the message of Land's Retinex theory (Land, 1977, 1986a), in which color is assumed to be defined by a triplet of receptor-specific lightness values. How lightness (or reflectance) is to be determined is another matter. Progress has been made (mainly in the machine domain) in making the problem amenable to computation. However, to what extent, and how, the algorithms in question may also be implemented in the human visual system, is still an open question (cf. Hurlbert, 1986).

Support for the (triple) lightness approach has been obtained in experiments employing the well-known "Mondrian" test pattern (McCann, McKee and Taylor, 1976; McCann and Houston, 1983). However, the Mondrian pattern has the disadvantage that it does not allow a straightforward computation of the local contrast between a test patch and adjacent, differently colored, patches. The importance of local contrast, which has only been recently acknowledged in the Retinex algorithm (Land, 1986b), has been demonstrated in various studies pointing out this neglected factor in the Retinex model (Shapley, 1986; Creutzfeldt, Lange-Malecki and Wortmann, 1987; Valberg and Lange-Malecki, 1990). Many studies have also demonstrated the importance of local changes, i.e. edges, as determinants of perceived color and brightness (Krauskopf, 1963; Arend, 1973; Walraven, 1973, 1977). Note that the contrast of a stimulus may

From Pigments to Perception, Edited by A. Valberg and
B.B. Lee, Plenum Press, New York, 1991

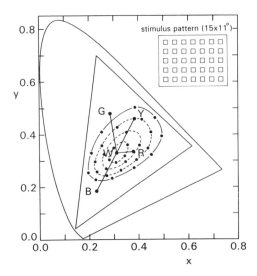

Fig. 1. CIE x,y chromaticity coordinates of simulated Munsell chips and illuminants. The latter were either blue (B), yellow (Y), red (R) or green (G). The samples are connected by lines of equal Chroma (8, 6, 4 or 2). The test samples are located on the line of highest Chroma (solid line). The achromatic samples coincide at the white point (W = D_{65}). The inset shows the geometry of the test pattern.

vary, since it depends on the luminance of the adjacent surround. Lightness, on the other hand, is invariant (at least when it is defined as the perceptual correlate of reflectance), since it is measured relative to a fixed (white) reference.

In the following we present data obtained with a stimulus pattern that, unlike the Mondrian pattern, allows easy calculation of border contrast. With that stimulus we could also create the stimulus conditions for chromatic induction. In a precursor to this communication we already discussed that chromatic induction and color constancy may have a common origin (Walraven, Benzschawel and Rogowitz, 1987). Here we present the data on which that conclusion was based, and also an analysis that is aimed at evaluating the role of (receptor) contrast in color constancy.

METHODS

Equipment

The stimuli were presented on a computer controlled 16-in color monitor (Hitachi). A box-shaped hood (65 cm long), with the inside painted matt black, was mounted in front of the display. The display was viewed through two holes in front of the box. Two patterns, a match and test pattern, were presented alternately (under the observer's control) to the left and right eye, respectively. The color of the matching stimulus could be manipulated by pressing keys controlling (with 8-bit precision) the RGB gun driving voltages.

Test stimulus. The test pattern consisted of 35 color samples (1°x1°), arranged in a 5x7 matrix, as shown in the inset of Fig. 1. The samples represented (colorimetric) simulations of 5 achromatic and 30 chromatic Munsell chips. The latter were distributed uniformly around the white point, at four different loci of equal Chroma (see Fig. 1). All samples "reflected" 50% of the incident light when "illuminated" by white light (D_{65}). Only part of the samples was used as test set, i.e. the 10 most saturated chromatic samples and one achromatic sample. The rest of the samples only served to provide a display with sufficient color variety.

The four colored "illuminants" (R, G, B and Y in Fig. 1) were defined as mixtures of the three phosphor primaries. The luminance produced by these "illuminants" (in combination with the 100% reflecting white grid) was not a variable. It was fixed at 13 cd/m², except for the blue light which, due to its low output, was set at 6.5 cd/m².

Since the stimuli were constrained to reflect only in the wavebands of the phosphors, we decided to constrain reflectance accordingly. Thus, object reflectance is defined by three coefficients, r_R, r_G and r_B, that determine how much of the "incident" light primaries (Y_R, Y_G and Y_B) is reflected. The color of the light source is similarly defined by three transmission coefficients, t_R, t_G and t_B that define a filter that transforms white light into colored light ($t_R Y_R + t_G Y_G + t_B Y_B$). Thus, in our "RGB world" the amount of light that the eye receives from a surface (s), illuminated by a test light (l), is given by

$$Y_i(s, l) = r_i(s).t_i(l).Y_i \qquad \text{with i = R,G,B} \tag{1}$$

Note that responding to $r_i(s)$ rather than to $r_i(s).t_i(l)$ is what color constancy is all about.

Match stimulus. The match pattern differed from the stimulus pattern in only two respects: it was always illuminated by white light, and all samples, except for the (adjustable) match stimulus, were shown as 50% reflecting achromats. The match stimulus was shown at a grid location corresponding to that of the test sample to be matched.

The rationale for presenting the match stimulus in (chromatic) isolation, was to make sure that the observer would concentrate on a single sample in the test pattern. He/she might otherwise try to match on a more cognitive level; that is, compare the color "position" of the sample relative to that of the other samples (in both match and test pattern), thereby ignoring its actual color. Such an attitude, whereby the observer recognizes a sample on a relative rather than an absolute basis may have a major effect on matching data (Arend and Reeves, 1986).

Visual scenarios. By manipulating the intervening space (the grid) between the samples, we created three different modes of color appearance. That is, surface colors, aperture or self-luminous colors, and induced colors. The surface and aperture mode result when using a white (100% reflectance) and dark grid (0% reflectance), respectively. The induced colors were produced by having the colored light selectively illuminate the (white) grid, whereas the samples were illuminated by white light. The latter configuration resembles the classical "colored shadows" demonstration (the samples are "shadowed" from the colored light source).

The task of the observer was to successively compare a sample under white and colored light, thereby adjusting the sample under white light (match sample) to match the one seen under colored illumination (test sample). The two stimuli were viewed in a self-paced sequence, that lasted as long as was necessary for a satisfactory match.

Three of the authors (B.R., J.W. and T.B.), all with normal color vision, good visual acuity and no abnormal ocular dominance, served as observers.

RESULTS

The results, averaged over the three observers, are plotted in CIE (x,y) coordinates in Fig. 2. Note that the physical color shift of the samples, due to the colored illumination (panel A), is not evident in the matches obtained for surface colors (panel B). The latter show that the perceived colors of the samples under colored light still resemble the colors seen under white illumination (solid line). In this condition, therefore, the visual system is capable of maintaining a fair degree of color constancy.

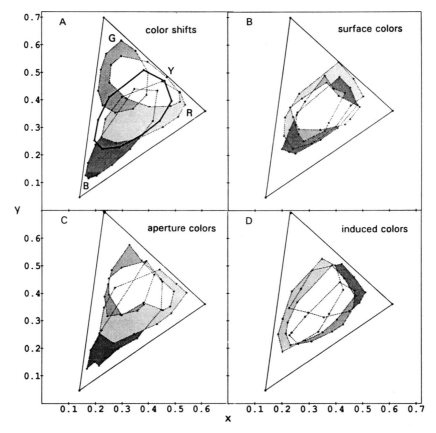

Fig. 2 A: Chromaticities of test colors under green (G), red (R), yellow (Y), and blue (B) illumination. The solid line represents the locus of the test samples under white light. B: Dichoptic color matches (in white light) to the shifted color sets shown in panel A. C: The same as B, for samples in the aperture color mode. D: The same as B, for samples in the induced color mode.

The data shown in panel C show, rather surprisingly, that color constancy is not very effective in the aperture color mode. This result has important theoretical implications, that will be addressed in the Discussion.

Panel D shows the familiar effect of chromatic induction. The matches indicate that the perceived color of the test set (which is illuminated by white light) are shifted in the direction of a color complementary to that of the grid. This induced color shift is opposite to the physical color shift (shown in panel A) that would be produced if the colored light illuminating the grid, would illuminate the entire test pattern. As pointed out before (Walraven et al., 1987), chromatic induction could thus be interpreted as a misdirected attempt at color constancy; misdirected, because the colored illumination incident on the surround is not incident on the samples.

ANALYSIS

We calculated for each sample and its match how the light (in cd/m^2) was distributed over the long-wave (L), middle-wave (M) and short-wave sensitive (S) photoreceptors. We used the Vos-Walraven cone fundamentals, as tabulated by Vos (1978), thereby normalizing the action spectra for equal sensitivity at equal-energy white (Walraven and Werner, 1986). Next, we determined the receptor-specific contrast, relative to the grid, for all test samples (under colored light) and their matches (under white light). This was not possible for the aperture colors (the background being absent), so these data were analyzed in terms of absolute rather than relative receptor inputs. The results, thus expressed in terms of receptor contrast, are plotted in Fig. 3. What is shown here is a comparison of the contrast produced by test (C_T) and match (C_M) samples, respectively.

The data shown in panel A of Fig. 3, those relating to the surface colors, indicate that matching a test sample implies matching the contrast it produces at the receptor level, at least for the L and M-cones. The S-cones are less well-behaved, indicating that, depending on the color of the test illuminant, the match contrasts may have to be increased or decreased by a certain factor. For the contrast range studied here, this factor seems to be fairly constant (for a given illuminant), since the S-cone data can be fitted, without too much stress, by straight lines.

The data shown in panel B of Fig. 3 do not suggest that the matches for induced colors follow a different pattern from that observed for surface colors. This provides the evidence for our earlier statement that chromatic induction and color constancy have a common origin (Walraven et al., 1987).

The data shown in panel C of Fig. 3 were obtained with samples on a dark background. So, in this condition, contrast can only be measured in terms of absolute receptor input. The values in question were multiplied by a factor 0.23 to enable plotting in the same scale units as used for the other data in Fig. 3. The receptor input is expressed in cd/m^2 per receptor system (as discussed above), but any other receptor-weighted unit might be used as well, as long as it is applied to both test and match samples. In as far as these data fall along the 45° line, they represent absolute matches, and hence, indicate the absence of color constancy. This appears indeed to be the general pattern, as was already apparent from Fig. 2C. Still, there are deviations from the physical identity match (particularly under blue light) that remain to be explained. One can think of local (chromatic) adaptation effects and/or a mechanism that operates beyond the borders of the samples.

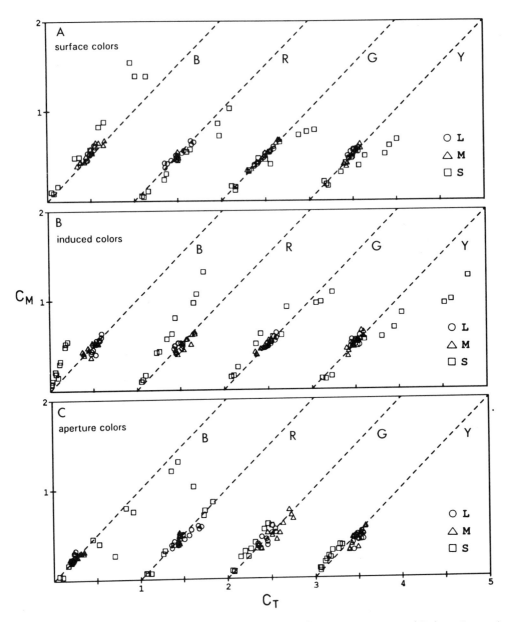

Fig. 3 Comparison of L, M and S (receptor-specific) contrast for test (C_T) and match (C_M) samples. The three panels show the data obtained for surface colors (A), induced colors (B) and aperture colors (C), respectively. The four data sets in each panel relate to the four different colors R(ed), G(reen), B(lue) and Y(ellow) of the light illuminating the test pattern. The sets are laterally shifted by 0, 1, 2 and 3 log units relative to the origin. In panel C the C_T values for the S-cones had to be halved in order to fit the scale format. In panel D, the contrast scale represents absolute rather than relative receptor input, as explained in the text.

DISCUSSION

The short-wave system is known to differ in a number of respects from the other cone systems (e.g. Mollon, 1977). In this study we showed, for the first time, that this is also true for its role in color constancy. The present data do not provide sufficient information for any firm conclusions, but an analysis, based on a more extensive data set (Lucassen and Walraven, 1990; Walraven and Lucassen, 1990), indicates that there is a second process involved, that possibly responds to the degree to which the short-wave system is selectively stimulated.

The results obtained for aperture colors (black grid condition), which show only a very weak color constancy effect, confirm earlier (qualitative) findings by Karp (1959), who used a similar stimulus pattern, but with a much finer grid (more like a network). The same result has also been obtained in a study by Tiplitz Blackwell and Buchsbaum (1988), in an entirely different experimental paradigm. It is noteworthy that the near absence of color constancy in the dark-grid condition, cannot be accounted for by the Retinex theory, and probably neither by models that are based on estimating the illuminant. The latter should be capable of estimating the color of the illuminant on the basis of the 35 samples in the stimulus pattern, irrespective of whether they are separated by a white or a black grid.

The main result of this study is that, except for the "short-wave factor", the data support the simple hypothesis that the color of a sample is fully determined by the sample/surround (receptor) contrast. We did not have to invoke a mechanism for estimating the illuminant. It is of interest in this respect that, in the chromatic induction scenario, the visual system could have treated the stimulus pattern as being illuminated entirely by white light, that is, samples displayed on a colored background. Why it did not do so, but instead, assigned "wrong" colors to the samples, is not self-evident. The phenomenon by itself, chromatic induction (simultaneous contrast) is familiar of course, but that does not explain its *raison d'être*. It does make sense, however, if we consider that exactly that same "mistake" (responding to contrast) provides a simple means for obtaining a fair degree of color constancy (see Fig. 2B).

Contrast has already been identified as the relevant variable for the perception of achromatic stimuli (e.g. Wallach, 1948; Shapley, 1986). This applies to lightness, the perceptual correlate of reflectance, rather than brightness (cf. Walraven, Enroth-Cugell, Hood, MacLeod and Schnapf, 1990). Thus, an increase in illumination, which does not affect reflectance (nor contrast), will be recognized as such, whereas lightness will tend to remain constant, consistent with the physical reality (Jacobsen and Gilchrist, 1988). Contrast processing can be achieved by a mechanism that performs a logarithmic transformation on the retinal image, like for example, the "Weber-machine" described by Koenderink, Van de Grind and Bouman (1971).

Obviously, our stimulus conditions did not allow the visual system to use the various cues that may be used in the natural visual environment. Still, it is not unreasonable to assume that the mechanism subserving the (constant) color response in our RGB world, i.e. contrast processing, will operate in the real world as well.

This study was performed while J. Walraven was on (summer) leave from the TNO Institute for Perception. His support by an IBM Science and Education fellowship is hereby gratefully aknowledged.

REFERENCES

Arend, L.E., 1973, Spatial differential integral operations in human vision: implications of stabilized retinal image fading, *Psychol. Rev.* 80:374-395.

Arend, L.E. and Reeves, A., 1986, Simultaneous color constancy, *J. Opt. Soc. Am. A* 3:1743-1751.

Buchsbaum, G., 1980, A spatial processor for object color perception, *J. Franklin Inst.* 310:1-26.

Creutzfeldt, O., Lange-Malecki, B., and Wortmann, K., 1987, Darkness induction, retinex algorithm and cooperative mechanism in vision, *Exp. Brain Res.* 67:270-283.

Horn, B.K.P., 1985, Robot Vision, McGraw-Hill, New York.

Hurlbert, A., 1986, Formal connections between lightness algorithms, *J. Opt. Soc. Am. A* 3:1684-1693.

Hurlbert, A. and Poggio, T.A., 1988, Synthesizing a color algorithm from examples, *Science (NY)* 239:482-483.

Jacobsen, A. and Gilchrist, A., 1988, The ratio principle holds over a million-to-one range of illumination, *Percept. Psychophys.* 43:1-6.

Krauskopf, J., 1963, Effect of retinal image stabilization on the appearance of heterochromatic targets, *J. Opt. Soc. Am.* 53:741-744.

Karp, A., 1959, Colour-image synthesis with two unorthodox primaries, *Nature* 184:710-712.

Koenderink, J.J., Grind, W.A. van de, and Bouman, M.A., 1971, Foveal information processing at photopic luminances, *Kybernetik* 8:128-144.

Land, E.H., 1977, The retinex theory of color vision, *Sci. Am.* 237:108-128.

Land, E.H., 1986a, Recent advances in retinex theory, *Vision Res.* 26:7-21.

Land, E.H., 1986b, An alternative technique for the computation of the designator in the retinex theory of color vision. *Proc. Nat. Acad. Sci. USA* 83:3078-3080.

Lucassen, M.P. and Walraven, J., 1990, A closer look at the role of the blue system in color constancy, *Perception* 19:332 A10.

Maloney, L.T. and Wandell, B., 1986, Color constancy: a method for recovering surface spectral reflectance, *J. Opt. Soc. Am. A* 3:29-33.

McCann, J.J., McKee, S.P., and Taylor, T.H., 1976, Quantitative studies in retinex theory. A comparison between theoretical predictions and observer responses to the color Mondrian experiments, *Vision Res.* 16:445-458.

McCann, J.J. and Houston, K.L., 1983, Color sensation, color perception and mathematical models of color vision, in "Color Vision" (eds: J.D. Mollon and T.L. Sharpe), 533-544, Academic Press, London.

Mollon, J.D., 1977, The oddity of blue, *Nature* 268:587-588.

Shapley, R., 1986, The importance of contrast for the activity of single neurons, the VEP and perception. *Vision Res.* 26:45-61.

Tiplitz Blackwell, K. and Buchsbaum, G., 1988, Quantitative studies of color constancy, *J. Opt. Soc. Am. A* 5:1772-1780.

Valberg, A., and Lange-Malecki, B., 1990, "Colour constancy" in Mondrian patterns; a partial cancellation of physical chromaticity shifts by simultaneous contrast, *Vision Res.* 30:371-380.

Vos, J.J., 1978, Colorimetric and photometric properties of a 2-degree fundamental observer, *Color Res. and Appl.* 3:125-129.

Wallach, H., 1948, Brightness constancy and the nature of achromatic colors, *J. Exp. Psychol.* 38:310-324.

Walraven, J., 1973, Spatial characteristics of chromatic induction; the segregation of lateral effects from straylight artefacts, *Vision Res.* 11:1739-1753.

Walraven, J., 1977, Colour signals from incremental and decremental light stimuli, *Vision Res.* 17:71-76.

Walraven, J., Benzschawel, T.L., and Rogowitz, B.E., 1987, Color-constancy interpretation of chromatic induction, Proc. AIC Interim Meeting "Stiles-Wyszecki Memorial", 269-273, Muster-Schmidt, Göttingen.

Walraven, J. and Lucassen, M.P., 1990, Color constancy in an RGB world, *Proc. Eurodisplay '90,* 112-115, VDE-Verlag, Berlin.

Walraven, J., Enroth-Cugell, Ch., Hood, D.C., MacLeod, D.I.A., and Schnapf, J.L., 1990, The control of visual sensitivity: receptoral and post-receptoral processes, in "Visual perception: the neuro-physiological foundations" (eds: L. Spillmann and J.S. Werner), 53-101, Academic Press, San Diego.

Walraven, J. and Werner, J.S., 1986, The invariance of unique white; implications for normalizing cone action spectra, *Perception* 15:A27.

ADAPTATION AND COLOR DISCRIMINATION

John Krauskopf and Karl Gegenfurtner

Center for Neural Science and Psychology Department

New York University

A central problem in color vision is how color discrimination varies over color space. MacAdam (1942) attacked the problem in an elegant set of experiments. He developed a special color mixer which produced a bipartite disc consisting of a fixed half field and a variable half field. Both halves were mixtures of lights derived from the same source and which passed through the same two color selective filters. The experimenter determined the relative amount of the two lights in the fixed half and the observer repeatedly adjusted the relative amount of the lights in the variable half. A large number of color filters sets were selected and trimmed by additional neutral filters so as to produce equally luminous primaries at many different points in the CIE diagram. By using a number of carefully chosen sets of these filters, fixed lights at the same point in the diagram could be obtained with several pairs of filters allowing the matching half of the field to be adjustable along different lines through that point.

The basic data were the standard deviations of the matches made along the different lines. These were used to construct equal discriminability ellipses about a number of points in the CIE diagram. The variation of the orientation and the lengths of the major and minor axes of these MacAdam ellipses over the diagram is complex and is not compatible with any simple theory of color discrimination. A further problem is that other investigators have reported different patterns of ellipses (Wyszecki and Fielder, 1971).

The differences may be due to the way adaptation was controlled in the experiments. MacAdam (1942) presented his 2-degree diameter bipartite field in the center of a 42-degree uniform surround having a chromaticity that approximated CIE illuminant C and that was about half the luminance of the matching stimuli. Since the observer was concentrating on the matching stimuli, his central retina may be have been presumed to have been adapted primarily to the matching stimuli with some marginal contribution of the surround. In other experiments (Wyszecki and Fielder, 1971) the observers were encouraged to move their gaze over the field including the neutral surround so as to control adaptation.

The powerful effect state of adaptation has on brightness discrimination was shown by Craik (1938). In his experiments the test, a target, modulated in luminance at 2 Hz around a test luminance level, briefly replaced a steady field whose luminance determined the adaptation level. The threshold amplitude of the modulation was typically a minimum when the adaptation and test luminances were equal.

Therefore, we decided to study the effects of adaptation on color discrimination under carefully controlled conditions. In measuring the variation of color discrimination

From Pigments to Perception, Edited by A. Valberg and
B.B. Lee, Plenum Press, New York, 1991

over color space we can define two useful conditions with respect to adaptation. In one condition, the observer is adapted to a fixed stimulus throughout the experiment with the test stimuli being briefly presented at various points in the color space. In the second condition, the observer is adapted throughout the experiment to the same point at which the discriminations are made. Both conditions are of both theoretical and practical interest. We were particularly interested in finding out whether any factors which influenced color discrimination under either of these conditions operated independently with respect to the mechanisms selectively responsive along the cardinal directions of color space (Krauskopf, Williams and Heeley, 1982).

METHODS

The stimuli were displayed on a Tektronix 690SR color television monitor driven by an Adage 3000 frame buffer controller which generates images on the monitor by reading through the picture memory in a raster scan and interpreting the numbers in each location as a color defined in a 256-element color lookup table. The intensity of each of the three primaries was controlled by 10-bit digital-to-analog converters. The display was generated at 120 Hz interlaced. All of the stimuli in the present experiments had a luminance of 50 cd m^{-2}.

The color space we use to describe our stimuli (Derrington, Krauskopf and Lennie, 1984), is an extension of the chromaticity diagram suggested by MacLeod and Boynton (1978). The origin of this space is an equal-energy white. There are two chromatic axes which lie within the isoluminant plane and run through the origin. Along one of these axes, the Constant B axis, the excitation of the short wavelength (B) cones is constant while the long wavelength (R) and middle wavelength (G) cone excitations covary to keep luminance constant. The other chromatic axis, the Constant R&G axis, is one along which only the B cone excitation varies. The third axis, the Luminance axis, is one along which the excitations of all three classes of cones vary in proportion to their value for equal-energy white.

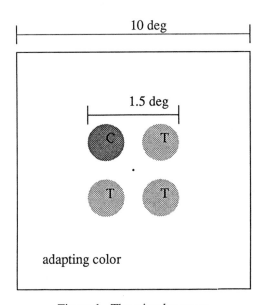

Figure 1. The stimulus array.

The general scheme of the experiments is depicted in Fig. 1. The observer fixated the center of the TV screen which was about 10 degrees square, which for most of the time was uniformly illuminated in the current adaptation color. On an experimental trial 4 discs 36 minutes in diameter were presented as a gaussian pulse of plus and minus 3 sigmas of one second total duration. The centers of the discs were located at the corners of a 54 minute square centered on the fixation point. Three test discs were the same color. The fourth comparison disc, randomly located at one of the corners on any trial, differed slightly in color from the test discs. The observers job was to report which of the discs was different from the others by pressing one of four buttons. A three to one staircase procedure was used to find the difference at which the test and comparison stimuli were just discriminable. In some of the experiments the test vector was of zero length. In these experiments color discrimination thresholds were measured at the adaptation point. In other experiments discrimination at points removed from the adaptation point were measured.

There are no agreed-upon units for representing the magnitude of stimuli along the Constant B and Constant R&G axes. We decided to use the mean thresholds for detecting changes from equal-energy white in both directions along each cardinal direction as units. The results of these measurements for one observer are shown in Fig. 2. On the left the results are given in machine units, and on the right in threshold units. The threshold unit in the Constant B direction for this observer is 12 machine units which is the mean of the two threshold measurements in the positive and the two threshold measurements in the negative direction along this axis. The threshold unit for the Constant R&G axis is 42 machine units which is the mean of the two threshold measurements in the positive direction along this axis and the two threshold measurements in the negative direction along this axis. In all subsequent plots of the results for this observer the threshold units based on detection relative to the equal-energy white point will be used.

Additional threshold measurements for stimuli modulated along directions intermediate between the cardinal directions are plotted. All of these points were fitted, using the method of least squares, by the ellipse drawn through them. One the right of this figure the same data are plotted in threshold units. The best fitting ellipse is now a circle. There are some asymmetries in the thresholds in different directions but these are probably a manifestation of experimental error.

MACHINE UNITS   THRESHOLD UNITS

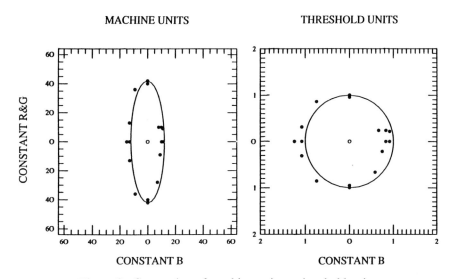

Figure 2. Conversion of machine units to threshold units.

RESULTS

Discrimination thresholds at locations removed from the adaptation point in color space

The general scheme for those experiments in which thresholds were measured for discrimination at points removed from the adaptation point is presented in Fig. 3. In some experiments both the test and the comparison vectors lay along the same cardinal axis. In these cases the comparison vector was either longer (as depicted) or shorter than the test vector. For example, the observer had to choose which of the four discs was a more (or less) saturated red. In other cases the difference between the test and comparison vectors was orthogonal to the test vector. For example, the observer might have to choose the disc that appeared yellower or bluer among a group of generally greenish discs.

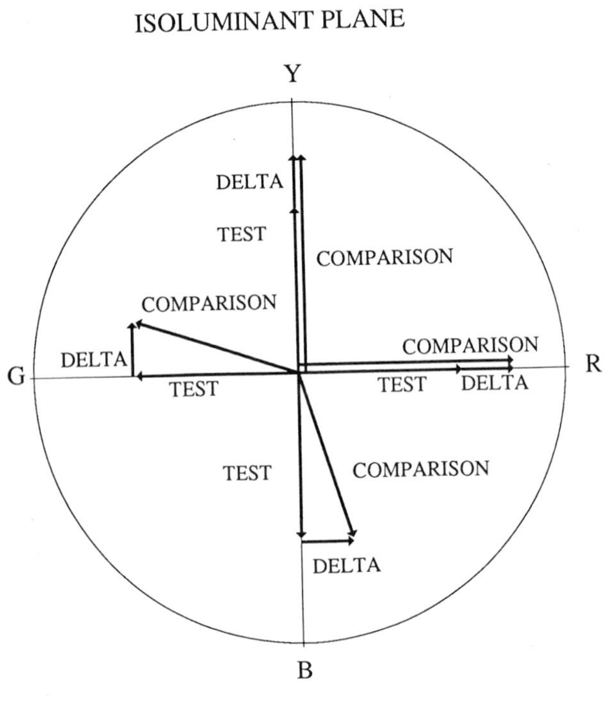

ISOLUMINANT PLANE

Figure 3. Schematic representation of test, comparison and difference vectors.

The results for the case when the test and comparison vectors lie along the same cardinal axis are illustrated in Fig. 4. Thresholds rise linearly with the length of the test vector.

The results when the difference between the test and comparison vectors are orthogonal to the test vector is illustrated in Fig. 5. Thresholds are independent of the length of the test vector.

CONSTANT BLUE

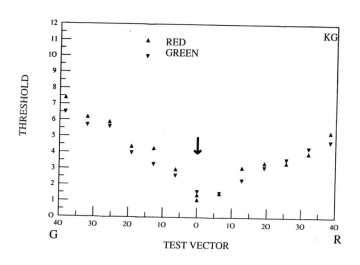

CONSTANT R&G

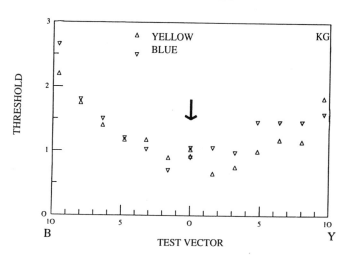

Figure 4. Thresholds when test and comparison stimuli are both on cardinal axes.

CONSTANT BLUE

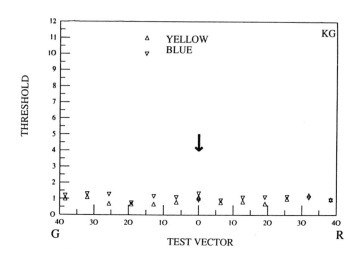

CONSTANT R&G

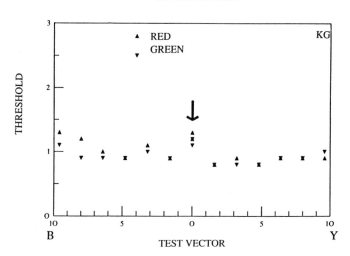

Figure 5. Thresholds when test and comparison stimuli are orthogonal.

These results suggest that discrimination of color differences may be mediated independently by mechanisms selectively responsive along the cardinal axes. To examine this proposition we measured thresholds when the test and difference vectors were in non-cardinal directions. The results of these experiments are shown in Fig. 6. If thresholds were mediated independently by such mechanisms we would expect the thresholds in different directions relative to test vectors in directions 45 degrees from the cardinal axes to plot as circles. This prediction was well satisfied in the lower right and upper left quadrants of Fig. 6 but clearly fails in the other two quadrants.

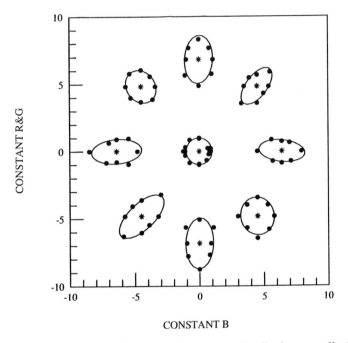

Figure 6. Thresholds in various directions from test stimuli along cardinal axes and directions 45 degrees from cardinal axes. Asterisks mark the tip of the tests vectors.

Discrimination thresholds at the adaptation point in color space

The results of the previous experiments demonstrate that discrimination thresholds are strongly effected by the state of adaptation. Therefore, in the second set of experiments adaptation was maintained at the point in color space at which the discriminations were made by presenting the stimuli to be detected on a continuously viewed background.

The variation in the thresholds for detecting changes in the reddish or greenish directions are plotted in Fig. 7 as a function of the location of the adaptation point along the Constant B axis. The adaptation point varies more than 30 threshold units in the red or green directions from the equal-energy white point without any measurable effect on the thresholds.

Similarly, the variation in the thresholds for detecting changes in the yellowish and bluish directions are plotted in Fig. 8 as a function of the location of the adaptation point along the Constant R&G axis. In this case, thresholds are linearly related to the excitation of the short wavelength sensitive (blue) cones.

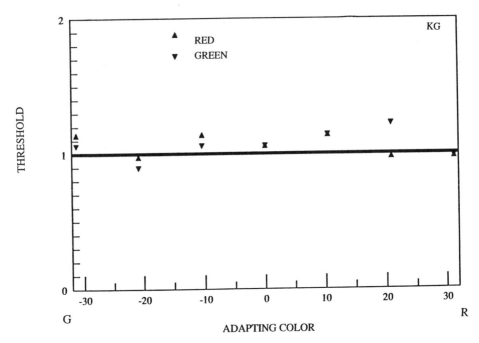

Figure 7. Thresholds for detecting changes in the reddish or greenish directions while adapted at various points along the Constant B axis.

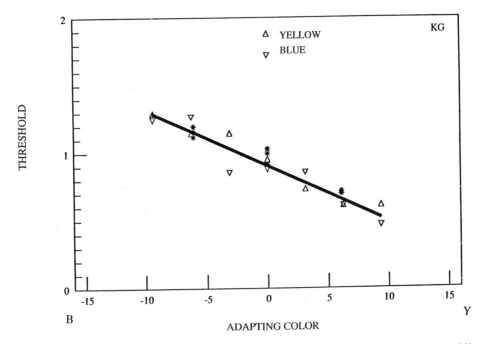

Figure 8. Thresholds for detecting changes in the yelowish and bluish directions while adapted at various points along the Constant R&G axis.

Thus, the effect of variation of the adaptation point on the thresholds differ along the two cardinal axes. The question arises are these effects independent. To help answer this question we measured the thresholds in several directions about nine different adaptation points as shown in Fig. 9. The crosses mark the location of the adaptation points. Stimuli that are just detectably different are indicated by the circular symbols. Open and closed circles are used to avoid confusion. The distance between the adaptation points and the threshold stimuli has been multiplied by three. The points about the central row of adaptation points are well fit by a circle of unit diameter. The rest of the data are best fit with ellipses. The axes of these ellipses are the same length in the Constant B direction in all cases. The axis in the Constant R&G direction is smaller when the adaptation point is moved in the yellow direction and larger when it is moved in the blue direction. There is no indication of interaction when the adaptation point is moved in both directions simultaneously.

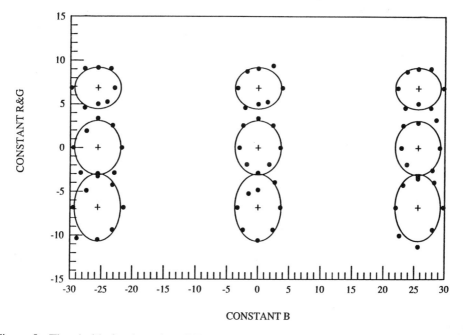

Figure 9. Thresholds for detecting differences in various directions from different adaptation points. Crosses mark actual adaptation points in threshold units. Magnitude of plotted thresholds plotted at four time actual values for clarity.

DISCUSSION

When adaptation is carefully maintained at the point in color space at which the discrimination is made our results provide a simpler picture of the variation of color discrimination over color space than those of MacAdam (1942). Using coordinates normalized with respect to thresholds for detecting changes from an equal-energy white, the results are well fitted by a set of ellipses whose major and minor axes are along the cardinal directions. Only one parameter, the length of the axis along the Constant R&G axis, must be varied to fit all the data.

The variation of the thresholds with the level of excitation of the B cones is, of course, consistent with the results of Stiles' (1978) increments threshold experiments and are in accord with the LeGrand (1949) analysis of MacAdam's (1942) results. However, our results differ from the LeGrand analysis in that we find no variation of discrimination thresholds with changes in the discriminanda along the Constant B axis.

The variation of detections thresholds with the level of B cone excitation is consistent with the notion that the signal detected in this case is proportional to $\Delta\frac{B}{B}$. It might be similarly hypothesized that the detection of changes involving only the R and G cones might be predicted by:

$$\frac{|\Delta R|}{R} + \frac{|\Delta G|}{G} = C$$

where R and G are the excitations of the red and green cones, respectively and C is a constant. It is useful to follow the approach of MacLeod and Boynton (1978) and normalize the cone excitations with respect to luminance given by $R + G$.

$$\frac{|\Delta r|}{r} + \frac{|\Delta g|}{g} = C$$

and thus,

$$g\,|\Delta r| + r\,|\Delta g| = Crg$$

but since $|\Delta r| = |\Delta g|$ for isoluminant stimuli:

$$2\Delta r\,(r + g) = Crg$$

and since $r + g = 1$:

$$|\Delta r| = |\Delta g| = \frac{Crg}{2}$$

For the results depicted in Fig. 7, r ranged from about 0.6 to 0.7 which predicts about a 10 % decline in threshold from the green to the red end of the abscissa. Such a difference should be evident given the reliability of our measurements but is not seen. We have extended the range to from for r values ranging from about 0.5 to about 0.85 in preliminary experiments. Under these conditions, this analysis predicts almost a two fold variation in thresholds but the thresholds remained essentially constant.

That control of adaptation is of great significance is demonstrated by the large increases in thresholds when the stimuli to be discriminated are remote from the adaptation point. We therefore believe that the difference between our results and those of MacAdam (1942) are likely to have been due to subtle variations in adaptation in his procedure.

There are practical consequences of the results presented here. Discrimination thresholds are smallest when the eye adapted in the vicinity of the stimuli to be discriminated. Therefore, critical comparisons are best made, and the largest number of discriminable colors can be attained under this regime. Since the results are more orderly when adaptation is carefully controlled, it would seem appropriate to use measurements made under such conditions in developing standards. However, it should be realized that such data would not correctly predict performance when the eye is adapted to one point in color space and monitoring lights in other places in color space, e.g., in a situation in which colored light flashes are used for signals.

The number of discriminable colors throughout color space is smaller when the adaptation point is held constant. By applying the Fechnerian method one might calculate this number for different conditions of adaptation. If adaptation must be held fixed the largest number of steps would be available in an isoluminant plane when one is adapted at the center of the plane.

REFERENCES

Craik, K.J.W., 1938, The effect of adaptation on differential brightness discrimination. *J. Physiol., 92,* 406-421.

Derrington, A.M., Krauskopf, J. and Lennie, P., 1984, Chromatic mechanisms in lateral geniculate nucleus of macaque. *J. Physiol. 357,* 241-265.

Krauskopf, J. Lennie, P. and Sclar, G., 1990, Chromatic mechanisms in striate cortex of macaque. *J. Neuroscience, 10,* 649-669.

Krauskopf, J., Williams. D.R. and Heeley, D.W., 1982, Cardinal Directions of Color Space, *Vision Res., 22,* 1123-1131.

Krauskopf, J., Williams, D.R., Mandler, M.B. and Brown, A.K. (1986, Higher order color mechanisms. *Vision Res., 26,* 23-32.

LeGrand, Y., 1949, Les seuils differentiels de coleurs dans ls theorie de Young., *Revue D'Optique Theorique et Instrumentale, 28,* 261-278.

MacAdam, D.L., 1942, Visual Sensitivities to color differences in Daylight. *J. Opt Soc. Am., 32,* 247-274.

MacLeod, D.I.A. and Boynton, R.M., 1978, Chromaticity diagram showing cone excitation by stimuli of equal luminance. *J. Opt. Soc. Am., 69,* 1183-1186.

Stiles, W. S., 1978, "Mechanisms of Colour Vision," Academic Press, London.

Wyszecki, G. and Fielder, G. H., 1971, New color-matching ellipses. *J. Opt. Soc. Am., 61,* 1135-1152

STUDIES ON COLOUR CONSTANCY IN MAN

USING A "CHECKERBOARD - MONDRIAN"

Annette Werner

Institut für Neurobiologie
Freie Universität Berlin
Königin-Luise-Str. 28-30
D-1000 Berlin 33

INTRODUCTION

Colour constancy is referred to as the ability of colour vision systems to code object colours so that they are perceived approximately constant despite changes in illumination. This phenomenon has been observed in both vertebrates e.g. man (Land, 1977) and invertebrates e.g. the honeybee (Werner et al., 1988), and is of general importance for investigations on colour coding. Chromatic interactions between different visual field areas were found to be a significant feature of information processing underlying colour constancy in both species.

Local border contrasts are thought to be an essential part of spatial integration (Land, 1977; Shapley, 1986), and for this reason it would be interesting to know how closely colour constancy is related to local contrasts. The failure of colour constancy i.e. the actual invariance of the contrast signals, is also discussed.

METHODS AND EXPERIMENTAL PROCEDURE

Colour constancy in man was investigated by means of a "Checkerboard-Mondrian", in which 13 differently coloured glass plates were mounted in a vertical plexiglass frame with black areas acting as separators (Fig. 1). Each unit square subtended 2 x 2 degree visual angle. The apparatus was shielded from daylight and homogeneously illuminated from behind by a mixture of 3 different lamps of short-, middle- and long-wavelength (S, M, and L respectively); the chosen wavelengths corresponding to the 3 spectral types of human cones i.e. λ max = 430 nm, 540 nm, 660 nm. The wavelenth-band intensity of each lamp could be varied individually. Spectral distribution of the emitted light was measured for each of the 13 colour squares and colour stimuli were quantitatively determined by calculating the related tristimulus values X, Y, and Z, that correspond to the relative amount of light absorbed in the 3 types of cones. Test subjects viewed the Mondrian against a black background.

Five colour constancy experiments with different reference colours and different illumination changes are described; the number of test persons per test is 10. Colour constancy was tested in a forced choice paradigm, candidates marking the position of the supposed reference plate on a black and white diagram.

From Pigments to Perception, Edited by A. Valberg and
B.B. Lee, Plenum Press, New York, 1991

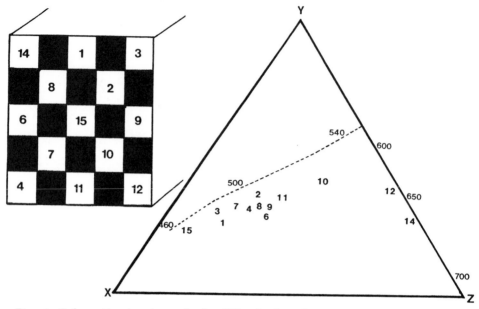

Fig. 1. Schematic drawing of the "Checkerboard-Mondrian" and Mondrian
stimuli for the standard illumination in a physiological colour triangle.
X, Y, and Z, represent the chromaticity values which are related to
each stimulus.

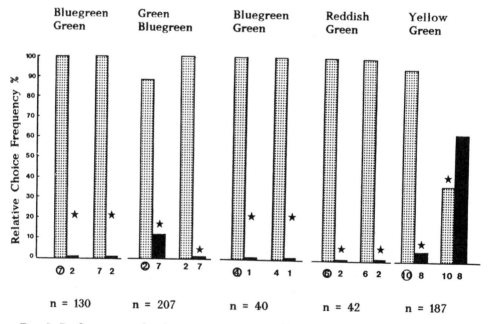

Fig. 2. Performance of colour constancy in different spectral regions and for
different illumination changes. Relative choice frequency gives the
choice distribution of choices between the reference plate (encircled
number) and the matching alternative. n = number of decisions,
★ indicates significant differences in choice frequency.

This task was performed both under a standard illumination and then, in further tests, with different colour matching illuminations. For each colour match, the illumination of the whole Mondrian was changed in such a way that only one of the alternative plates (the so-called "matched alternative") now emitted exactly the same light flux (i.e. resulting in the same tristimulus values) as the reference plate emitted previously during the standard illumination conditions. This colour match was performed 10 times for each test subject, and the position of the plates within the Mondrian altered for each test. Results are given as the distribution of decisions made for the reference plate and the matching alternative (rel. choice frequency % - see Fig. 2). Colour constancy was determined by comparing the choice behaviour in the discrimination and colour constancy tests. In the case of perfect colour constancy, one would expect the distribution of choice behaviour between the training and the alternative matching plate to remain the same. Alternatively, if normalization of receptor signals does not occur, perception would be determined by the absolute spectral light fluxes of the plates. Therefore, under the changed illumination one would then expect the matched alternative plate to be chosen with the same frequency as the training plate during the discrimination test.

RESULTS

1. Coding Contrasts

Contrasts are an appropriate code to carry constant information under conditions of varying illumination as they represent the relationships between reflections of different surfaces. However, this is only true for situations where surface reflections change accordingly i.e. illumination changes act simultaneously and similarly on all surfaces (as is the case in Land's Mondrian). In the checkerboard arrangement presented here, the black seperators between the colour plates ensure that border contrasts between the equally illuminated colour fields are limited to the corners of each plate. In other words, if the illumination is varied, the majority of local contrast signals do not remain constant since illumination changes effect the colour plates only (as opposed to the black fields). If colour constancy is primarily based on short range interactions then no colour constancy would be expected under the checkerboard condition.

Colour constancy was tested for 5 different colour matches in different spectral regions. In Fig. 2, the first double block ("train.") of every experiment shows the distribution of choices between the test plate and alternative plate during standard illumination conditions and the second double block ("match.") shows the distribution of choices during the changed illumination. The choice behaviour of the observers does not change in 4 of the 5 colour matches, and this indicates that there is a compensation in the physical colormetric shift. The lack of constant border contrasts is, therefore, not a limiting factor for colour constancy - at least under the given experimental conditions. In experiment 10/8 colour constancy is not observed and this indicates that the experimental procedure is sensitive enough to detect for situations where the colour vision system cannot cope with illumination changes.

II. A possible reason for the failure of colour constancy

The effects of illumination changes on receptor signals were determined for experiment 10/8 (failure of constancy) and experiment 2/7 (successful compensation for illumination change). In order to detect differences which can be traced back to the coding of these signals in relation to their surround, i.e. contrast coding, a simple spatial normalization procedure is applied for calculating the receptor signals (Fig. 3a). This computation can be described as the weighting of each signal by the chromatic average of all the colour signals of the surrounding background, and separately for each chromatic channel. It is, therefore, comparable to a von Kries transformation (von Kries, 1905). Receptor signals are derived from

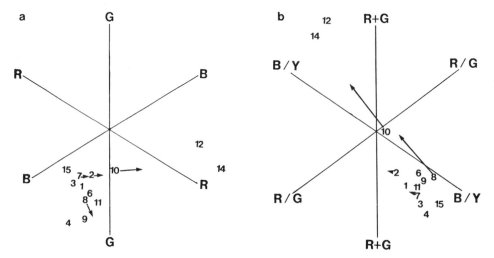

Fig. 3a. Effect of illumination change on the level of receptor signals after
normalization. Colour loci are presented in a 2-dim. projection of a
vector space whose axes represent the quantities of receptor signals
R, G, B. Arrows indicate the occurring shifts of the colour loci of
Nos: 2 and 7 in experiment 2/7, and Nos: 10 and 8 in experiment 10/8.
Δ gives the Euclidian distance between the loci of the reference and
alternative matching plate under matching illumination, and the locus
of the reference stimulus under standard illumination respectively:
Δ (2-2) = 1.17; Δ (2-7) = 0.62; Δ (10-10) = 1.26; Δ (10-8) = 7.4.

Fig. 3b. Effect of illumination changes on the level of opponent signals after
normalization. The diagram represents a 2 dim. projection of a vector
space whose axis are given by the normalized spectral opponent signals
R/G, B/Y and R+G. Δ (2-2) = 1.22; Δ (2-7) = 2.21; Δ (10-10) = 6.48;
Δ (10-8) = 5.7.

the cone absorption values i.e. tristimulus values by a logarithmic transformation.
In Fig. 3a the loci under standard illumination are indicated by the numbers of
the colour plates 2, 7, 10, and 8, and arrowheads give the position of the loci of
the same plates under the matching illuminations. As can be seen by the distances
Δ between the reference loci under standard illumination and the colour loci under
matching illumination, the normalization procedure predicts that plate No: 7
should be mistaken for plate No: 2, whereas plate No: 10 should be well
discriminated from No: 8. This is in contradiction to the experimental results
shown in Fig. 2.

The same normalization procedure was applied to the opponent signals,
which were calculated as a linear combination of the weighted receptor signals
(see Fig. 3b). Weighting factors were taken from Guth et al. (Guth, 1980). It
should be mentioned that in contrast to Worthy's calculations (Worthy, 1985) the
normalization process is applied directly to the opponent signals. Normalization
of the opponent signals predicts that plate No: 2 should be recognized as the
reference plate, whereas in experiment 10/8 plate No: 8 should be mistaken for
plate No: 10 (see loci differences Δ), and Fig. 2 shows that this is indeed the case.
Therefore, the observed lack of colour constancy in experiment 10/8 can be
interpreted as a normalization of opponent signals.

DISCUSSION

Since local contrasts in the "Checkerboard-Mondrian" do not provide a constant cue for coding colour signals under varying illumination conditions, the observed colour constancy cannot be directly related to local contrasts. Additional long-range interactions, as demonstrated by Gelb (1929), Land et al. (1983) and Pöppel (1986), contribute to spatial integration, and this is particularly so if the intermediating area is black (Valberg et al., 1985; see also this volume). However, the integration of long-range contrast signals cannot lead to colour constancy immediately since the colour plates contribute to only 50% of the total Mondrian area (independent of which radius around the chosen reference field is taken for the integrating area). I assume, therefore, that colour constancy is derived by repeated steps of contrast integration, where local contrasts are the initial input stage. This assumption takes into account the simultaneous effect of longe-range interactions and the role of distance-weighted local contrasts. Indeed, algorithms such as Blake's modified Horn's Algorthim, which involve iterative differentiation of contrast signals, can successful be applied "for recovering reflectance in the Mondrian world" (Blake, 1985). However, it should be mentioned that the complete failure of constant local contrasts may effect the performance of colour constany (Walraven et al., 1990).

Quantification of the effect of a spatial normalization of receptor and spectral opponent signals suggests that colour constancy is not appropriately described at the level of receptor signals, but is rather the result of continuous steps of contrast integration in the on-going processing of chromatic information between the retina and the areas of the visual cortex.

REFERENCES

Blake, A., 1985, On lightness computation in the Mondrian world, in: "Central and Peripheral Mechansims of Colour Vision", T. Ottoson and S. Zeki, ed., MacMillan, New York.

Gelb, A., 1929, Die Farbkonstanz der Sehdinge, in: "Handbuch der normalen und pathologischen Physiologie", Vol. 12, A. Bethe, G. V. Bergmann, G. Embden and A. Ellinger, ed., Springer, Berlin.

Guth, S. L., Massof, R. W., and Benzschawel, T., 1980, Vector model for normal and dichromatic colour vision, J. opt. Soc. Am., 70(2):197.

Kries, J. v., 1905, Die Gesichtsempfindungen, in: "Handbuch der Physiologie des Menschen", Vol. 3, W. Nagell, ed., Vieweg, Braunschweig.

Land, E. H., 1977, The retinex theory of colour vision, Sci. Am., 108.

Land, E. H., Hubel, D. H., Livingstone, M. S., Perry, S. H., and Burns, M. M., 1983, Colour generating interactions across the corpus callosum, Nature, 303:616.

Pöppel, E., 1986, Longrange colour generating interactions across the retina, Nature, 320(10):523.

Shapley, R., 1986, The importance of contrast for the activity of single neurons, the VEP and perception, Vision Res., 26(1):45.

Valberg, A., Lee, B. B., Tigwell, D. A., and Creutzfeldt, O. D., 1985, A simultaneous contrast effect of steady remote surrounds on responses of cells in macaque lateral geniculate nucleus, Exp. Brain Res., 58:604.

Walraven, J., Enroth-Cugell, C., Hood, D. C., MacLeod, D. I. A., and Schnapf, J. L., 1990, The control of visual sensitivity: Receptoral and postreceptoral processes, in: "Visual Perception: The Neurophysiological Foundations", L. Spillmann and J. S. Werner, ed., Academic Press, San Diego.

Werner, A., Menzel, R. and Wehrhahn, Chr., 1988, Colour constancy in the honeybee. J. Neurosci. 8:156.

Worthy, J. A., 1985, Limitations of colour constancy, J. opt. Sci. Am. A. 2(7):1014.

DISCUSSION: POST-RECEPTORAL PROCESSES II

Dick Cavonius

Institut für Arbeitsphysiologie
Dortmund
F.R.G.

Cavonius: I'd like to comment briefly on a matter related to the session on evoked potentials. When I was very young, before I knew better, I spent a great deal of time recording human electroretinograms and visual evoked potentials. Averaging methods had just become generally available, and were being applied blindly to try to find 'objective' measures of almost any psychophysical event that you could think of. This resulted in many weak papers: a favorite of mine was one in which the size of a certain 'component' - i.e., a bump - in VEPs elicited by flashes was correlated across subjects with 20 different traits in a conventional pencil-and-paper personality inventory. The authors were very excited by the fact that one of the 20 traits correlated significantly, at $p=0.05$.

This dreadful enthusiasm for trying to find evoked potential correlates for every conceivable psychophysical event has now largely died out. I was therefore distressed to hear Henk Spekreijse say that many people who used to record evoked potentials are now lusting after the means to record magnetic potentials. The only fitting comparison that I can think of is the classic description of second marriages: they are the triumph of hope over experience.

The present session is rather more heterogeneous than some of the earlier sessions. Jan Walraven described the effects of adaptation on colour constancy, and Mary Hayhoe described a method by means of which one can distinguish between subtractive and gain-control aspects of adaptation. I feel that this study is very timely, since recent ARVO meetings have been burdened with models of colour constancy that describe in increasing detail some chosen data set; often as if only a single mechanism were involved. In the last paper, John Krauskopf turned from appearance to discrimination, and described the influence of the adaptive state on hue discrimination. In his paradigm the adapting and test hues can be varied independently, which is a considerable methodological advance on the classic studies of David MacAdam and W.D. Wright, in which adaptation was either uncontrolled, or had (because of the available equipment) the same chromaticity as the test.

In thinking about these studies, I'd like to ask for comments on an issue that was raised earlier: where does adaptation occur, is it retinal and monocular or is it central and potentially binocular? In this regard I'd like to mention an unpublished study that Edgar Auerbach performed at my department several years ago. He studied dark-adapted thresholds for the monocular detection of light flashes either while the other eye was dark-adapted or while it was adapted to an intense, homogeneous, 656 nm light. The test was presented at several locations, from 0 deg to 16 deg in the peripheral field. From 0 to 4 deg, contralateral light adaptation caused a uniform loss of sensitivity to the test; whereas from 7 to 12 degrees, adaptation improved sensitivity by an average of 44%. Data beyond 12 deg tend to be noisy, but show no clear effect of

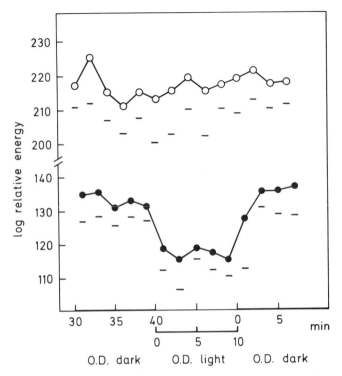

Fig. 1. Sensitivity at 7 deg to a 5 cycle/deg, 502-nm square-wave grating (open symbols) and to a 1 deg, 502-nm homogeneous field (filled symbols) before (O.D. dark), during (O.D. light) and after (O.D. dark) contralateral light adaptation. Bars show 1 s.d.

contralateral adaptation. These results suggest a binocular rod-cone interaction, in which the detection is performed by rods, but (because the rods are so insensitive to 656 nm) the adaptation results from cone activity. The evidence for rod detection appears in Fig. 1, which shows thresholds when the non-test eye is initially dark-adapted (the last 10 min of a 40 min adaptation). After this it is light adapted for 10 min, and finally dark-adapted again. When the task is detection of a 1 deg unstructured field, sensitivity improves during contralateral light adaptation (filled symbols), but when the task is detection of a grating that is too fine for the peripheral rods to resolve, thresholds remain constant during contralateral light adaptation (open symbols). I'd appreciate any suggestions as to possible physiological bases for such an interaction.

Hayhoe: I was in my own little world up here.

Shapley: In a paper with Enroth-Cugell we describe in cat ganglion cells slow adaptation processes with strong (about 2 log unit) background changes. But for small background changes, things are over very fast.

Zaidi: Jan Walraven makes a distinction between colour constancy and colour induction. It seems that all your experiments are colour induction experiments.

Walraven: You could say we are studying colour constancy and induction under similar conditions.

Shapley: What is the difference between your and Arendt's work?

Walraven: He did not see much colour constancy in his experiments. I am not sure what causes the differences, though perhaps it is his simultaneous comparisons. In our case, our subjects used successive comparison. The subject matched chromaticity or reflectance or whatever. We used about 50% reflectance, and this may have improved the matching technique in comparison to matches of colours of higher relative luminance.

van Essen: The rapid alternation you used seems rather unusual, and a not quite natural way of doing matches. What would happen with forced blank intervals between matches?

Walraven: Such memory matching does not give good data.

Krauskopf: I want to comment on Krauskopf's paper, which seems heretical to me. For years we've believed that Stiles' π mechanisms adapt independently. In my experiment we present a S-cone stimulus on a yellow background. This should be a way of isolating π_2, but independence failed. Specifically, we could vary our adaptation so that the S-cone input was constant and the ratio of M/L-cone inputs varied; or so that the M/L ratio was constant and only the S-cone input varied. In each case, luminance was held constant. When the cone excitations were expressed in terms of their values relative to the white point, thresholds for the detection of a test that changed S-cone stimulation varied in the same way for each of the background variations. The result of changing the S-cone excitation is no surprise, since it's just another case of cone-specific adaptation. But the effect of changing the M- and L-cone inputs to the adapting field is a different matter, and implies that the effect of adaptation occurs in a mechanism that combines signals from all three types of receptor.

The fact the M- and L-cones can control detection of short-wavelength stimuli was in fact shown by Stiles; the present experiment should be a more sensitive way to study the elusive π_2 mechanism. I think that our result is important, and I have no idea why Quasim Zaidi failed to find similar effects, other than his experimental conditions were different.

Walraven: I am not surprised that you did not find much selective adaptation of L- and M-cones, because your display did not allow much differential adaptation of the two cone systems. If I recall directly, Brown (JOSA, 1952) did find such effects in MacAdam ellipses, so it could be just a matter of improving the experimental conditions.

Krauskopf: Although it is difficult to vary the M to L excitation ratio over a wide range, we have shown cone-specific effects. When you vary both signals you just don't get the effects you expect from independent von Kries adaptation.

Vienot: Oscar Estevez put forward the hypothesis, based on colour matching functions, that there could be self-adaptation in the receptors, which could explain the variability in Stiles and Burch's results. We set up an experiment to test this, and I thought it possible observers were equating ratios of cone excitation, rather than equating quantum catches.

Zaidi: We measured S-cone thresholds on adaptation backgrounds varying only in L- and M-cone catches and found a straight line; we think this is Stiles π_1. One distinction between our tests and those that Krauskopf used is that ours have sharp edges and are brief. This may have something to do with it or perhaps Krauskopf's experiments also involve hue judgements and not simply detection.

Richter: MacAdam ellipses are smallest at the achromatic point, which is an indication that either the red or the green process is the most sensitive at threshold and that they do not interact. In scaling of larger colour differences, both processes interact.

Krauskopf: We don't see this at threshold. Thresholds are constant in the constant S-direction. Anyway, you can bet that Lennie and I are going to do neurophysiological experiments on adaptation in the near future.

Cavonius: Let me ask about the distances over which constancy and similar processes can operate. Annette Werner tested colour-constancy with a chequerboard-like pattern in which there were dark lines between the coloured areas. She concludes that calculation of local border contrasts could not handle the results. Can these effects jump over borders or short, or even long, distances?

Walraven: In my experiment in a dark grid situation, it seems that when just the chips are illuminated there is minimal colour constancy. With a closer apposition of chips, Annette

Werner found bigger effects, so there seems to be a change with distance between constancy and no constancy.

Creutzfeldt: Physiological evidence for long range effects is shown in the poster of Arne Valberg and in ours. Also psychophysical evidence suggests the effects can go over distances but the gap must be dark.

ORIGIN OF PERCEPTUALLY MEASURED PHASE SHIFTS IN THE VISUAL SYSTEM

Vivianne C. Smith[*]

Visual Sciences Center
The University of Chicago
Chicago, IL 60637, U.S.A.

INTRODUCTION

De Lange (1958) first remarked that a pair of heterochromatic lights in temporal sinusoidal counterphase modulation show residual flicker after adjustment to the minimum flicker percept consistent with equiluminance. He used long wavelength ("red") and mid wavelength ("green") lights and suggested that the residual flicker could be cancelled by adjusting the relative physical phase of the "red" and "green" component lights. De Lange ascribed the phenomenon to a latency difference between the cones. The phase adjustment for minimal flicker required that the "green" light lead the "red" light at the heterochromatic transition. De Lange concluded that the long wavelength sensitive (LWS) cones had shorter latencies than the middle wavelength sensitive (MWS) cones. De Lange's findings and subsequent data of Walraven & Leebeck (1964) and Vos & Walraven (1965) were subsequently attributed to rod intrusion in the flicker match and to a difference between rod and cone latencies rather than to latency differences between LWS and MWS cone types (von Grunau, 1977; van der Berg & Spekreijse, 1977). However, Cushman & Levinson (1983) working at high luminance levels and with high temporal frequencies were able to confirm the existence of residual flicker which could be cancelled by phase adjustment of one of the component lights. Their data also required a phase adjustment to "green" leads "red" at the heterochromatic transition. Thus they established that perceptual phase shifts can be mediated within LWS and MWS cone pathways.

PSYCHOPHYSICAL STUDIES OF PERCEPTUAL PHASE SHIFTS

We developed a different psychophysical procedure to measure perceptual phase shifts. We measured modulation thresholds to a high luminance heterochromatic stimulus with equiluminant "red" and "green" components as a function of the relative physical phase of the component lights. The equipment and techniques are described in Lindsey, Pokorny & Smith, (1986) and Swanson, Pokorny & Smith (1987). As the physical phase difference between the lights was varied, the modulation thresholds showed a characteristic U-shaped function. An example of a typical data plot is shown as threshold sensitivity (1/modulation threshold) in Figure 1, using data from the Swanson, Pokorny & Smith (1987) paper. The "red" and "green" components were at 564 nm and 625 nm giving a time average luminance of 900 tds and time average chromaticity metameric to 600 nm. Sensitivity is plotted as a function of the physical phase of the long

From Pigments to Perception, Edited by A. Valberg and
B.B. Lee, Plenum Press, New York, 1991

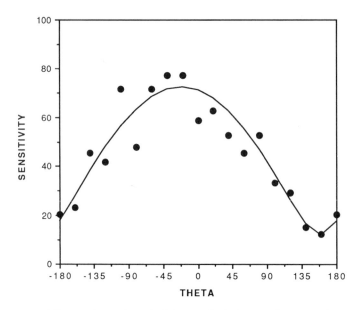

Figure 1. Threshold sensitivity (1/modulation)
as a function of relative physical phase, θ.
Modulation thresholds were measured for an 8 Hz, 2°
field using 564 nm and 625 nm equiluminant components
with time average luminance of 900 tds and time
average chromaticity metameric to 600 nm.

wavelength light and in this format shows an inverted U-shape.

This figure is for an 8 Hz temporal frequency and a 2° field. The
types of phase lags measured by De Lange or Cushman and Levinson are
revealed by asymmetry of the inverted U-shaped function from 0° and 180°
on the physical phase axis. The data are shown as solid circles. The
smooth line is an equation derived by Lindsey, Pokorny & Smith (1986):

$$1/M \ = \ \{\cos^2[(\theta-\phi)/2]/M_{LT}^2 + \sin^2[(\theta-\phi)/2]/M_{CT}^2\}^{0.5} \qquad (1)$$

where M is modulation threshold, θ is the relative physical phase of the
"red" light, φ is the physiological phase shift and M_{LT} and M_{CT} represent
threshold modulation at physical phase values which isolate pure
luminance and pure chromatic response. For the data of Figure 1, the
value of φ was -21, and values for $1/M_{LT}$ and $1/M_{CT}$ were 72.277 and 12.050
respectively. The equation describes detection as a vector sum of
achromatic and chromatic components. The meaning of the "physiological"
phase shift, φ is undefined. At 8 Hz the minimum sensitivity occurs at
positive theta values or "red-leads-green". These data confirmed the
Lindsey, Pokorny & Smith (1986) study with "red-leads green" for
frequencies from 3 Hz to 15 Hz. Cushman & Levinson's (1983) high
frequency minimum occurred at negative theta values or "green-leads-red".
This finding was also replicated by Swanson, Pokorny & Smith (1987), who
found a shift of the phase minimum from positive to negative theta values
near 18 Hz.

Subsequently Swanson, Pokorny & Smith (1987) defined an
atheoretical method of summarizing the data, by determining the axis of

402

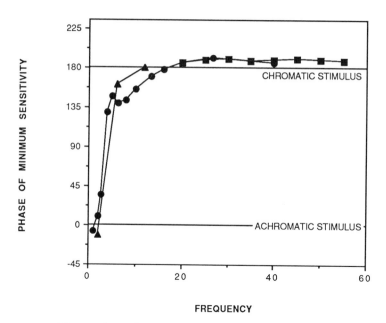

Figure 2. The phase of least sensitivity as a function of temporal frequency. Data are shown from three studies. Solid circles show the average data of Swanson Pokorny & Smith (1988). Solid triangles show the data of Lindsey, Pokorny & Smith (1986). Solid squares show the data of Cushman & Levinson (1983).

symmetry of the data. From similar data plots at other frequencies, Swanson, Pokorny & Smith (1987) derived the physical phase at which modulation sensitivity was least (180° from the phase where sensitivity as best) as a function of temporal frequency. The results are shown in Figure 2 together with data of Lindsey, Pokorny & Smith (1986) and Cushman & Levinson (1983). At low frequencies, modulation sensitivity was best when the lights had a physical phase near 180° (counterphase, heterochromatic stimulation) while at high frequencies, modulation sensitivity was best when the lights had a physical phase near 0° (inphase, luminance stimulation). The function relating the phase of least sensitivity to temporal frequency rose rapidly from near 0° at 1 Hz, crossed 180° near 20 Hz and reached 190° at higher frequencies. These data were collected under conditions where rod participation in flicker detection is unlikely. We assume that thresholds at 1-3 Hz were mediated by chromatic channels while those above 3 Hz were mediated by achromatic channels.

The temporal phase shifts appeared to be related to LWS and MWS cone activity. Swanson, Pokorny & Smith (1988) showed that phase shifts were affected by chromatic adaptation. The largest phase shifts occurred for adaptation conditions that render the LWS and MWS cones most similar in sensitivity. The smallest phase shifts occurred for test and adaptation conditions which are thought to isolate one or other cone type. The phase shifts were robust among observers and correlated with the observer's heterochromatic flicker photometric match (Swanson, Pokorny & Smith, 1986). These findings seemed to suggest that the phase shifts were determined in LWS and MWS cone pathways early in the visual system.

De Lange's initial idea was that the phase shifts reflected latency

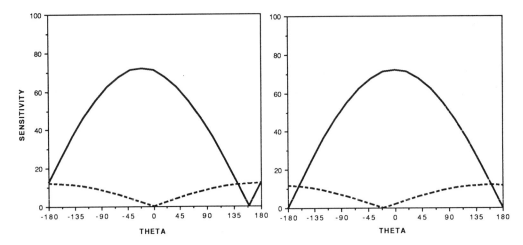

Figure 3. Calculated threshold sensitivity as a
function of relative physical phase, θ. The
achromatic mechanism is shown as six times more
sensitive than the chromatic mechanism. Panel 1 shows
an achromatic meachanism with a 20° phase shift and a
chromatic mechanism with zero phase shift. Panel 2
shows an achromatic mechanism with zero phase shift
and a chromatic mechanism with a 20° phase shift. The
phase shift is revealed only in the more sensitive
mechanism.

differences in the LWS and MWS cones as they summed in a "luminance
channel". This idea was predicated on two assumptions. The first is
that the flicker photometric match was mediated by a luminance channel.
The second is that the "red" light excited the LWS cone and the "green"
light excited the MWS cone. Within the framework of channel models, the
mechanism mediating detection at equiluminance is usually thought to be a
chromatic mechanism, since the sensitivity of an achromatic mechanism
would be zero at equiluminance. None-the-less, de Lange's first
assumption was shown to be true by Lindsey, Pokorny & Smith (1986), at
least for frequencies above 3 Hz where luminance sensitivity exceeds
chromatic sensitivity. This result is demonstrated in Figure 3. The
achromatic and chromatic sensitivities are defined as:

$$S_A = (\cos[(\theta-\phi)/2]/M_{LT}) \tag{2}$$

$$S_C = (\sin[(\theta-\phi)/2]/M_{CT}) \tag{3}$$

where the symbols are as defined above with $1/M_{LT}$ at 72 and $1/M_{CT}$ at 12.
The demonstration depends on the phase-dependant inverted U-shaped
sensitivity profile. At the phase of least sensitivity, detection may be
mediated by the least sensitive channel but the phase asymmetry will be
determined by the descending limbs of the most sensitive channel. Phase
shifts in the least sensitive channel are unlikely to be resolved by the
technique.

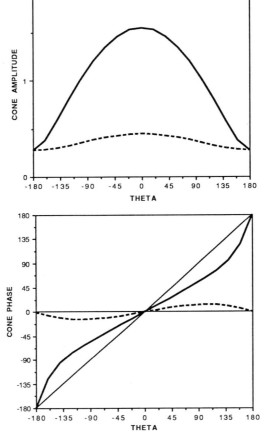

Figure 4. Calculated cone amplitude and
relative phase as a function of relative physical
phase, θ.

De Lange's second assumption however is incorrect. Both cone types
are excited by both lights. Figure 4 shows the calculated amplitude and
phase of the LWS and MWS cones to equi-luminant "red" and "green"
component lights of 553 nm and 636 nm. Each cone of course has higher
amplitude (greater sensitivity) to the inphase condition. The phases of
the cones differ, the LWS cone follows more closly to the "red" component
while the MWS cone follows the "green" component. A linear achromatic
channel should sum these amplitudes with regard to phase. Swanson,
Pokorny & Smith (1986) noted that an achromatic channel characterized by
the linear sum of LWS and MWS sensitivities, would show minimal phase
shifts even if the latencies of the cone types differed considerably.
This result is demonstrated in Figure 5, which shows predicted cone
amplitude functions for a linear achromatic channel summing LWS and MWS
cone sensitivities with a cone latency difference. At 8 Hz temporal
frequency, the LWS cone was advanced by 90° (i.e. a 31.25 msecs latency
difference) in the calculation in order to obtain a phase minimum near
the 160° typical of the psychophysical data (Swanson, Pokorny & Smith,
1986). Distal cone latency differences of this amount have not been
observed physiologically. Therefore it seemed that the phase shifts must
be determined by post-receptoral channels, not by the cones themselves.

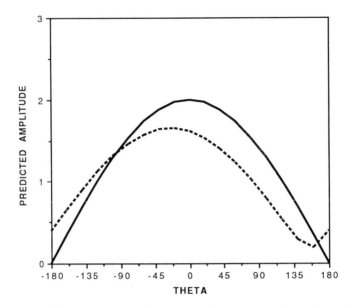

Figure 5. Predicted amplitude as a function of
relative physical phase, θ. Calculations are for an
achromatic channel summing LWS and MWS activity at 8
Hz, allowing for possible cone latency differences.
Solid line is a zero latency difference and dotted
line is a 31.25 msec latency difference between the
cones with the mWS cone delayed.

PHYSIOLOGICAL BASIS OF PERCEPTUAL PHASE SHIFTS

In an attempt to resolve the question, we have worked with Lee,
Martin and Valberg to measure retinal ganglion cell responses of macaque
retinal ganglion cells. Recording and cell classification methods have
been detailed in Lee, Martin & Valberg, 1988 and the equipment,
essentially identical to that of Swanson, Pokorny & Smith (1986), is
described in Lee, Martin, Pokorny, Smith & Valberg (in press). Ganglion
cell responses were measured to a pair of equiluminant 636 nm and 553 nm
lights in temporal sinusoidal modulation as a function of the physical
phase difference of the lights. Equiluminance was that of the human
experimenters, giving a time average luminance of 2000 tds and time
average chromaticity metameric to 594.6 nm. We used a fixed suprathreshold physical contrast and a 4.6° field centered on the receptive
field. We recorded about 6 secs data per phase condition and frequency.
The amplitude and phase of the first Fourier harmonic were derived from
the peri-stimulus histograms and plotted as a function of physical phase.
The amplitude plots showed the inverted U-shaped functions characteristic
of the psychophysical sensitivity data. The technique of Swanson,
Pokorny & Smith (1987) was used to derive the phase of least sensitivity.

We recorded from 25 P-pathway, red-green chromatic cells. Red-
green chromatic cells showed a phase of least sensitivity near 0° at low
frequencies. As frequency increased, there were frequency-dependent

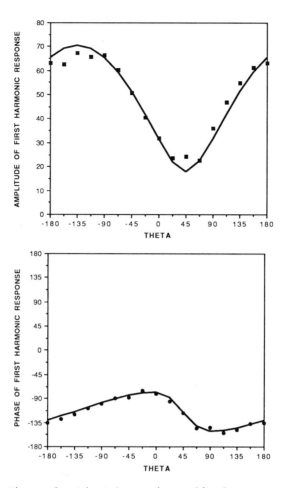

Figure 6. First harmonic amplitude as a
function of relative physical phase, θ. Solid symbols
are data are for a "red-on" cell measured at 50%
physical contrast; the solid line is a prediction
based on a linear model of center-surround behavior
(see text). Amplitudes were measured for a 9.76 Hz,
4.6° field using 553 nm and 636 nm equiluminant
components with time average luminance of 2000 tds and
time average chromaticity metameric to 594.6 nm.

phase shifts. Typical amplitude and phase data (solid symbols) for a P-
pathway ("red-on") cell is shown in Figure 6. Cells with a presumed LWS
cone center ("red-on" and "red-off") showed a minimum which shifted into
the positive theta region as frequency increased (as in Figure 6). Cells
with a presumed MWS cone center ("green-on" and "green-off") showed a
minimum which shifted into the negative theta region as frequency
increased. The phase of minimum amplitude for average cell data is shown
in Figure 7. The data were consistent with the concept that LWS and MWS
cone types are segregated in center and surround and that the surround
has a small phase delay relative to the center. A linear model summing
LWS and MWS cones was developed, described in Lee, Smith, Pokorny, Martin
& Valberg (in preparation). The solid line in Figure 6 represents the
implementation of this model. The LWS cone center showed a phase of -
119°; the MWS cone surround showed a phase of -315°. The center-surround
cone weighting was 0.38, a value that is consistent with the time-average

407

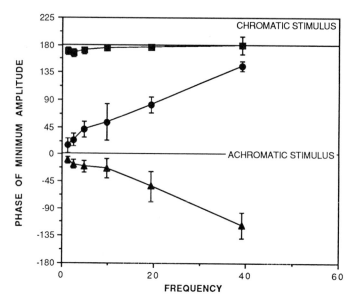

Figure 7. The phase of least amplitude as a
function of temporal frequency. Data are shown for
red-green and blue-yellow chromatic cells. Solid
circles show average data for "red-on" and "red-off"
cells. Solid triangles show average data of "green-on"
and "green-off" cells and solid squares show average
data for "blue-on" cells.

chromaticity of the stimulus. The center-surround phase difference of
$194°$ represents $180°$ phase difference between center LWS cone and
surround MWS cone plus the center-surround phase difference of $14°$ (about
a 4 msec delay in a latency model).

   We also recorded from 15 P-pathway, blue-yellow chromatic cells.
These cells are presumed to have SWS cone centers with mixed LWS and MWS
cone surrounds. Since SWS cones are not appreciably excited by the 553
nm and 636 nm stimuli, these cells revealed a pure surround response.
Blue-yellow cells did not show resolvable phase shifts at any frequency.
The average phase of minimum amplitude is shown in Figure 7. The data
were consistent with the concept that LWS and MWS cone types are mixed in
the surrounds and that there are minimal differences in the temporal
responses of the LWS and MWS cones feeding the surround. It is clear
that the phase behavior of the chromatic cells does not mirror the
psychophysical data. The change in minimum amplitude with frequency for
red-green chromatic cells is quite gradual compared with the
psychophysical data, and has not reached $180°$ by 39.04 Hz.

   M-pathway (phasic) cells, included a sample of 11 "on-center" cells
and 15 "off-center" cells. Again the amplitude data showed the
characteristic inverted U-shape. The amplitude behavior of the "on-
center" and "off-center" cells was very similar, though of course their
phase responses differed. M-pathway cells showed, as expected, a minimum
amplitude response near $180°$ at high frequencies. However, phase
asymmetries were observed at low frequencies. Figure 8 shows amplitude
and phase data for a "phasic-on" cell at 4.88 Hz, where large phase
shifts could be noted. These data were collected at 20% contrast. The
solid line was derived from a linear model of cell behavior described in
Smith, Lee, Pokorny, Martin & Valberg (in preparation). The flattening

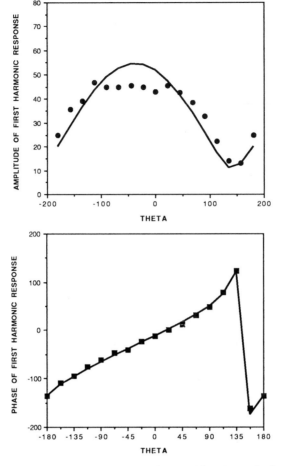

Figure 8. First harmonic amplitude and phase as
a function of relative physical phase, θ. Solid
symbols are data are for a "phasic-on" cell measured
at 20% physical contrast; the solid line is a
prediction based on a linear model of center-surround
behavior (see text). Amplitudes were measured for a
4.88 Hz, 4.6° field using 553 nm and 636 nm
equiluminant components with time average luminance of
2000 tds and time average chromaticity metameric to
600 nm.

of the data points compared to the solid line probably reflects
saturation of the contrast gain response of the cell.

The phase of least sensitivity ranged from 45°-135° at 1 Hz, rose
to 180° near 25 Hz and for some cells reached 190° at 40 Hz. Average
data for "on-center" and for "off-center " cells is shown in Figure 9.
This behavior is strikingly similar to the psychophysical data, both for
"on-center" and for "off-center" cells. There are large phase shifts
into the positive theta region at low frequencies. These are not evident
by 20 Hz. A large selection of cells show small phase shifts into the
negative theta region above 30 Hz. These results suggested strongly that
the origin of the psychophysically measured phase shifts lay in the
magnocellular cell response. These results reinforce a series of studies
from Lee's laboratory, in which the M-pathway (phasic) cells have been

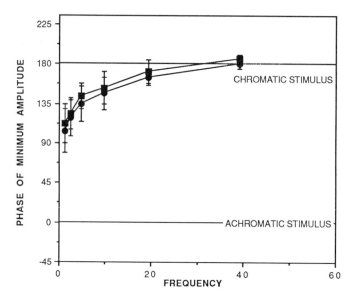

Figure 9. The phase of least amplitude as a
function of temporal frequency. Data are shown for
M-pathway (achromatic) cells. Solid circles show
average data for "phasic-on" cells. Solid squares show
average data of "phasic-off" cells.

identified as the probable physiological substrate of the psychophysical
luminance channel tapped by heterochromatic flicker photometry (Lee,
Martin & Valberg, 1988) and minimally distinct border (Kaiser, Lee,
Martin & Valberg, 1990).

A consideration of the data of Figure 9 shows that the common text
book models of M-pathway cells cannot explain the data. Usually the
centers and surrounds of M-pathway cells are considered to have the same
spectral sensitivity, given by mixed LWS and MWS activity, similar to the
surrounds of "blue-on" cells. Both center and surround would be most
sensitive to in-phase modulation and their phase responses would be 180°
out of phase for all values of physical phase, θ. A frequency-dependent
surround delay would increase the amplitude response, but the amplitude
maximum would always occur to in-phase stimulation. Even if the cone
weightings of center and surround were unequal, it is impossible to
obtain behavior like that of Figure 9. Once again, both center and
surround would be most sensitive to in-phase stimulation, as in Figure 4.
A small surround delay could only introduce small frequency-dependant
shifts in the amplitude minima as in Figure 5. For example the change in
phase of minimum amplitude from 180° to 190° at 40 Hz could be explained
by surround delays and a center and surround whose cone weightings were
different. We think that the behavior of the M-pathway cells is
consistent with the idea that a small component of the M-pathway surround
is opponent in origin. A model incorporating such opponency has been
developed and can account for the amplitude and phase responses of the M-
pathway data. The solid line on Figure 8 represents the implementation
of this model. The net response from the center is derived from a summed
LWS and MWS cone response with a phase of 86°. The surround is derived
from a difference of LWS and MWS cone response and has a phase of -35°.
The center-surround weighting is 0.45. The phase difference between
center and surround of 121° reflects the differential amount of phase

advance of the summing and differencing components, about 110°, plus a small frequency-dependant center-surround phase difference, about 5-15° at 4.88 Hz. The center-surround weighting decreases with frequency and the opponency is most evident at low temporal frequencies, suggesting that it is only a minor component of the phasic surround. It is important to note that the low frequency phase shifts occur becuse the quadrature stimulus introduces a 90° phase shift between the non-opponent and the opponent components of the cell.

SUMMARY

Psychophysical studies of equiluminant heterochromatic lights in temporal sinusoidal modulation have shown that modulation sensitivity is least when the relative phase of the component lights is at a value other than 180° (counterphase). This result is robust and cannot be attributed to inherent latencies in the distal signals originating from the cone photoreceptors. Physiological studies in macaque retina using identical stimulus paradigms reveal a parallel behavior in the amplitude responses of M-pathway (phasic) retinal ganglion cells. The data can be explained by postulating a small color-opponent input to the surround of the M-pathway retinal ganglion cell.

FOOTNOTES

*Various aspects of this work were performed in collaboration with Barry B. Lee, Delwyn T. Lindsey, Paul R. Martin, Joel Pokorny, William H. Swanson & Arne Valberg. Preliminary reports of the physiological data were given at the annual meetings of ARVO, 1989 and OSA, 1989.

ACKNOWLEDGEMENTS

This research was supported by USPH NEI grant EY00901 to Joel Pokorny and by NATO grant 0909/87 to Barry B. Lee.

REFERENCES

van den Berg, TJTP and Spekreijse, H: Interaction between rod and cone signals studied with temporal sine wave stimulation. J. Opt. Soc. Am. 65:1210-1217, 1977.

Cushman, WB and Levinson, JZ: Phase shift in red and green counterphase flicker at high temporal frequencies. J. Opt. Soc. Am. 73:1557-1561, 1983.

von Grunau, MW: Lateral interactions and rod intrusion in color flicker. Vision Research. 17:911-916, 1977.

de Lange, H: Research into the dynamic nature of the human fovea-cortex systems with intermittent and modulated light. II Phase shifts in brightness and delay in color perception. J. Opt. Soc. Am. 48:784-789, 1958.

Lindsey, DT, Pokorny, J and Smith, VC: Phase-dependent sensitivity to heterochromatic flicker . J. Opt. Soc. Am. A. 3:921-927, 1986.

Kaiser, PK, Lee, BB, Martin, PR, and Valberg, A: The physiological basis of the minimally distinct border demonstrated in the ganglion cells of the macaque retina. J. Physiol. (London). 422:153-183, 1990.

Lee, BB, Martin, PR and Valberg, A: The physiological basis of heterochromatic flicker photometry demonstrated in the ganglion cells of the macaque retina. J. Physiol. (London). 404: 323-347, 1988.

Lee, BB, Pokorny, J, Smith, VC, Martin, PR & Valberg, A: Luminance and chromatic modulation sensitivity of macaque ganglion cells and human observers. J. Opt. Soc. Amer. (in press).

Lee, BB, Smith, VC, Pokorny, J, Martin PR & Valberg, A: Responses of tonic ganglion cells of the macaque retina on changing the relative phase of two flickering lights. (in preparation)

Smith, VC, Lee, BB, Pokorny, J, Martin PR & Valberg, A: Responses of phasic ganglion cells of the macaque retina on changing the relative phase of two flickering lights. (in preparation)

Swanson, WH, Pokorny, J and Smith, VC: Phase-dependant sensitivity to heterochromatic flicker: subject differences in color-normal males, J. Opt. Soc. Am. A. 13:P27, 1986.

Swanson, WH, Pokorny, J and Smith, VC: Effects of temporal frequency on phase-dependent sensitivity to heterochromatic flicker. J Opt. Soc. Am. A. 4: 2266-2273, 1987.

Swanson, WH, Pokorny, J and Smith, VC: Effects of chromatic adaptation on phase-dependent flicker. J.Opt. Soc. Am. A 5: 1976-1982, 1988.

Vos, JJ & Walraven, PL: Phase shift in the perception of sinusoidally modulated light at low luminances. In Perception of the Eye at Low Luminances: Proceedings of the Colloquium in Delft, 1965, MA Bouman and JJ Vos eds, Excerpta Medica Foundation, NY 1966.

Walraven, PL and Leebeek, HJ: Phase shift of alternating coloured stimuli. Doc. Ophthalmol. 18:56-71, 1964.

PSYCHOPHYSICAL CORRELATES OF PARVO CHANNEL FUNCTION

Carl R. Ingling, Jr.

Department of Zoology, The Ohio State University
Columbus, Ohio 43210

INTRODUCTION

The identification of magnocellular and parvocellular channels within the primary visual system has prompted attempts to correlate the results of psychophysical experiments with the properties of these channels. Despite the relatively clear differences in channel properties, there is currently disagreement about the assignment of various tasks to the different channels.

A single channel cannot be optimized for detection of stimuli on many dimensions because the requirements for one task conflict with those of another. The large axons and receptive fields of the magno channel optimize speed of response but sacrifice acuity and color vision. The separate fibers for each receptor found in the parvo channel are necessary for color vision and acuity, for which cone signals must be kept separate, but sacrifice speed and sensitivity. Hence the need for different channels.

This obvious separation of function is of course the first line of attack on the problem of deciding which channel does what. By using such well-documented differences between the parvo and magno channels as spectral sensitivity, photometric additivity, temporal and spatial resolution and other properties to be developed, we attempt to determine which channel is used to perform particular psychophysical tasks.

1. ABSOLUTE CONE THRESHOLD

Which channel is most sensitive at absolute cone threshold? To decide, note the difference in spectral sensitivity of the parvo and magno channels. The parvo channel weights R and G cones in the ratio 2R:3G, the weighting needed for the R and G cones to have equal sensitivities around 580 nm. The magno channel weights R and G cones in the ratio 5R:3G, the ratio needed to fit V_λ, the luminosity curve. Hsia and Graham (1957) measured foveal absolute threshold curves (42' test field, 4 msec flash) for normal observers, protanopes and deuteranopes. To fit the spectral sensitivity curves for the normal observers the protan and deutan curves must cross at 580 nm. Apparently, the channel detecting the threshold test flashes weights cones in the ratio 2R:3G. Further, for such stimuli, there is no discernible photochromatic interval (excepting spectral locations near the tritanopic and protan-deutan neutral points). If seen, long-wavelength flashes look red and short-wavelength flashes look green, not achromatic.

From Pigments to Perception, Edited by A. Valberg and
B.B. Lee, Plenum Press, New York, 1991

However, other experiments show evidence of magno channel input, but apparently not enough to significantly move the 580 nm crosspoint. Massof (1981) measured the slopes of frequency-of-seeing curves at absolute cone threshold as a function of wavelength (20', 50 msec test flash). He found that the slopes were steeper in the middle of the spectrum than at the ends. This result is explained (Ingling, Martinez-Uriegas and Lewis, 1983) if a channel with a steeper psychometric function - namely, the magno channel - contributes to the detection of mid-spectral lights. Furthermore, Guth and colleagues (e.g., Guth et al., 1980) have shown, using both spectral test flashes and mixtures of spectral lights at foveal threshold, that an achromatic channel is needed to account for detection at absolute threshold. However, to obtain these results requires mixtures of lights. Mixtures tend to stimulate both R and G cones (as do mid-spectral lights, particularly near 580 nm), which in the parvo channel inhibit one another, thus reducing their sensitivity and enabling the less sensitive magno channel to contribute to detection.

In summary, for pure spectral lights which at threshold stimulate mostly either one cone or the other (excepting at the 580 nm region), Hsia and Graham's spectral sensitivities show that the parvo channel is the most sensitive channel; although there is evidence for a magno contribution in the region where the R and G cones inhibit one another, it is too small to significantly alter the spectral sensitivity.

2. INCREMENT THRESHOLD

Which channel detects test flashes on backgrounds? Stiles's (1978; Wyszecki and Stiles, 1967) extensive data on two-color thresholds gives the same answer as Hsia and Graham's threshold data. For Stiles's data, the wavelength of the background that raises the threshold for π_4 and π_5 equally is again near 580 nm, not the 490 nm magno channel crosspoint. However, for increment threshold conditions the question of parameters becomes crucial. We have shown earlier (Ingling and Martinez-Uriegas, 1981) that the parvo channel input to detection can be modified by altering the temporal characteristics of the test flash. More recent (unpublished) observations show that when Stiles's 200 msec test flash is shortened to 5 msec., the intensity of a yellow (580 nm) background that raises a red test flash 10 times above its absolute threshold no longer also raises a green test flash 10 times. For 5 msec test flashes, the 580 background must be increased by a factor of about 2.5 to cause the same 10 times increase for a green test flash. For the channel that detects 5 msec flashes, the G-cone sensitivity is 2.5 times less at 580 than the red; for a 200 msec test flash the sensitivities of the mechanisms that detect the red and green test flashes are equal. The background wavelength that would produce equal threshold elevations for 5 msec flashes is near 500 nm, the crosspoint for the magno channel. Thus the answer to the question, "Which channel detects increments?" is parameter-dependent; for short flashes the magno channel dominates, for long flashes the parvo.

3. ACUITY

Which channel resolves high-frequency gratings? To answer this question, it is useful to review a few experiments that have addressed this issue. Because photometric additivity is an important property of the psychophysical achromatic channel, it has been widely used to decide whether or not the achromatic channel mediates a given visual task. Consider Guth's threshold additivity paradigm. Determine threshold for a red test flash and a green test flash. Mix them in half-threshold amounts and ask if they are still at threshold. They are not; they are markedly subadditive, meaning that it takes a great deal more than half a threshold unit of the red plus half a unit of the green to sum to one threshold unit of the mixture. The result is explained by pointing to the opponent sensitivity of the parvo channels. For stimuli detected by the parvo channel, not only will half a unit of red plus half a unit of green not add to one unit of the mixture, they may well add to zero units of the mixture. One concludes from this experiment that both parvo and magno channels detect the test flash (see (1) above; the use of mixtures recruits magno input) but the chromatically opponent parvo channels cause the subadditivity.

Repeat the experiment exactly but with one change; place a grating in the test field and require the observer to detect the orientation of the grating instead of the presence of the test flash. Otherwise the experiment is identical. Now, half a threshold unit of red grating plus half a unit of green grating adds to precisely one threshold unit of the mixture grating (Guth and Graham, 1975; Myers et al., 1973). Given the rule that subadditivity is a parvo channel signature and photometric additivity a luminance channel signature, the straightforward conclusion from this experiment is that the conventional photometrically additive luminance channel resolves the gratings (Guth and Graham, 1975). However, informed by the parallel channel literature, we reject this conclusion, logical and parsimonious though it be (Myers et al., 1973). The grounds for rejection are simply that the parvo channel has the proper characteristics for subserving acuity and the magno channel does not. Mere recitation of parvo characteristics convinces: parvo units dominate the central fovea, the region of highest acuity; they have the smallest receptive fields; acuity is unaffected by image stabilization, which eliminates magno contributions (Kelly and Martinez-Uriegas, 1990), etc.

Attributing detection of the gratings to the parvo channel requires an explanation of how the parvo channel can be both subadditive (when detecting spatially uniform test flashes; i.e., low spatial frequencies) and also photometrically additive (when detecting gratings; i.e., high spatial frequencies). Figure 1 summarizes an analysis given elsewhere (Ingling and Martinez-Uriegas, 1985, 1983a,b). As the figure shows, the parvo channel transmits both an opponent-color (r-g) signal, which is subadditive, and also an additive luminance signal. The opponent-color signal is transmitted at low spatial frequencies (e.g., for spatially uniform test flashes) whereas the luminance signal is transmitted at high (e.g., for gratings).

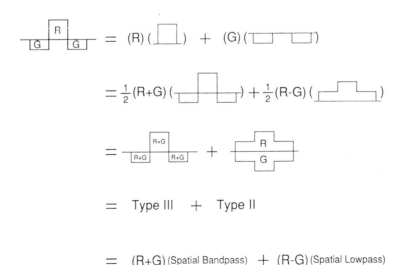

Figure 1. The parvo channel r-g simple-opponent receptive field is identically equal to a combination of both an achromatic (R + G) cell with spatial bandpass sensitivity and a chromatic (R - G) spatially lowpass cell.

Although Fig. 1 complicates the picture of the achromatic channel, inasmuch as it is no longer conceived of as a single structure, now having both magno and parvo components, it is supported by much physiological evidence. Wiesel and Hubel (1966) were the first to report that parvo cells respond to achromatic contrasts, and De Valois and Pease (1971) long ago recorded, in convincing fashion, both types of responses from chromatically opponent LGN units.

As reasonable as it seems to attribute acuity to the parvo channel, a discrepancy remains. If acuity be mediated by the parvo channel, then spectral sensitivity measured with an acuity criterion should be fit by R and G cones combined in the ratio 2R:3G, not the 5R:3G proportion of the magno channel. However, lights of equal luminance produce equal acuities, which means that an acuity criterion isolates a V_λ -like or magno spectral sensitivity. This finding has been obtained for at least two different paradigms; determination of the highest acuity as a function of luminance for different wavelengths (Pokorny, Graham and Lanson, 1968), and measurement of the contrast to resolve a fixed spatial frequency as a function of wavelength (Ingling et al., 1988). Ingling and Tsou (1988) have advanced an explanation of how an acuity criterion, believed to tap the parvo channel, unexpectedly yields a magno channel sensitivity. At threshold, for high-spatial-frequency stimuli, the receptive field centers of parvo units are more sensitive than their surrounds. Surrounds, of larger spatial extent, have lower cut-off frequencies and are less sensitive than centers at all spatial frequencies (Burbeck and Kelly, 1980). On the hypothesis that the centers of parvo receptive fields are distributed in the same ratio as the cone populations that feed the magno channel, we expect to obtain approximately the magno channel sensitivity with any criterion that taps only cone centers. Only a suprathreshold acuity criterion that stimulates both centers and surrounds would be expected to produce the parvo sensitivity.

This question of which channel mediates acuity illustrates the problems confronting attempts to establish correlations. As discussed, the acuity criterion is both phototmetrically additive, and furthermore has the magno channel sensitivity. Both facts argue for a magno channel origin for acuity. Nonetheless we conclude that the parvo channel subserves acuity, and advance other explanations for these discrepant facts. In this case as well as others that follow arguments can be marshalled on both sides.

4. MINIMALLY DISTINCT BORDER (MDB)

Which channel is used to minimize the contrast of a border? As a photometric criterion, MDB produces the same spectral sensitivity as heterochromatic flicker photometry (Wagner and Boynton, 1972). It also passes another test required of an additive luminance channel; it is photometrically additive (Boynton and Kaiser, 1968). Finally, Kaiser et al. (1990) measured the responses of parvo and magno units to MDB stimuli and conclude that the magno unit supports detection of minimally distinct borders.

However, there are arguments for believing that MDB is mediated by the parvo channel. Evidence favoring a parvo channel origin derives from the De Vries form of the additivity test. De Vries (1948), in his investigation of the properties of heterochromatic flicker photometry, matched a pair of lights at successively increasing intensities. By the De Vries test, flicker photometry is not an additive photometric procedure. It is additive for an octave or two at most, around an intensity of 100 trolands. The cause of the additivity failure is well-understood. Red lights cause greater differential adaptation in the R and G cones than do green lights. Thus as intensity is increased when measuring a De Vries function, the R cone is adapted more than the G cone, and therefore the intensity of the red light must be more than doubled, say, to match a doubled green light. This is a simple consequence of the shape of the spectral sensitivity curves and the choice of the matching lights. Adaptation in this context produces a non-linearity. The ideal achromatic channel implements the solution of a simple equation; (R-cone response + G-cone response) for the green light equals (R-cone response + G-cone response) for the red light. If the adaptation site precedes the summing site and the matching lights be chosen to asymmetrically adapt the cones, then the channel must show the De Vries function signature. If the MDB criterion taps the magno channel, it presumably does so by equating the sum of the R- and G-cone responses on one side of the field to the sum of the R- and G-cone responses on the other. If so, then the MDB criterion ought to have a characteristic De Vries function signature. It does not. The De Vries function for MDB is essentially flat, as if it were measured on a monochromat (Ingling et al., 1978). Given this result, we propose a parvo channel mechanism that produces a flat De Vries function.

Figure 2 shows the response of the parvo unit to three edges (Ingling and Martinez-Uriegas, 1983b). In general, the parvo unit has two responses

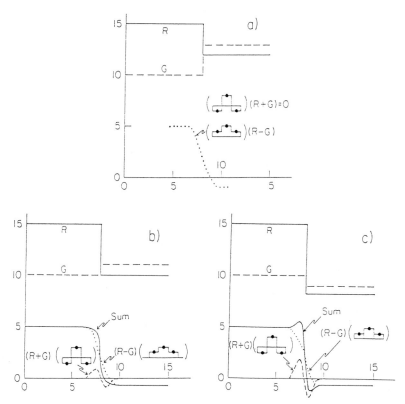

Figure 2. Response of the two components of the (r-g) receptive field of Fig. 1 to an edge. (R and G only in this particular Figure represent cone responses.) (a) An equal luminance edge; (R + G) cone responses are constant across the border; the response is lowpass. (b) Luminance contrast is added by decreasing the green side. Response remains low-pass, but a masked achromatic component is present. (c) A higher-contrast edge which satisfies the inequalities given in the text produces a bandpass response, the criterion required for stereopsis for Lu and Fender's stereograms.

to an edge, a chromatic response and an achromatic response. (Fig. 1 shows
the two receptive field components that produce the achromatic and chromatic
responses, which when combined form the parvo receptive field). For edges
having both chromatic and luminance contrast both signals are present.
Although both signals may be present, the bandpass achromatic component can
be masked by the chromatic component until a certain criterion contrast is
reached, as shown in Fig. 2. The value of the achromatic contrast needed to
make the composite signal switch from lowpass to bandpass is given by a pair
of inequalities. One inequality applies to the response of the receptive
field center; the other inequality applies to the response of the surround.
Both must be satisfied in order to produce a bandpass signal. The
inequalities are:

$$I_1 R_1 > I_2 R_2, \quad \text{and}$$
$$I_1 G_1 > I_2 G_2 \quad .$$

In these equations, $I_1 R_1$ is (Intensity)(Sensitivity) = R-cone response for
Side 1 (the left) and $I_2 R_2$ the analogous quantity for the right; similarly
for $I_1 G_1$. For Fig. 2, the first inequality - the center inequality - is
satisfied even at equal luminance. Now dim the right side until the second
(surround) inequality is satisfied, that is, until the response of the
G cone is greater to the red side of the field than it is to the green side.
When this condition is met, the composite signal switches from lowpass to
bandpass. When the inequalities are satisfied the response profile shows a
Mach-band type enhancement at the edge, which we refer to as a bandpass
signal. The inequalities simply state that for enhancement, both the R-cone
and the G-cone response on one side of the field must be greater than the R-
and G-cone responses on the other. For the converse situation, in which we
make the right side of the field brighter, it will be found that the
surround inequality is first met, and instead the right side of the field
must be brightened until the first or center inequality is met.

We suppose that an observer in minimizing the distinctness of a border
alternates between these two inequalities (or between two intensities
proportional to the intensities which satisfy the inequalities) as he dims
and then brightens the test field, looking for that point which lies midway
between the points of enhancement defined by the inequalities. Because the
inequalties are between cones of the same type, their state of adaptation
is irrelevant. Cone responses on one side of the field are not summed to
equal cone responses on the other side of the field. Instead, the point
midway between two enhancement points is found by equating R-cone responses
across the boundary at one of the limits and G-cone responses across the
boundary at the other limit.

The major effect producing the De Vries signature for a conventional
additive luminance channel, namely adaptation of receptors at a site which
precedes summation of cone signals, has no effect on the parvo channel
inequalities because cones are not summed to satisfy the inequalities.

It needs to be shown that this model for mimimizing border distinctness
also produces a magno channel sensitivity. At this time, simply note that
the spectral sensitivity curve for detecting enhancement when the test side
(right side) of Fig. 2 is brighter is that of the G cone; similarly, the
spectral sensitivity for detecting enhancement when it is dimmer is that of
the R cone. In order for the MDB criterion to produce a magno spectral
sensitivity, the observer must set the midpoint at a position between the R
and G spectral sensitivities which is proportional to the numbers of cones
that enter into each inequality.

Regarding the conclusions of Kaiser et al. based on electrophysiological recording, their rejection of the parvo unit as a possible
mediator of MDB settings seems premature. We agree that magno units might
appear more sensitive than parvo units. However, the case here is similar to
the acuity case. It has also been occasionally concluded that parvo units

are too insensitive compared to magno units to subserve acuity. A point made by Derrington and Lennie (1984) bears repeating: "When one takes account of differences in threshold criterion, a single magnocellular unit seems to be slightly more sensitive than any detector inferred from psychophysical experiments, while a parvocellular unit is perhaps one-third as sensitive; a gap that could be removed by probability summation across as few as ten units."

Although it may not be readily apparent, and may be difficult to extract from physiological recordings, there is an achromatic response buried in Kaiser et al.'s parvo unit recordings. Pursuing Derrington and Lennie's argument, it may well be that processing in the brain is capable of extracting this achromatic response by techniques not available to or tried by electrophysiology. Kaiser et al. note that for high contrasts such signals are present. Parvo sensitivity has been underestimated or discounted in other contexts. In this regard we note that Burbeck and Kelly (1980) model detection of the entire chromatic <u>and achromatic</u> spatiotemporal sine-wave sensitivity surface <u>with a single receptive field of the parvo type</u>. Thus it seems plausible that the parvo unit is sensitive enough to mediate contrast detection across MDB's. Until electrophysiology is certain of being able to process parvo signals as efficiently as the brain, it seems premature to dismiss a role in minimizing border distinctness.

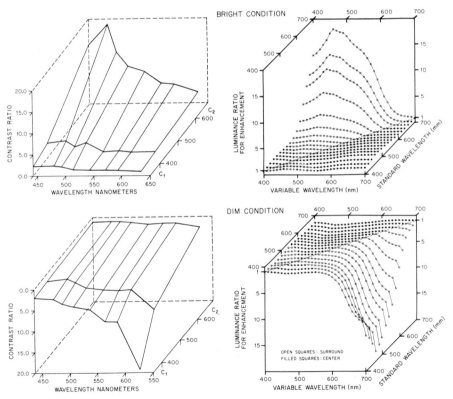

Figure 3. Left side; Lu and Fender's measurements of the contrast (ordinate) needed to produce stereopsis for pairs of lights shown on abscissae. There are two surfaces because dimming one member of a pair until stereopsis occurs takes a different contrast than brightening it. Right side: computation of the contrast needed to produce enhancement on the response profile of the r-g parvo cell.

5. STEREOPSIS

Which channel is responsible for stereopsis? Figure 3 shows the
surfaces measured by Lu and Fender (1972) for isoluminant random-dot
stereograms, say red and green dots. The surfaces of Fig. 3 show the
luminance contrast required for stereopsis for different wavelength pairs.
For such stereograms, stereopsis is absent or greatly reduced at
isoluminance; achromatic contrast must be added in order to see depth.

This finding has been interpreted to mean that the parvo system does
not support stereopsis. However, the amount of achromatic contrast that must
be added to produce stereopsis for different wavelength pairs is not
constant; it is a function of wavelength. Russell (1979), by using the
inequalities given above, showed that the amount of contrast needed in order
to see depth in a random-dot stereogram is predicted by finding the contrast
needed to produce the transition in the response of the parvo unit from
lowpass to bandpass. To explain this theory in reference to Fig. 3 and the
inequalities, stereopsis appears when, say, the green dots have been dimmed
enough to satisfy the surround inequality ($I_1G_1 > I_2G_2$); at this point, the
achromatic or bandpass response of the parvo unit (also see Fig. 2) just
starts to appear upon the total response. That is, the response profile
becomes enhanced. It is at this enhancement point that stereopsis appears.
Similarly, stereopsis also appears when the red dots have been dimmed enough
to satisfy the center inequality. Figure 3 shows the surfaces calculated
from the inequalities (from Ingling, 1982); i.e, the luminance ratios
required for the parvo unit to enhance, or acquire a bandpass-type profile.
Comparison of these surfaces with Lu and Fender's data leaves little doubt
that the parvo unit must be the major channel mediating stereopsis for
Julesz random-dot stereograms.

Again, as in the MDB case, we are aware that physiologists have come to
different conclusions (Livingstone and Hubel, 1988). As for MDB, it appears
that the achromatic response component of the parvo unit has either been
ignored or masked by magno channel responses.

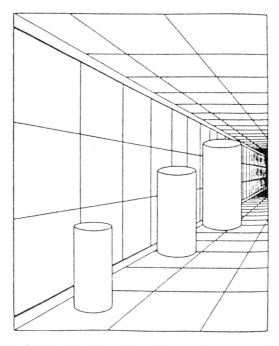

Figure 4. Illusory depth stimulus. When the contours are isoluminant, the
illusion of depth is absent. The illusion of depth appears as above in the
positive afterimage, which depends solely upon the parvo channel.

6. MONOCULAR DEPTH AND FORM PERCEPTION

Which channel mediates the perception of monocular depth and form? Isoluminant stimuli have been used in attempts to answer this question, in the belief that such stimuli isolate chromatic mechanisms, namely, the parvo channel. Ideally, color-blind or luminance channels are blind to isoluminant stimuli, thus allowing other channels to be studied in isolation. The problem with using isoluminant stimuli to study the properties of the parvo channel (assuming that the magno channel is silent for such stimuli; cf. Schiller and Colby, 1983) is that isoluminant stimuli also silence the achromatic signal from the parvo channel. Thus, should a particular percept be absent for an isoluminant stimulus - for example, monocular depth - the only conclusion permissible is that monocular depth is mediated either by the magno channel, the achromatic signal of the parvo channel, or by both. What is needed are stimulus formats that eliminate the magno channel without affecting either of the parvo channel contributions. That is, we require a stimulus with luminance contours that, nonetheless, cannot be seen by the magno-luminance channel.

Because of the transient nature of the temporal responses of the magno channel, it appeared likely, and has since been confirmed (Kelly & Martinez-Uriegas, 1990), that stabilized images or afterimages would be mediated solely by the parvo channel.

The striking depth apparent in Fig. 4 is absent in the isoluminant format. As discussed above, no conclusions can be drawn from this fact. Either the magno or the parvo channel, or both, might mediate the percept of monocular depth.

However, when Fig. 4 is visualized as an afterimage, the perceived depth is as great or greater than that seen in free viewing. Given that the afterimage mode eliminates magno channel signals, we conclude that the achromatic signal of the parvo channel mediates monocular depth and form perception (Ingling and Grigsby, 1990; but cf. Livingstone and Hubel, 1988).

7. PARVO MULTIPLEXING OF HUE AND LUMINANCE

As Fig. 1 shows and electrophysiology confirms, the parvo channel is not a univariant channel, in the sense that it encodes both luminance and hue signals and ostensibly transmits both within a single channel. In general, of course, one cannot transmit two independent variables over one channel (Ingling and Martinez-Uriegas 1983b). It would be impossible to decide if the channel output represented a value for the hue, a value for the luminance, or some combination of both. Two solutions of this problem have been advanced. Ingling and Martinez-Uriegas (1983b) point out that although both signals are in the same channel, they are sent at different spatial frequencies. This suggests that a demultiplexer in the brain could recover hue and luminance. The luminance signal is spatially bandpass whereas the hue signal is lowpass. Were the parvo fibers to project to lowpass and bandpass filters in the brain which matched the character- istics of the lowpass and bandpass retinal filters, to within certain limits set by the tuning and overlap of the filters the hue and luminance signals could be demultiplexed and recovered. This is a true multiplexing scheme. It takes advantage of a dimension - the spatial frequency spectrum - along which signals can be encoded and thereby saves bandwidth by obviating separate channels for luminance and hue. (This multiplexing scheme does not allow complete recovery of luminance and hue for all points in the scene, inasmuch as luminance is only transmitted at edges or high-spatial-frequency regions of the scene.)

The second method for recovering hue and luminance signals from the parvo channel has been advanced independently by several people (Martinez-Uriegas (1985) also in Kelly (1989); Billock (personal communication); Lennie (1988) and others.) This second method is not a form of multiplexing.

It does not take advantage of the spatial-frequency dimension for differentially encoding hue and luminance, nor for extracting such signals at the cortex with matching filters. Instead, it requires the addition of extra channels. In one form it postulates that the r-g channel, say, is composed of two <u>independent</u> subchannels, +R - G and +G - R. Although these channels might seem to be sending the same kind of information and for that reason not be independent, for the proposed recovery scheme they must be truly independent channels. Two variables require two channels, regardless of how similar they may appear. Given this independence, luminance and hue are recovered by summing and differencing these two channels in the brain. Fig. 1 pictures an r-g receptive field as [(R)(+ center) + (G)(- surround)], which can be rewritten as [(R + G)(bandpass filter) + (R - G)(lowpass filter)]. Suppose a complementary independent channel composed of receptive fields [(G)(+ center) + (R)(- surround)]. Adding these two fields gives [(R + G)(bandpass filter)], or luminance, and subtracting gives [(R - G)(lowpass filter)], or hue. Although this scheme might be said to "recover" hue and luminance, they were never lost, but simply transformed for transmission over two channels. In fact, there is no need to ever recover them. (Certainly the brain does not trouble itself to invert the opponent transformation from R- and G-cone signals to (R + G) and (R - G) channel signals in order to recover information about quantum absorptions in cones.) However, because observers seem to classify stimuli along hue and luminance dimensions, in this case they may well be recovered. This scheme is entirely analogous to the transformation of R- and G-cone signals into (R + G) and (R - G) channel signals, and demands two channels. Assuming the y-b channel functions similarly, 5 independent channels are required in the optic nerve; two r-g subchannels, two y-b, and the magno channel.

Both of the suggested solutions appear plausible, and both might be used. We comment that the Martinez-Lennie-etc. scheme demands considerable spatial precision. On the other hand, the Lu and Fender result described above shows that the filters needed for demultiplexing are already in place. That is, to account for the Lu and Fender result requires matching cortical filters to extract luminance and hue from the composite parvo signal; i.e., central demultiplexing filters.

8. CONCLUSIONS

We have considered absolute cone thresholds, increment thresholds, the minimally distinct border, stereopsis, monocular depth and form perception, and physiological mechanisms for analyzing the composite signals of the parvo channel. We conclude that the parvo channel contributes significantly to the tasks considered; some, of course, more certainly than others. At this time the role of the parvo channel in most of these tasks is controversial. The root of the controversy seems to stem from difficulties in recognizing and measuring the luminance component of the parvo channel code.

REFERENCES

Boynton, R. M. and Kaiser, P. K. , 1968, Vision: The additivity law made to work for heterochromatic photometry with bipartite fields, Science, 161, 366-368.
Burbeck, C. A. and Kelly, D. H., 1980, Spatiotemporal characteristics of visual mechanisms: Excitatory-inhibitory model, J. Opt. Soc. Am., 70, 1121-1126.
Derrington, A. M. and Lennie, P., 1984, Spatial and temporal contrast sensitivities of neurones in lateral geniculate nucleus of macaque. J. Physiol., 357, 219-240.

De Valois, R. W. and Pease, P. L., 1971, Contours and contrast: Responses of monkey lateral geniculate nucleus cells to luminance and color figures, Science, 171, 694-696.

De Vries, HL, 1948, The luminosity curve of the eye as determined by measurements with the flickerphotometer, Physica, 14, 319-348.

Guth, S. L., Massof, R. W. and Benzschawel, T., 1980, Vector model for normal and dichromatic vision, J. Opt. Soc. Am., 70, 197-212.

Guth, S. L. and Graham, B. V., 1975, Heterochromatic additivity and the acuity response, Vision Res., 15, 317-319.

Hsia, Y. and Graham, C. H., 1957, Spectral luminosity curves for protanopic, deuteranopic, and normal subjects, Nat. Acad. Sci., 43, 1011-1019.

Ingling, C. R. Jr. and Grigsby, S. S., 1990, Perceptual correlates of magnocellular and parvocellular channels: Seeing form and depth in afterimages, Vision Res., 30, 823-828.

Ingling, C. R. Jr. and Tsou, B. H.-P., 1988, Spectral sensitivity for flicker and acuity criteria, J. Opt. Soc. Am. A., 5, 1374-1378.

Ingling, C. R. Jr., Long, R. and Grigsby, S. S., 1988, The spectral sensitivity of normal observers using heterochromatic flicker photometry and an acuity criterion, Lighting Research Institute Final Report (Contract No. 85:DR:1) New York, N.Y. 10001.

Ingling, C. R. Jr. and Martinez-Uriegas, E., 1985, The spatiotemporal properties of the r-g X-cell channel, Vision Res. ,25, 33-38.

Ingling, C. R. Jr. and Martinez-Uriegas, E., 1983a, The relationship between spectral sensitivity and spatial sensitivity for the primate r-g X-cell channel, Vision Res., 23, 1495-1500.

Ingling, C. R. Jr. and Martinez-Uriegas, E., 1983b, The spatiochromatic signal of the r-g channel, in "Colour Vision," J. D. Mollon and L. T. Sharpe, eds., Academic Press, London.

Ingling, C. R. Jr., Martinez-Uriegas, E. and Lewis, A. L., 1983, Tonic-phasic channel dichotomy and Crozier's law, J. Opt. Soc. Am., 73, 183-189.

Ingling, C. R. Jr., 1982, The transformation from cone to channel sensitivities, Color Res. and Application, 7, 191-196.

Ingling, C. R. Jr. and Martinez-Uriegas, E., 1981, Stiles's mechanism: Failure to show univariance is caused by opponent-channel input, J. Opt. Soc. Am., 71, 1134-1137.

Ingling, C. R. Jr., Tsou, B. H.-P., Gast, T. J., Burns, S. A., Emerick, J. O. and Riesenberg, L., 1978, The achromatic channel - I. The nonlinearity of minimum-border and flicker matches, Vision Res., 18, 379-390.

Kaiser, P. K., Lee, B. B., Martin, P. R. and Valberg, A., 1990, The physiological basis of the minimally distinct border demonstrated in the ganglion cells of the macaque retina, J. Physiol., 422, 153-183.

Kelly, D. H. and Martinez-Uriegas, E., 1990, Are parvocellular pathways the sole source of non-bleaching afterimages?, Invest. Ophthal. and Visual Science (Suppl.), 31, 88.

Kelly, D. H., 1989, Spatial and temporal interactions in color vision, J. Imaging Technology, 15, 82-89.

Lennie, P. and D'Zmura M., 1988, Mechanisms of color vision, CRC Critical Reviews in Neurobiology, 3, 333-400.

Lindsey, D. T. and Teller, D. Y., 1989, Influence of variations in edge blur on minimally distinct border judgments: A theoretical and empirical investigation, J. Opt. Soc. Am. A, 6, 446-458.

Livingstone, M. S. and Hubel, D. H., 1988, Segregation of form, color, movement, and depth: Anatomy, physiology, and perception, Science, 240, 740-749.

Lu, C. and Fender, D. H., 1972, The interaction of color and luminance in stereoscopic vision, Invest. Ophthal., 11, 482-490.

Martinez-Uriegas, E., 1985, A solution to the color-luminance ambiguity in the spatio-temporal signal of primate X cells, Invest. Ophthal. and Visual Science (Suppl.), 26, 183.

Massof, R. W., 1981, Wavelength dependence of the shape of foveal absolute threshold probability of detection functions, Vision Res., 21, 995-1004.

Myers, K. J., Ingling, C. R. Jr. and Drum, B. A., 1973, Brightness additivity for a grating target, Vision Res., 13, 1165-1173.

Pokorny, J., Graham, C. H. and Lanson, R. N., 1968, Effect of wavelength on foveal grating acuity, J. Opt. Soc. Am., 58, 1410-1414.

Russell, P. W., 1979, Chromatic input to stereopsis, Vision Res., 19, 831-834.

Schiller, P. H. and Colby, C. L., 1983, The responses of single cells in the lateral geniculate nucleus of the rhesus monkey to color and luminance contrast, Vision Res., 23, 1631-1641.

Stiles, W. S., 1978, Mechanisms of Color Vision, Academic Press, New York.

Wagner, G. and Boynton, R. M., 1972, Comparison of four methods of heterochromatic phototmetry, J. Opt. Soc. Am., 62, 1508-1515.

Wiesel, T. N. and Hubel, D. H., 1966, Spatial and chromatic interactions in the lateral geniculate body of the rhesus monkey, J. Neurophysiol., 29, 1115-1156.

Wyszecki, G. and Stiles, W. S., 1967, Color Science, Wiley, New York.

ON THE PHYSIOLOGICAL BASIS OF HIGHER COLOUR METRICS

Arne Valberg and Thorstein Seim

Department of Physics, Section of Biophysics
University of Oslo, Box 1048 Blindern
0316 Oslo 3, Norway

INTRODUCTION

We explore here the predictive power of a physiological model of colour vision that combines the outputs of six types of colour-opponent cells of the parvocellular pathway. This model has been demonstrated to account well for the equidistant spacing of chromaticities in the Optical Society of America's Uniform Color Scale and in the Munsell System. Chroma, in planes of constant luminance, is related to response magnitude after subtraction of an achromatic component, and hue to the relative activities of cell systems. Here, we demonstrate that the model is also capable of predicting the Bezold-Brücke phenomenon and the change of saturation with intensity without auxiliary assumptions. Implications of the model for colour discrimination of dichromats are also presented.

When the luminance of foveal mid-spectral lights increases from zero, there is an achromatic interval between detection of light and the identification of its hue (Abney, 1913; Graham and Hsia, 1969). As luminance further increases above chromatic threshold, all chromatic stimuli show a shift in the perceived hue. At relative high intensities long wavelength red light becomes yellowish and short-wavelength violet light more bluish. This latter phenomenon is known as the Bezold-Brücke phenomenon (Bezold, 1873; Brücke, 1878; Purdy, 1931a; Judd, 1951; Hurvich and Jameson, 1955; Hurvich, 1981).

The Bezold-Brücke hue shift is accompanied by strong changes in the perceived chromatic content of the stimulus (saturation or chromatic strength). When luminance of chromatic lights increases, saturation first increases to reach a maximum and then decreases again, in a way that is wavelength specific (Haupt, 1922; Purdy, 1931; Vimal, Pokorny, and Smith, 1987; Valberg, Seim, and Lee, 1987b). For high luminances, most lights become whitish in appearance.

These colour changes of stimuli of constant chromaticity with intensity are related to the Abney effect of hue changes of spectral lights in colour mixtures with white (Abney, 1910, 1913). Such non-linearities of colour vision occur, for instance, in self-luminous colour displays of various kinds where luminance ratios relative to a background can vary over a much larger range than for surface colours that reflect the incident light.

From Pigments to Perception, Edited by A. Valberg and
B.B. Lee, Plenum Press, New York, 1991

There is a need to explain these phenomena within a new theoretical framework. Whereas the Bezold-Brücke hue shifts have been explained by saturating intensity-response curves of the cones themselves (Cornsweet, 1978; Walraven, 1961; Vos, 1986), or by relative activities of chromatic and achromatic signals (Judd, 1951; Hurvich and Jameson, 1955; Guth and Hovis, 1986), it is hard to see how such intensity-response curves can explain the non-monotonic relationship that is observed between luminance and chromatic strength (saturation). Explanations in terms of compressive non-linearities of receptor excitations (e.g. Helmholtz, 1911) are rather general and have not been demonstrated to predict the behaviour of individual chromatic lights. For instance, the great differences in luminance for maximum chromatic strength for different wavelengths is not expected from such models. This behaviour is more readily accounted for at a postreceptoral level by the non-monotonic intensity-response curves of cone-opponent P-cells of the parvocellular pathway (Valberg, Lee, Tigwell, and Creutzfeldt, 1983; Valberg, Lee, Seim, and Tryti, 1986; Valberg, Lee and Tryti, 1987; Lee, Valberg, Tigwell, and Tryti, 1987).

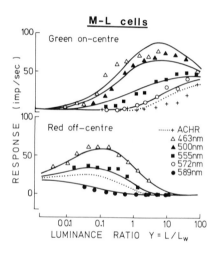

Fig. 1. Luminance-responses for chromatic stimuli covering the receptive field of opponent cells. Symbols represent recorded impulses/s, and curves are simulations by the opponent model. These ON and OFF-centre M-L cells belong to the weakly- and strongly inhibited P-cells of the LGN. Cells with the opposite L-M cone inputs can also be divided into similar ON- and OFF-groups.

From the assumption that both the chromatic content of a stimulus and its hue are related to the relative activities of opponent cells, it follows that the intensity-dependence of these percepts must also be predictable from the same processes. The results presented here show that chromatic scaling and the combined changes in chromatic appearance with intensity find a common physiological interpretation in the non-linear and non-monotonic responses of P-cells in the retina and the lateral geniculate nucleus (LGN) of the primate. Preliminary results have been presented elsewhere (Valberg, Seim and Lee, 1987b; Valberg and Lee, 1989).

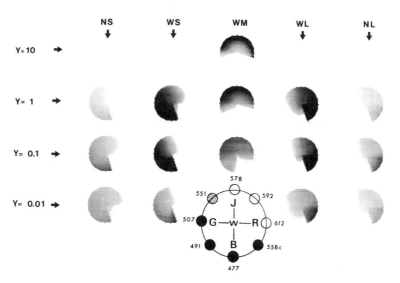

Fig. 2. Response density plots of five colour-opponent cells in equiluminous chroma-hue planes of different luminance ratios (Y). Purity increases from centre (achromatic point) towards the periphery. These plots have been derived from simulation of responses to different wavelengths as shown in Fig. 1. ON-cells (WS, WM, WL) respond better to bright colours than OFF-cells (NS, NL). The blue ON-cell is not shown here.

PHYSIOLOGICAL MODEL AND CHROMATIC SCALING

Valberg et al. (1986) showed that Munsell chroma is closely related to the response magnitude of primate colour-opponent cells, and that hue is related to the relative activities of the same cells. This model also implied a dependence of these attributes on intensity, although this was not tested.

The model was based on the behaviour of typical opponent P cells of Macaca fascicularis, where responses were measured to stimuli darker and brighter than a white adaptation field (Valberg et al., 1983, 1986; Lee et al., 1987; Valberg et al., 1987a;). In the primate retina and LGN one finds weakly and strongly inhibited P-cells with the same cone inputs. They correspond to the ON- and OFF-centre cells of Wiesel and Hubel (1966). For instance, a red ON-centre cell and a green OFF-centre cell both have L-M cone input. The inhibition from the M-cones is stronger for the green OFF-cell, and such cells are therefore activated by increments of long-wavelength light at low luminances and by decrements of all spectral lights at high intensity. We have found six major groups of P-cells: two with L-M cone input, two with M-L cone input (these are the ON and OFF-cells), and one ON cell type with M-S and one with S-M(-L) cone input.

Examples of luminance-response curves of the ON- and OFF M-L cell types to 4 deg spots are shown in Fig. 1. Basing on quantitative evidence, we have argued that both types share the task of coding for colour, but for different ranges of relative luminance, L/L_w, between a stimulus and an adapting field. They thus share the intensity dimension between them; ON-cells are best suited for coding the brighter colours of self-luminous objects, and OFF-cells for coding the colours of darker, reflecting surfaces, with some overlap between them. The resultant "chromatic response" is derived from the summed responses of ON and OFF-cells with the same cone input.

The diagrams of Fig. 2 illustrate the responses of five P cells to stimuli within hue planes of constant luminance, where purity (saturation) varies with distance from the centre achromatic stimulus. Vertically, the responses of ON- and OFF-cells are shown for different luminance ratios relative to a white adaptation field. The darker the shading, the higher is the response. The difference between ON and OFF-cells is clearly seen. OFF-cells respond better to dark colours and ON-cells to brighter colours.

For stimuli covering the whole receptive field, Figs. 1 and 2 show how response behaviour of opponent-cells can be modelled mathematically by taking sums and differences of (usually) two cone inputs to an opponent cell (Valberg et al., 1985, 1987). The cone signals are hyperbolic functions of cone excitations. The same response equation applies to changes in all three dimensions of colour space (intensity, wavelength, and saturation).

Figs. 3A and B demonstrate how a colour space made up of a combination of the six different types of P-cells represent colour scaling. In A, six typical geniculate P-cells are combined to form orthogonal axes, where the positive and negative x-axis (p_1 and $-p_1$) are made up of the summed response of L-M ON- and OFF-cells and of M-L ON-and OFF-cells, respectively. Distances along the positive y-axis (p_2) are proportional to the response of a yellow M-S ON-cell, and along the negative y-axis ($-p_2$) to the activity of a blue S-L ON-cell. This leads to an elliptical form of constant Munsell chroma curves, and close to straight hue lines.

In Fig. 3B, more circular chroma curves have been obtained by a linear transformation of the p_1 and p_2-axes of A:

$$F1 = 1.5p_1 - 0.5p_2 \tag{1}$$

$$F2 = 0.7p_2$$

CHANGES OF HUE AND CHROMATIC STRENGTH WITH INTENSITY

Whereas the scaling data of Fig. 3 were representative of the literature, in studying the Bezold-Brücke phenomenon, we relate the results of our own psychophysical scaling experiments with neurophysiological experiments done on the same apparatus, and under the same stimulus conditions.

The human observers participating in the scaling experiments viewed 40 stimuli of increasing luminance flashed on a tangent screen for 300 ms in alternation with a white 100 cd/m² adaptation field of 1.2 s duration. The presentation was monocular and foveal. Luminance of the chromatic stimuli was changed over a 5 log unit range in steps of 0.3 log unit from 0.1 to 10 000 cd/m². The surround was dark. Luminances were calculated using the CIE 10° V_λ luminosity curve (Valberg et al., 1987). All other colorimetric parameters were determined using a Spectrascan spectrophotometer (Photo Research, USA).

Human observers were asked to estimate two attributes of each stimulus, the hue and the chromatic strength, but only one attribute in each session. Hue was judged as proportions of two of the neighbouring elementary (unique) hues yellow and red, red

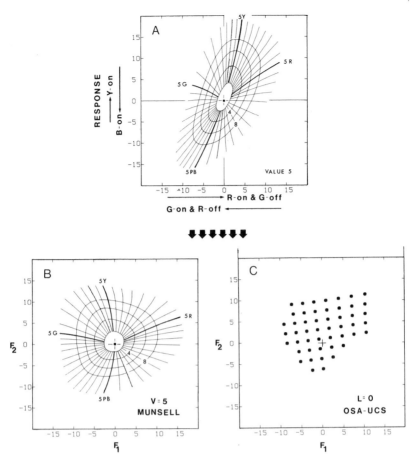

Fig. 3 Uniform colour space as derived from a combination of the responses of six different types of P-cells. In A, the axes (p_1, p_2) are made up of LGN colour-opponent
cells. In B and C, these outputs have been transformed to new coordinates $(F1, F2)$
by the linear transformation of Eq.(1) which may be assumed to take place after
the LGN.

and blue, blue and green, or green and yellow, such that the numbers added up to
10. Elementary hues obeyed the usual neither-nor criterion (e.g. elementary yellow is
neither reddish nor greenish). In this way a particular reddish orange light could be
characterized as 3Y and 7R.

In scaling the chromatic content of a stimulus, the observer was instructed not to
pay attention to the intensity changes of the lights, but to estimate only "the perceived
chromatic difference relative to an equally bright achromatic stimulus". This is essentially the definition of Munsell chroma (Wyszecki and Stiles 1982), but in this context
we will call it chromatic strength.

An overview of the combined Bezold-Brücke phenomenon and the saturation
changes is shown in Fig. 4. Luminance increases from the origin as is indicated by the
chevrons. At low luminance ratios, orange and greenish, or greenish-yellow hues dominate. Violet (454 nm) and purple (11a) stimuli seem to asymptote at some reddish-blue

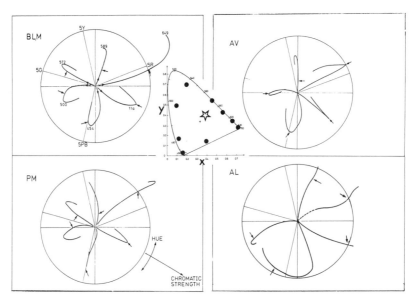

Fig. 4. The Bezold-Brücke hue shifts and the saturation changes of six near monochromatic lights are here combined in polar plots for four observers. Radius vector is proportional to chromatic strength and angle corresponds to hue. All data are normalised so that chromatic strength of the 649 nm stimulus is equal for all observers at a luminance ratio L/L_w of 0.1 (shown by the circle). Luminance ratio L/L_w equal to 0.1 is also indicated by arrows on the other curves. The Munsell hues 5R, 5PB, 5G, and 5Y serve as orientation. Luminance increases from the origin. The magnitude of the perceived chromatic strength of a stimulus was rated by the observer on a scale from 0 to 100.

hue at low luminance. When the luminance ratio increased from the minimum value, several of the curves of Fig. 2 display hue shifts first in one direction and then back again. For instance, over the first 1.5-2 log units of intensity above chromatic threshold, the 649 nm light gets redder, and only at higher intensity it changes towards yellow. As luminance increases from threshold, short wavelength violet light become more bluish (and finally bluish-green for AV and AL). All observers found that the green, 500 nm stimulus became more bluish as luminance increased, and all but one found that the purple stimulus no. 11 got reddish. Over a 4-5 log unit luminance range above chromatic threshold, hardly any stimulus is hue invariant, except perhaps for 589 nm and 11a (which are not elementary hues) for observer AL.

The directions of hue shifts are in general agreement with earlier observations reported in the literature for comparable luminance ranges. However, several studies of the Bezold-Brücke hue shift have concluded that the elementary hues are invariant with intensity (Judd, 1951; Linksz, 1964; Hurvich, 1981), whereas others have provided evidence that generally they are not (Boynton and Gordon, 1965; Savoie, 1973; Nagy, 1979; Purdy, 1931a). None of the stimuli of Fig. 3A and B show invariance of hue for all observers. Obviously, both the position of the stimuli on an individual hue scale and the relative hue shifts with intensity is subject to inter-observer variation.

Short-wavelength stimuli reach maximum chromatic strength at low luminances, at a ratio L/L_w of about 0.1 or below. For the 500 nm light, maximum chromatic strength is found at a luminance ratio close to 0.3 for all observers, and for mid-spectral yellow of 589 nm, the maximum is at about 10 times the luminance for the white reference. The greatest variability between observers was found for the red, 649 nm stimulus. Whereas observer AV show maximum chromatic strength at a low luminance ratio close to that for short-wavelength light, the maximum was at a 10 times greater ratio for the other observers.

Despite the obvious inter-individual differences in the colour estimates, clear similarities are present.

A PHYSIOLOGICAL INTERPRETATION

A polar plot of the combined response magnitudes of the model of Figs. 2 and 3 are shown in Fig. 5 for stimuli of constant chromaticity but increasing luminance. These plots are comparable to Fig. 4 and confirm that chromatic changes as a function of intensity are predicted by the model, without auxiliary assumptions. In summary, the most important assumptions of the model were:

a) that responses of ON- and OFF-cells with the same cone inputs add their inputs,

b) that an achromatic component can be subtracted,

c) orthogonality between the L-M and M-L cell outputs on one hand, and M-S and S-L outputs on the other,

d) vector summation between the orthogonal components, and

e) combinations of the coordinate axis of b) to transform equi-chroma ellipses into circles about the origin (Eq.(1)).

The predictions of the model for the the combined Bezold-Brücke and chroma effects of Fig. 5 are very similar to the plots of Fig. 4. Luminance ratio relative to the white adaptation field increases from the origin over 4 log units, and the curves in the diagram are projections of these trajectories in space down on a plane. Radius vector corresponds to chromatic strength (response magnitude), and its orientation relative to the coordinate axes is related to hue (ratio of responses between opponent processes). Colour strength thus initially grows as luminance increases from zero, reaches a maximum for a luminance ratio that is characteristic for each hue, and then decreases again. The metric of this figure (i.e. the relationship between distances, angles, and colour differences) is the same as that Valberg et al. (1986) used to account for Optical Society of Americas Uniform Color Scale (UCS) and the Munsell System.

In accordance with the Bezold-Brücke hue shifts of Fig. 4, the model predicts that a stimulus which appears reddish at low luminance, will turn more orange and finally yellowish at very high relative intensity. At some intermediate intensity, chromatic strength reaches a maximum, and for still higher intensities, the light becomes more whitish. Short wavelength stimuli that appear violet at low luminances, turn blue and finally bluish-green after maximum chromatic strength has been reached at higher luminances. Thus, the typical directions of hue shifts of Fig. 4 are well predicted, with one exception for the 500 nm stimulus. Fig. 5 thus shows that the Bezold-Brücke hue shift, and the related change of chroma (saturation), may have a common origin in the non-monotonic responses of P-cells with intensity.

The 'cell-opponent stages' of Fig. 3A and B allow the outputs of ON- and OFF P-channels to converge on cortical mechanisms, and combine them in opponent ways.

Theoretically, this is one possibility of arriving at an equidistant colour space (Valberg et al., 1986). However, considering the great variability in response properties among cells, another strategy could be a vector combination of cell types with their best responsiveness distributed throughout colour space.

This physiological model is successful in predicting salient features of observers' estimates of hue and chroma at equal luminance (Fig. 3B and C), and when relative luminance changes (Fig. 5). These phenomena, and the Abney effect, can therefore be accounted for by non-linear and non-monotonic responses of opponent cells in the retina and the LGN, and their possible combination at higher levels.

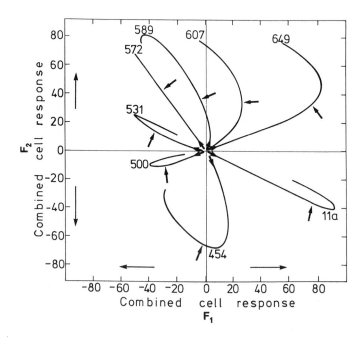

Fig. 5. Polar plot of the chromatic response magnitude of the model to the stimuli of Fig. 4 and the two additional wavelengths 531 and 607 nm on increasing the luminance. As in Fig. 4, radius vector is proportional to chromatic strength, and angles relative to the orthogonal axes represent hues. Arrows indicate the same luminance ratio of 0.1. Unit of cell response F corresponds to Munsell chroma. The predictions of the model show the same general features as subjects' estimates of hue and chromatic strength in Fig. 4.

In Fig. 6 we demonstrate that, for dichromats, curves of constant chroma collapse onto straight lines with orientation that is dependent on which type of cone is missing. For protanopes or deuteranopes, when L-cones replaces M-cones, or vice versa, the model predicts optimal (saturation-) discrimination in the same direction of violet and greenish-yellow hues. For tritanopes, two axes result dependent on which cone types take over the function of S-cones. If S-cones are replaced by M-cones, discrimination is in the direction of either constant red or constant bluish-green hues, and if S-cones are replaced by L-cones, discrimination is in the direction of bluish-red and green. These hue directions are in general agreement with those seen by the defective eye as reported by patients with unilateral acquired colour vision deficiencies (Trendelenburg, 1961).

The axes of Fig. 6 are also maintained when luminance increases. Therefore, the model postulates that these dichromats do not show a Bezold-Brücke hue shift, although the dependency of chromatic strength on luminance is non-monotonic.

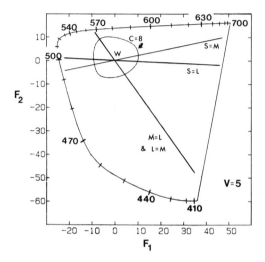

Fig. 6. Spectrum locus and model predictions of hues seen by dichromats if we assume that one cone type is replaced by another, leaving the number of cones constant, and their input to opponent-cells intact. The plot is for Munsell Value 5 (C = chroma).

DISCUSSION

In our scaling experiments, the Bezold-Brücke hue shifts varied in magnitude among observers, but they were consistent in direction. The way chromatic strength changed as a function of luminance was qualitatively similar to the results of Haupt (1922) and Purdy (1931b). Individual differences caused results to differ in detail, but there was reasonably consistency in the important features of the results.

The desaturation of chromatic stimuli at high luminances has interesting theoretical implications but has long been neglected. Purdy (1931b), and later Judd (1951) and Hurvich and Jameson (1955), suggested that it was caused by the relative activity of chromatic and achromatic processes, the achromatic signal being dominant at high intensities. Our results, however, indicate a new and unifying explanation of the Bezold-Brücke phenomenon and the related changes of chromatic strength. These two phenomena, as well as chromatic scaling and the Abney effect, can be accounted for by combining non-monotonic responses of retinal and LGN opponent cells (Valberg et al. 1986). The non-monotonic responses of the model of Fig. 5 follow the same functions of luminance as do chromatic strength and hues in Fig. 4, and this behaviour is caused solely by combining excitatory and inhibitory cone inputs to opponent P-cells. There is no need to introduce a separate achromatic mechanism to explain these results. Both chromatic and achromatic response components can, in principle, be derived from combinations of signals from opponent P-cells (Valberg et al., 1987a). Once a chromatic threshold is overcome at low luminance, it is the response magnitude of a combination of opponent cells, that alone set the correlate for chromatic strength and hue, regardless the response of an unknown achromatic mechanism. Of course, an achromatic response component could be derived from comparing the responses of different P-cells to the achromatic stimulus.

The excitatory profile across colour-selective cells contains the code for lightness and brightness attributes as well as for the chromatic content of a stimulus. Since both colour and intensity attributes are implicit in the activity of colour-opponent cells, cortical mechanisms may derive several perceptual attributes by analyzing multiple converging and diverging pathways. Here, we only show a few examples.

The response to achromatic light is, in the model, subtracted to yield a pure chromatic component. This rests upon the assumption that such a process indeed takes place after the LGN, where inputs from LGN cells are compared. Removing the 'white-response' seems necessary because retinal and LGN P-cells are generally responsive to changes in the intensity of achromatic light (ON-cells are activated by light increments and OFF-cells by light decrements). We did not find W- and Bl-cells, the purely non-opponent sustained cells of DeValois (1967; Lee et al., 1987), except as very weakly inhibited ON-cells and very strongly inhibited OFF-cells, thus confirming similar findings of Padmos and van Norren (1975), and Derrington and Lennie (1984).

At threshold, however, the achromatic interval between detection of light and discrimination of its hue is probably determined by the relative sensitivities of the non-opponent, phasic M-cells and cone-opponent, tonic P-cells of the retina (Graham and Hsia, 1969; Valberg and Lee, 1989).

Judd (1951), and also Hurvich and Jameson (1955), have referred the correlate of hue and the Bezold-Brücke phenomenon to the relative activity of two neighbouring 'primary processes', where these processes were understood to lead to Hering's elementary hues. Such a relationship is not easy to establish, since retinal and geniculate opponent cells responses cannot be directly linked to the perception of elementary hues (Valberg, 1984; Valberg et al., 1986), or to the achromatic percepts of white, neutral grey and black.

The physiological model is general in that it also accounts for empirical data on colour discrimination (Fig. 3). The spacing of chromaticities in the two most representative uniform colour systems is well accounted for without auxiliary assumptions (Valberg et al., 1986).

Individual differences. The relation between cone proportions and hue perception is by no means clear (Pokorny, this volume). The optimal responses of typical retinal and geniculate opponent cells correspond to purple (approximately Munsell 5RP) for L-M cells, to bluish-green (10G) for M-L cells, to greenish-yellow (2.5GY) for M-

S cells, and to violet (2.5P) for S-L cells (Valberg et al., 1986, Fig. 6). Thus, in the retinal and geniculate opponent processes, L-cones seem to contribute to bluish-red percepts, whereas M-cones can mediate different hues such as bluish-green in M-L cells and greenish-yellow in M-S cells. S-cones seem to mediate reddish-blue via their excitatory inputs to S-L cells. Hue and chromatic content of a stimulus may be more related to individual differences in the proportions of different cell types than to the proportions of cones.

REFERENCES

Abney, W., 1910, On the changes in hue of spectrum colours by dilution with white light. *Proc. Roy. Soc. A, 83*: 120-127.

Abney, W., 1913, "Researches in Colour Vision and Trichromatic Theory" Longmans-Green; London.

Bezold, W. von, 1873, Ueber das Gesetz der Farbmischung und die physiologischen Grundfarben. *Pogg. Ann. d. Physik u. Chemie*, 150: 221-247.

Boynton, B. and Gordon, J., 1965, Bezold-Brücke hue shifts measured by color-naming techniques. *J. Opt. Soc. Am.*, 55: 78-86.

Brücke, M. E., 1878, Ueber einige Empfindungen im Gebiet der Sehnerven. *Sitz. Ber. d.K.K. Akad. d. Wissensch., Math. Nat. Wiss. Classe*, 77: 39-71.

Burns, S. A., Elsner, A. E., Pokorny, J. and Smith, V. C., 1984, The Abney effect: Chromatic coordinates of unique and other constant hues. *Vision Res.*, 24: 479-489.

Cornsweet, T. N., 1978, The Bezold-Brücke effect and its complement, hue constancy, in "Visual Psychophysics and Physiology,", pp 233-244, Eds. J. C. Armington, J. Krauskopf, and B. R. Wooten, Academic Press, New York.

De Valois, R. L., Abramov, I., and Jacobs, G. H., 1966, Analysis of response patterns of LGN cells. *J. Opt. Soc. Am.*, 56: 966-977.

Derrington, A. M., Krauskopf, J., and Lennie, P., 1984, Chromatic mechanisms in lateral geniculate nucleus of macaque. *J. Physiol.*, 357: 241-265.

Graham, C. H. and Hsia, Y., 1969, Saturation and the foveal achromatic interval. *J. Opt. Soc. Am.*, 59: 993-997.

Guth, L. and Hovis, J. K., 1986, Chromatic and achromatic contributions to Bezold-Brücke effects. *Invest. Ophthalmol. Vis. Sci. (Suppl.)*, 27: 206.

Haupt, I. A., 1922, The selectiveness of the eye's response to wave-length and its change with change of intensity. *J. Exp. Psychol.*, 5: 347-379.

Helmholtz, H. von, 1911, "Handbuch der Physiologischen Optik. Third Ed." pp. 154-155, Leopold Voss, Hamburg.

Hurvich, L. M., 1980, "Color Vision" p. 73, Sinauer, Sundeland; Massachusetts.

Hurvich, L. M. and Jameson, D., 1955, Some quantitative aspects of opponent-colors theory. II Brightness, saturation, and hue in normal and dichromatic vision. *J. Opt. Soc. Am.*, 45: 602-616.

Ingling, C. R. and Tsou, B. H.-P., 1988, Spectral sensitivity for flicker and acuity criteria. *J. Opt. Soc. Am., A* 5: 1374-1378.

Judd, D. B., 1951, An Essay on Color Vision, in "Experimental Psychology" ed. A. Linksz, 1964, Grune and Stratton, New York.

Lee, B. B., Valberg, A., Tigwell, D. A. and Tryti, J., 1987, An account of responses of spectrally opponent neurones in macaque lateral geniculate nucleus to successive contrast. *Proc. Roy. Soc. B*, 230: 293-314.

Nagy, L: A:, 1979, Unique hues are not invariant with brief stimulus durations. *Vision Res.*, 19: 1427-1432.

Padmos, P. and van Norren, D., 1975, Cone systems interactions in single neurones of the lateral geniculate nucleus of the macaque. *Vision Res.*, 15: 617-619.

Purdy, D. McL., 1931a, Spectral hue as a function of intensity. *Am. J. Psychol.*, 43: 541-559.

Purdy, D. McL., 1931b, On the saturations and chromatic thresholds of spectral colours. *Brit. J. Psychology*, 21: 283-313.

Purdy, D. M., 1937, The Bezold-Brücke phenomenon and contours for constant hue. *Am. J. Psychol.*, 49: 313-315.

Savoie, R. E., 1973, Bezold-Brücke effect and visual non-linearity. *J. Opt. Soc. Am.*, 63: 1253-1261.

Swanson, W. H., Ueno, T., Smith, V. C., and Pokorny, J., 1987, Temporal modulation sensitivity and pulse detection thresholds for chromatic and luminance perturbations. *J. Opt. Soc. Amer.*, 4: 1992-2005.

Trendelenburg, W., 1961, "Der Gesichtsinn", p.104, Springer, Berlin.

Valberg, A. and Seim, T., 1983, Chromatic induction: responses of neurophysiological double opponent units?. *Biol. Cybernetics*, 46: 149-158.

Valberg, A., 1984, Die physiologische Verarbeitung von Licht- und Farbreizen. *Farbe + Design*, 31/32: 7.

Valberg, A., Lee, B. B., and Tryti, J., 1985, Computation of responses of opponent-cells in the macaque lateral geniculate nucleus to light stimuli varying in luminance, wavelength and purity, *Report 85-29, Inst. of Physics, University of Oslo, Norway.*

Valberg, A., Lee, B. B. and Tryti, J., 1987a, Simulation of responses of spectrally-opponent neurones in the macaque lateral geniculate nucleus to chromatic and achromatic light stimuli. *Vision Res.*, 27: 867-882.

Valberg, A., Seim, T., and Lee B.B., 1987b, A three-stage model of colour perception. *Die Farbe*, 34: 229-234

Valberg, A., Seim, T., Lee, B. B. and Tryti, J., 1986, Reconstruction of equidistant color space from responses of visual neurones of macaques. *J. Opt. Soc. Am.*, 3: 1726-1734.

Valberg, A. and Lee, B. B., 1989, Detection and discrimination of colour, a comparison of physiological and psychophysical data. *Physica Scripta*, 39: 178-186.

Vimal, R. L. P., Pokorny, J. and Smith, V. C., 1987, Appearance of steadily viewed lights. *Vision Res.*, 27: 1309-1318.

Vimal, R. L. P., Pokorny, J., Smith, V. C., and Shevell, S. K., 1989, Foveal cone thresholds. *Vision Res.*, 29: 61-78.

Vos, J. J., 1986, Are unique and invariant hues coupled? *Vision Res.*, 2: 337-342.

Walraven, P. L., 1961, On the Bezold-Brücke phenomenon. *J. Opt. Soc. Am.*, 51: 1113-1116.

Wiesel, T. N. and Hubel, D. H., 1966, Spatial and chromatic interactions in the lateral geniculate body of the rhesus monkey. *J. Neurophysiol.*, 29: 1115-1156.

Wyszecki, G. and Stiles, W. S., 1982, "Color Science: Concepts and Methods Quantitative Data and Formulae" Wiley and Sons, New York.

NEURAL DECODING

J. A. Hertz[1], B. J. Richmond[2], B. G. Hertz[1] and L. M. Optican[3]

[1]Nordita, Blegdamsvej 17
2100 Copenhagen, Denmark

[2]Laboratory of Neuropsychology, NIMH
Bethesda MD 20892 USA

[3]Laboratory of Sensorimotor Research, NEI
Bethesda MD 20892 USA

SUMMARY

A simple neural network is used to decode spatial pattern information in measured temporal firing patterns of a cell in the primary visual cortex of a macaque monkey. In this way one can quantify how much information beyond that contained in the average spike count is coded in the temporal modulation of the cell's activity. Most of the results obtained here are for simple discriminations between spike trains generated by two Walsh-pattern stimuli; some are for a task of identifying the correct one out of 4 such stimuli. A simple information-theoretic analysis of the trained networks indicates that significant extra information, of the order of 50% of that contained in the spike count, is carried in the temporal modulation.

INTRODUCTION

Richmond, Optican and their coworkers (Richmond et al., 1987; Richmond and Optican, 1987; Optican and Richmond, 1987; Richmond et al., 1989; McClurkin et al., 1990a,b; Gawne et al., 1990; Richmond et al., 1990; Richmond and Optican, 1990) have analyzed the temporal structure of neural spike trains in several parts of the visual system, from the LGN to inferior temporal cortex. They found that at every level the information about the stimulus transmitted in the spike trains is measurably greater than that transmitted by the spike count alone, indicating that temporal structure in the firing pattern is relevant. Its relative importance appears to increase with successive processing stages. For example, in the LGN the spike count contains about 3/4 of the total transmitted information (McClurkin et al., 1990b), while in the primary visual cortex it only contains half (Richmond and Optican, 1990).

Here we will report some preliminary findings on the problem of decoding these signals. We will not be concerned with how downstream parts of the visual system actually process the data. Rather, we will take the point of view that it is interesting to see how well the decoding can be done by a simple algorithm. This will at least give an indication of how hard a job the downstream parts of the system have.

From Pigments to Perception, Edited by A. Valberg and
B.B. Lee, Plenum Press, New York, 1991

The algorithm we use is a simple neural network, with a feed-forward architecture, "trained" by backpropagation (Rumelhart et al., 1986). The properties of this type of net are relatively well-understood, so we do not have problems of trying to understand the network confounded with those of trying to understand the data.

We will see that these simple networks can be trained to decode the spike trains. The decoding is not perfect, of course, because a single cell transmits only limited information, but its quality is consistent with the results of previous studies (Richmond and Optican, 1990) on information transmission by cells like the one used in this work. We will make the measure of decoding quality quantitative by computing the *equivocation* of the neuron, regarding it as a noisy encoding channel whose output is the spike train. We will see that for the tasks we consider, the equivocation is reduced by 30-40% when the receiver can use the temporal structure in the spike train in addition to the spike count.

Our approach to this problem is similar to that taken by Bialek et al. (1990; see also de Ruyter et al., 1988), who decoded spike trains from a motion-sensitive neuron in the blowfly, obtaining an accurate estimate of the velocity of the image in the visual field. In our problem, however, there is no correspondingly obvious choice of stimulus property to decode. Here we have chosen simple pattern discriminations, but information on other qualities than pattern might also be coded by the cells we study.

The next section is a brief introduction to these networks for the benefit of readers who may be unfamiliar with them. In the following section we then describe how a network is used to decode the spike trains. The section after that summarizes the results obtained so far, and a final section contains a brief discussion of the results and where one can go from here.

NETWORKS

For purposes of this discussion, a neural network is just a formula for approximating a function. To make things concrete, suppose we want to approximate a scalar function of N variables $(x_1, \ldots x_N) = \mathbf{x}$. Call the true function f and the approximation z. The simplest kind of fit is a linear one, a weighted sum of the input variables:

$$z(\mathbf{x}) = \sum_{j=1}^{N} w_j x_j. \tag{1}$$

A modest generalization is one where the fit is to a fixed nonlinear function of the weighted sum:

$$z(\mathbf{x}) = g(\sum_{j=1}^{N} w_j x_j). \tag{2}$$

If the function to be fit is a Boolean one (as, for example, in the binary classification problems we will be interested in here), then a natural choice for $g(\ )$ is the unit step function $g(u) = \theta(u)$. Other g's are appropriate in other situations.

We frequently think of the approximate function as a simple model "neuron" or processing unit that receives the different inputs x_j on different input lines that multiply them by "connection strengths" w_j, summing these up and running them through the (generally) nonlinear g function (Fig. 1a). (This is done physically in VLSI in real neural hardware like the Intel 80170 chip, but we still use this mechanistic language even when we are just talking about software, as in the computations to be described here.) Such a system is called a *simple perceptron*.

Simple perceptrons are severely limited in the kinds of functions they can represent accurately (Mynsky and Papert, 1969). However, if we have a set of M such units

$$y_i(\mathbf{x}) = g(\sum_{j=1}^{N} w_{ij} x_j) \tag{3}$$

and feed their outputs into another unit which treats these they way they did the original inputs, i.e. (Fig. 1b)

$$z(\mathbf{x}) = g(\sum_{i=1}^{M} v_i y_i(\mathbf{x})), \tag{4}$$

then very general kinds of functions can be approximated (Cybenko, 1989). This procedure can be generalized to more internal ("hidden") layers, but we won't need more than one hidden layer here. If f is a binary decision function and $g(\ )$ is a threshold function, we can think of the extra units as representing the truth or falsehood of various intermediate hypotheses, which are then weighed with strengths v_i in arriving at the final decision. The fitting parameters at our disposal are both the v_i and the w_{ij}, there are $M(N+1)$ of them in all. In the fitting procedure we want to avoid both under- and overfitting by selecting M judiciously.

In the kind of problem we have at hand, we want to determine these fitting parameters from a number of examples of the true mapping $f(\ )$. Everyone would know how to do this for the linear approximate map (1); it is a standard regression problem, minimizing the squared difference between the true and approximate functions, summed over the example data points $\mathbf{x}^{\mu}$:

$$E(\mathbf{w}) = \frac{1}{2} \sum_{\mu} [f(\mathbf{x}^{\mu}) - \sum_{j} w_j x_j^{\mu}]^2. \tag{5}$$

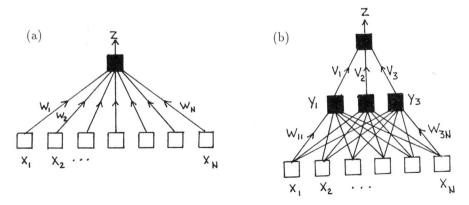

Figure 1. (a) a simple perceptron. (b) a two-layer perceptron.

A simple iterative way to minimize (5) with respect to the fit coefficients w_j is *gradient descent* – making small corrections so one "slides downhill" on the surface $E(\mathbf{w})$:

$$\Delta w_j \propto -\frac{\partial E}{\partial w_j} = \sum_\mu [f(\mathbf{x}^\mu) - z(\mathbf{x}^\mu)]x_j^\mu, \tag{6}$$

continually reducing the error until one is satisfied with the quality of the fit. In the context of neural networks, this is called "learning" or "training". More specifically, it is *supervised learning*, since the corrections in (6) can be viewed in terms of example-by-example comparison of the network's output z with the correct value f, supplied by a "teacher".

The same simple method (with a few variations) works for learning the network weights v_i and w_{ij} in the more complicated parametrized fit (4) if $g(\ )$ is a differentiable function. (In the computations we will report below, we used the common choice $g(u) = (1 + e^{-u})^{-1}$.) The error measure is now

$$E(\mathbf{v}, \mathbf{w}) = \frac{1}{2} \sum_\mu [f(\mathbf{x}^\mu) - g(\sum_{i=1}^M v_i y_i(\mathbf{x}^\mu))]^2, \tag{7}$$

and we adjust the weights by analogous downhill steps in $\mathbf{v} - \mathbf{w}$ space. Thus

$$\Delta v_i = -\alpha \frac{\partial E}{\partial v_i} = \alpha \sum_\mu [f(\mathbf{x}^\mu) - z(\mathbf{x}^\mu)]g'(h^\mu)y_i(\mathbf{x}^\mu), \tag{8}$$

where $g'(\ )$ is the derivative of $g(\ )$, $h^\mu \equiv \sum_i v_i y_i(\mathbf{x}^\mu)$ is the net input to the output unit, and α is called the learning rate. Similarly, using the chain rule,

$$\Delta w_{ij} = \alpha \frac{\partial E}{\partial w_{ij}} = \alpha \sum_\mu [f(\mathbf{x}^\mu) - z(\mathbf{x}^\mu)]g'(h^\mu)v_i g'(h_i^\mu)x_j, \tag{9}$$

where $h_i^\mu \equiv \sum_i w_{ij}x_j^\mu$ is the net input to hidden unit number i. Except for the factor of g', (8) is in the form (6) of a sum over examples of the product of an input y_i to a connection weight and the error measured for this example at the output unit. The adjustment (9) of the input-to-hidden weights can be put in the same form, with an *effective error* associated with a hidden unit obtained by weighting the output error proportionally to the connection from that unit to the output (and with another derivative factor) – i.e. propagating the error measured at the output backward through the network. This way of assigning blame on hidden units gives the algorithm the name "back propagation" (Rumelhart et al., 1986).

Note that both (8) and (9) involve sums of error-input products for the individual examples. In practice, it is more effective to go through the examples one at a time, updating the weights at each step with a fading window average of such single-example terms over around 10 recent examples. Furthermore, in the version used here, an up-dating is made only if the error at the output exceeds a preset tolerance value. (This leads to a more uniform fit quality.)

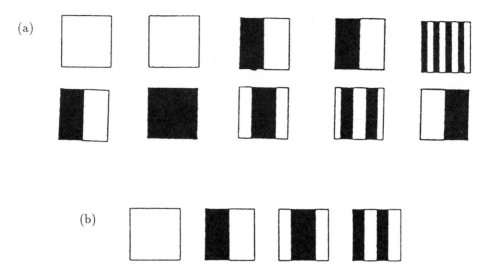

Figure 2. (a) the pairs of Walsh patterns used for the 1–of–2 discrimination experiments. (b) the 4 Walsh patterns used in the 1–of–4 discrimination experiment.

DECODING NEURONAL SIGNALS

We have trained several networks to decode the spike trains recorded from a single cell in the primary visual cortex of a macaque monkey. The monkey saw many presentations of a complete set of eight 1-dimensional black and white patterns based on Walsh functions at 7 contrast levels, and the neuronal signals produced by these stimuli were collected (Richmond et al., 1990). One set of calculations was done as follows: We selected a pair of one-dimensional Walsh patterns (vertical stripes), each presented about 20 times to the monkey at each of two different contrast levels, and trained a 2-layer perceptron as described above to give an output of 1 if the signal was produced by the first of the patterns and 0 if it was produced by the second one. Thus we are doing a simple psychophysics experiment on the neural network. (Note, however, that the monkey did not perform any such task – all it had to do was to fixate on the pattern.)

Because of the intrinsic noise in the signal, it is impossible to train the network to make this discrimination perfectly (unless one uses many more hidden units than the size of the data set warrants). The output of the network, trained as well as it can be, then admits a natural interpretation as the *probability* that the stimulus was pattern number 1.

This was done for the 5 different pairs of Walsh patterns shown in Fig. 2a. The corresponding experiments were then done for the same pairs of Walsh patterns, but each presented at 4 different contrast levels. This naturally made the task harder to learn.

Finally, a similar experiment was done for a 1–of–4 discrimination task. The signals were produced by 4 different Walsh patterns (the 4 lowest-"frequency" ones: Fig. 2b). The network here had 4 output units, each of then giving the inferred probability that the input was produced by a particular one of the patterns. Only 2 contrast levels were used in this experiment.

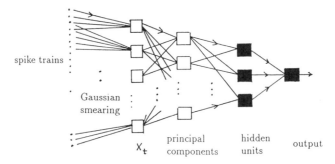

spike trains

Gaussian
smearing

x_t

principal
components

hidden
units

output

Figure 3. 4-layer effective network structure including preprocessing.

Preprocessing

The input to the network described above was not a raw spike train. The data were transformed in two steps. In the first of these, each spike was smeared out into a Gaussian of unit integrated weight. The widths of the Gaussians were chosen small for times (relative to stimulus onset) where there tended to be many spikes, and large where there were few. These Gaussians were then added up and sampled at 5-msec intervals up to 320 msec after onset. Thus the signal was represented by a vector of 64 components $\{x_t\}$. We omit some details here; see Richmond and Optican (1987) to learn more.

In the second step, the 64-dimensional input vectors were compressed into 5-dimensional ones in the following way. The covariance matrix

$$C_{tt'} = \langle (x_t - \langle x_t \rangle)(x_{t'} - \langle x_{t'} \rangle) \rangle \tag{10}$$

(where the average is over all data collected) was computed and diagonalized. The input vectors were then re-expressed in terms of their components $\{x'_\alpha\}$ ("principal components") in the basis where $\mathbf{C}$ is diagonal. Ordering these according to the magnitude of the corresponding eigenvalues, all but the first 5 were discarded, leaving 5-dimensional input vectors. (Again, more details about the transformation to principal components can be found in Richmond and Optican, 1987.)

To check that we were not throwing away essential information here, we did a few experiments in which we compared the network's learning from the 5 principal components with that from all 64 inputs. No sign of extra information in the omitted principal components was seen. This is also consistent with the calculations of Richmond and Optican (1990), who found that almost all the information in the signals was contained in the first 5 principal components.

There is also another reason for making this data compression. It means we need fewer fitting parameters w_{ij}, so we run less of a risk of overfitting. For example, if we had 10 hidden units, the network would have 650 weights, so we would need a number of examples of at least this order to feel safe against overfitting. On the other hand, with only the 5 principal components as inputs, there are only 60 weights. As we sometimes have only about 70 examples, the restriction to 5 principal components seems prudent.

These preprocessing steps can themselves be thought of as layers in a neural network, so really we have a 4-layer structure like that shown in Fig. 3. However, in this work only the part corresponding to the 2-layer perceptron which gets the 5 principal components as inputs is trained; the preprocessing parts are fixed in advance.

The main aim of the experiments was to see how much it helped the network to have knowledge of the temporal structure of the signal. Now we also know from the work of Richmond and Optican (1990) that the first principal component x_0' is very strongly correlated with the spike count. Thus it is the rest of the input principal components that contain the temporal modulation information. So all the experiments were done in two ways: once, as described above, with the first 5 principal components as inputs to the learning perceptron, and again with only the first principal component. The difference in training and in performance after training then reflects directly the extra information that was temporally encoded.

RESULTS

The most striking effect of the addition of temporal modulation information in the training was the fact that it increased the speed of learning by a factor of two or more. It also increased the accuracy of learning, i.e. the network got a significantly higher fraction of the examples "correct" in the sense that the output was within the 10% tolerance limit of the target (1 or 0) output for that example. This was true for both the experiments with 2 contrast levels and those with 4.

This difference was much smaller, however, if one simply interpreted the output as meaning "pattern 1" if it exceeded $\frac{1}{2}$ and "pattern 2" if it was less than $\frac{1}{2}$. That is, the network was "surer" of its output when trained on 5 principal components than when trained on only the first one, but in most cases the spike count alone was enough information to make the output fall on the right side of $\frac{1}{2}$.

There was one exception to this general pattern: one of the 1–of–2 discriminations (the last one in Fig. 2a) couldn't be done with just one principal component. This failure occurred for both the 2- and 4-contrast-level experiments.

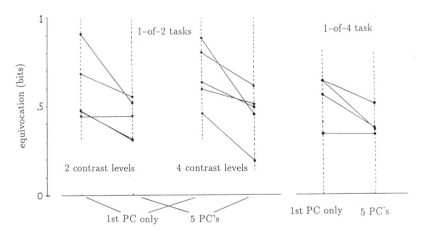

Figure 4. Equivocation, without and with temporal modulation taken into account, for the two 1–of–2 discriminations (2 and 4 contrast levels, left and center, resp., and the 1–of–4 discrimination (right).

To quantify the greater certainty exhibited by the networks trained with temporal modulation information, we computed the equivocation (Abramson, 1963) of the channel from the stimulus to the neuronal signal:

$$e = -\langle [P(\mathbf{x}) \log P(\mathbf{x}) + (1 - P(\mathbf{x})) \log(1 - P(\mathbf{x}))] \rangle_{\mathbf{x}}, \tag{11}$$

where $P(\mathbf{x})$ is the network output when signal $\mathbf{x}$ is the input (i.e. the best estimate of the probability that signal $\mathbf{x}$ was produced by pattern number 1). Confident outputs (P near 0 or 1) lead to a low equivocation, which is the same as a high information transmission, since

$$T + e = S \tag{12}$$

where T is the transmitted information and S is the entropy of the pattern distribution. For two patterns presented equally frequently, $S = 1$, so $T = 1 - e$.

Figure 4 shows the results. There is considerable quantitative variation from task to task, but the tendency toward lower equivocation when temporal modulation information is used is clear. The average results are summarized in this table:

task	equivocation with temporal information	equivocation with spike count only
1–of–2, 2 contrasts	0.38	0.59
1–of–2, 4 contrasts	0.45	0.67
1–of–4, (2 contrasts)	0.39	0.53

The trend seen in the figure is evident – the equivocations with spike count only are 40-50% higher than those with the temporal modulation information included. We conclude that our simple feed-forward neural network algorithm can extract significant amounts of this information from the signals.

DISCUSSION

Neurons encode information about the form of a stimulus in both the strength and temporal waveform of their responses. This work shows that the extra temporally-encoded information can be decoded using a simple neural network. This suggests that it would be easy for neurons in the brain to decode temporally modulated information about stimulus form. Another implication is that the discrimination of other qualities, such as color and texture, may also be improved by utilizing temporally encoded information.

Particular pathways in the visual system have been hypothesized to carry information about distinct qualities. However, as has been discussed at this conference, we now see that the different functional pathways are not so cleanly separated as we might once have believed. Perhaps further mixing of the kinds of information in different pathways might become evident when the temporal multiplexing of messages suggested by these results is taken into account.

A real animal often has to react more quickly than the half-second or so it would take to collect and analyze the data from a neural signal like those we have analyzed. It would be interesting to repeat experiments like ours, but varying the length of the signals used as inputs to see how long an observation is necessary to make simple discriminations.

We can not determine whether the extra temporally-encoded information we have identified is actually used by later processing stages in the brain. This must depend on weighing the benefits to the animal of having this information against the cost of extracting it. Nonetheless, we have now shown that decoding this extra information is relatively simple. Furthermore there are benefits to the animal in using it, such as reducing the number of axons in neural pathways. It is interesting to note that engineers designing neural chips have independently proposed the idea of temporal multiplexing as a way of beating the "too many wires" problem (Bailey and Hammerstrom, 1988). This problem may also be relevant in the brain.

This work on understanding the use of temporally-encoded information is just a beginning. One neuron carries very little visual information, and thus many neurons must be able to work together in visual perception. Our approach can be extended to interpreting the responses of many neurons recorded simultaneously. Thus, this method provides a way to understand how neurons cooperate in transmitting enough visual information for behavioral purposes.

Acknowledgement: We would like to thank Mogens Aalund, NEuroTech A/S, for computing resources.

REFERENCES

Abramson, N., 1963, "Information Theory and Coding," New York, McGraw-Hill.

Bailey, J. and Hammarstrom, D., 1988, Why VLSI implementation of associative VLCNs requires connection multiplexing, *Proc. IEEE Conf. on Neural Networks* II:173-180, San Diego.

Bialek, W., Rieke, F., de Ruyter, R. R., van Steveninck and Warland, D., 1990, Reading a Neural Code,*in* "Advances in Neural Information Processing Systems" 2, D. S. Touretzky, ed., San Mateo CA, Morgan Kaufmann.

Cybenko, G., 1989, Approximation by superpositions of a sigmoidal function. *Math. of Controls, Signals, and Systems* 2:303-314.

Gawne, T. J., McClurkin, J. W., Richmond B. J. and Optican, L.M., 1990, Lateral geniculate neurons in behaving monkeys: III. Predictive power of a multi-channel model. Submitted to *J. Neurophysiol.*

McClurkin, J. W., Gawne, T. J., Richmond, B. J., Optican, L. M. and Robinson D. L., 1990, Lateral geniculate neurons in behaving monkeys: I. Responses to two-dimensional stimuli. Submitted to *J. Neurophysiol.*

McClurkin, J. W., Gawne, T. J., Richmond, B. J. and Optican, L. M., 1990, Lateral geniculate neurons in behaving monkeys: II. Encoding of visual information in the temporal modulation of the spike train. Submitted to *J. Neurophysiol.*

Minsky, M. and Papert, S., 1969, "Perceptrons," Cambridge MA, MIT Press.

Optican, L. M. and Richmond, B. J., 1987, Temporal encoding of two-dimensional patterns by single units in primate inferior temporal cortex. III. Information theoretic analysis. *J. Neurophysiol.*, **57**: 162-178.

Richmond, B.J. and Optican, L. M., 1987, Temporal encoding of two-dimensional patterns by single units in primate inferior temporal cortex. II. Quantification of response waveform. *J. Neurophysiol.*, **57:** 147-161.

Richmond, B.J. and Optican, L. M. 1990, Temporal encoding of two-dimensional patterns by single units in primate primary visual cortex: II. Information transmission. *J. Neurophysiol.*, **64:**370-380.

Richmond, B. J., Optican, L. M. and Gawne, T.J., 1989, Neurons use multiple messages encoded in temporally modulated spike trains to represent pictures. *in* "Seeing Contour and Colour", J. J. Kulikowski and C. M. Dickinson, eds., Oxford, Pergamon Press.

Richmond, B.J., Optican, L. M., Podell, M. and Spitzer, H., 1987, Temporal encoding of two-dimensional patterns by single units in primate inferior temporal cortex. I. Response characteristics. *J. Neurophysiol.*, **57:**132-146.

Richmond, B. J., Optican, L. M. and Spitzer, H., 1990, Temporal encoding of two-dimensional patterns by single units in primate primary visual cortex: I. Stimulus-response relations. *J. Neurophysiol.*, **64:**351-369.

Rumelhart, D. E., Hinton, G. E. and Williams, R. J., 1986, Learning internal representations by back-propagating errors. *Nature,* **323:**533-536.

de Ruyter, R. R., van Steveninck S. and Bialek, W., 1988, Real-time performance of a movement-sensitive neuron in the blowfly visual system: coding and information transfer in short spike sequences. *Proc. Roy. Soc. B,* **234:**379-414.

EFFECTS OF PHASE SHIFTS BETWEEN CONE INPUTS

ON RESPONSES OF CHROMATICALLY OPPONENT CELLS

William H. Swanson

Retina Foundation of the Southwest and Dept. Ophthalmology, UT Southwestern
8230 Walnut Hill Lane #414
Dallas, TX 75231

Weighted differences of the responses of long-wavelength-sensitive (LWS) and middle-wavelength-sensitive (MWS) cones have been successfully used to model both red-green opponent cells in the primate parvocellular pathway (P-pathway) and psychophysical LWS-MWS chromatic mechanisms. However, relatively little attention has been paid to how changes in the relative phases of the LWS and MWS cone inputs affect predictions of these models. The purpose of this paper is to use quantitative evaluation of the effects of phase shifts to show how a delay between center and surround responses can account for apparently contradictory results between published studies of of P-pathway cells which used different paradigms for evaluating chromatic opponency.

Let ω be the temporal frequency (in Hz), t be time (in sec), d be the delay (in sec) between center and surround, W_L and W_M be the weights of cone inputs ($W_L + W_M = 1$), and R_L and R_M be the cone quantal catches. Then the phase shift between cone inputs (in degrees) is $\rho = 360\omega d$.

The response of a LWS-MWS cell is

$$W_L R_L \sin(\omega t) - W_M R_M \sin(\omega t - \rho) = [W_L R_L - W_M R_M \cos(\rho)]\sin(\omega t) + [W_M R_M \sin(\rho)]\cos(\omega t) \qquad (1)$$

The amplitude of this response is

$$\sqrt{\{[W_L R_L - W_M R_M \cos(\rho)]^2 + [W_M R_M \sin(\rho)]^2\}} = \sqrt{\{(W_L R_L)^2 + (W_M R_M)^2 - 2 W_L R_L W_M R_M \cos(\rho)\}}. \quad (2)$$

SPECTRAL TUNING FUNCTIONS

Spectral tuning functions of LWS-MWS opponent cells typically show a sharp elevation of threshold in a mid-spectral region where the LWS and MWS cone inputs are approximately equal (and of opposite sign). The absence of a mid-spectral region of minimal sensitivity often is interpreted as indicating a loss of chromatic opponency. Spectral tuning functions of LWS-MWS opponent retinal ganglion cells have a region of minimal sensitivity at low temporal frequencies but not at high temporal frequencies (Gielen, van Gisbergen & Vendrik, 1982; Gouras & Zrenner, 1979). For the data of Gouras & Zrenner (1979) sensitivity at 540nm increased relative to 660nm by ~15 dB between 1 Hz and 33 Hz, while for the data of Gielen et al. sensitivity at 548nm increased relative to 656nm by ~19 dB between 1 Hz and 33 Hz. These studies obtained similar results despite the use of quite different luminances (4.5 vs. 2.5 log td), indicating that this phenomenon is not due to a particular state of adaptation.

Fig. 1 shows the effect of phase shifts between cones on responses of an LWS-MWS mechanism to spectral lights, for a mechanism with $W_L R_L = W_M R_M$ at 560nm. Note that for values of ρ from 0° to 5° there is a sharp elevation of threshold near 560nm, but as ρ increases the elevation becomes less distinct; for $\rho = 30°$ there is no threshold elevation. Therefore phase shifts as small as 30° (delays of 2.5 msec and greater) could account for the loss of a mid-spectral region of increased threshold at 33 Hz.

From Pigments to Perception, Edited by A. Valberg and
B.B. Lee, Plenum Press, New York, 1991

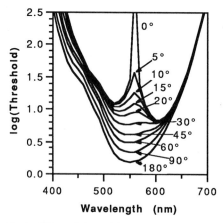

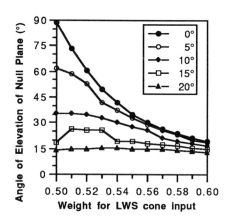

Fig. 1. Effects of phase shifts on responses of a LWS-MWS opponent mechanism to spectral lights. Arrows indicate the phase shifts (ρ).

Fig. 2. Effects of phase shifts on the angle of elevation of the null plane of a LWS-MWS opponent mechanism, where the weights of the cone inputs are normalized so they sum to 1.0.

ELEVATION OF THE NULL PLANE IN CONE EXCITATION SPACE

Derrington et al. (1984) measured responses of P-pathway thalamic cells to modulation in a range of directions in cone excitation space, and derived the *angle of elevation of the null plane*. The null plane is the plane passing through white which contains all lights that can be exchanged without the cell responding. The angle of elevation of the null plane is the angle between it and the equiluminant plane. Derrington et al. found that increasing the temporal frequency of a uniform field from 4 to 15 Hz caused only a small decrease in the angle of elevation of the null plane.

All but one of the cells Derrington et al. studied at 15 Hz had angles of elevation greater than 20°. The angle of elevation of the null plane is a function of the phase shift and of the relative weight of the LWS and MWS cone inputs, as shown in Fig. 2. Examination of this figure reveals that a phase axis greater than 20° can only be achieved if the absolute value of any phase shift is less than 20°. This means that Derrington et al.'s 15 Hz data require delays of less than 3.7 msec.

Examination of equation (2) shows that for ρ other than 0° or 180°, the amplitude of the response will be zero only when $W_L R_L = W_M R_M = 0$. This means that when there is a phase shift there will not be a null plane *per se*, yet Derrington et al. obtained good fits to their data with an analysis assuming the existence of a null plane. To determine the effects of phase shifts on the estimates of null plane elevations obtained by Derrington et al.'s analysis, responses of LWS-MWS mechanisms to the 27 stimuli they were computed for a range of values for W_L and ρ. These responses were fit with their method to obtain a putative elevation of the null plane. Fits obtained in this way were quite good; an example is shown in Fig. 3 for a cell with a phase shift of 30°. Therefore the fact that their model gave good fits to their data does not rule out the existence of small phase shifts.

COMPARISONS WITH OTHER STUDIES

The preceding analysis is in terms of phase shifts between cone inputs due to a delay between center and surround responses. The analysis requires that the delay be at least 2.5 msec (to account for the spectral tuning data) and less than 3.7 msec (to account for the data on angle of elevation of the null plane in cone excitation space). Two recent studies of LWS-MWS opponent P-pathway retinal ganglion cells obtained comparable estimates for the delay (Lee, Martin & Valberg, 1989; Smith, Lee, Pokorny, Martin & Valberg, 1989). Lee et al. measured amplitude and phase of P-pathway retinal ganglion cell responses to flickering spectral lights, and derived an estimate of 3-8 msec for the center-surround delay of red-green chromatic cells at 1400 td. Smith et al. measured amplitude and phase of P-pathway retinal ganglion cell responses to heterochromatic flicker of red and green lights modulated in a range of phases, and estimated that the delay was ~5 msec (see also Smith, this volume).

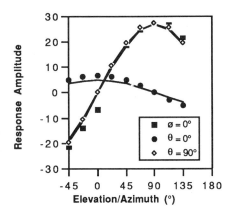

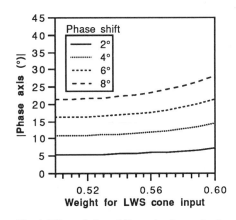

Fig. 3. Responses of a LWS-MWS opponent mechanism with $\rho = 30°$ to Derrington et al.'s stimuli (symbols) and fits obtained with their analysis (curves).

Fig. 4. Effects of phase shifts on the phase axis of a LWS-MWS opponent mechanism, for the primaries used by Lindsey et al. (1986) and Swanson et al. (1987).

Lindsey, Pokorny & Smith (1986) and Swanson, Pokorny & Smith (1987) measured psychophysical modulation thresholds for red-green flicker as a function of the relative phase of red and green lights. They found that thresholds at low temporal frequencies were symmetric around phase axes other than 0° or 180°. The phase axes for low frequencies ranged from -15° to +21°, indicating phase shifts between LWS and MWS cone inputs to mechanisms mediating threshold. Derivation of the phase axis from the phase shift requires a modification of equation (2). The amplitude of the response is

$$\sqrt{([R_{L1} + R_{L2}\cos(\theta) - R_{M1}\cos(\rho) - R_{M2}\cos(\theta+\rho)]^2 + [-R_{L2}\sin(\theta) + R_{M1}\sin(\rho) + R_{M2}\sin(\theta+\rho)]^2)} \quad (3)$$

where θ is the relative phase of the two primaries and R_{L1}, R_{L2}, R_{M1}, R_{M2} are the responses of the LWS and MWS cones to the two primaries. Setting the derivative of response amplitude to zero,

$$\text{phase axis} = \arctan\{ [(R_{L1}R_{M2} - R_{L2}R_{M1})\sin\rho] / [(R_{L1}R_{L2} + R_{M1}R_{M2} - (R_{L1}R_{M2} + R_{L2}R_{M1})\cos\rho] \} \quad (4)$$

The phase axis is a function of the phase shift and of the relative weights of the LWS and MWS cone inputs, as shown in Fig. 4. Examination of this figure shows that phase axes with an absolute value of 21° or less require phase shifts with absolute values less than 8°. While psychophysical thresholds probably reflect the responses of many cells at different levels in the visual system, it appears that red/green psychophysical mechanisms, like P-pathway retinal ganglion cells, have small phase shifts between LWS and MWS cone inputs.

MIXED CONE INPUTS TO THE SURROUND

The analysis presented above assumes that the center and the surround of a given cell each receive input from only one cone type. If the surround has mixed cone inputs (that is, both LWS and MWS cones contribute to the surround), then smaller phase shifts could account for both types of data. The response of a LWS - (MWS+LWS) cell is

$$W_{L1}R_L\sin(\omega t) - (W_M R_M + W_{L2}R_L)\sin(\omega t-\rho)$$

$$= [W_{L1}R_L - (W_M R_M + W_{L2}R_L)\cos(\rho)]\sin(\omega t) + [(W_M R_M + W_{L2}R_L)\sin(\rho)]\cos(\omega t), \quad (5)$$

where W_{L1} is the LWS cone weight for the center and W_{L2} is the LWS cone weight for the surround. The amplitude of this response is

$$\sqrt{\{[W_{L1}R_L - (W_M R_M + W_{L2}R_L)\cos(\rho)]^2 + [(W_M R_M + W_{L2}R_L)\sin(\rho)]^2\}}. \quad (6)$$

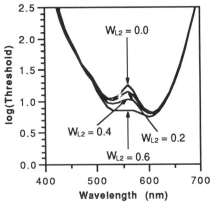
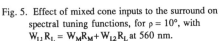
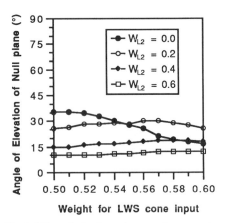

Fig. 5. Effect of mixed cone inputs to the surround on spectral tuning functions, for $\rho = 10°$, with $W_{L1}R_L = W_M R_M + W_{L2} R_L$ at 560 nm.

Fig. 6. Effect of mixed cone inputs to the surround on the angle of elevation of the null plane for $\rho = 10°$. W_{L2} is expressed as a fraction of W_{L1}.

Fig. 5 shows the effect of W_{L2} on spectral tuning functions for $\rho = 10°$, where W_{L1} varies with W_{L2} to keep $W_{L1}R_L = W_M R_M + W_{L2} R_L$ at 560nm. The primary effect of W_{L2} on spectral tuning functions is to increase sensitivity near 560nm. Therefore mixed cone inputs to the surround would allow the data of Gouras & Zrenner to be accounted for by even smaller phase shifts than 30°. Fig. 6 shows the effect of W_{L2} on the angle of elevation in cone excitation space for $\rho = 10°$, where W_{L2} is expressed as a fraction of W_{L1}. For $W_{L2} = 0.2$ the angle of elevation is decreased at $W_{L1} = 0.5$, but increased for $W_{L1} > 0.55$, so no change in phase shift is required to account for Derrington et al.'s data. For $W_{L2} \geq 0.4$, the angle of elevation tends to decrease, and for these values smaller phase shifts would be required to account for Derrington et al.'s data. A model with mixed cone inputs to the surround could allow shorter latencies than the 2.5 - 3.7 msec derived above.

More complex models could be constructed, with different amplitude and phase spectra for the center and surround. Such models would introduce additional parameters, allowing shorter or longer latencies. The purpose of the current model is simply to show how a reasonable latency, or other sources of phase shifts between center and surround, can account for the seemingly quite different effects of temporal frequency on spectral tuning functions and on modulation thresholds in cone excitation space.

REFERENCES

A. M. Derrington, J. Krauskopf & P. Lennie (1984) Chromatic mechanisms in lateral geniculate nucleus of macaque. *J. Physiol.* **357**, 241-265.

C. C. A. M Gielen, J. A. M van Gisbergen & A. J. H. Vendrik (1982) Reconstruction of cone-system contributions to responses of colour-opponent neurones in monkey lateral geniculate. *Biol. Cybern.* **44**, 211-221.

P. Gouras and E. Zrenner (1979) Enhancement of luminance flicker by color-opponent mechanisms. *Science* **205**, 587-589.

B. B. Lee, P.R. Martin, & A. Valberg (1989) Amplitude and phase of responses of macaque retinal ganglion cells to flickering stimuli. *J. Physiol.* **414**, 223-244.

D. T. Lindsey, J. Pokorny and V. C. Smith (1986) Phase-dependent sensitivity to heterochromatic flicker. *J. Opt. Soc. Am. A* **3**, 921-927.

V. C. Smith, B. B. Lee, J. Pokorny, P. R. Martin & A. Valberg (1989) Response of macaque ganglion cells to changes in the phase of two flickering lights. *Invest. Ophthalmol. Visual Sci. Suppl.* **30**, 323.

W. H. Swanson, J. Pokorny and V. C. Smith (1987) Effects of temporal frequency on phase-dependent sensitivity. *J. Opt. Soc. Am. A* **4**, 2267-2273.

DIFFERENT NEURAL CODES FOR SPATIAL FREQUENCY

AND CONTRAST

Mark W. Greenlee, Svein Magnussen and James P. Thomas

Dept. of Neurophysiology, Freiburg, Germany
Institute of Psychology, Oslo, Norway
Dept. of Psychology, UCLA, Los Angeles, U.S.A.

S. S. Stevens (1957) argued that there are two different types of sensory dimensions, extensive and intensive, and that they have different types of neural codes. Stevens based his arguments on the results of psychophysical scaling experiments. However, if the two types of dimensions are coded in fundamentally different ways, the difference should be reflected in other tasks as well, including short-term memory tasks. In this paper, we present evidence that the short-term memory for spatial frequency, an extensive dimension, does not degrade over periods up to 10 sec, whereas the memory for contrast, an intensive dimension, does. In our experiments, the viewer must discriminate between two, successively presented stimuli. To respond correctly, the viewer must have an accurate memory of the first stimulus to compare with the second. If the memory degrades over time, discrimination performance must decline as the interval between the stimuli (ISI) lengthens. We find that spatial frequency discrimination, as measured by the difference threshold, is independent of ISI for intervals up to 10 sec, whereas the threshold for contrast discrimination progressively increases over this same period. This difference occurs at all contrast levels and has the same magnitude when the viewer judges both dimensions simultaneously as when the dimensions are judged separately.

Methods

Sinewave gratings truncated in space and time by Gaussians were presented briefly (300 msec) with a defined ISI. On each trial the subject was presented two stimuli, one stimulus had the reference frequency and contrast, the other differed from the reference either in spatial frequency or contrast and could be either higher or lower. In the *single-judgment task*, the subject knew along which dimension the stimuli would differ and had only to respond which interval contained the grating with the higher value. In the *dual-judgment task*, on half of the trials stimuli differed in spatial frequency and on half they differed in contrast. In this task, the subject had to make two responses on each trial, one signalling the dimension along which the stimuli differed and another indicating which interval contained the grating with the higher value. Both responses had to be correct for the trial to be scored as correct. Ten interleaved, independent staircases estimated discrimination thresholds for frequency and contrast at five reference contrast levels (0.02 - 0.32) using a maximum-likelihood algorithm (Best-PEST, Lieberman and Pentland, 1982). Thresholds were set at 75% correct response for the single-judgment task and 62.5% for the dual-judgment task.

Results

Results for one subject are presented. The difference thresholds for frequency and contrast were analyzed separately by analysis of variance and the results are presented in Table 1. The major finding is that the difference threshold for contrast rises with increasing ISI, but the threshold for spatial frequency does not. This finding is illustrated in Fig. 1 where the Weber fractions for contrast discrimination (part a) and frequency discrimination (part b) are shown as a function of ISI. For this figure, the data have been averaged across contrast levels. The second finding is that the differential effect of ISI is the same for both single-judgment and dual-judgment tasks, although thresholds are higher in the latter case. (See Fig. 1 and Table 1). Finally, the differential effect of ISI is the same at all contrast levels, although the thresholds themselves vary with contrast level. (See Table 1).

Table 1.   Analysis of Variance

frequency discrimination

Source	df	MS	F	p
ISI	2	.001	2.54	n.s.
Task Complexity	1	.002	8.73	0.044
Contrast	4	.006	34.88	0.0001
ISI x Task	2	.00021	0.12	n.s.
ISI x Contrast	8	.001	2.96	0.007
Task x Contrast	4	.00037	0.206	n.s.
ISI x Task X Con.	8	.00025	1.39	n.s.

contrast discrimination

Source	df	MS	F	p
ISI	2	.197	23.56	0.0001
Task Complexity	1	.938	8.731	0.0001
Contrast	4	.254	30.42	0.0001
ISI x Task	2	.031	1.54	n.s.
ISI x Contrast	8	.009	1.09	n.s.
Task x Contrast	4	.015	1.84	n.s.
ISI x Task X Con.	8	.012	.49	n.s.

452

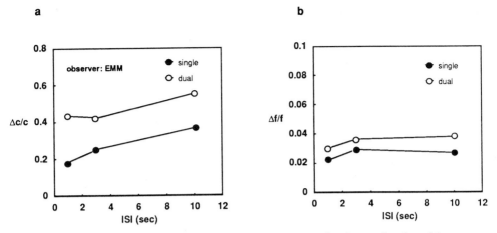

Fig. 1. Discrimination thresholds averaged over contrast levels as a function of the inter-stimulus interval (ISI).

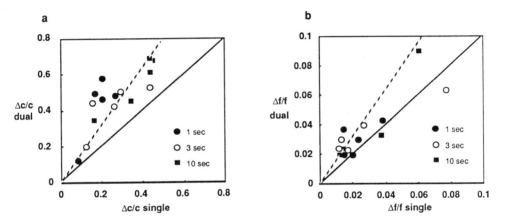

Fig. 2. Relationship between discrimination thresholds in single and dual-judgment tasks.

Discussion

As measured by the threshold technique, the accuracy of memory for spatial frequency does not decline over intervals up to 10 sec long. This finding replicates and extends the results of Magnussen et al. (in press). However, the accuracy of memory for contrast does decline over these same periods of time. The differential effect of ISI argues that information about these two perceptual dimensions is represented and stored differently in the nervous system. This difference is unaffected by the contrast levels used or by whether the viewer judged one or both dimensions. These latter variables manipulate task difficuly and/or processing demands, and the fact that they do not interact with the effect of ISI argues that the differential effect of ISI stems from a basic coding difference rather than from some limitation of processing resources.

Thresholds for both spatial frequency and contrast were higher in the dual-judgment task than in the single-judgment task by amounts which can be predicted on the basis of stimulus uncertainty (Thomas, Gille, and Barker, 1982; Thomas and Olzak, in press). In Fig. 2, the thresholds obtained in the dual task are plotted against thresholds from the single task. Each point represents the two types of thresholds for a given contrast level and ISI. The solid diagonal shows the locus of the points expected if the nature of the task had no effect; the broken line shows the locus predicted on the basis of uncertainty. When the frequency and contrast data are combined to minimize the effects of response bias, there is no systematic deviation from the uncertainty prediction. The uncertainty prediction assumes that the two judgments are made independently, i.e. they do not interfere with each other and do not compete for limited processing resources. Thus, this result also argues that the differential effect of ISI arises from differences in the neural codes for frequency and contrast, rather than from a competition for limited processing resources.

These findings are particularly interesting given that early neurons in the visual system, up to at least V1, confound information about spatial frequency and contrast. That is, the response of any given neuron is jointly determined by frequency and contrast. Were the responses of these neurons stored in memory, it would be hard to see how differential memory effects could arise. Apparently, the information about frequency and contrast is decoded and placed in separate codes at a later stage, and it is these later codes which are stored and used in the discrimination task.

References

Lieberman H.R. and Pentland A.P. (1982) Microcomputer-based estimation of psychophysical thresholds: The best PEST. **Behavoral Research Methods and Instrumentation**, 14, 21-25.

Magnussen, S., Greenlee, M.W., Asplund, R. and Dyrnes, S. (1990) Perfect visual short-term memory for periodic patterns. **European Journal of Cognitive Psychology**, (in press)

Stevens, S.S. (1957) On the psychophysical law. **Psychological Review**, 64, 153-181.

Thomas, J.P., Gille, J. and Barker, R.A. (1982) Simultaneous detection and identification: theory and data. **Journal of the Optical Society of America**, 72, 1642-1651.

Thomas, J. P. and Olzak, L.A. (1990). Cue summation in discrimination of simple and compound gratings. **Vision Research** (in press).

Supported by: Deutsche Forschungsgemeinschaft, SFB 325, B4, the Deutsche Akademische Austauschdienst (DAAD), and the National Eye Institute EY360

DISPLACEMENT ESTIMATION, STEREO MATCHING AND 'OBJECT' RECOGNITION:

A COMPUTER SIMULATION APPROACH WORKING WITH REAL WORLD IMAGERY

Georg Zimmermann

Fraunhofer-Institut für Informations- und Datenverarbeitung (IITB)

Fraunhoferstr. 1, D-7500 Karlsruhe 1, FRG

Introduction

Most of the work done in computer vision is intended for industrial applications. The methods applied there are predominantly based on mathematical notions (gradients, curvatures) of the gray value images obtained by optical sensors. Many of them are designed for real-time implementation on computers but not for the explanation of existing biological vision systems (Rosenfeld 1990). With Neural Networks, a closer connection to biological circuitry has been found. But experimenting with them is tedious, especially when using large image sequences, and interpreting the resulting weights between the cells after training is almost as hard as in real biology.

This paper presents a method for displacement estimation in image sequences, for stereo image matching and the recognition of image 'objects' that has been developed for industrial real-time application, but with biological notions in mind. The first stages of the method are compatible with neuronal connections in biology (lateral interactions, thresholding), well understood in terms of signal processing, robust in the presence of noise or inadequate signal conditions and work with almost every type of real world imagery. This method might therefore serve as a hypothesis of how early vision could work.

Simple features (blobs) obtained by nonlinear filtering

The basis for displacement estimation, stereo matching and object recognition are features obtained from the gray value image by applying a band pass filter (linear stage) and then detecting regions of local maxima and minima (nonlinear stage). Bandpass filtering is performed by a very simplified version of the Difference-Of-Gaussian approach. The detection of local extrema is done by comparing in a window eight surrounding pixels with the central one. If all surrounding pixels have a gray value that is less than the central one, the central pixel is classified as belonging to a region of a local maximum. This procedure results in a binary image with blobs marking the local maxima. For further computations, the centroid of the blob is taken as a representative for the blob. Fig. 1 shows the method in a one-dimensional implementation. In two dimensions, averaging is done successively in two stages horizontally and vertically. Layer 3 gets input from four pixels, two in both directions. Layer 4 and 5 contain eight elements arranged in a square. More details and further properties of the method are given in Zimmermann & Kories 1989.

The center spatial wavelength of the band pass filter is the only parameter as all geometrical distances between pixels are connected directly with the wavelength. Using large

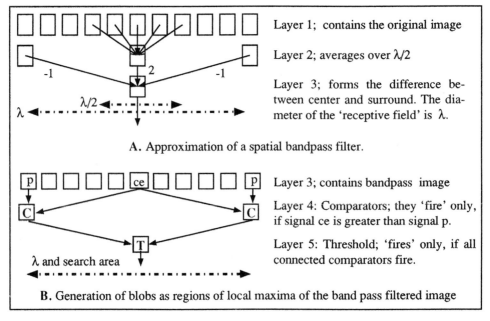

Layer 1; contains the original image

Layer 2; averages over λ/2

Layer 3; forms the difference between center and surround. The diameter of the 'receptive field' is λ.

A. Approximation of a spatial bandpass filter.

Layer 3; contains bandpass image

Layer 4: Comparators; they 'fire' only, if signal ce is greater than signal p.

Layer 5: Threshold; 'fires' only, if all connected comparators fire.

B. Generation of blobs as regions of local maxima of the band pass filtered image

Fig. 1. Homogeneous layers for the generation of simple features, depicted for one dimension. All cells within one layer have the same connections. For better clarity, only one example is given. The output layer of Fig. 1A is the input for Fig. 1B.

(small) wavelengths results in the extraction of coarse (fine) structures from the image. There are no thresholds at all that have to be adjusted from outside.

Estimation of displacement vector fields using blobs

For the estimation of displacement vector fields, searching for corresponding features in the next image is a crucial task. By virtue of the definition of the blobs used here, this task is trivial <u>if</u> the displacement is less than λ/2 . In this case, the phase information of the filtered image is preserved and it is (in the ideal case) impossible, that within ± λ/2 of a maximum in the first image there is a maximum in the second image that does not emerge from the same gray value structure. The search space for the corresponding blob in the next image is therefore simply ±λ/2 (see Fig. 1). If it is additionally demanded, that a blob must be tracked through four consecutive image pairs without interruption (low pass in time), a robust method for displacement estimation is defined. This has been tested extensively with TV-sequences from indoor and outdoor scenes using stationary and moving cameras (Kories & Zimmermann 86, Zimmermann & Kories 88, 89). Fig 3 gives an example of a displacement vector field of the scene depicted in Fig 2.

Stereo matching using complex features (constellations)

If displacements between two images become larger than ±λ/2, then the method described above degrades gracefully as it delivers no displacement vectors at all (Zimmermann & Kories 89). To overcome this situation, one can use larger wavelengths in order to increase the search area. For very large disparties however, only very coarse structures can be matched. A better method is therefore to construct complex features by combining simple ones, so increasing the 'individuality' of the features. In the present implementation, this is done by using three wavelengths (coarse, medium, fine) within one octave and taking into account the mutual arrangement of the respective blobs. As these arrangements resemble the constellations in the night sky, the complex feature is named a constellation. The coarse blob is taken as the reference point of the constellation. In Fig. 3, constellations

Fig. 2. Stereo pair from an image sequence taken from a moving car.

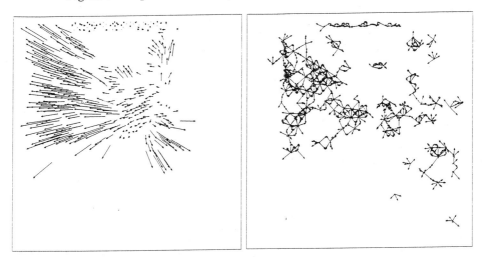

Fig. 3 left. Displacement vector field of those blobs, that persist uninterrupted over 24 consecutive image pairs, i.e. one second. **Right:** Constellations of the right image of Fig. 2.

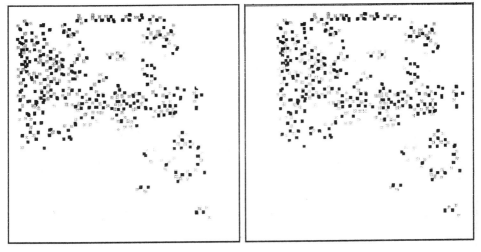

Fig. 4. Synthesized stereo pair from matched constellations of the images in Fig. 2.

457

of one image are displayed. The lines connect coarse blobs with medium and fine ones. In parts of the image, the constellations overlap giving so a natural tesselation of the image.

Stereo matching is achieved by comparing all constellations of the left image with all of the right image that lie within a very large search area of about half of the image size (which is 512^2 pixel^2). That pair of constellations, that fits together with the highest number of blobs is taken as a candidate pair for a match. The procedure is repeated starting from the right image and a match is assumed only if both candidate pairs prove to be the same pair. In Fig. 4 a synthesized stereo pair is shown. Each square represents a blob found by automatic constellation matching. Dark blobs represent minima, light ones maxima, regardless of λ. In order to enable fusion for the human observer, vertical disparity has been set to zero. If one fuses the two images, the emerging tramway can be clearly seen.

'Object' recognition using constellations

Constellations are useful for describing parts of the image almost uniquely. This holds even when one compares the constellatons in 500 images taken from a moving car with those of 4 images taken 10 min later in the same streets (Zimmermann 89). Such, constellations might serve as a representation for real objects imaged from a similar point of view.

Discussion

If one assumes, that the circuitry described above is implemented in the retina as sustained on- (off-) center cells and their output signal is transmitted via LGN to the cortex, some useful properties of the method become evident: 1) Only few dendritic connections are needed leaving space for multiple evaluation of the same receptor area. 2) The receptive field width defines the spatial wavelength. 3) The output signal is binary and therefore immune against noise and crosstalk which might occur heavily in nerve nets. 4) A bundle of neighbouring axons transmit the same signal, provided that the retinotopic organization is preserved. This enhances fault tolerance, as the degeneration of single axons is not deletrious. 4) Despite the simplicity of the signals, useful information can be extracted automatically (motion e.g, for eye movement control and tracking; stereo fusion) or intentionally using memorized data ('object' recognition).

It is clear however, that the present implementation simulates only a tiny amount of what is happening in a real retina. To mention a few deficits: Color is left out at all, time is treated discretely and the poor signal to noise ratio of neural spike sequences is not yet taken into account in the algorithms for stereo matching and object recognition.

References

Rosenfeld, A. (1990): Image Analysis and Computer Vision: 1989. Computer Vision, Graphics, and Image Processing. Vol 50, No 2, May 1990, pp. 188-240.

Kories, R., Zimmermann, G. (1986): A Versatile Method for the Estimation of Displacement Vector Fields. Workshop On Motion, Kiawah Island, Charleston SC, USA, May 1986, pp. 101-106.

Zimmermann, G., Kories, R. (1988): What an Autonomous Robot can Learn about its Environment Starting from Scratch. In: Real-Time Object Measurement and Classification; A. Jain ed.; NATO ASI Series, Vol. F42, Springer-Verlag Berlin, Heidelberg, pp. 215-226.

Zimmermann. G., Kories, R. (1989): Eine Familie von nichtlinearen Operatoren zur robusten Auswertung von Bildfolgen. In Fortschrittberichte VDI, Reihe 10: Informatik/ Kommunikationstechnik, Nr. 114: Ausgewählte Verfahren der Mustererkennung, W. Schwerdtmann (Hrsg.), VDI-Verlag Düsseldorf, pp. 96-119.

Zimmermann, G. (1989): Creating and Verifying Automatically a Representation of Three Streets in Karlsruhe City Using Complex Image Features. Proc. Stichting Int. Conf. on Intelligent Autonomous Systems · 2, Amsterdam The Netherlands, Dec. 1989, pp. 705-714.

SCALING AND THRESHOLDS OF COLOR AND LIGHT DESCRIBED BY AN

OPPONENT MODEL OF COLOR VISION BASED ON PSYCHOPHYSICAL DATA

Klaus Richter

Federal Institute for Materials Research and Testing
(BAM 5.44), Laboratory "Color Reproduction"
Unter den Eichen 87, D–1000 Berlin 45, Germany

SUMMARY

A newly developed model for vision of color and light, based on psychophysical results and some physiological experiments with monkeys, defines equations to describe scaling and threshold data over a large range of luminances and chromaticities The luminance of central and surround fields vary between 1 and 10,000 cd/m^2.

One basic result is a new relationship between scaling and threshold data, e.g. for a red-green series of colors. At threshold either the red D-process (D=Deuteranop) or the green P-process (P=Protanop) is the more sensitive along the red-green color series of equal luminance and defines the threshold. For large appearance differences, both processes are far above threshold and the model has to use both processes to describe chromaticness of the red-green series. Similar results describe the yellow-blue and white-black directions of the color space, e.g. two processes are active in each of these directions. the following 3 figures show the basic ideas.

RECEPTOR SENSITIVITIES AND CHROMATIC SATURATION

We have looked to the fundamentals of the visual system with the aim of calculating the two processes in the red-green direction from the known receptor sensitivities. In Fig. 1 two of three receptor sensitivities (P=Protanop, D=Deuteranop, T=Tritanop), and a photopic luminous efficiency function U are shown as function of wavelength. The curves show simple approximations in form of symmetric parabolas, differing in maxima on a wavelength scale. There are genetic reasons why the P-sensitivity may have a second hump through a small contribution of (S-cone) T-sensitivity with a maximum near 450 nm. A contribution of 2% is assumed in the figure, which leads to opponent red-green saturations on the wavelength scale as compared in Fig. 2.

Fig. 2 shows the ratio of the cone sensitivities and the luminous efficiency function U. The ratios P/U and D/U, which are normally labeled "saturation", look like results of threshold experiments in the red-green direction (Richter, 1985). Similar results are presented by Krauskopf (1990) in this book. Either the red P-process or the green D-process is the more sensitive and defines threshold. Above threshold, the sum of both processes (Richter, 1985) describes scaling experiments of the red-green color series.

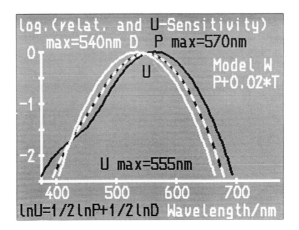

Fig. 1. Log. of relative cone sensitivities P and D (P corrected by 0.02*T) and log. luminous efficiency U as function of wavelength W.

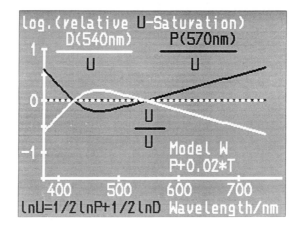

Fig. 2. Log. U-saturation P/U, and D/U of cone sensitivities D and P (P corrected by 0.02*T) with ln U = 1/2 ln D + 1/2 ln P (U=luminous efficiency).

DESCRIPTION OF LUMINANCE CONTRAST L/△L BY TWO OPPONENT PROCESSES

In Fig. 3 the contrast L/△L (L=central field luminance) is plotted as function of L for a presentation time of 0.1 s of the central bipartite field in a grey steady background of differing luminance. The contrast reaches a maximum at the surround field luminance, Lu, for all surround field luminance levels. The horizontal relationship near the maximum indicates that over a small region the Weber-Fechner law (L/△L=1) is valid. There are large changes with presentation time which are not discussed here.

The curves for short presentation times (0.1 s) can be approximated in the left (dark) part by one function and the in right (light) part by another function (Richter, 1988). These two parts correspond to the sigmoid firing rates on a log luminance scale of two cell types described by Valberg et al. (1986). The two cell types respond to achromatic colors either darker or lighter than the surround. Again, threshold and scaling experiments can be described by the two separate functions (firing rates) or the sum of both. The model seems the first based on *two* opponent mechanisms in red-green, yellow-blue, and black-white direction. The two processes in the three opponent directions allow a description of psychophysical data for scaling and thresholds.

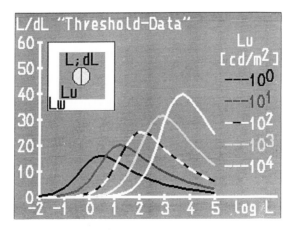

Fig. 3. Luminance (contrast) discrimination L/△L for central field colors of different luminance L in a grey surround. Parameter: surround luminance.

REFERENCES

Krauskopf, J., 1990, Chromatic descrimination and adaptation, In this Book: Advances in understanding the visual process ..., Plenum.

Richter, K., 1985, Farbempfindungsmerkmal Elementarbuntton und Buntheitsabstände als Funktion von Farbart und Leuchtdichte von In- und Umfeld, Wirtschaftsverlag, Postfach 101110, D–2850 Bremerhaven 1, 119 pages.

Richter, K., 1988, Beschreibung des Leuchtdichte-Unterscheidungsvermögens durch einen schmal- und breitbandigen Schwarz-weiß-Gegenfarben-Prozeß, Tagungsberichte Licht '88, 2:421–433, LiTG, Berlin 1988.

Valberg, A., Seim, T. Lee, B.B., and Tryti, J., 1986, Reconstruction of equidistant color space from responses of visual neurones of macaques, J. opt. Soc. Am. 3:1726–1734.

DISCUSSION: MODELS AND NEURAL NETS

Ivar Giæver

Dept. of Physics
University of Oslo
Norway

Giæver: I have enjoyed this conference although when I came here Monday I knew virtually nothing of what you people are talking about. Despite or because of this, I have enjoyed the conference. My vision of the eye was that the eye was a television camera, with a little guy up in my head watching television. Now I only understand in gross outline what you are worried about, but I do not understand the details. This is difficult for a discussant, so I propose to give a few philosophical remarks about computers and neural networks and then throw the discussion open.

We can ask if computers can think or not? Most of you may say no, but nowadays even cheap computers can beat me at chess. How can we decide if computers think? One classic way is the Turing test. Using a keyboard, you communicate with a computer or a person in the next room. If you cannot decide if a person or computer is on the other end, then by pragmatic definition the computer can think; I think we are only a few years away from this. Roger Penrose in his book "The Emperor's New Mind" thinks that even if a machine passes the Turing test it cannot think. I feel differently; if God could do it again he would have used copper wires and silicon, being more reliable and robust. In that context, you can imagine trying to record from a PC, drilling a hole in the case, putting in a voltage probe and maybe tipping in a little acid. One can see how difficult the problem is and I admire your progress. Also, all IBM PCs are wired up alike, but we are probably wired up very differently. If you look at the information in DNA, there is not enough information to code for all the connections in the brain; we are self-organizing and all different; those of us who are amblyopic can testify to that.

I have learned this week that one cannot believe one's eyes; you are interested in the details of this system. As an outsider, I am interested in how robust the system is; you can put stuff in a monkey's eye and it still can see. Destroying one transistor can stop a PC. With 'nural' networks (misspelling deliberate, since neurophysiologists don't think they have anything to do with neurones), one also has a very robust system. Elements can be removed and inserted with little effect. I believe nural networks can tell us something about the brain, however. Finally, we build nural networks for parallel processing, to find ways of improving computer processing speed. Let me finally talk about the back-propagation model. People have used this model for many real problems, for example, to generate credit ratings. The network generalizes from past experience to produce a result. The network thus takes on a life of its own. Still, John Hertz told me that a nural network is viewed as the second best way to do anything. Another model of a network, of Hopfield, is an associative memory system. Hearing the word Røros, you will think of Norway. With an algorithm able to connect simultaneously excited cells in a network, you have an associative memory. When we recognise faces, we use associative memories. I have no idea how it works, but it is a simple, usable model.

From Pigments to Perception, Edited by A. Valberg and
B.B. Lee, Plenum Press, New York, 1991

Zrenner: Can Vivianne Smith say a bit more about P-cells?

Spillmann: Also, if your stimulus covered the whole receptive field, centre plus surround, you might confound an effect from a M-/L-cone temporal difference with an effect due to the spatial antagonism.

Smith: We used full-field stimuli and our data are consistent with L- and M-cone antagonism in the red-green cells with a phase delay from the surround. The red and green on-cells are very symmetrical, so any M/L cone latency difference has to be less than two milliseconds. There will be a change in spectral sensitivity with frequency, but its going to be a very gradual change, and even by 40 Hz we seldom had a 180 deg phase shift. If you want to interpret the phase delay as a latency, its about 3-6 msec, so you need substantially more than 40 Hz to make L/M cells have the spectral sensitivity of a luminance system.

Lennie: Why do you put the opponency in magno cells in the surround?

Smith: Partly because Bob Shapley tells us that centres are well-behaved, partly because in pilot studies we tried smaller fields and the phase shifts disappeared, so the phase shifts seem to be associated with the surround.

Kaplan: I would like to respond to Carl Ingling's paper. Most of what you say is predicated upon the plus, that the centre and surround add linearly. We don't have strong reasons to know if this is really true.

Ingling: That is true, in order for the identity to hold the cell has to be separable and reasonably linear. Some people have recorded responses of this sort.

Walraven: Valberg's modelling seems to work quite well. There are, however, many models that use a non-linearity somewhere, causing responses to saturate earlier. At high intensity, all cells will signal white, or, in the red-green system they will signal yellow.

Valberg: One reason for testing this neurophysiologically based model on psychophysical effects is, as with any model, to find out where it goes wrong in describing the details of psychophysical phenomena. Only such information will help us understand the functional significance of the different types of cells in colour vision and to improve the model.

Walraven: There is a recent model of Hans Vos where he extrapolates the Bezold-Brücke phenomenon to the achromatic domain as well. Which kind of point would your model converge on at high intensities; on white or on something else?

Valberg: We have not extrapolated so far beyond our data, but it can easily be done, and I anticipate it would converge on white.

Dow: Do M cells not contribute to the Bezold-Brücke effect?

Valberg: It is not necessary to include M cells to account for it. P cells just do fine on their own.

Spekreijse: If you take the Hertz machine, and train it, and put in the actual spike responses, can the machine perform as well as the monkey?

Hertz: The monkey wasn't doing that so it is difficult to compare. We did not average over many runs but the machine did do one trial recognition.

Kaplan: Can you say why the principal component analysis is better than the Fourier spectrum?

Hertz: If you keep all components, it's the same. The advantage of principal components is that their importance falls off sequentially, and you can make a sensible truncation.

Van Essen: You have established that the information is there; how do you establish that it is actually used by the real nervous system?

Hertz: That is a very hard question, all I can say is, if the networks in the brain can easily learn extraction of temporal information, then it is a plausible scheme. It is difficult to experimentally test it.

Lennie: How do you decide how long to sample a spike train?

Hertz: We have not tested this, but my guess is that by 100 or 150 msec you have most of the information.

Lennie: Is that too long for a real visual system?

Hertz: Some discriminations we make quickly and some we make slowly. You can ask whether in each case the spike count alone is enough for the decision.

Krauskopf: One has to think in terms of cost benefit of the apparatus; you presumably have to build some elaborate apparatus to do principle component analysis in the brain. It might be cheaper in the brain to simply get neurons to do spike counting rather than any other way.

Hertz: I have no concrete way of testing that possibility more specifically.

Giæver: Any last minute hesitations.

Creutzfeldt: Commenting on what you said, where computers differ from brains is that there is a certain isomorphy between what is out there and what action the organism must take. This is not so for a computer.

Giæver: If you talk about whether a computer can think, I still think the Turing test is a good one. Many years ago some jokester made a program where a computer behaved like a psychologist. Some people came to prefer that to a real psychologist. To you, it feels absolutely real even though the computer knows noting about what it does. This is one of Penrose's objections to the Turing test.

Schiller: Another important difference is that computers were designed by a God-like creature to a particular plan, but most of us believe evolution is blind, and we are the product of non-planned development. This makes comparisons questionable.

Zaidi: Penrose's central thesis is that if you can define all mental functions in algorithmic form, then computers can think, because you can program the computer to perform the algorithm. If this is not possible for all mental functions, then computers cannot think. A lot of what we have been talking about concerns very early parts of vision, and this might be defined in terms of simple algorithms. Computers may not be so bad at this level, but what goes on later is anybody's guess.

Giæver: In a neural network you do not really work in terms of algorithms, the system self-organizes, but basically I agree with you.

Zaidi: My view of neural networks is more cynical. You take an input and an output you want to train it. If, for example, you take edges and have an output of Mach bands, you get a network with lateral inhibition. But really you can use your own wetware like Ernst Mach did to get this result without programming.

Schiller: Emphasizing this fallacy of thinking the brain is like a computer, I guess we are limited in our imagination and always try to figure out how the brain works in terms of some kind of analogy. You remember in Descartes day people tried to think about the brain in terms of gases going through tubes, in the industrial revolution with the development of clocks or other machinery people thought of the brain as a clock, thereafter they invented the telephone and the brain was switching circuits, then came electronics and the brain was a bunch of resistors and transistors. Nowadays we are foolish enough to think of the brain as a computer; I assure you it's not.

GENERAL DISCUSSION

Barry B. Lee

Department of Neurobiology
Max Planck Institute for Biophysical Chemistry
Göttingen, F.R.G.

Lee: I would first like to select some specific topics, partly related to the posters presented. A first set of questions have to do with what the retina might be doing. These may be listed as follows;

1) How is cone selectivity maintained? We know that there are gap junctions between the cones of the primate's retina and we know that horizontal cells seem to collect indiscriminately from all cones. So there seems to be something of an anatomical problem here; perhaps opponency must be pushed into the inner plexiform layer.

2) At the next stage, we have the bipolar cells. Paul Martin in his poster shows counts of the total number of bipolars, and could also specifically stain the rod bipolars, to give about 1.5 cone bipolar cells per ganglion cell.

3) How distinct is the S-cone system, anatomically or otherwise? Bob Rodieck identified the stratified ganglion cell as a type II cell, but there is another group of ganglion cells projecting to the parvocellular laminae and those are the ones which get S-cone input. These are about of the same proportion as the stratified cells.

4) Another issue is the relationship between M and P cells at different retinal eccentricities, and if their projections M-scale in the same kind of way.

5) How adequate are current models of P and M cells? An interesting question is how linear these systems are, and what is the nature of the non-linearity in the M-pathway particularly.

Kaplan: In relation to Drasdo's poster on retinal eccentricity, Lisa Kroner is now looking at M and P cells at various eccentricities as to whether or not in the periphery the differences are maintained. At least from the point of view of contrast sensitivity this seems to be so.

Drasdo: There has been a long history of suggestions that the central fovea was perhaps under-served by M or Pα cells. Recently Livingstone and Hubel looked at 4cα and 4cβ and found a rather even distribution of neurones with eccentricity. This also coincides with the paper from Perry and Silveira who claim an anatomically even distribution in the retina, though there are experimental difficulties. It is very difficult to show the strongest effects which are expected in the central 5 degrees, due to the foveal excavation. There is also an artefact due to displaced amacrines which tend to be classed as P cells rather than M cells. My paper of course is purely based on psychophysics, and I will be very interested to argue with anybody about it.

Martin: We are presently working on the idea that there is nothing really special about the fovea, and as far as the M scaling is concerned, Heinz Wässle's work suggests that previous estimates which maintained that the fovea was over-represented in the cortex were probably incorrect, mostly because of this artefact of the displaced amacrine cells which become relatively more numerous in peripheral retina. There are other artifacts associated with counting in the fovea that tend to produce underestimates there, so we are working on the idea that the fovea is really essentially the same as other parts of the retina and that the ganglion cell density gradient determines the cortical magnification factor. Here, we are demonstrating that there are enough bipolar cells present to allow for the existence of separate streams converging on to separate streams of ganglion cells. In other words, different types of ganglion cells don't have to share their input from one population of bipolar cells.

Schiller: The question I want to pose has to do with eccentricity and receptive field size. Several people showed pictures today of the relative differences in sizes of the M and P cell receptive fields in the retina, and these sizes for both these cells gradually increases from the fovea to the periphery. For P cells this seems in conflict with the anatomy, according to which a single cone makes up the center mechanism as far as 10°, so their centres should not show gradual increase, at least much less than it appears. Is there an parallel increase in receptive field center of M and P cells in both systems or do they diverge? This is an important question in relation to the sampling these two systems carry out.

Walraven: With increasing eccentricity cone density varies, falling off roughly with the square of the distance. So you get a more diluted retina and to compensate, receptive field size increases. Lots of models increase in proportion to cone density or cortical magnification. Then you find that acuity, colour discrimination and whatever do not vary. So only the scale of the system changes. I guess that all functional units change at the same rate.

Lee: We have attempted to measure receptive field diameters of M and P cells at different eccentricities with area summation curves, and though this is technically difficult, our impression was that they were increasing at roughly the same rates. Which ever way you look at it, cone diameter, and presumably the cone's sampling aperture doesn't increase at the same rate that visual acuity decreases. So there is a real problem in there somewhere which I think has yet to be resolved.

Shapley: Because there is a possibility of cone coupling, even if all the P cells get single cone input through a single midget bipolar (which need not be the case), coupling conceivably could make continuous variation in a P pathway possible. We just don't know.

Lee: But how would you maintain cone specificity in that kind of framework?

Shapley: I think there is much about cone specificity that we don't understand. Our data tell us it is there, we have to figure how it's done.

MacLeod: The idea that the midget bipolars and the midget ganglion cells serve one cone each does not seem to be consistent with the numbers of the cells involved. We have several million cones in the retina and about one million ganglion cells which include at least on and off types, and so if you subdivide the ganglion cell population into various categories, you have an average convergence of several cones per midget ganglion cell over the whole retina. So the one to one connection cannot be the rule over the entire retina even if it is so over a substantial parafoveal region.

Martin: There is a little confusion about, among the population of so-called P cells, what are midget cells. It appears that there is a continuum with midget cells having larger receptive fields contacting several cones past about 20 degrees. Whether these cells then selectively find the correct cones to maintain chromatic specificity is still unsure, but at 5 mm which is equivalent to about 20 degrees, one can still find classically appearing midget cells. One can also find larger appearing cells which project to the parvocellular layers.

Lee: What about the way the S cone-system fits into this anatomical pattern?

Zrenner: There are nice data from Arnelt on this. The inner segment of the S-cone is longer than that of the other cones, so although it looks like a cone, it sticks out more distally in the retina. The outer segment of S-cones is shorter than that of the other cones. If one looks in the mosaic the S-cone seems thicker in the inner segment, and it is usually connected with another cone. Secondly, how distinct is this cone system? One difference might arise because it is coded on a different chromosome. As far as function is concerned, if you look at the blue-cone monochromat the visual acuity of the S-cone is approximately 0.2, much worse than that the other cones, probably because of sampling. However, it's much better than the rods although the S-cones are much sparser. The temporal frequency response of these cones is higher than the literature says, for there is now much data that S-cones and their ganglion cells transmit temporal frequencies up to about 35 Hz. Susceptibility to toxic agents is another point. A number affect the S-cones to a much larger degree than the long-wavelength systems, such as antiepileptic drugs.

Lee: I would like to ask a question specific to your paper. I am very ignorant about this, but do your blue-cone monochromats have no other cones present, or are they filled with rhodopsin?

Zrenner: The blue-cone monochromat has rods and S-cones only, and they do not have the genes which determine the M- and L-cones.

Lee: Is there anatomical evidence as to cone and ganglion cell densities?

Zrenner: No, there's not. So it seems as if the normal rod and the normal S-cone retinae are maintained in the monochromats, and that the S-cones interact in an opponent manner with the rods, so that blue-cone monochromats are able to discriminate colours as well as normals in the short-wavelength part of the spectrum in a very narrow range of adaptation. If the adapting light is too bright, rods are saturated and colour discrimination drops very fast. If the light is too dim, S-cones are not sensitive enough, and again colour discrimination is lost.

Smith: There are certain families of blue-cone monochromats in which there are residual cones, and additionally there are families of blue-cone monochromats in which there is a preserved 'green-cone' gene. It's not clear what that gene does. I didn't see on your poster specific evidence for opponency. You can get discrimination without opponency.

Zrenner: From Nathan's paper, he has so-called blue-cone monochromats which I wouldn't consider true blue-cone monochromats because the genes show that they have remnant middle-wavelength sensitive cones, or even they have both the other cone pigments but inactivated for some reason. Are those strictly blue-cone monochromats? I think we dealing with the real ones, as far as the genetics are concerned. As far as opponency is concerned, the short-wavelength equivalent of 'Sloan's notch', between rods and S-cones, can be manipulated as you would expect from an opponent system.

Shapley: I'm curious what vision is like for these blue-cone monochromats, what pattern vision is like and if they can do shape recognition under scotopic conditions.

Zrenner: In the dark they behave like we do; Kathy Mullen has studied this. We may help them with blue filters, to minimise light seen by the rods, so they can utilize both systems.

Creutzfeldt: What is exactly the light adaptation level at which the rods completely stop functioning. Is there vision remaining in the rod system during normal daylight?

Zrenner: Between around 80 to 800 trolands rod vision deteriorates very strongly, spectral sensitivity is dominated by S-cones in ranges above this.

Creutzfeldt: I would like to ask Barry Lee to comment a little bit on this morning's session.

Lee: I thought there was fair degree of agreement. I think the main point of disagreement that we personally have with Schiller's and Merigan's lesion work is the difficulty we see in developing from P cell activity some sort of achromatic, luminance signal. You can get a brightness signal out of P cells, and I am not surprised that animals with M cell lesions can make a brightness discrimination as Schiller described. That is quite different from photometric tasks. It is very difficult from our physiological responses to derive from P cells which could play a role in, for example, the minimally distinct border. This is an unresolved problem where our physiology and your lesion experiments are not really in concordance.

Kaplan: To expand on this, there is a problem for several reasons. One is, if it is really possible to design a motion task or a stereo task or a shape discrimination task in which no other cue could be used. I think it's hard to exclude the possibility that there are other cues the animal uses. As Kulikowski says, if there are no big fish we have to use little fish. That might be one explanation why if you take away the magno system, the animal can still do something with the parvo system and *vice versa*. There are other conceptual difficulties with lesion experiments. Taking the eyes of a horseshoe crab out means it doesn't walk but we wouldn't want to suggest that they walk with their eyes. There are also technical problems with lesions in the brain about localization. The perigeniculate nucleus is extremely sensitive to ibotinic acid. If you put it anywhere near the animal, the perigeniculate nucleus is lesioned, let alone if you put it in the skull. However, these may all be excuses, and we just don't understand the disagreement between the results that Peter and Bill described, and the physiology that Barry Lee and Bob Shapley and I measure.

Schiller: I feel that my physiology is in good agreement with my findings, so we have a physiological disagreement. For example, the idea of an overlap between the center and the surround mechanism I find almost outrageous.

Shapley: Peter *(Schiller)*, as we were talking with Gilbert, you said sure it's true that you can't measure an overt surround with achromatic stimuli in a geniculate cell. At the same time it's very clear that you can measure cone antagonism in the same neurons. How can it possibly be explained expect in terms of spatial overlap or lack of it.

Schiller: In single unit recordings in paralyzed animals probably we see these things overlapping because we do not have the resolution. Remember that the cones in the central retina are only 3 μm across, and if a single cone makes up the center, even an alert monkey fixating is going to have a lot of trouble providing you with the spatial segregation. We know from Williams' work that the true constraint in acuity is not the cone grid but rather the optics. Bypassing the optics, your spatial frequency limit goes up to beyond 60 cycles per degree. If you and I could bypass the optics one might find very nice spatial segregation of the center and surround mechanisms.

Kaplan: I don't think that your lesions have so far challenged the details of the receptive field, but to me it is especially confusing that even in your own work magnocellular cells are more sensitive to luminance contrasts.

Schiller: That's true, but your P system is much more numerous than your M system. Only 8% of the cells in the retina or geniculate are M cells, but the whole organism looks at brightness discrimination or whatever. If M cells have much larger receptive field center, so their center mechanism is probably receiving as many cones as an aggregate of the P cells in a particular part of the retina. Where does the extra sensitivity come from? From a single cell point of view, you are correct. However, the brain or the mind or whatever is not looking at the visual scene with one M cell and one P cell. It's looking at any particular locus say with one M cell and 8 P cells. That has to be taken into account if you are talking about differential sensitivity. I was surprised that after a magnocellular lesion there was no tremendous decrease in contrast sensitivity. I didn't find it. You only find a big difference in contrast sensitivity when you do motion.

Kaplan: We had the same reaction, we were also surprised.

Lee: Perhaps for flicker you may just need a few M cells which escape somehow the lesion or are at its edge. However, you might need many more to provide a matrix for a motion task.

Schiller: In our case the motion is always confined to less than the area of the lesion. It still doesn't exclude the vague possibility that a few cells remained within the M system. Motion is highly affected by the magno lesion at low contrast, and at high contrast it takes much longer to process. I think the flicker and motion results are complementary.

Kulikowski: It's difficult to escape the conclusion that the systems complement each other. If anyone reports achromatic contrast sensitivity which is less than 100 (a 1% percent threshold) it's very difficult to rule out the parvo system because P cells might have contrast sensitivity around 10-12 but probability summation might contribute a factor of 3, so a contrast sensitivity of 50 might be supported by the parvo system. So in these cases where contrast sensitivity is not very high, I am not surprised that the effect of M lesions might not be very big. So unless we go to a task which requires very high contrast sensitivity, I don't see how really we can be confident in distinction between parvo and magno contributions.

Lee: I would like to make a last comment on this point. In terms of tasks requiring a high degree of resolution of fine spatial detail, if you postulate these are due to the P cell system, you can either increase the sensitivity of the system by combining cells, or you use the whole the sampling matrix to process the fine spatial detail. You cannot do both; you cannot have your cake and eat it too.

Drasdo: There was a statement earlier that M scaling would equate all thresholds. This has been disproved; M scaling will not equate many thresholds.

Lee: As a second group of topics, I have subsumed under central mechanisms. Three posters have to do with colour constancy or adaptive constancy, and an interesting question is how far do we expect these kinds of constancy to be generated perhaps at the ganglion cell level, and how far are they going to be developed at later stage. Secondly, with higher level mechanisms, such as colour scaling, motion and development, detection and discrimination, can we relate any of these tasks to physiological mechanisms?

Shapley: I think that there is some evidence both in the posters and elsewhere that certain aspects of brightness and colour constancy have peripheral components. Some of these effects, both behavioral and neurophysiological, may be related to these retinal adaptive, controlling constancies.

Lee: An important question is how important is the role of scattering in the optical media.

Walraven: Scattering in the ocular media is one artifact you have to consider when you are dealing with colour constancy and chromatic induction, which are actually very closely related. I hope to come to that in my talk. I would like to ask the reverse question, how much evidence is there that it is not peripheral. With binocular stimulation and dichoptic presentation, many people have not found any effect. It seems that simultaneous contrast is a largely retinal phenomenon. I think that colour constancy is a very primitive system, meaning that you are reacting to a stimulus which depends on reflectance. You respond to contrast, and if you don't respond to contrast immediately, you have a very slow system. A wise system would immediately start converting everything to ratios, and so escape the problem of having to consider changing illuminance or light levels. Ganglion cells may be trying to compute such a contrast by trying to null out the surround. For instance, if you have a gain box that divides by its own input, you get something like a log transformation. Once you have this, you have a contrast detector or a reflectance machine and you are in business. I would prefer to have this done early in the system.

MacLeod: Normally I would argue that everything happens peripherally, but since Jan Walraven has just done that, I will argue for the opposite position, and mention some experiments that show a role of central processes. Thomson and Lashford showed that when you generate a McCulloch effect, and you trick the subject by changing the colour of the illumination during induction, then the effect depends upon the peripherally presented stimulus, rather than the colours seen by the subject during the induction phase. That is evidence that the corrections that affect perception are actually being applied after the place where the McCulloch effect is generated and therefore in the cortex somewhere.

Creutzfeldt: In our dichoptic experiments with simultaneous contrast we came to about 30% of the value in the monocular situation. Everyone knows with dichotic colour contrast, the effects are very unsaturated.

MacLeod: These are the results that Jan Walraven was referring to, and it is indeed difficult to reconcile those with the results I mentioned. One straw that one could clutch at, though not particularly plausible, is binocular rivalry in the normal dichoptic experiment.

Thomas: There is another way to get at this question as to whether there is a peripheral transform such as log transform, and that involves discrimination. In the normal viewing situation the illumination is multiplied times the reflectance, and if both the reflectance pattern and the illumination pattern are periodic, you are going to get the Fourier spectrum of the two convolved with each other. If you choose the conditions properly you can set up a situation where that convolution should increase the bandwidth of your reflectance pattern, and make discrimination more difficult. We have done this experiment, and you do not get the fall in discrimination you would expect. That's evidence supporting the idea that there is an early log-like transform which prevents that convolution from happening.

Walraven: There is a vast literature on chromatic adaptation where they use the other eye for a reference. You can have glaring fields in one eye and the other eye simply doesn't know. These experiments have been done assuming that the eyes are pretty independent. If you bleach with one colour at one spot, you will notice that the corresponding spot in the other eye will become more sensitive to the other colour. But these are small effects. In general we are all aware that the eyes are pretty independent from each other.

Lee: Lastly let us consider some of these more elaborate tasks. David Burr's poster is perhaps most obviously related to physiology. Are you suggesting that two different physiological systems develop at different rates in the infant, and is there any type of anatomical evidence for that?

Burr: The data is certainly consistent with that, but I don't think you could decide that, say, chromatic opponency in one system develops after achromatic opponency in another. As far as I know, there is no agreement among anatomists as to whether the magno- or the parvo-system develops earlier. In a recent conference I polled the anatomists, and got a roughly distribution of 1/3 magno, 1/3 parvo and 1/3 didn't know.

Swanson: I don't know the anatomy very well, but my understanding is that in the fovea the cones are still moving around during the first 6 month to a year of life. I wonder what that has to do with getting proper wiring of the opponent cells. Is the wiring done before the cones stop moving or after?

Shapley: I want to ask particularly whether the equal luminant point varies within an individual with age from infancy through childhood. Is that point moving or is it fixed at an early age?

Burr: From memory, the isoluminance point was fairly similar for all infants. It did shift a bit with spatial frequency, but you could easily explain that on chromatic aberration.

Fiorentini: To expand on that, we didn't find any systematic change. I have a question myself about the development of the two systems. Is it possible that peripherally one system develops first. Take for instance the M pathway, would you conceive that at retinal level you have an earlier development for the M cells, but those cortical functions that are subserved mainly by the M system may develop later than the ones that are subserved by the P system. Is that conceivable or not?

Kulikowski: I remember reading an old paper by Hickey in which he argues that the M system might start earlier in development, but finishes later. That would clearly imply later development in the visual cortex.

Fiorentini: There seems to be some evidence from Wiesel and Hubel that there is a difference period of plasticity in the two systems, with the M system stopping earlier than the P system.

Spillmann: This is a question that leads us elsewhere, but I was wondering if elevated intraocular pressure in glaucoma effects the P and M systems differently. If so, would one have a model for either M or P in such patients?

Schiller: I don't have a definitive answer. People are studying this quite furiously. One problem is that, under conditions of glaucoma, the tendency is for the larger fibers to go first, and there is a tremendous change in fiber size from center to periphery, which goes beyond the difference between parvo and magno. So the deficit following glaucoma may not be purely damage to the parvo- or magno-systems. The general belief is that the magno or fast fibers are more affected.

Swanson: That is one story, but a group in Germany did retrograde filling after inducing glaucoma in one eye, and then looked at the same patch in both eyes and compared the distribution of different cells sizes. They found equal loss across all different cell sizes, which was unexpected. Earl Smith in Houston did electrode penetrations in the LGN, and at any given location there seemed to be equal loss across different cell sizes. This means it's still up in the air.

Kaplan: Bob Shapley and I have recorded in our lab from monkeys prepared with one glaucomatous, and we get the same result as Earl Smith. With an electrode in your hand, it's very difficult to tell that the magnos have been more compromised than the parvos in these animals.

Shapley: A sample of glaucoma patients showed definite flicker deficits which I think is the strongest psychophysical result.

Swanson: Quigley also showed the same in ocular hypertensive patients. Those patients, most of whom will not develop glaucoma, have flicker deficits, so we don't know if it's a reversible deficit due to high pressure or if it's actual cell damage.

Lee: I would like to raise some points for a general discussion.
1) From contributions so far, the neuron doctrine which was described by Barlow in 1972 has not received a great deal of support. Has the grandmother cell passed away?
2) It is difficult enough to link physiology and psychophysics with rather simple stimuli, for example with flicker photometry or threshold measurements. Some tasks we have heard discussed require a much higher level and much more complex sort of process. Several years ago Davida Teller published a rather crushing review of the attempts to interpret Mach bands in terms of neuronal activity in the retina. Like the retinal Mach band explanation, induced brightness or lightness effects are also very much class B hypotheses. Is it possible, and what criteria should we use, to link in these cases class B psychophysics with physiology?
3) Can we really generate selective stimuli which activate specific sub-systems, either spatially or by considering thresholds?
4) What physiology can psychophysicists trust, a question of Joel Pokorny's. His implication is that the physiology gets more difficult to interpret the further into the brain you go.
5) Finally, can we make do with a random cone distribution and random cone connectivity, or do we have to postulate a great deal of preprogrammed connectivity within the visual system.

Rodieck: There is at least one group hard at work on developing an immunocytochemical distinction between the red and the green pigments, and I expect that soon we are going to know the answer.

Swanson: One way of evaluate random connectivity might be to take a function that may be sensitive to it, and then test patients who you know to have a more random than usual retina. Carriers of protanopia and protanomaly appear to have in peripheral retina random inactivation of one of the genes, so they would have a more disordered receptor mosaic than normal.

Smith: I want to make two comments. First, the fact that female squirrel monkeys have colour discrimination would suggest that you do not *a priori* require an ordered mosaic and cells that know their destinations. The second point is that human carriers are not a very good model because it depends on what you decide the possible size of abnormal patches of cones might be. If they are large, the carrier will also have normal patches which could mediate discrimination, so you need to tap specifically the abnormal patches, which could be a very hard thing to do. I don't think however there is even very good evidence that there are such patches.

Drasdo: Just a brief comment on the first question. I suggest an answer might come from the report that we can record VEPs related to a face recognition response. This suggests that a very large number of neurons are involved in such a recognition process rather than one very specific one. If this is true, I suspect that the grandmother cell doesn't exist, and we have a more diffuse representation, proceeding towards the older concept of a holographic type of representation.

Rodieck: I am surprised in this meeting that so little emphasis has been given to the notion of hypothesis testing as a mental activity, and in that context a number of the questions that have been asked don't really seem to me to be questions that really refer to the visual system. A good example was the question as to where all the signals in different pathways come together to unite various features and so forth. A possible answer is that they don't come together anywhere, but they are made available so that the the brain could check them out, and see whether they violate or select between various alternative hypotheses. That's another way of looking at things other than the grandmother cell, and it doesn't have so many problems associated with it.

Kaplan: About question 3, and whether we could cook up a stimulus that would tap only one group of cells. From our data, we could prescribe such a stimulus only for the M cells, and such a stimulus will have low contrast let's say below 10 or 8%, and it would help if you lowered the mean luminance because at low luminance and low contrast M cells are the only ones which respond to patterns. I am not sure that any other stimulus will isolate other cell groups.

Lee: An isoluminant stimulus modulated along a tritanopic confusion line will isolate cells with S-cone input, a predominantly parvocellular group.

Kaplan: There is always a bias in sampling with electrodes and I am prepared to believe that there are many cells that have not yet been recorded.

Schiller: We are obviously not talking about two completely separate systems but we have two systems which probably sample different aspects of the environment with a considerable overlap. Given that caveat, you can probably isolate the parvocellular system, however heterogeneous it may be, by using high spatial frequency stimuli which in my opinion are not seen well by the magnocellular system. Conversely I believe low-contrast stimuli at high velocity will be better seen by the magnocellular system. Using stimuli only in those two domains very much restricts what one can test. I do not believe that isoluminant stimuli for example can discriminate the two systems. There may be other methods, however, such as afterimages, which can also be used.

Kulikowski: We cut our teeth on trying to find correlates of parvocellular activity by using very fine gratings. There are problems because many other factors affect such stimuli. You are never certain that your subjects are completely normal refractively. I wonder how many among those here have a visual resolution over 50 cycles per degree. Our experience is that to be really certain about sole stimulation of parvo-layers you have to go to 30 cycles per degree or above, and this is a very difficult task. On the other hand low contrast, low temporal and spatial frequency, isoluminant gratings, can at least isolate part of the parvocellular system. At high contrast we have to restrict ourselves to tritanopic confusion lines because high contrast red-green gratings stimulate magno neurons very well. Only with such stimuli have we any hope of keeping away from activity of magno cells.

Schiller: I neglected to say that to separate these system on the basis of spatial frequency, you have to use equal retinal eccentricities, preferably not the fovea. I don't think it's a very good experiment to let the eye wander about to inspect the figure and to use 60 cycles a degree to do this.

Russ DeValois: To make an obvious point about specific systems in the cortex, it is possible to isolate many dozens of separate systems by use of patterns with different orientations and spatial frequency.

Lee: Is such isolation specific or is there much overlap?

Russ DeValois: I think you can get total isolation.

Dow: I second that. It is difficult to appreciate how difficult it is to find a good stimulus for cortical cells. With gratings or some sorts of complex figures, it's very easy to get no response at all. If you find the right stimulus, the cells start to respond very vigorously. So it's very easy to isolate specific systems, the only trouble is that there are so many of them.

Lee: This is very much the argument used 20 years ago in developing the concept of grandmother cells. At this meeting there seems to be much greater emphasis on distributed network processing. Yet Bruce Dow has just expressed a viewpoint that could have been put forward 20 years ago.

Rodieck: I don't think there's much evidence for distributed processing in the sense that a variety of different signals are carried by the same set of neurons, and demultiplexed somewhere else. I think that there is a good deal of specificity, as shown by the work of Charley Gilbert. It is a muddling of the distinction between what a cell responds to and what it codes for. Below the cortex you can get any cell to respond to anything. The idea of distributed processing of, say, motion or texture, maybe means that you haven't asked the right question or you characterized your feature in too broad a sense.

Lee: On the other hand, in the experiments of Bob Desimone you can modify the properties of V4 cells according to the kind of task which an animal is performing. That's an example of a flexible network which is adopting different modes according to the task.

Rodieck: Farther away from V1 and V2, you come up against the issue which I referred to as hypothesis testing, and the more this becomes true, the poorer our rationale is for single unit recording in attempting to delineate functions of these areas, particularly when the receptive fields become very large. We may simply be missing the boat farther and farther away from V1 in terms of the adequate stimuli, and the adequate stimuli may not be static, anatomical feature, but a very dynamic one.

Spekreijse: I would advocate that indeed at a cortical level a distributed network seems a proper model, and not a static one but a dynamic one. You can show it directly by making simultaneously single cell recordings, and at the same time surface recordings from circles of electrodes, up to 1 cm away from the cell you are recording. If you use the spike response as a trigger for averaging the EEG, the correlation with the spike response changes when you start to activate the cell. Its responses are influenced by what is happening in quite a large network of cells surrounding it. If such a dynamic network would not exist it would probably be impossible to make a record from the surface of evoked activity, because cells working independently would cancel each other out at the surface.

Drasdo: If we are concerned with summated responses of neurones in a psychophysical or evoked potential experiment, and if we are concerned with the threshold, it is only the system that gets there first that matters. And therefore one doesn't have to have a totally exclusive stimulation but simply a superior stimulation.

Gilbert: When talking about the possible segregation of functions for a different cell populations there is a tendency to use as parsimonious a stimulus as possible in order to silence one system. What you learn from that may be a bit deceptive. When you are dealing with supra-threshold conditions or natural visual environments, these various populations may be both contributing to the percept and interacting in a way that wouldn't be necessarily evident when you are working at threshold with these very parsimonious stimuli.

Merigan: Although it is easy to stimulate the P- and M-pathways independently, I think one should ask in this context if it is interesting or not.

Kulikowski: This is an important point. We have to distinguish between two questions. What are the stimuli that isolate artificially the systems, and secondly, what is the interaction between the systems in response to supra-threshold stimuli.

Russ DeValois: I would agree that most natural stimuli activate very large numbers of cells. The question posed was whether we can devise stimuli which isolate specific systems, and in the cortex one can. It's important to point out that these are very artificial types of stimuli specifically designed to isolate individual cells. As I mentioned earlier, one might have simultaneously active different varieties of cells, some of which are showing assimilation and some are responding to contrast and thereby one can get unstable percepts.

Spillmann: To your fourth point, I would like to ask what physiology psychophysicists should welcome, not trust. There is, for instance, the finding of Zeki that colour constancy happens high up, in the neurones of V4. However, Jan Walraven suggested here that colour constancy happens at very early levels. What is the status of this controversy? The question is relevant to psychophysics and I would welcome much more physiological research in this area.

Participants

J.K. Bowmaker, Institute of Ophthalmology, University of London, Judd Street, London WC1H 9QS, UK.

D.C. Burr, Istituto di Neurofisiologia CNR, Via S. Zeno 51, 56100 Pisa, Italy.

P. Cavanagh, Institute of Psychology, Harvard University, 33 Kirkland Street, Cambridge, MA 02138, USA.

C.R. Cavonius, Institut für Arbeitsphysiologie, Ardeystraße 67, D-4600 Dortmund 1, FRG.

O.D. Creutzfeldt, Neurobiology Department, Max Planck Institute for Biophysical Chemistry, Postfach 2841, D-3400 Göttingen, FRG.

L.P. Csernai, Physics Department, University of Bergen, Allégaten 55, N-5007 Bergen, Norway.

C.M.M. de Weert, Psychological Laboratory, University of Nijmegen, Montessorilaan 3, NL-6525 HR Nijmegen, The Netherlands.

R. DeValois, Section of Optometry, University of California, Berkeley, CA 94720, USA.

K. DeValois, Section of Optometry, University of California, Berkeley, CA 94720, USA.

B.M. Dow, Department of Anatomy and Neurobiology, University of California, Irvine; California College of Medicine, Irvine, CA 92717, USA.

N. Drasdo, Department of Vision Sciences, Aston University, Aston Triangle, Birmingham B4 7ET, UK.

A. Fiorentini, Istituto di Neurofisiologia CNR, Via S. Zeno 51, 56100 Pisa, Italy.

C.D. Gilbert, The Rockefeller University, 1230 York Avenue, New York, NY 10021, USA.

I. Giæver, Department of Physics, University of Oslo, P.O. Box 1048 Blindern, N-0316 Oslo, Norway.

M. Hayhoe, Center for Visual Science, University of Rochester, 274 Meloria Hall, Rochester, NY 14627-9999, USA.

P. Heggelund, Institute of Neurophysiology, University of Oslo, Karl Johansgate 47, 0162 Oslo 1, Norway.

J.A. Hertz, NORDITA, Blegdamsvej 17, DK-2100 Copenhagen 1, Denmark.

C.R. Ingling, Jr., Ohio State University Research Center, Ohio State University, 1314 Kinnear Road, Colombus, Ohio 43212, USA.

J. Jansen, Physiological Institute, University of Oslo, Karl Johansgate 47, 0162 Oslo 1, Norway.

P.K. Kaiser, Department of Psychology, Faculty of Arts, York University, 4700 Keele Street, North York, Ontario M3J 1P3, Canada.

E. Kaplan, Biophysics Laboratory, Rockefeller University, 1230 York Avenue, New York, NY 10021, USA.

S. Kastner, Neurobiology Department, Max Planck Institute for Biophysical Chemistry, Postfach 2841, D-3400 Göttingen, FRG.

J. Krauskopf, Department of Psychology, New York University, 6 Washington Place, 8th Floor, New York, NY 10003, USA.

J. Kremers, Neurobiology Department, Max Planck Institute for Biophysical Chemistry, Postfach 2841, D-3400 Göttingen, FRG.

J.J. Kulikowski, Visual Science Laboratory, UMIST Box 88, Manchester M6O 1QD, UK.

B.B. Lee, Neurobiology Department, Max Planck Institute for Biophysical Chemistry, Postfach 2841, D-3400 Göttingen, FRG.

P. Lennie, Center for Visual Science, University of Rochester, Rochester, NY 14627, USA.

D.I.A. MacLeod, Department of Psychology C-009, University of California, La Jolla, CA 92093-0109, USA.

S. Magnussen, Institute of Psychology, University of Oslo, P.O. Box 1094 Blindern, 0317 Oslo 3, Norway.

P.R. Martin, Department of Neuroanatomy, Max Planck Institute for Brain Research, Deutschordenstraße 47, D-60 Frankfurt - Niederrad, FRG.

W.H. Merigan, University of Rochester Medical Center, Box 314 Rochester, NY 14642, USA.

J. Mollon, Department of Experimental Psychology, Downing Street, Cambridge CB2 3EB, UK.

J. Pokorny, The Visual Sciences Center, University of Chicago, 939 East Fifty-Seventh Street, Chicago, Illinois 60637, USA.

K. Richter, Laboratory "Color Reproduction", Federal Institute for Materials Research and Testing (BAM 5.44), Unter den Eichen 87, D-1000 Berlin 45, FRG.

R.W. Rodieck, Department of Ophthalmology RJ-10, University of Washington, Seattle, WA 98195, USA.

P.H. Schiller, Department of Brain and Cognitive Science, Mass. Institute of Technology, 77 Mass. Avenue, Cambridge, MA 02139, USA.

T. Seim, Department of Physics, Section of Biophysics, University of Oslo, Box 1048 Blindern, 0316 Oslo 3, Norway.

R. Shapley, Center for Neural Science, Department of Psychology, New York University, 6 Washington Place, New York, NY 10003, USA.

L.T. Sharpe, Abteilung Klinische Neurologie und Neurophysiologie, Klinikum der Albert-Ludwigs Universität, Hansastraße 9, 7800 Freiburg i. Br., FRG.

S.K. Shevell, The Visual Sciences Center, University of Chicago, 939 East Fifty-Seventh Street, Chicago, Illinois 60637, USA.

V.C. Smith, The Visual Sciences Center, University of Chicago, 939 East Fifty-Seventh Street, Chicago, Illinois 60637, USA.

H. Spekreijse, The Netherlands Ophthalmic Research Institute, P.O. Box 12141, 1100 AC Amsterdam-Zuidoost, The Netherlands.

L. Spillmann, Abteilung Klinische Neurologie und Neurophysiologie, Klinikum der Albert-Ludwigs Universität, Hansastraße 9, 7800 Freiburg i. Br., FRG.

W.H. Swanson, Retina Foundation of the Southwest, 8230 Walnut Hill Lane No 414, Dallas, Texas 75231, USA.

J.P. Thomas, Abteilung Klinische Neurologie und Neurophysiologie, Klinikum der Albert-Ludwigs Universität, Hansastraße 9, 7800 Freiburg i. Br., FRG.

A. Valberg, Department of Physics, Section of Biophysics, University of Oslo, Box 1048 Blindern, 0316 Oslo 3, Norway.

D.C. van Essen, California Institute of Technology, Biology Division 216-76, Caltech, Pasadena, CA 91125, USA.

F. Viénot, Muséum National d'Histoire Naturelle, Laboratoire de Physique Apliquée, 43 rue Cuvier, 75231 Paris Cedex 05, France.

J. Walraven, Institute for Perception, P.O. Box 23, 3769 ZG Soesterberg, The Netherlands.

A. Werner, Institut für Neurobiologie, Freie Universität Berlin, Königin-Luise-Straße 28-30, D-1000 Berlin 33, FRG.

P. Whittle, Department of Experimental Psychology, University of Cambridge, Downing Street, Cambridge CB2 3EB, UK.

D.R. Williams, Center for Visual Science, University of Rochester, Rochester, New York 14627, USA.

Q. Zaidi, Columbia University in the City of New York, Department of Psychology, New York, NY 10027, USA.

G. Zimmermann, Fraunhofer-Institut für Informations- und Datenverarbeitung (IITB), Fraunhoferstraße 1, D-7500 Karlsruhe 1, FRG.

E. Zrenner, Universitäts-Augenklinik, Abteilung II, D-7400 Tübingen 1, FRG.

Observers

T. Arestov, Department of Psychology, Moscow University, Marx Prospekt 18/5, Moscow, 103009 USSR.

L.R. Bjørnevik, Department of Physics, Section of Biophysics, University of Oslo, Box 1048 Blindern, 0316 Oslo 3, Norway.

S. Dyrnes, Institute of Psychology, University of Oslo, Box 1094 Blindern, 0317 Oslo 3, Norway.

D. Enarun, I.T.U. Elektrik-Elektronik Fak., 80191 Gümüssuyu, Istanbul, Turkey.

E. Hartveidt, Institute of Neurophysiology, University of Oslo, Karl Johansgate 46, 0162 Oslo, Norway.

L. Hyvärinen, Institute of Occupational Health, Topeliuksenkatu 41aA, SF-00250 Helsinki, Finland.

B.T. Olsen, Norwegian Telecom., Instituttvn. 23, Box 83, 2007 Kjeller, Norway.

R. Watten, Institute of Psychology, University of Oslo, Box 1094 Blindern, 0317 Oslo 3, Norway.

J.H. Wold, Department of Physics, Section of Biophysics, University of Oslo, Box 1048 Blindern, 0316 Oslo 3, Norway.

T. Yeh, The Visual Sciences Center, University of Chicago, 939 East Fifty-Seventh Street, Chicago, Illinois 60637, USA.

Organizing Committee

Barry Lee, Max Planck Institute for Biophysical Chemistry, Göttingen, FRG.
Svein Magnussen, University of Oslo, Oslo, Norway.
Joel Pokorny, University of Chicago, Chicago, USA.
Steve Shevell, University of Chicago, Chicago, USA.
Arne Valberg, University of Oslo, Oslo, Norway.

Session Chairmen

D.C. Burr
L.P. Csernai
J. Jansen
S. Magnussen
P. Martin
J. Pokorny
K. Richter
T. Seim
L. Spillmann

Technical Assistance

Lars-Rune Bjørnevik (sound)
Thorstein Seim (recording)
Stein Dyrnes (projection)
Jan-Henrik Wold (recording)

Secretariat

Anne-Sophie Andresen
Anne-Grethe Gulbrandsen
Sissel Knutsen

Sponsors

NATO
National Science Foundation (USA)
Norwegian Council for Science and
the Humanities the Fridjof Nansen
Foundation

University of Oslo
Norwegian Physical Society
Norwegian Telecom
Scandinavian Airline System

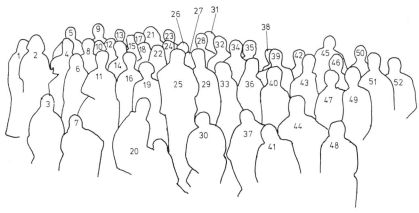

1. P.H. Schiller	19. M. Hayhoe	37. C.R. Ingling Jr.
2. C. Zrenner	20. E. Zrenner and son	38. B.B. Lee
3. W.H. Swanson	21. L.T. Sharpe	39. L. Spillmann
4. H. Spekreijse	22. K. DeValois	40. A. Fiorentini
5. R. DeValois	23. S. Magnussen	41. G. Zimmermann
6. T. Yeh	24. J. Krauskopf	42. N. Drasdo
7. E. Kaplan	25. I. Giæver	43. W. Merigan
8. P.K. Kaiser	26. J. Hertz	44. L.R. Bjørnevik
9. E. Hartveidt	27. J. Kremers	45. C.R. Cavonius
10. J. Mollon	28. P. Lennie	46. J. Pokorny
11. D. Enarun	29. B.T. Olsen	47. V.C. Smith
12. T. Seim	30. T. Arestov	48. A. Valberg
13. C. de Weert	31. K. Richter	49. L. Hyvärinen
14. P.R. Martin	32. D.R. Williams	50. D.C. Burr
15. J. Bowmaker	33. A. Werner	51. S.K. Shevell
16. F. Viénot	34. L.P. Csernai	52. C. Gilbert
17. S. Dyrnes	35. J.H. Wold	
18. J. Kulikowski	36. S. Kastner	

INDEX